Chemical Analysis in Extractive Metallurgy

Chemical Analysis in Extractive Metallurgy

Roland S. Young

M.Sc., Ph.D., C.Eng., F.R.I.C., F.I.M., M.I.Chem.E., M.I.M.M.

Department of Mines and Petroleum Resources
Victoria, B.C., Canada

Griffin London

Charles Griffin & Company Limited
42 Drury Lane, London, WC2B 5RX

First published . 1971

229 × 152 mm, vii + 427 pages
ISBN: 0 85264 198 2

SET AND PRINTED IN GREAT BRITAIN BY
CHORLEY & PICKERSGILL LIMITED LEEDS

CONTENTS

PREFACE

Many years have elapsed since a book devoted to general chemical analysis for the mining and extractive metallurgical industries has appeared in English. Long experience in this field has convinced me of the need for a handbook outlining current procedures for the elements which are usually encountered in works practice.

The rapid and welcome trend towards instrumentation in analytical chemistry should not obscure the fundamental importance of the basic principles of analysis, the separation of groups and elements, the effect of interferences, and the selection of a final method of determination. Possession of a thorough knowledge of these subjects must always be mandatory for the analytical chemist.

This book is intended to be less comprehensive and more selective than the excellent multi-volume treatises on analytical chemistry, to be more attuned to the routine industrial tempo than is desirable in a university textbook, and to contain more discussion of theory and interferences than is found in the valuable analytical manuals of companies and trade associations.

Most of the procedures are in vogue for rapid routine work in the laboratories of the mineral industries. Some older, more time-consuming methods have been included, however, for their reliability in occasional analyses or in standardization and checking. It is hoped that this book will prove helpful to all concerned with the analytical chemistry of extractive metallurgy.

R. S. YOUNG

Victoria, B.C.
February 1971

1 Aluminium

Because aluminium is not only the most abundant metal on the earth's surface but is also one of the most important engineering metals, its determination is of frequent occurrence in nearly all inorganic analytical laboratories, particularly in the mining and metallurgical industries. Analysis for aluminium is required in rocks, minerals, ores, concentrates, tailings, slags, flue dusts, fluxes, metals and alloys, abrasives, refractories, soils, and many other materials; percentages may vary over a wide range from a few tenths in some rocks to nearly 100 per cent for aluminium metal.

Aluminium is an important constituent of so many rocks and minerals that it is encountered in at least trace amounts in all ores. For the production of metallic aluminium or of pure alumina, the principal ore is an impure hydrated aluminium oxide known as bauxite, which usually contains about 40–50% Al_2O_3.

The accurate determination of aluminium in a complex product is one of the most difficult in an analytical laboratory. After removal of silica and metals of the hydrogen sulphide group, aluminium may be precipitated by ammonia in the presence of ammonium chloride, and be thereby separated from the alkaline earths and alkali metals.

Unfortunately, about one-third of the elements in the periodic table are likewise precipitated by ammonia, and the isolation of aluminium from other members of the ammonium sulphide or ammonium hydroxide groups may be lengthy and difficult.

In many materials, however, the precipitate obtained on treatment with ammonia and ammonium chloride consists almost entirely of hydroxides of iron and aluminium, other elements occurring only in very small quantities. For technical analyses, this mixture of oxides is weighed and the iron content, obtained volumetrically or colorimetrically, is subtracted from this total to give Al_2O_3. If titanium is present to a significant extent, it is determined colorimetrically and likewise deducted. Sometimes phosphorus is the only other element found in weighable quantity with Fe_2O_3, in which case it is determined colorimetrically and the weight of P_2O_5 subtracted from the total [7, 8, 9, 26, 33, 36, 47].

For many ores and similar materials, the analysis for aluminium, calcium, and magnesium is carried out on the same sample; frequently silica also is determined initially on this portion.

ISOLATION OR SEPARATION OF ALUMINIUM

Transfer a 0·5–2 g sample to a beaker and decompose with 10 ml hydrochloric acid, 10 ml nitric acid, and 10 ml 1 : 1 sulphuric acid. For high sulphides add a few ml of bromine and allow to stand for several minutes prior to addition of acids. Evaporate on the hot plate to strong fumes of sulphuric acid, remove and cool.

Add 5 ml hydrochloric acid and 100 ml hot water, boil, filter, wash with hot water, scouring the beaker thoroughly with a rubber policeman. Retain the filtrate. Place the paper in a platinum crucible, dry, ignite, cool, add approximately 5 g sodium carbonate, and fuse the contents of the crucible for at least five minutes at the full heat of a gas burner. Cool, place the crucible in the beaker containing the dilute acid filtrate, and when the melt has dissolved, wash and remove the crucible. Evaporate the solution to dryness on asbestos at the side of the hot plate. Moisten twice with hydrochloric acid to dehydrate the silica, and evaporate to complete dryness each time. Take up in 5 ml hydrochloric acid, 100 ml hot water, boil, filter using paper pulp, and scrub the beaker thoroughly with a rubber policeman. For accurate work, this filtrate will have to be evaporated to dryness again, dehydrated twice with hydrochloric acid as before, and the small quantity of residual silica filtered off and added to the first precipitate.

Pass hydrogen sulphide into the filtrate from the silica separation, after adjusting the acid content to 5–10% by volume of hydrochloric acid, to remove elements of Group 2, i.e. As, Sb, Sn, Cu, Cd, Bi, etc. Filter and wash thoroughly with acidulated hydrogen sulphide water; discard the precipitate. Boil the filtrate until all hydrogen sulphide is removed, add 5 ml of 3% hydrogen peroxide to oxidize iron, and again boil to eliminate excess hydrogen peroxide. Add a few drops of methyl red, then ammonium hydroxide cautiously until the colour of the solution changes to a distinct yellow. If much iron is present the colour can be seen by adding enough ammonium hydroxide to precipitate the iron and allowing the latter to settle after a short boil; the indicator can be observed in the supernatant liquid.

Boil for a minute or two and filter through a Whatman No. 40, 41, or 541 paper, using pulp; wash thoroughly with hot 2% ammonium chloride. Reserve the filtrate for the calcium and magnesium determinations, if required. Transfer the precipitate to the original beaker, dissolve with hydrochloric acid any adhering to the filter paper, and add more acid if necessary to bring the hydroxides into solution. Re-precipitate the aluminium with ammonium hydroxide as before, using methyl red as indicator. Filter the solution through the original paper, and wash the precipitate thoroughly with hot 2% ammonium chloride. Scrub the beaker thoroughly with a policeman; small quantities of aluminium hydroxide are often tenaciously retained on glassware. Combine this filtrate with the one previously retained for the calcium and magnesium determinations, if required.

The aluminium in the form of hydroxide has now been separated from members of preceding and succeeding groups, but it may be contaminated by one or more of the following elements: Fe, Be, Ti, Cr, Th, V, Zr, U, Ga, In, P, the rare earths and traces of Ni, Co, Nb, Ta, and W.

It is difficult to remove nickel and cobalt from aluminium entirely by repeated separations with ammonium hydroxide. A large excess of ammonia, which would minimize the retention of these elements, cannot be used, for then appreciable re-solution of aluminium hydroxide occurs. For routine purposes, the small quantities of these elements adsorbed or occluded by aluminium hydroxide after two ammonia separations are usually disregarded; in analyses of the highest accuracy, other methods of removal must be employed or the content of nickel and cobalt in the mixed oxides may be determined colorimetrically.

The decomposition and separation steps detailed above may be altered when the composition of the sample is known. For instance, if the material is a silicate containing no members of the hydrogen sulphide group, and silica is not required, treatment with hydrofluoric and sulphuric acids in a large platinum dish, evaporating twice to strong fumes of sulphuric acid, and taking into solution with hydrochloric acid and water, will usually yield a filtrate ready for an ammonia separation.

Calcined alumina, and some natural forms of corundum, are very refractory and require prolonged fusion with 2 parts anhydrous sodium carbonate and 1 part anhydrous sodium tetraborate. Aluminium and some of its alloys can be dissolved in hot 20% sodium hydroxide solution; this separates aluminium at once from iron, one of its commonest interfering elements.

The procedure from this point depends on the presence of other elements, and on the speed and accuracy demanded for the analysis.

1. Gravimetric Determination

A. In the Presence of Iron and other members of the Ammonium Hydroxide Group

(1) *Weighing as Oxide*

If the sample contains only iron and traces of other elements, precipitate with ammonium hydroxide as described above. Determine aluminium by subtracting the iron, calculated as Fe_2O_3, from the weight of the combined oxides. In this method the precipitate of hydroxides of aluminium + iron is placed in a weighed platinum crucible, dried, charred, ignited at 1200°C to constant weight, and weighed as crude Al_2O_3. If iron has not been determined on a separate portion of the same sample it may be determined by fusing the ignited oxides with potassium pyrosulphate and carrying out the usual volumetric or colorimetric procedure. If titanium is present in more than traces it may be determined colorimetrically, and the amount of TiO_2 deducted from the weight of the mixed oxides. If a significant quantity of phosphorus is present it should be determined colorimetrically, and its value as P_2O_5 also subtracted from the final residue.

$$Al_2O_3 \times 0{\cdot}5292 = Al$$

(2) *Precipitation with 8-hydroxyquinoline*

Small quantities of aluminium may sometimes be advantageously determined with 8-hydroxyquinoline; the precipitate has a favourable weight factor [21, 26, 30].

Remove silica by acid dehydration, or by volatilization with hydrofluoric acid. Separate iron and other heavy metals by cupferron or a mercury cathode as described later. Acidify the sample solution, containing 75–150 ml, with 1 : 1 hydrochloric acid and add 5 drops in excess. Add sufficient 8-hydroxyquinoline solution in 5–10% acetic acid to precipitate the aluminium, plus an excess, and heat to 80°C. Add 30% ammonium acetate solution until a precipitate forms, and then 15 ml in excess.

Cool, adjust the *p*H to 5·7 with ammonium hydroxide; allow to stand several hours or overnight. Filter through a medium-porosity, fritted-glass crucible and wash with a neutral 0·02% solution of 8-hydroxyquinoline. Dry at 135°C for 2 hours, cool in a desiccator and weigh.

$$Al(C_9H_6NO_3)_3 \times 0{\cdot}0587 = Al$$

A procedure for the 8-hydroxyquinoline precipitation in ammoniacal cyanide–EDTA solution has been described; there is no interference from iron and many other heavy metals [22].

(3) *Other Gravimetric Procedures*

Precipitation of aluminium as the hydroxide or hydroxyquinolate is used in nearly all laboratories of the mining and metallurgical industries, but other gravimetric procedures may be employed [26, 33, 47]. Precipitation of sodium fluoaluminate, without prior separation of Fe, Cu, Mo, Mn, Ti, V, Pb, Sn, Cr, and Zr, has been described [40].

Formerly, a popular procedure for aluminium in mining laboratories was the precipitation of this element as phosphate in the presence of iron [52]. When ammonium phosphate is added to a weak acetic acid solution of aluminium and iron, containing sodium thiosulphate, aluminium phosphate slowly precipitates on boiling whereas iron, being in the ferrous state, remains in solution. The variables in this method tend to be rather critical, and for most work it is preferable to separate aluminium from iron before precipitating as hydroxide, or to subtract the iron value from the final mixed oxides.

B. Removal of Iron, etc.

If the precipitate obtained by the ammonium hydroxide + ammonium chloride precipitation contains considerable quantities of other elements, further separations must be made to isolate the aluminium, or to remove those elements which interfere in the final determination of the procedure which is selected. Common separations are listed below.

(a) Iron may be removed by an ether separation, the details of which are given in the chapter on cobalt. Because nearly all other elements are left in the acid fraction with aluminium, this separation has limited value in aluminium determinations except in the analysis of some steels and ferrous alloys.

(b) Cobalt, nickel, and manganese may be separated from aluminium by a zinc oxide precipitation; details are given in the chapter on cobalt. Almost all other elements of the ammonium hydroxide group accompany aluminium in the precipitate, however, hence its use is somewhat restricted.

(c) Treatment with sodium hydroxide will leave aluminium, with Be, V, Zn, and W in the filtrate; the precipitate contains Fe, Ti, Zr, Cr, U, Ni, Co, Th, and Mn. If sodium peroxide is also added, chromium and uranium will remain in solution with aluminium.

(d) Aluminium, together with Be, Ti, Th, U, V, Nb, Ta, and Zr, is not precipitated by hydrogen sulphide in ammoniacal tartrate solution and can thereby be separated from Fe, Mn, Ni, Co, and Zn [30].

(e) Cupferron in cold 10% sulphuric acid solution will precipitate Fe, Ti, Zr, V, Ga, Nb, Ta, U(IV), and W; the filtrate contains Al, Be, P, Mn, Ni, Co, Zn, In, Ge, B, and Cr. Th and the rare earths are partially precipitated. Instead of separation by conventional filtration, the cupferrates of Fe, etc. can be extracted into chloroform and thereby separated from aluminium.

(f) Electrolysis with a mercury-cathode cell, in 0·3M sulphuric acid will deposit Fe, Ni, Co, Cr, Ga, Ge, In, Tl, and Zn, leaving Al, B, Be, Ti, Zr, P, Th, V, Nb, Ta, U, and W in solution; manganese is partly deposited. This useful separation is frequently employed in industrial laboratories because a number of mercury-cathode units can be operated simultaneously with little attention [39, 42].

(g) Aluminium is precipitated by 8-hydroxyquinoline and separated from phosphates, fluorides, and borates in ammoniacal solution; from V, Ti, Nb, and Ta in ammoniacal solution containing hydrogen peroxide; and from beryllium in acetic acid [30].

(h) Chromium can be separated from aluminium by repeated treatment with boiling hydrochloric and perchloric acids to volatilize chromyl chloride. Chromium may also be left in the filtrate if it is first oxidized to chromate with sodium peroxide or perchloric acid, before precipitation of aluminium by ammonia.

(i) Addition of an excess of sodium carbonate to a solution of aluminium leaves the latter, with V, U, and W, in the filtrate, whereas Fe, Ti, Co, Ni, Be, Mn, Zr, and Cr are precipitated.

(j) Uranium can be separated from aluminium by treatment with ammonium carbonate, uranium remaining in solution.

(k) Phosphorus can be removed by fusing with sodium carbonate and extracting the melt with ammonium carbonate; aluminium is precipitated while sodium phosphate passes into solution.

(l) Vanadium in sodium hydroxide solution may be precipitated by the addition of barium chloride, filtered, and thus removed from aluminium. Excess barium chloride can be removed from the aluminium filtrate by addition of sulphuric acid and filtration.

(m) In strong hydrochloric acid solution, iron and many other elements in their highest valence state such as Mn(VII), Cr(VI), V(V), U(VI), Zn, Zr, and W are adsorbed on anion exchange resins like Dowex 1 [27], Dowex 1–X10 [35], and Amberlite IRA 400 [13], while Al, Mn(II), Cr(III), V(IV), Ni, Ti, and Be are not retained.

When separating aluminium with ammonium hydroxide, from a solution from which no elements have been removed, it is useful to remember that the following common elements are usually not precipitated, assuming an excess of ammonia and ammonium salts are present, and a double precipitation is employed: Ag, Cu, Mo, Cd, Pt, Ni, Co, Zn, Tl(ous), V(ate), Mn, Ca, Sr, Ba, Mg, and W. The following elements are precipitated with aluminium by ammonia under these conditions: Hg, Pb, Sn, Bi, Sb, Au, Rh, Os, Fe, Cr, Ti, U, Be, Zr, Th, Ce, In, Ga, Tl(ic), V(vanadyl), Nb, and Ta. The above lists must be used with caution, for often the presence of other elements in certain ratios alters the composition of the precipitate. Furthermore, Cu, Ni, and Co are strongly occluded with precipitated aluminium hydroxide, and traces of these metals are always present, even after a double precipitation.

The Cupferron–Chloroform Separation

The cupferron–chloroform separation of aluminium from Fe, Ti, V, Zr, etc. is convenient and reliable for nearly all materials in the mining and metallurgical industries. To an ice-cold 10% sulphuric or hydrochloric acid solution of the sample in a separatory funnel add sufficient cold 6% aqueous solution of cupferron to precipitate all the iron as a flocculent brown cupferrate. A slight excess of cupferron is indicated by the temporary appearance of a white precipitate, which re-dissolves. Theoretically, 0·84 g of cupferron is required for 0·1 g iron, but one should add at least 16 ml of 6% solution of cupferron for each 0·1 g of iron.

Shake the separatory funnel to mix the contents, add 5–10 ml of cold chloroform and shake vigorously for 20–30 seconds. Allow to settle, draw off and discard the lower brown chloroform layer containing the ferric cupferrate. Add another 5–10 ml chloroform, shake, and again discard the lower brown chloroform portion. Continue the chloroform additions and extractions until the lower layer is colourless. Transfer the upper, acid layer to a beaker, adding the rinsings from the separatory funnel. Place on the edge of the hot plate until the traces of chloroform have evaporated, add 10 ml nitric acid, 10 ml 1 : 1 sulphuric if the cupferron separation was not carried out in sulphuric acid, and take to strong fumes of sulphuric acid. Cool, wash down the inside of the beaker, add another 10 ml of nitric acid, and again evaporate to strong fumes of sulphuric acid to eliminate all organic matter. If traces of the latter still persist, add more nitric acid and a few drops of perchloric acid, and again evaporate to strong sulphuric fumes.

When silica, tungsten, and members of the hydrogen sulphide group have been removed initially, and Al, Fe, etc. have been isolated by an ammonia separation after oxidation of chromium to chromate, the only elements found with aluminium in the cupferron–chloroform extraction will be phosphoros, germanium, indium, boron, beryllium, and a part of the thorium and rare earths originally present. Usually, phosphorus will be the only element present in significant quantities. It is customary, for ores and similar materials, to proceed with the precipitation of aluminium + phosphorus with ammonia, as described earlier, weighing the combined oxides, fusing the latter with sodium carbonate, dissolving in 1 : 1 nitric acid, and determining phosphorus colorimetrically by the molybdovanadophosphoric acid method.

The Mercury Cathode Separation

As an alternative to the cupferron–chloroform solvent extraction for the separation of aluminium, described above, another popular isolation procedure in mining and metallurgical laboratories utilizes the mercury cathode cell. By this device the following are deposited on the mercury: Fe, Cr, Ni, Co, Cu, Sn, Mo, Ag, Zn, Hg, Pb, Ga, Ge, In, Pt, Pd, Rh, Ir, Bi, Au, Tl, Cd and part of the manganese. In the solution are left Al, Ca, Ba, Sr, B, Ti, Zr, V, U, Th, F, Be, Mg, W, Nb, Ta, and rare earths. After the usual removal of the silica and hydrogen sulphide groups, and the treatment of the ammonium hydroxide group in a mercury cathode cell in about 0·3M sulphuric acid, the only elements left in solution with aluminium will be P, B, Be, Ti, V, Zr, Th, and rare earths.

Many types of mercury cathode cells have been described [8, 26, 30, 33, 39, 42]. The anode is usually a flat spiral of platinum wire, and the sulphuric acid concentration in the sample of 50–100 ml is preferably less than 0·25M for faster and more complete separations. Perchloric acid may be substituted for sulphuric. Stirring of the 200 g of mercury is done mechanically or by a stream of air. Current density is generally 0·1–0·2 A/sq. cm, and the temperature 25–60°C; sometimes cooling is recommended. High concentrations of ammonium salts are undesirable; it is better to reduce the acid concentration by evaporation rather than by neutralization with alkali. Separations are quite rapid; in 15 minutes 99·99% of the iron in 1 gram of steel can be deposited, and in 1 hour 99·99% of the zinc in a 5-g zinc-base alloy can be amalgamated with the mercury. For routine work a bank of cells can provide separations with a minimum of time and attention. At the conclusion of the electrolysis, the solution and washings are readily separated from the mercury and deposited metals, and the mercury can be cleaned by dilute acid or distillation before re-use.

For complex ores, the mercury cathode separation may be followed by the cupferron–chloroform isolation, or vice versa [24]. Assuming the usual prior separations, the only elements then left with aluminium will be beryllium, phosphorus, boron, and part of any thorium and rare earths originally present. Boron can be completely eliminated initially by evaporating with sulphuric and hydrofluoric acids. The insolubility of the fluorides of thorium and the rare earths furnishes a ready means of separating these elements. Beryllium can be separated from aluminium by precipitation of the latter with 8-hydroxyquinoline in feebly acid solution as described earlier. Phosphorus is usually determined colorimetrically in the final weighed precipitate of $Al_2O_3 + P_2O_5$; larger quantities of phosphorus are separated as ammonium phosphomolybdate in nitric acid solution, discussed in the section on Phosphorus.

2. Volumetric Determination

From time to time, volumetric procedures for determing aluminium in certain products have been published. Typical of these are preliminary separations with cupferron and benzoate, and titration with ethylenediaminetetraacetic acid [38], or separation on Dowex 1–X10 anion exchange resin and titration with (ethylenedinitrilo) tetraacetic acid [35]. Another method depends on the formation of soluble sodium aluminate and precipitation of hydroxides of iron and titanium above *p*H 10, and on the fact that, when aluminium reacts with fluoride the hydroxide which combines with aluminium can be released for titration with standard acid. The titration can be performed in the presence of the precipitated hydroxides; interference from calcium can be eliminated by the precipitation of calcium oxalate, and, by limiting the sample size, aluminium can be determined in the presence of silica [14, 49, 50].

A volumetric procedure for aluminium in complex alloys, using 1,2-cyclohexydenedinitrilo tetraacetic acid (CDTA), has been published [18]. Removal of interferences by fuming with a perchloric-hydrochloric acid mixture, sodium chlorate digestion, mercury cathode electrolysis, and extraction with tri-n-octylphosphine oxide in cyclohexane, is followed by back titration of excess CDTA with standard zinc solution, using xylenol orange indicator.

As a general rule, for ores and metallurgical products, volumetric methods for aluminium have limited application.

3. Colorimetric Determination

A. Aluminon

For small quantities of aluminium, a colorimetric method using aluminon is frequently employed [17, 31, 44, 48]. Aluminon, the ammonium salt of aurin tricarboxylic acid, is a dye which forms a bright red lake $(C_6H_3OHCOONH_4)_2{:}C{:}C_6H_3COONH_4{:}O$ with aluminium. The lake is obtained by adding a solution of ammonium carbonate in dilute ammonium hydroxide to a solution containing the dye and free acetic acid. If the quantit of aluminium is less than 0·1 mg in 50 ml of solution, a clear faint-pink to a deep-red colour is obtained; with larger quantities, the red lake precipitates out. Iron and beryllium are the chief interfering elements. Basic ion-exchange resins, such as Dowex 1 can be used to remove most interfering metals in strong hydrochloric acid, except Ni, Be, and Ti [32]. The effect of nickel can be masked by pyridine; titanium does not interfere unless it exceeds the amount of aluminium present [32].

To the solution, which has been passed through an anion exchange column, if necessary, containing 5 ml hydrochloric and 5 ml acetic acid, add 5 ml of a 0·2% aqueous solution of aluminon. Add ammonium hydroxide which has been saturated with ammonium carbonate, until the solution is alkaline to litmus. Acidify with 5 ml of acetic acid, allow to stand 10 minutes, again neutralize with the ammonium hydroxide-carbonate solution, and add 5 ml excess. Cool and compare the colour against standards similarly prepared, or measure the optical density at 525 nm on a spectrophotometer.

B. Eriochrome Cyanine R

There is scope for further applications of a direct photometric determination of aluminium which has been in use on certain iron ores for some years [28, 29]. After fusion with sodium carbonate and solution in hydrochloric acid, sodium mercaptoacetate is added to form a colourless complex with iron, and EDTA is added to the blank. Ammonium acetate buffer and stabilized Eriochrome Cyanine R dye are added to both solutions. The red Eriochrome Cyanine R–aluminium complex which forms in the sample solution is measured against the blank at 535 nm. Beryllium, the background colour of the dye, and other interfering elements are compensated for in the blank. EDTA inhibits interferences by all complexes, except those of Be and Zr.

C. Chromazurol S

In another procedure for aluminium in iron ores, sinters, and slags, the sample is fused in a nickel crucible with sodium peroxide, the cooled melt dissolved in water and hydrochloric acid, boiled, cooled, and made to volume. An aliquot is evaporated with perchloric acid. Ascorbic acid, an acetate buffer, and Chromazurol S solution are added and the colour intensity read at 545 nm [10].

D. 8-Hydroxyquinoline

For micro-analysis of silicate rocks, aluminium can be determined spectrophotometrically as the 8-hydroxyquinolate after ligand exchange in the organic phase with aluminium acetylacetonate extracted into benzene from aqueous medium at *p*H 6–7 [20]. Interfering elements such as Fe, Ti, V, and Zr are removed by extraction into *o*-dichlorobenzene from 1M hydrochloric acid, but beryllium is not removed by this procedure.

E. Other Colorimetric Procedures

Aluminium in soil extracts has been determined spectrophotometrically with xylenol orange, iron interference being eliminated with EDTA [41].

4. Optical Spectrography

Small amounts of aluminium are readily determined by emission spectrography in many materials by the sensitive lines 3082·2, 3092·7, 3944·0, and 3961·5 Å. Detailed information is available for metals [8], geological materials [2], plants [7], and alloy steels [37]. The spectrographic analysis of bauxite for 20–60% Al_2O_3, 0–60% Fe_2O_3, 0–50% SiO_2, and 0–10% TiO_2 has been described [12].

5. X-Ray Spectrography

Aluminium, like all other elements having an atomic number lower than that of titanium, cannot be determined by conventional X-ray fluorescence. With a vacuum spectrograph and special techniques it is possible to determine aluminium, but the procedure is not yet applicable to routine analysis [1].

6. Atomic-Absorption Spectrophotometry

Small quantities of aluminium can be rapidly and accurately determined by atomic absorption; the most sensitive line is a doublet at 3093 Å. Like other metals which form very stable oxides, aluminium is best determined with a nitrous oxide–acetylene flame. Some workers have reported interferences in the presence of large quantities of Cu, Zn, Fe, Ca, and Cl, whereas others observed only slight effects which could be overcome by matching the composition of the standards and unknowns. The use of atomic absorption in geology, metallurgy, and related fields is fully covered in recent reference books and papers [6, 19, 25, 34, 43, 45, 46].

7. Miscellaneous Determinations

(a) For very small quantities of aluminium, for instance in boiler waters where it may occur in the range of one part in 10^9, the fluorescence of the aluminium-morin complex may be utilized [51].

(b) Determinations of other elements present in aluminium and its alloys, and in bauxite and various aluminas, are given in numerous references [3, 4, 5, 8, 15, 16, 23, 26, 47].

(c) An X-ray diffraction method for Al_2O_3 in bauxite exploration samples has been published [11].

For assistance in any special problem in aluminium analyses, a chemist may seek advice from:

Alcoa Research Laboratories,
New Kensington, Pa., U.S.A.

The British Aluminium Company Ltd.,
Norfolk House, St. James's Square,
London, S.W.1

Alcan Research and Development Laboratories Ltd.,
Kingston, Ont., Canada

REFERENCES

1. ADLER, I., *X-Ray Emission Spectrography in Geology*, New York, Elsevier, 1966.
2. AHRENS, L. H., and TAYLOR, S. R., *Spectrochemical Analysis: A Treatise on the D.C. Arc Analysis of Geological and Related Materials*, 2nd Ed., Reading, Mass., Addison-Wesley, 1964.
3. ALCOA RESEARCH LABORATORIES, *Analysis of Bauxite, Calcined Alumina, Hydrated Alumina*, New Kensington, Pa., Alcoa Research Laboratories, 1960.
4. ALCOA RESEARCH LABORATORIES, *Analysis of Activated Alumina, Tabular Alumina, and Miscellaneous Materials*, New Kensington, Pa., Alcoa Research Laboratories, 1960.
5. ALUMINUM COMPANY OF AMERICA, *Chemical Analysis of Aluminum*, New Kensington, Pa., Aluminum Company of America 1950.
6. ANGINO, E. E., and BILLINGS, G. K., *Atomic Absorption Spectrometry in Geology*, New York, Elsevier, 1967.
7. A.O.A.C., *Official Methods of Analysis of the Association of Official Analytical Chemists*, Washington, D.C., Association of Official Analytical Chemists, 1965.
8. A.S.T.M., *Chemical Analysis of Metals. Sampling and Analysis of Metal Bearing Ores*, Philadelphia, Pa., American Society for Testing and Materials, 1969.
9. BENNETT, H., and HAWLEY, W. G., *Methods of Silicate Analysis*, 2nd Ed., New York, Academic Press, 1964.
10. BHARGAVA, O. P., and HINES, W. G., *Anal. Chem.* **40**, 413–5 (1968).
11. BLACK, R. H., *Anal. Chem.* **25**, 743–8 (1953).
12. BLACK, R. H., and LEMIEUX, P. E., *Anal. Chem.* **29**, 1141–44 (1957).
13. BOK, L. D. C., and SCHULER, V. C. O., *J. South African Chem. Inst.* **XI**, No. 1, 1–8 (1958).
14. BORUN, G. A., *Anal. Chem.* **34**, 1166–7 (1962).
15. BRITISH ALUMINIUM COMPANY, *Chemical Analysis of Aluminium and its Alloys*, London, The British Aluminium Company, 1961.
16. BRITISH ALUMINIUM COMPANY, *Spectrochemical Analysis of Aluminium and its Alloys*, London, The British Aluminium Company, 1961.
17. BRITISH DRUG HOUSES, *The B.D.H. Book of Organic Reagents*, Poole, England, British Drug Houses, 1958.
18. BURKE, K. E., and DAVIS, C. M., *Anal. Chem.* **36**, 172–5 (1964).
19. CAPACHO-DELGADO, L., and MANNING, D. C., *Analyst*, **92**, 553–7 (1967).
20. CHALMERS, R. A., and BASIT, M. A., *Analyst* **93**, 629–32 (1968).
21. CLAASSEN, A., and BASTINGS, L., *Analyst* **92**, 614–7 (1967).
22. CLAASSEN, A., BASTINGS, L., and VISSER, J., *Analyst* **92**, 618–21 (1967).
23. COMMITTEE OF CHEMISTS CONVENED BY ALAR, *Modern Methods for the Analysis of Aluminium Alloys*, London, Chapman and Hall, 1949.
24. CORBETT, J. A., and GUERIN, B. D., *Analyst* **91**, 490–8 (1966).
25. ELWELL, W. T., and GIDLEY, J. A. F., *Atomic-Absorption Spectrophotometry*, 2nd Ed., Oxford, Pergamon, 1966.
26. FURMAN, N. H., ed., *Scott's Standard Methods of Chemical Analysis*, 6th Ed., Vol. 1, Princeton, N.J., Van Nostrand, 1962.
27. GILFRICH, J. V., *Anal. Chem.* **29**, 978–80 (1957).
28. HILL, U. T., *Anal. Chem.* **28**, 1419–24 (1956).

29. HILL, U. T., *Anal. Chem.* **38**, 654–6 (1966).
30. HILLEBRAND, W. F., LUNDELL, G. E. F., BRIGHT, H. A., and HOFFMAN, J. I., *Applied Inorganic Analysis*, 2nd ed., New York, Wiley, 1953.
31. HOPKIN and WILLIAMS LTD., *Organic Reagents for Metals*, Chadwell Heath, England, Hopkin and Williams, Vol. 1, 1955, Vol. 2, 1964.
32. HORTON, A. D., and THOMASON, P. F., *Anal. Chem.* **28**, 1326–8 (1956).
33. KOLTHOFF, I. M., and ELVING, P. J., *Treatise on Analytical Chemistry*, Part 11, Vol. 4, 367–439, New York, Interscience Publishers, 1966.
34. LAFLAMME, Y., *Atomic Absorption Newsletter* **6**, 70–1 (1967).
35. LEWIS, L. L., NARDOZZI, M. J., and MELNICK, L. M., *Anal. Chem.* **33**, 1351–5 (1961).
36. MAXWELL, J. A., *Rock and Mineral Analysis*, New York, Wiley–Interscience, 1968.
37. MCKAVENEY, J. P., BALDWIN, H. E., AND VASSILAROS, G. L., *J. Metals* **20**, 54–9 (1968).
38. MILNER, G. W. C., and WOODHEAD, J. L., *Anal. Chim. Acta* **12**, 127–37 (1955).
39. PARKS, T. D., JOHNSON, H. O., and LYKKEN, L., *Anal. Chem.* **20**, 148–51 (1948).
40. PENDER, H. W., *Anal. Chem.* **31**, 1107–9 (1959).
41. PRITCHARD, D. T., *Analyst* **92**, 103–6 (1967).
42. RABBITTS, F. T., *Anal. Chem.* **20**, 181–2 (1948).
43. RAMIREZ-MUNOZ, J., *Atomic-Absorption Spectroscopy, and Analysis by Atomic-Absorption Flame Photometry*, New York, Elsevier, 1968.
44. ROBERTSON, G., *J. Science Food and Agriculture* **1**, 59–63 (1950).
45. ROBINSON, J. W., *Atomic Absorption Spectroscopy*, London, Edward Arnold, 1966.
46. SLAVIN, W., *Atomic Absorption Spectroscopy*, New York, Interscience Publishers, 1968.
47. SNELL, F. D., and HILTON, C. L., *Encyclopedia of Industrial Chemical Analysis*, Vol. 5, New York, Interscience Publishers, 1967.
48. STRAFFORD, N., and WYATT, P. F., *Analyst* **72**, 54–6 (1947).
49. WATTS, H. L., *Anal. Chem.* **30**, 967–70 (1958).
50. WATTS, H. L., and UTLEY, D. W., *Anal. Chem.* **28**, 1731–5 (1956).
51. WILL, F., *Anal. Chem.* **33**, 1360–2 (1961).
52. YOUNG, R. S., *Industrial Inorganic Analysis*, London, Chapman and Hall, 1953.

2 Antimony

Antimony is found in small quantities in many minerals, ores, concentrates, smelter and refinery products, and it is an important constituent of many non-ferrous alloys.

The principal antimony mineral is stibnite, Sb_2S_3, but oxidized forms of the element are also found in nature. Traces of antimony are found in many ores of copper, lead, zinc, nickel, cobalt, silver, and other metals. It is an objectionable impurity in some refined metals, especially in copper, and sensitive methods for its determination are important in many laboratories of the mining and metallurgical industry.

Moderate or large amounts of this element are nearly always determined volumetrically; for small quantities colorimetric, spectrographic, atomic absorption or similar procedures are employed.

The chlorides of antimony are volatile in hydrochloric acid at low temperatures, $SbCl_3$ at 110°C and $SbCl_5$ at 140°C. Great care must accordingly be taken in decomposition to avoid losses, by using a bisulphate fusion, or mixtures of nitric, sulphuric, and hydrofluoric acids, followed if necessary by fusion of insoluble material in sodium carbonate or sodium hydroxide. Do not use boiling hydrochloric acid when bringing materials into solution, and in subsequent operations do not evaporate hydrochloric acid solutions to dryness or to a low volume. Another point to bear in mind is that antimony compounds tend to hydrolyse readily at low acidities; the acid concentration in solutions must be carefully watched.

ISOLATION OR SEPARATION OF ANTIMONY

Antimony belongs to the hydrogen sulphide group, and precipitation with this gas in 5–10% sulphuric or hydrochloric acid is often employed in preliminary separations [1, 12, 13, 14, 24, 28]. Antimony, with As, Sn and Mo, is soluble in alkaline sulphide solution, and can thereby be separated from Cu, Cd, Bi, and Pb which are insoluble. Arsenic is precipitated with hydrogen sulphide in strong hydrochloric acid, such as 2 : 1 HCl, whereas antimony remains in solution. Arsenic is frequently separated from antimony by distillation of arsenious chloride from hydrochloric acid at a temperature of 108–112°C. Antimony can be precipitated and separated from molybdenum by boiling with high purity lead in 20% hydrochloric acid, or it can be collected with 20–30 times its weight of iron in an ammonia separation, and re-precipitated to remove occluded molybdenum. Tin can be separated from antimony by precipitation of the latter with metallic iron in a hot solution of 10% by volume concentrated hydrochloric acid, adding more acid as the iron dissolves.

Antimony can also be separated from arsenic and tin by distillation. To the mixed sulphates in a distilling flask add 100 ml hydrochloric acid and a reducing agent such as hydrazine sulphate or sulphurous acid. Distill the arsenic at 108–112°C, adding more hydrochloric acid if necessary, until the receiver contains all the arsenic in a volume of 50–100 ml. Change the receiver, add 7 ml of 85% phosphoric acid, and introduce hydrochloric acid at the rate of 30–40 drops per minute, keeping the temperature between 155° and 165°C. Add more hydrochloric acid if necessary until the receiver contains all the antimony distilled over in a volume of 100 ml. Cool the solution to 140°C, change the receiver, and add a mixture of 3 volumes hydrochloric acid to 1 volume hydrobromic acid at the rate of 30–40 drops per minute. Continue distilling the tin until a volume of about 75 ml has been collected [1, 12, 13].

It is often convenient to collect a small quantity of antimony from a sample containing a high content of copper, nickel, cobalt, or molybdenum by a ferric hydroxide separation [12, 13]. If there is not at least 20 times as much iron as antimony in the material, add sufficient ferric iron, make ammoniacal, boil, allow to stand 30–60 minutes, filter, and wash. Transfer the precipitate to the original beaker, dissolve in hydrochloric acid, re-precipitate with ammonia, boil, and filter through the original paper, washing thoroughly to eliminate all Cu, Ni, Co, or Mo. Antimony, with As, Sn, P, Bi, Pb, Se, and Te, is in the precipitate.

Another useful separation method for antimony is co-precipitation with manganese dioxide [1, 15, 16, 20, 21]. To a 10% nitric acid solution add 10 ml of 2% potassium permanganate solution and 20 ml of 10% manganese nitrate solution. Boil, allow to stand for 30 minutes, filter, and wash with hot 10% nitric acid. Antimony, with As, Sn, Se, Te, and Bi, is in the precipitate.

Occasionally it may be helpful in separations to remember that antimony, like arsenic, does not form a dithizonate [23]. Extraction in dilute acid with a solution of dithizone in carbon tetrachloride or chloroform will separate the dithizonates of Cu, Hg, Ag, Au, Pt, and Pd; the remainder of the hydrogen sulphide group, i.e. Bi, Cd, Pb, and Sn, can be extracted as dithizonates from a weakly basic solution. Antimony, with arsenic, is left in the upper aqueous phase.

The separation of antimony from lead merits a brief discussion. Fortunately, lead does not interfere in the volumetric determinations for antimony by potassium permanganate or potassium bromate. The latter titrant, with its sharper end point, is preferred when a large amount of lead is present. If antimony is to be determined colorimetrically, or by atomic absorption, in a sample containing an appreciable quantity of lead there is the problem of obtaining a clear solution, containing all the antimony, from the residue following decomposition in sulphuric acid. Lead sulphate occludes a significant quantity of antimony, and various alternatives to conventional filtration must be employed:

(a) Antimony can be separated from large amounts of lead by the manganese dioxide collection in dilute nitric acid, mentioned above. In this way, all the antimony, arsenic, tin, and thallium, with some of the gold, and a little lead and manganese, can be collected in a small precipitate from a 50-g sample of lead.

(b) Antimony can be distilled at about 200°C as antimony trichloride from a solution of hydrochloric acid containing lead, as discussed later.

(c) Antimony sulphide can be dissolved in a solution of sodium hydroxide and sodium sulphide, lead sulphide remaining insoluble; this is described subsequently.

(d) Trivalent antimony is extracted by cupferron–chloroform in 10% by volume hydrochloric acid, whereas lead is not.

(e) The amount of antimony occluded by lead sulphate is considerably decreased if the solution contains 1% tartaric acid. Some lead remains in solution but this is generally well below the interference level, and traces of occluded antimony may usually be disregarded in technical analyses of ores and metallurgical products.

(f) Solution and re-precipitation of lead sulphate will liberate most of the antimony, but the operation is rather lengthy. Dissolve in saturated ammonium acetate solution, add sulphuric acid, evaporate to strong fumes of the latter, cool, dilute, boil, allow to stand several hours, and filter off the lead sulphate. As an alternative procedure, boil the lead sulphate precipitate, containing occluded antimony, with an excess of ammonium hydroxide. Pour, with stirring, into enough nitric acid to leave an excess of 10 ml acid per 100 ml of solution. To the clear solution add sufficient sulphuric acid to precipitate the lead, evaporate to fumes, cool, dilute, boil, allow to stand, and filter as usual.

(g) A substantial quantity of lead sulphate can be kept in solution by the use of hydrochloric acid and sodium chloride. For example, 1 g of lead sulphate will not precipitate at 20°C if it is dissolved in a hot solution of 50 ml hydrochloric acid and 25 g sodium chloride in a total volume of 200 ml.

1. Volumetric Determination

After careful decomposition to avoid loss from the volatility of its chloride, antimony is separated with hydrogen sulphide in an acid or alkaline solution, arsenic is removed, and antimony is reduced and titrated with a standard oxidant.

A. Titration with Standard Potassium Permanganate

Weigh out 0·5–2 g sample into an 8-oz Pyrex pear-shaped flask, often called a "copper flask", add 2–4 g anhydrous sodium sulphate, 8–15 ml concentrated sulphuric acid, and $\frac{1}{8}$ of a 9 cm filter paper. Heat on a hot plate until fumes appear, place on a wire gauze over a burner in a fume hood, and slowly increase the temperature until all carbonaceous matter has been destroyed, the flask contents are clear, and all sulphur has been driven off the walls of the flask. Other fusion mixtures can be used, for example 10 g potassium sulphate, 10 ml sulphuric acid, and $\frac{1}{8}$ of a 9 cm filter paper, or 2·5 g ammonium sulphate, 0·5 g potassium sulphate, 8 ml sulphuric acid, and 0·2 g sulphur.

Swirl the flask when cooling to distribute the melt around the sides near the bottom of the flask. Add 50 ml warm water, 20 ml 1 : 1 sulphuric acid, warm until all soluble salts are in solution, then boil for a few minutes to remove sulphur dioxide, and cool.

This method of decomposition will bring into solution all antimony in nearly every type of ore and metallurgical product; a few materials may require further treatment:

(a) If it is evident that the residue in the flask is not merely silica, filter it through a small paper and return to the flask with 5–10 ml nitric acid, and the same quantities of sodium sulphate and sulphuric acid which were used originally. Repeat the heating over a flame until a colourless or white melt is obtained.

(b) Filter off the material which is insoluble in the original bisulphate fusion, wash, dry, ignite at a low temperature in an iron crucible and fuse with sodium hydroxide. Cool, dissolve in dilute sulphuric acid, and add to the original filtrate.

(c) Filter the residue from the bisulphate fusion, wash, ignite at a low temperature in a platinum crucible, and fuse at a low heat with a mixture of equal parts of sodium carbonate and sulphur. Dissolve the solidified melt in water, acidify with sulphuric acid, and filter. If a residue still remains, treat with potassium hydroxide and hydrogen peroxide, acidify with sulphuric acid, evaporate to fumes, dissolve in water, and add to the original filtrate.

An alternative to the fusion procedures given above is to decompose the sample in a platinum dish with sulphuric, hydrofluoric, and nitric acids, and evaporate to strong fumes of sulphuric.

Because arsenic is frequently associated with antimony, it must usually be removed at this stage. This may be conveniently done by precipitating arsenic with hydrogen sulphide in strong acid solution; antimony and other members of the acid sulphide group remain in solution. Adjust the acid strength to 9 N hydrochloric acid, i.e. 2 volumes of hydrochloric acid to 1 volume water, and pass in hydrogen sulphide at room temperature until all arsenic is precipitated. Filter and wash with the same strength of acid. Dilute the filtrate with double its volume of warm water, and saturate with hydrogen sulphide. Filter and wash thoroughly with 10% hydrochloric acid saturated with hydrogen sulphide.

Treat the sulphides on the filter paper with 15 ml of 20% sodium hydroxide and 10 g sodium sulphide crystals. Dilute to 100 ml and warm gently on a steam bath for an hour. Filter and wash with a dilute solution of sodium hydroxide–sodium sulphide. Antimony and tin are in the filtrate, free from Cu, Bi, Pb, Cd, etc. Boil to remove all hydrogen sulphide.

Acidify with sulphuric acid and evaporate to fumes of this acid. Cool, adjust acidity roughly to 1 : 2 sulphuric acid, add 2–3 ml saturated sulphur dioxide water, and boil until all sulphur dioxide is expelled. Cool, adjust the acid concentration to 10 ml sulphuric acid and 10 ml hydrochloric acid per 100 ml. Cool to 10°C and titrate with standard potassium permanganate solution until a faint pink tint persists for 5 seconds. The end point is sharp, but the presence of hydrochloric acid causes the subsequent gradual fading of the pink colour.

If the strength of permanganate is 0·1 N, or approximately 3·16 g per litre, then 1 ml = 0·00609 g Sb. Weaker solutions are frequently used in industrial practice. The titration is stoichiometric and the value of the permanganate solution, determined with a primary standard such as sodium oxalate, can be used for calculation. Metallic antimony in a high state of purity is readily available, and may also be used to standardize the potassium permanganate solution.

$$5Sb_2(SO_4)_3 + 4KMnO_4 + 16H_2SO_4 \rightarrow 5Sb_2(SO_4)_5 + 4MnSO_4 + \\ + 2K_2SO_4 + 16H_2O$$

$$5SbCl_3 + 2KMnO_4 + 16HCl \rightarrow 5SbCl_5 + 2MnCl_2 + 2KCl + 8H_2O$$

In Presence of Tin. When tin is desired on the sample, it may be determined iodimetrically on the same solution after the permanganate titration of antimony.

In Presence of Arsenic. If arsenic is required on the sample, it can be conveniently determined initially by the volatilization of arsenic trichloride, keeping the temperature of the vapour below 108°C; the analysis for antimony is carried out on the residual liquid in the distilling flask. Ferrous sulphate, hydrazine sulphate, or sodium sulphite should be used instead of cuprous chloride as the reducing agent for arsenic, to avoid the introduction of large quantities of copper which would later require removal from antimony. After the volatilization of arsenic, transfer the contents of the distilling flask to a beaker, dilute with water until a concentration of about 10% v/v hydrochloric acid is obtained, and pass in hydrogen sulphide until all antimony and other metals of this group are precipitated. Dissolve antimony in sodium hydroxide and sodium sulphide, and complete the determination as described above.

If arsenic is not required on the sample, it can be volatilized, without distilling, as follows. Dissolve the cooled fusion in the Pyrex flask with a small amount of water, transfer to a 400 ml beaker with several granules of silicon carbide to prevent bumping. Add 100 ml of hydrochloric acid and 10 ml of a saturated sodium sulphite solution. Cover the beaker, heat to gentle boiling, and continue boiling until the contents have evaporated to about 50 ml. This should expel arsenic and the excess of sulphite; samples high in arsenic may require a repetition of this reduction and volatilization step [1, 12]. There is no danger of volatilizing significant quantities of antimony unless the temperature of the hot plate is high, boiling is very vigorous, and the volume of solution is reduced below that specified above.

In Absence of Arsenic and Copper. Many samples will have only traces of arsenic or copper, and an initial precipitation with hydrogen sulphide in acid solution will separate antimony from its principal interferences in titrimetry. Return the filter paper containing the washed sulphides to the Pyrex fusion flask, add sulphuric acid and anhydrous sodium sulphate or other sulphates, and fuse as described previously. Add more sulphuric acid, if necessary, until all carbonaceous matter has been destroyed, sulphur has been completely volatilized, and the melt is clear. Allow the flask to cool, swirling to distribute the melt around the walls, add water and sulphuric acid if necessary to bring its concentration to about 10% by volume, add 10 ml of hydrochloric acid per 100 ml of solution, and titrate as described.

Isolation of Antimony by Volatilization. If desired, antimony can be isolated by volatilization, from the residue following distillation of arsenic at 108°C, by adding hydrochloric acid and zinc chloride, raising the temperature to about 200°C and evaporating to a low volume. The residual reducing agent in the arsenic volatilization may be sufficient to keep the antimony in the trivalent state; if not, add a little more hydrazine sulphate, because some quinquevalent antimony compounds are not readily volatile at this temperature. Catch the distillate in a beaker of cold water, and carry out the titration with standard potassium permanganate or potassium bromate.

In Presence of Selenium and Tellurium. When a sample contains selenium and tellurium above the trace level, these elements tend to accompany antimony in most procedures, even in those involving distillations of arsenic and antimony. It is preferable to remove them initially by sulphur dioxide in 30–40% hydrochloric acid by volume, containing 10–20 ml of a 15% solution of hydrazine hydrochloride. Filter off the precipitated selenium and tellurium, wash with 2% hydrochloric acid, and evaporate the filtrate gently to remove sulphur dioxide. Proceed with the elimination of arsenic by distillation, and the subsequent determination of antimony.

B. Titration with Standard Potassium Bromate

Many chemists, including the writer, prefer to use potassium bromate as the oxidizing titrant instead of potassium permanganate [1, 4, 8, 12, 24, 28]. Methyl orange is used as indicator in a hot solution; the dyestuff is irreversibly bleached at the end point by the action of free bromine. In addition to giving a sharper end point, potassium bromate has the advantage of being a primary standard which can be weighed directly, and it is stable in solution for a long period. The reactions may be represented:

$$3SbCl_3 + KBrO_3 + 6HCl \rightarrow 3SbCl_5 + KBr + 3H_2O$$
$$KBrO_3 + 5KBr + 6HCl \rightarrow 6KCl + 3H_2O + 3Br_2$$

Decompose the sample of ore or other material by a bisulphate fusion in a Pyrex flask as previously described, or by any other convenient procedure to avoid the use of hydrochloric acid. Separate the antimony with other members of its group, by precipitation in 5–10% sulphuric acid with hydrogen sulphide. If arsenic, or more than traces of copper, are present, remove them by one of the separation methods discussed earlier.

Heat the sample containing antimony in the trivalent state, in 5–10% by volume sulphuric acid and 20% by volume hydrochloric acid, to 80–90°C. Add 3–4 drops of an aqueous 0·1% solution of methyl orange, and titrate with a standard potassium bromate solution to the disappearance of the red colour. The end point is a little slow; the final additions of potassium bromate should be made drop by drop with vigorous stirring, to avoid over-titration. A 0·1 N solution of potassium bromate contains 2·7835 g per litre, and has a value of 1 ml = 0·00609 g antimony.

2. Colorimetric Determination

Very small quantities of antimony can be determined by the reaction of rhodamine B with quinquevalent antimony, extraction of this compound with benzene, and its photometric measurement at 565 nm [7, 9, 11, 12, 15, 16, 19, 21, 23, 25, 26]. The method is sensitive and satisfactory, but the colour intensity depends upon strict adherence to amount of reagents, order of addition, time, temperature, and other variables for the preparation of the calibration curve and the determination of unknowns.

Decompose the sample by bisulphate fusion in a Pyrex flask, treatment in a platinum dish with hydrofluoric, sulphuric, and nitric acids, or by any other suitable method which avoids evaporation with hydrochloric acid. For a very low content of antimony, a convenient initial enrichment can be effected by co-precipitation with ferric hydroxide for metals or oxides of copper, cobalt, nickel, or molybdenum; co-precipitation with oxides of manganese in dilute nitric acid serves the same purpose for lead, brasses, and bronzes.

The method described below is a general one, suitable for ores or materials which do not contain appreciable amounts of metals precipitated by hydrogen sulphide in acid medium. Antimony is initially precipitated as sulphide, copper being used as collector [23]. The following metals can be tolerated in quantities at least 100 times that of antimony: As, Sn, Bi, Pb, Cd, Hg, Mo, Au, Ag, Pt, Pd, Se, and Te.

To about 50 ml of 0·5N sulphuric or hydrochloric acid solution containing 0·5–2·5 μg of antimony, add 2 g tartaric acid and 3 ml of 10% hydroxylamine hydrochloride; the latter is sufficient to reduce 0·1 g of ferric iron. Boil for 5–10 minutes, cool, and add 1 ml of copper sulphate solution containing 0·40 g $CuSO_4{\cdot}5H_2O$ in 100 ml of water. Pass in hydrogen sulphide for 10 minutes, heat to nearly boiling, and continue the passage of gas for another 10 minutes. Allow to stand for 3 hours or preferably overnight.

Filter through a small porcelain filter crucible, and wash 3–4 times with 0·5N sulphuric acid saturated with hydrogen sulphide. Discard the filtrate and washings. To the crucible add 3 ml hydrochloric acid and 1 ml nitric acid; after several minutes draw the acid through the crucible and wash with

5 ml of 1 : 1 nitric acid and finally with water. Evaporate the combined solutions in a small covered beaker on a steam bath or low-temperature hot plate to a volume of a few drops. Add 0·5 ml of 1 : 2 sulphuric acid and heat just to fumes of sulphuric acid. Cool and add 5 ml of sulphurous acid solution. Cover the beaker and boil gently for 10 minutes or until the volume has been reduced to 3–4 ml. Gold, if present, is precipitated with sulphurous acid and may be filtered off at this point.

Cool and transfer the solution to a small short-stemmed separatory funnel, and add 5 ml of hydrochloric acid to make the mixture 7 N in hydrochloric acid. Cool to room temperature, add 5 ml of isopropyl ether and shake for 30 seconds. Allow the phases to separate, run the aqueous layer into another separatory funnel, rinse with 1 : 1 hydrochloric acid, and shake with another 5 ml portion of ether. Draw off the aqueous layer into a separatory funnel, rinse with 1 : 1 hydrochloric acid, wash the combined ether phases with 1 ml or less of 1 : 1 hydrochloric acid and add the latter to the aqueous phase.

To the aqueous solution containing the antimony add enough ceric ammonium sulphate solution to give a yellow colour, about 2 ml being usually required; this is made by dissolving 64 g $(NH_4)_4Ce(SO_4)_4 \cdot 2H_2O$ in 1N sulphuric acid and diluting to 1 litre. Mix, and after exactly 1 minute blow air into the funnel for a few seconds to remove any free chlorine that may be present. Immediately add 5 drops of 1% hydroxylamine hydrochloride, stopper the funnel and shake. The yellow colour of ceric sulphate should disappear. Again blow air into the funnel for 15–20 seconds. Without delay add 3 ml of 85% phosphoric acid, mix, and follow with 1·0 ml of 0·2% aqueous solution of rhodamine B solution. Phosphoric acid is added to overcome the effect of traces of iron which may have contaminated the copper sulphide precipitate.

Mix and add 5·0 ml of reagent-grade benzene. Shake vigorously for 1 minute to extract rhodamine B chloroantimonate. Allow the funnel to stand for about 5 minutes for droplets of water to separate from the benzene phase. Drain off and discard the aqueous layer. Dry the stem of the funnel with a roll of filter paper. Filter the benzene through a small plug of glass wool into a 1-cm cell. Measure the absorbance at 565 nm.

Prepare the calibration curve by taking 0, 0·5, 1·0, 2·0, and 3·0 μg of antimony, adding 10 ml of 6N hydrochloric acid and 6·5 ml of ceric ammonium sulphate, and proceeding as described above. Run a blank on the reagents.

The rhodamine B method for low concentrations of antimony is deservedly popular in the mining and metallurgical field; detailed procedures are available for copper [9, 11], nickel [15], zinc [26], arsenic [25], germanium [19], and lead [16].

3. Optical Spectrography

Small quantities of antimony may be determined by emission spectrography, using the lines 2311·5 or 2598·1 [2]. Procedures have been described for antimony in copper and copper alloys in the range of about 7 ppm to 0·1% [9], for 0·001–0·03% in pig lead [1], 0·001–1% in tin alloys [1], 2–100 ppm in nickel [15], and 0·4–0·8% in aluminium alloys [6].

4. X-Ray Spectrography

Antimony can be determined by X-ray fluorescence over a wide range of concentration, generally using the K_α lines of 0·4719 or 0·9437 [22, 24].

A novel procedure for trace amounts of antimony consists of co-precipitation with hydrogen sulphide, cupferron, or phenylfluorone, filtration on a paper disc, and determination by X-ray spectrography [18].

5. Atomic-Absorption Spectrophotometry

Small to moderate amounts of antimony can be determined by atomic absorption [10]. The most sensitive absorption line is 2175·9; lines 2068·4 and 2311·5 may also be used.

When appreciable lead is present a complication arises, because the usual filtration step to obtain a clear solution for atomic absorption results in the retention of some antimony in the lead sulphate precipitate. Methods to overcome this error have been discussed earlier, but they do detract from the normally rapid and selective atomic absorption procedure.

The sensitivity and selectivity of atomic absorption methods for antimony can be enhanced by extraction of the chloride complex into n-amyl acetate [27], or methyl isobutyl ketone [29].

6. Miscellaneous Determinations

(a) A polarographic procedure for $< 0·5\%$ antimony in aluminium alloys has been published [5]. Antimony is distilled as trichloride in an inert atmosphere, and the distillate is polarographed in acid solution, using gelatin as suppressor.

(b) Although the selectivity of rhodamine B favours its use for most colorimetric determinations, a few are based on the yellow colour which antimony in acid solution gives with potassium iodide. Methods have been outlined for aluminium alloys [5], lead [3], and miscellaneous materials [12, 17, 23].

REFERENCES

1. A.S.T.M., *Chemical Analysis of Metals. Sampling and Analysis of Metal Bearing Ores*, Philadelphia, Pa., American Society for Testing and Materials, 1969.
2. A.S.T.M., *Methods for Emission Spectrochemical Analysis*, 5th ed., Philadelphia, Pa., American Society for Testing and Materials, 1968.

3. Bassett, J., and Jones, J. C. H., *Analyst*, **91**, 176–9 (1966).
4. Bradshaw, G., *Analyst* **88**, 599–602 (1963).
5. British Aluminium Co., *Chemical Analysis of Aluminium and its Alloys*, London, British Aluminium Co., 1961.
6. British Aluminium Co., *Spectrochemical Analysis of Aluminium and its Alloys*, London, British Aluminium Co., 1961.
7. British Drug Houses, *The B.D.H. Book of Organic Reagents*, Poole, England, British Drug Houses, 1958.
8. Brown, E. G., Forshaw, I. P., and Hayes, T. J., *Metallurgia* **55**, 45–7 (1957).
9. Dozinel, C. M., *Modern Methods of Analysis of Copper and its Alloys*, 2nd ed., London, Elsevier, 1963.
10. Elwell, W. T., and Gidley, J. A. F., *Atomic-Absorption Spectrophotometry*, 2nd ed., Oxford, Pergamon, 1966.
11. Elwell, W. T., and Scholes, I. R., *Analysis of Copper and its Alloys*, Oxford, Pergamon, 1967.
12. Furman, N. H., ed., *Scott's Standard Methods of Chemical Analysis*, 6th ed,. Vol. 1, Princeton, N.J., Van Nostrand, 1962.
13. Hillebrand, W. F., Lundell, G. E. F., Bright, H. A., and Hoffman, J. I., *Applied Inorganic Analysis*, 2nd ed., New York, Wiley, 1953.
14. Kolthoff, I. M., and Elving, P. J., *Treatise on Analytical Chemistry*, Part 11, Vol. 10, New York, Interscience Publishers, 1970.
15. Lewis, C. L., Ott, W. L., and Sine, N. M., *The Analysis of Nickel*, Oxford, Pergamon, 1966.
16. Luke, C. L., *Anal. Chem.* **31**, 1680–2 (1959).
17. Luke, C. L., *Anal. Chim. Acta* **39**, 447–56 (1967).
18. Luke, C. L., *Anal. Chim. Acta* **41**, 237–50 (1968).
19. Luke, C. L., and Campbell, M. E., *Anal. Chem.* **25**, 1588–93 (1953).
20. Ogden, D., and Reynolds, G. F., *Analyst* **89**, 538–43 (1964).
21. Portmann, J. E., and Riley, J. P., *Anal. Chim. Acta* **35**, 35–41 (1966).
22. Rossouw, A. J., *S. African Ind. Chemist* **14**, No. 3, 44–6 (1960).
23. Sandell, E. B., *Colorimetric Determination of Traces of Metals*, 3rd ed., New York, Interscience Publishers, 1959.
24. Snell, F. D., and Hilton, C. L., *Encyclopedia of Industrial Chemical Analysis*, Vol. 6, New York, Interscience Publishers, 1968.
25. de Souza, T. L. C., and Kerbyson, J. D., *Anal. Chem.* **40**, 1146–8 (1968).
26. Van Aman, R. E., Hollibaugh, F. D., and Kanzelmeyer, J. H., *Anal. Chem.* **31**, 1783–5 (1959).
27. Walker, C. R., Vita, O. A., and Sparks, R. W., *Anal. Chim. Acta* **47**, 1–7 (1969).
28. Wilson, C. L., and Wilson, D. W., *Comprehensive Analytical Chemistry*, Vol. 1C, 252–64, London, Elsevier, 1962.
29. Yanagisawa, M., Suzuki, M., and Takeuchi, T., *Anal. Chim. Acta* **47**, 121-6 (1969).

3 Arsenic

Arsenic is widely distributed in nature, ranging from traces to substantial quantities in many ores of antimony, copper, nickel, cobalt, iron, lead, zinc, silver, gold, and platinum metals. Unlike antimony, arsenic is not an important constituent of non-ferrous alloys, occurring nearly always in low concentrations as an undesirable impurity. The high toxicity of arsenic makes it imperative to periodically check the amounts in gaseous and liquid effluents, and in solid wastes, from mining and metallurgical operations. The element may also be encountered occasionally in mining laboratories in wood preservatives, fungicides, herbicides, etc.

ISOLATION OR SEPARATION OF ARSENIC

When determining arsenic it must be kept in mind that volatilization of arsenious chloride occurs readily in hot hydrochloric acid solutions. Arsenic can be determined in nearly all materials by an oxidizing decomposition, followed by a reducing step and distillation as arsenious chloride from hydrochloric acid solution. If the temperature of the vapour is kept below 108°C the only interference in this procedure is the rare element germanium [7, 8, 9]. When the latter is present it may be separated by precipitation of arsenic with hydrogen sulphide in hydrochloric acid solution containing oxalic acid; germanium forms a complex which is not precipitated [15].

A common initial separation for arsenic is a hydrogen sulphide precipitation in 5–10% hydrochloric or sulphuric acid. It is accompanied by copper, bismuth, cadmium, lead, antimony, tin, gold, molybdenum, silver, and the platinum metals.

Arsenic may be separated from other members of the hydrogen sulphide group in several ways:

(a) Treatment of the mixed sulphide precipitate with alkaline sulphide dissolves As, Sb, Sn, and Mo from the precipitate of Cu, Bi, Cd, and Pb.

(b) Arsenic can be separated from Sb and Sn, by hydrogen sulphide precipitation in 9N hydrochloric acid, i.e. 2 parts concentrated hydrochloric acid to 1 part water.

(c) Arsenic, together with Sb, Bi, P, Sn, Pb, Se, and Te, can be separated from copper, nickel, cobalt, and molybdenum by collection in an iron precipitate with ammonia.

(d) Another co-precipitation separation for arsenic, with Sb, Sn, and Bi, is the addition of potassium permanganate and manganese nitrate to a dilute nitric acid solution of the sample, and filtration of the manganese oxides which have collected arsenic, etc.

(e) Arsenic may be separated from other members of the hydrogen sulphide group by the extraction of the latter with dithizone and carbon tetrachloride or chloroform. In dilute acid the dithizonates of copper, gold, mercury, palladium, platinum and silver are extracted; in weakly basic solution bismuth, cadmium, lead, and tin can be removed. Arsenic, with antimony, does not react with dithizone.

1. Volumetric Determination

In industrial laboratories, amounts of arsenic above trace levels are usually determined chemically by volumetric procedures. Oxidizing decomposition is followed by distillation of arsenious chloride in hydrochloric acid, in the presence of a reducing agent. The reduced arsenic in the distillate is titrated with a standard oxidant such as iodine or potassium bromate [1, 2, 6, 7, 8, 9, 16, 17]. Sometimes distillation is replaced by isolation with hydrogen sulphide in strong hydrochloric acid solution.

Bring the sample into solution with nitric acid, or a nitrochlorate mixture. Add ten ml of 1 : 1 sulphuric acid and evaporate the sample to strong fumes to eliminate all nitric acid. If it is suspected that arsenic may be present in any insoluble residue which remains at this stage, filter it off, ignite at a low heat, fuse with 9 : 1 mixture of sodium carbonate and potassium nitrate, and add the contents of the dissolved melt to the bulk of the sample in sulphuric acid solution.

Alternatively, and preferably for most materials in extractive metallurgy, decompose 0·5–2 g sample by placing it in a Pyrex flask with 2–4 g anhydrous sodium sulphate, 8–15 ml concentrated sulphuric acid and $\frac{1}{8}$ of a 9-cm filter paper, and heat on a wire gauze over a flame in a hood. The heating should be conducted slowly at first, then over full heat until all carbonaceous matter has been destroyed, copious white fumes have been evolved for several minutes, and all sulphur has been driven off the walls of the flask. Other fusion mixtures may be employed, for instance 10 g potassium sulphate, 0·2 g sulphur, and 15 ml sulphuric acid.

When decomposition is complete, swirl the flask when cooling to distribute the melt around the sides near the bottom of the flask. Add 50 ml water and warm on the hot plate until all soluble salts are in solution.

Add 50–150 ml hydrochloric acid, transfer to a 250 or 500 ml flask fitted with a thermometer and an outlet tube attached to a vertical condenser; the lower end of the latter dips into a beaker or small flask filled with cold water and kept cool by standing in ice or cold water. It is usually not necessary to have an elaborate distillation assembly. A more efficient absorption system consisting of a stoppered flask with a guard tube, and the end of the condenser outlet tube drawn to a point to give small bubbles passing through the absorbent, are illustrations of refinements which may occasionally be employed. Sulphur-free rubber stoppers must be used. Several types of all-glass distilling equipment are more suitable when many determinations are required. A few pieces of silica brick in the distillation flask will prevent bumping.

To reduce the arsenic to the arsenious condition, one of several reducing agents may be used: ferrous sulphate, cuprous chloride, sodium sulphite, or hydrazine sulphate. About 1–10 g of a reducing agent are usually added, and distillation continued until 50–100 ml of distillate have been collected. When antimony or other elements are to be subsequently determined in the residual liquid in the distilling flask, use sodium sulphite or hydrazine sulphate. A favourite reducing agent is 1 g of hydrazine sulphate and 1 g potassium bromide. If large quantities of arsenic are present, a further 25–50 ml hydrochloric acid are added and distillation continued to a low volume.

Remove the distilling flask and retain the contents for the determination of antimony or any other element desired. Wash down the condenser and add the washings to the distillate in the beaker or flask.

A. Titration with Iodine

Add a few drops of methyl orange to the distillate, followed by ammonium hydroxide until the indicator turns yellow, then add hydrochloric acid carefully until it just turns red. Cool in ice water and add 5–10 g sodium bicarbonate. Add 10 ml of 0·5% starch solution and titrate with 0·1N or similar standard iodine solution to a permanent blue. Subtract the titration blank obtained by carrying through a distillation on acid and reagents.

A 0·1N iodine solution, containing 12·7 g iodine per litre, is made by dissolving 25 g potassium iodide in a little water, adding the iodine, and, when the latter has completely dissolved, diluting to one litre with water. This is standardized against pure arsenious oxide. Dissolve 0·1320 g of the latter, equivalent to 0·1000 g arsenic, in a little dilute sodium hydroxide solution, dilute with water to about 100 ml, add several drops of phenolphthalein indicator and make slightly acid with hydrochloric acid. Cool, add 3-4 g sodium bicarbonate and 5 ml of 0·5% starch solution, and titrate with the iodine solution to a permanent blue colour. One ml of 0·1N iodine solution = 0·003746 g arsenic.

$$2H_3AsO_3 + 4NaHCO_3 + 2I_2 \rightarrow As_2O_5 + 4NaI + 5H_2O + 4CO_2$$

B. Titration with Potassium Bromate

Adjust the acidity of the distillate so that it contains 20 ml hydrochloric acid and 10 ml sulphuric acid per 100 ml [1, 2, 7, 8, 9, 16, 17]. This may be done by neutralizing with sodium hydroxide pellets, followed by the additions of acid. Heat the solution to about 80°C, add 2–3 drops of an aqueous 1% solution of methyl orange, and titrate with a standard potassium bromate solution to the disappearance of the red colour. Make the final additions of potassium bromate drop by drop with vigorous stirring, to avoid over-titration.

A 0·1N solution of potassium bromate contains 2·7835g/litre, and has a value of 1 ml = 0·003746 g arsenic. Methyl orange acts as the indicator in hot solution, being bleached at the end point by the action of free bromine. Potassium bromate is an excellent standard; the salt can be weighed out directly to give the theoretical titre, and the solution remains stable for a long period. If desired, potassium bromate can be checked against pure arsenious oxide by dissolving the latter in a little sodium hydroxide solution, adjusting the acidity and titrating in the manner specified above.

$$3AsCl_3 + KBrO_3 + 6HCl \rightarrow 3AsCl_5 + KBr + 3H_20$$
$$KBrO_3 + 5KBr + 6HCl \rightarrow 6KCl + 3H_2O + 3Br_2$$

SEPARATION OF ARSENIC IN 9N HYDROCHLORIC ACID

Instead of isolating arsenic by distillation, it may be separated from other members of its sub-group by precipitation with hydrogen sulphide in strong hydrochloric acid solution. After preliminary separation of arsenic with other members by hydrogen sulphide in 5–10% hydrochloric acid, the arsenic-antimony-tin subgroup is dissolved by treatment for an hour on the steam bath with 15 ml of 20% sodium hydroxide and 10 g sodium sulphide crystals in 50 ml water. Filtration gives As, Sb, Sn, and Mo in the filtrate; Cu, Bi, Cd, and Pb remain as precipitates. Boil out all hydrogen sulphide from the filtrate, acidify with hydrochloric acid until the solution is about 9N, i.e. 75% hydrochloric acid by volume. Precipitate the arsenic, free of other elements, by passing in a rapid stream of hydrogen sulphide at room temperature and allowing to stand an hour or longer. Filter, dissolve in sodium hydroxide, acidify with 1 : 1 sulphuric acid and add 10 ml excess, add 3–4 g potassium bisulphate, and a little tartaric acid or filter paper. Evaporate on a wire gauze over a burner to strong fumes of sulphuric acid until all carbonaceous matter has been destroyed, no sulphur remains on the walls of the flask, and the melt is clear. Cool, dissolve in water and proceed with titration, using iodine or potassium bromate solution as described above.

It is sometimes useful to realize that arsenic is precipitated much more readily with hydrogen sulphide if it is first reduced with sulphur dioxide in dilute sulphuric acid solution. Boil out all sulphur dioxide, add hydrochloric acid, and pass in hydrogen sulphide, when it will be found that a clean precipitate is quickly obtained.

Arsenic in nearly all samples of rocks, minerals, ores, concentrates, mattes, slags, speisses, flue dusts, slimes and other materials in the mining industry can be brought into solution by fusion in a Pyrex flask with a bisulphate mixture, or by treatment with nitric and sulphuric acids, followed if necessary by a sodium carbonate-potassium nitrate fusion, as described earlier. For the refractory mineral sperrylite, $PtAs_2$, the following procedure can be used; the method should be borne in mind for other substances which are difficult to decompose. Fuse a suitable weight of the sample with 4–5 times its weight of a 1 : 1 mixture of sodium peroxide and sodium carbonate in a nickel or zirconium crucible. Cool, leach with hot water, boil the suspension with alcohol, cool, and dilute to 250 ml in a flask. Filter a 200 ml aliquot through dry paper, and acidify with nitric acid. Collect the arsenic by co-precipitation with ferric hydroxide as outlined later, filter, and wash. Dissolve the precipitate in hydrochloric acid, and separate the arsenic by distillation as described earlier. Finish the determination volumetrically or colorimetrically.

If selenium and tellurium are present in more than traces, separate them before the arsenic distillation step because it is difficult to prevent their volatilization with the arsenic. Boil the sample under a reflux condenser in 1 : 1 hydrochloric acid solution until selenium and tellurium are reduced to the quadrivalent state, and cool. Dilute so that hydrochloric acid is 30–40% by volume, add 10–20 ml of 15% hydrazine hydrochloride, and pass in a stream of sulphur dioxide to precipitate all selenium and tellurium. Set aside for several hours, filter, and wash with 2% hydrochloric acid. Precipitate arsenic, with other members of its group, by hydrogen sulphide in the filtrate as described earlier, and proceed with the distillation step as outlined previously.

2. Colorimetric Determination

A very low content of arsenic can be determined with confidence by two colorimetric procedures, particularly when the element is initially separated by volatilization.

A. Molybdendum Blue Metho

(1) *General Procedure.* When ammonium molybdate is added to pentavalent arsenic to form a heteropoly molybdiarsenate, in the presence of a reducing agent such as hydrazine sulphate, "molybdenum blue", a substance of undetermined composition, is formed and can be measured photometrically [2, 6, 7, 8, 9, 10, 12, 14, 16]. The interference of phosphorus, silicon, tellurium, and nearly all other elements is eliminated by first isolating arsenic as the volatilized trichloride.

The only remaining interferences are the rare element germanium, which will distil with the arsenic and likewise give a molybdenum blue, and substantial quantities of selenium. The latter element tends to be carried over to a small extent in an arsenic distillation, and may cause an interference in the colour measurement. In the presence of germanium, arsenic should be separated by precipitation with hydrogen sulphide in hydrochloric acid solution containing oxalic acid; the latter forms a soluble complex with germanium. Another procedure when germanium is high depends on the extraction of arsenic by chloroform and diethylammonium diethyldithiocarbamic acid from a hydrochloric–oxalic acid solution [12]. When selenium is high, its prior separation by sulphur dioxide before the molybdenum blue measurement will result in no loss of arsenic [13].

In the molybdenum blue method the acid concentration is important; if the acidity is too low, even molybdate alone will give a blue colour, and if it is too high the colour due to arsenic is decreased in intensity.

Decompose the sample and distil arsenic as the volatile trichloride in the manner described previously for the volumetric determination. Add a few ml of bromine water to the vessel receiving the distillate to oxidize the arsenic as it distils over. Because the quantity of arsenic in the sample will be small, all with be distilled over in 15 minutes or less, in a volume of 10–15 ml.

To the distillate add 10 ml of nitric acid and evaporate at a low heat to dryness. Heat at 130–135°C for 30–60 minutes to remove nitric acid. Add 10 ml of hydrazine sulphate–ammonium molybdate solution for each 30 μg of arsenic, and heat the sample on a steam bath for 15 minutes. Cool, and dilute to volume with hydrazine sulphate–ammonium molybdate solution. The final solution should not contain more than 3 μg of arsenic per ml. Transfer a suitable portion to an absorption cell of a spectrophotometer, and compare the transmittance with that of a series of arsenic standards treated in the same way as the unknown. Photometric measurement is usually made at approximately 660 nm, but can sometimes be carried out at 840 nm.

Hydrazine sulphate–ammonium molybdate. Dissolve (a) 1 g of ammonium molybdate in 100 ml of 5 N sulphuric acid, and (b) 0·15 g of hydrazine sulphate in water. Dilute 10 ml of solution (a) to 90 ml with water, add 1 ml of solution (b), and dilute to 100 ml with water. Prepare the mixed solution fresh daily. Alternatively, to avoid the frequent preparation of the mixed solution which is not stable, add appropriate volumes of the individual solutions separately to the sample. To prevent introduction of silica, keep the ammonium molybdate solution in a polyethylene bottle.

Standard arsenic solution. Dissolve 0·1320 g of arsenic trioxide in 2 ml of 1N sodium hydroxide, dilute with water, add 100 ml of hydrochloric acid, transfer to a litre flask, make to the mark with water, and mix. Transfer 100 ml of this solution to a litre flask, add 40 ml of hydrochloric acid, make to the mark with water, and mix. This second flask contains 0·01 mg arsenic per ml.

(2) *Special Procedures*

The general method outlined above can be applied to all samples of ores. For some products of the mining and metallurgical industries, the final molybdate photometric determination is more conveniently preceded by different dissolution and separation steps. The following are typical examples.

(*a*) *Blister or refined copper.* Phosphorus and silicon are low in these materials, and the sample can be dissolved in nitric acid, arsenic oxidized to the pentavalent state, ammonium molybdate added, and the resulting complex extracted with n-butanol and measured at approximately 400 nm [2].

Dissolve 5 g in 30 ml 1 : 1 nitric acid, boil to remove oxides of nitrogen, cool, and add 1 ml of fresh 5% ammonium persulphate solution. Boil for several minutes, cool, transfer to a separatory funnel and dilute to 50 ml. Add 25 ml of ammonium molybdate–nitric acid solution, mix, add 10 ml of chloroform–butanol mixture, and shake for 30 seconds. Drain off and discard the lower layer. Add a second 10-ml portion of chloroform–butanol mixture and again extract, discarding the lower layer. These preliminary extractions with chloroform–butanol remove any phosphomolybdic acid which may have formed.

Add 15 ml of n-butanol to the solution in the separatory funnel, shake for 30 seconds, and drain and discard the aqueous layer. Dry the funnel stem with a roll of filter paper, and transfer the n-butanol extract to a dry 25-ml flask. Rinse the separatory funnel with two small portions of water-saturated n-butanol, and add the washings to the extract in the 25 ml flask. Dilute to the mark with water-saturated n-butanol and mix. Filter a sufficient amount of the solution through a 9-cm rapid filter paper into an absorption cell. Compare the optical density with a series of standards in a spectrophotometer at approximately 400 nm.

Prepare a standard solution of arsenic containing 0·05 mg arsenic per ml in the following manner. Transfer 0·0661 g arsenious oxide to a polyethylene beaker, add one pellet of sodium hydroxide and 10 ml of water, and swirl to dissolve. Add about 90 ml of water and pour into a Pyrex beaker containing 5 ml of 1 : 1 nitric acid. Heat to boiling and add 1% potassium permanganate solution until a precipitate persists. Add 3% hydrogen peroxide until the precipitate dissolves, boil for several minutes, cool, and transfer to a 1-litre flask. Dilute to the mark, mix, and store in a polyethylene bottle. Transfer 1–12 ml portions of this solution to a separatory funnel, dilute to 50 ml, add 25 ml of ammonium molybdate–nitric acid solution and proceed as directed above for the unknown.

Ammonium molybdate–nitric acid solution. Dissolve 15 g of ammonium molybdate in 400 ml of water, add 100 ml of nitric acid, and mix. Store in a polyethylene bottle.

Chloroform–butanol mixture. Mix 150 ml of chloroform with 50 ml of n-butanol in a glass-stoppered bottle.

Water-saturated n-butanol. Add 50 ml of water to 200 ml of n-butanol in a 500-ml separatory funnel, and shake for 30 seconds. Discard the aqueous layer, and store the n-butanol in a glass-stoppered bottle.

(*b*) *Copper and its alloys.* Another useful procedure for copper and its alloys utilizes a dissolution in a hydrochloric acid–hydrogen peroxide mixture, reduction of arsenic with hypophosphorus acid, extraction of arsenic into chloroform, oxidation, and final photometric measurement of molybdenum blue [6]. In this method the following can be tolerated, in per cent: lead 4, nickel 30, selenium or tellurium 2, silicon 3, tin 10, and zinc 40. To obtain quantitative reduction of pentavalent arsenic, the amount of copper in solution must not be less than 0·4 g. In practice, a 1-g sample is used, and when the weight of copper in this is less than 1 g a compensating amount of copper is added.

Place the beaker containing the sample in a cold-water bath, add 5 ml hydrochloric acid, and 3 ml of 30% hydrogen peroxide. When the initial reaction has subsided, add another 7 ml of 30% hydrogen peroxide to completely dissolve the copper. Evaporate the solution to dryness on a steam bath, cool, add 50 ml of hydrochloric acid, 3 ml of 50% hypophosphorus acid, and allow to stand for 5 minutes. Wash the solution into a dry separatory funnel with 10 ml of hydrochloric acid, add 25 ml of chloroform, and shake for 1 minute. Transfer the lower layer into a second dry separatory funnel, and discard the aqueous layer.

Add 10 ml of hydrochloric acid, shake for 30 seconds, transfer the chloroform layer into another dry separatory funnel, and discard the aqueous layer. Add 20 ml of water, shake for 1 minute, discard the chloroform layer, and transfer the aqueous layer into a 50-ml flask, rinsing with about 5 ml of water. Add the following reagents in the order given, washing the neck of the flask and mixing well after each addition:

5 drops of 0·1 N iodine solution
5 ml of 1% ammonium molybdate in 14% sulphuric acid solution
2 ml of 0·15% hydrazine sulphate solution.

Stand the flask in a boiling-water bath for 10 minutes, cool to 20°C, and dilute to the mark. Measure the optical density at 840 nm, using 2-cm cells.

If tellurium is over about 0·1%, after the initial evaporation to dryness, dissolve the cold residue in 30 ml of hydrochloric acid instead of 50 ml. After adding 3 ml of 50% hypophosphorus acid and allowing to stand for 5 minutes, filter into a dry separatory funnel, and wash three times with 10 ml of hydrochloric acid. To the filtrate add 25 ml of chloroform and proceed as described earlier.

(c) *Lead and lead alloys.* A low content of arsenic in pig lead and high-lead alloys may be collected by the manganese co-precipitation method [2]. Antimony, tin, bismuth, selenium, and tellurium accompany arsenic.

Dissolve 5–50 g of the sample in 1 : 4 nitric acid, boil to expel brown fumes, dilute to about 300 ml, and heat to boiling. Add 10 ml of 2% potassium permanganate, heat to boiling, add 20 ml of 10% manganese nitrate, and boil for 1–2 minutes. Filter, wash with hot water, and reserve the precipitate. To the warm filtrate add 35 ml of 1 : 1 sulphuric acid, stir, filter through fine paper on a Buchner funnel with suction, wash once with water, and discard the precipitate of lead sulphate.

Neutralize the filtrate with ammonium hydroxide and add 15 ml in excess. Heat the solution to boiling, add 10 ml of 10% ammonium persulphate solution, and boil for one minute. Filter, wash 3–4 times with hot water, and discard the filtrate. Combine this paper and precipitate with that retained initially, transfer to a flask, and add 15 ml of sulphuric acid and 35 ml of nitric acid. Boil gently to destroy filter paper, adding more nitric acid if necessary. Evaporate to dense white fumes to expel all nitric acid.

Cool, add about 0·2 g of hydrazine sulphate, wash down the walls of the flask with 15 ml of water, and evaporate to dense white fumes until the volume has been reduced to about 10–15 ml.

Cool to room temperature, add 50 ml of hydrochloric acid, 10 g of sodium chloride, and distil the arsenic as arsenious chloride by the method described earlier. Determine arsenic in the distillate by the photometric molybdenum blue method outlined previously, measuring the optical density at about 660 nm.

(d) *Nickel, cobalt, and copper.* The colorimetric determination of minute quantities of arsenic, by the heteropoly molybdenum blue method, in nickel, cobalt, and copper is most reliably performed after an initial co-precipitation with ferric hydroxide, followed by distillation of arsenious trichloride. Arsenic, antimony, tin, bismuth, lead, phosphorus, selenium, and tellurium are quantitatively precipitated from acid solutions of nickel, cobalt, and copper when the latter are made ammoniacal, provided iron is present to the extent of 10–20 times that of the total weight of impurity elements. In this convenient manner, traces of arsenic can be collected in a small precipitate of ferric hydroxide from large samples of 10, 25, or 50 grams if necessary. Subsequent distillation of arsenic leaves antimony, tin, bismuth, lead, phosphorus, tellurium, and most of the selenium in the distilling flask. If selenium is high it is preferable to remove it before the photometric measurement by saturating an approximately 1 : 1 hydrochloric acid solution of the sample with sulphur dioxide, allowing to stand overnight, and filtering the precipitated selenium.

The following procedure, intended for arsenic in metallic nickel [10], is illustrative of methods for this impurity in metals, oxides, and salts of nickel, cobalt, and copper. Dissolve 1–5 g of nickel in 20–30 ml of 1 : 1 nitric acid. Add 1 ml of ferric ammonium sulphate solution, which is made by dissolving 9 g of $FeNH_4(SO_4)_2.12H_2O$ in 100 ml of 3·6 M sulphuric acid. Dilute to 150–200 ml, make ammoniacal, and add 10 ml excess ammonium hydroxide. Boil, and allow to stand in a warm place for 30 minutes to coagulate the precipitate. Filter through a 9-cm Whatman No. 541 paper, and wash the precipitate with dilute ammonium hydroxide solution; discard the filtrate.

Transfer the precipitate from the paper to the original beaker, using a fine jet of water. Wash the paper with hot 4N hydrochloric acid and hot water, catching the washings in the original beaker. It is not necessary to do a re-precipitation, because the traces of nickel, cobalt, or copper occluded in the ferric hydroxide will not be distilled with the arsenic in the next step. Evaporate the solution carefully to 10 ml, transfer to a distillation flask, using hydrochloric acid to rinse the beaker. Add 2 ml of hydrobromic acid, 0·3 g of hydrazine sulphate, and 10 ml of hydrochloric acid. Immerse the end of the condenser into about 10 ml of cold water in a small beaker surrounded by ice water. Pass a stream of nitrogen or carbon dioxide slowly through the distillation apparatus, and boil gently until the temperature of the vapour reaches 108°C and the volume in the distillation flask is reduced to about 10 ml.

To the distillate add 10 ml of nitric acid and evaporate to dryness on a hot plate which does not exceed 130°C. Heat in an oven at 130°C for 30–60 minutes. Add 10 ml of hydrazine sulphate–ammonium molybdate solution for each 30 μg of arsenic present, and heat on a water bath for 15 minutes. Cool and dilute to 25 ml, or more if necessary, with the hydrazine sulphate–ammonium molybdate solution. Read the absorbance of the solution in a 1- or 2-cm cell at 650 mμ. Prepare a calibration curve, using aliquots of a standard arsenic solution containing 0·01 mg arsenic per ml.

Hydrazine sulphate–ammonium molybdate solution. Dilute 50 ml of 1% ammonium molybdate solution in 5N sulphuric acid to 450 ml with water; add 5 ml of 0·15% hydrazine sulphate solution, dilute to 500 ml and mix. Prepare this solution daily.

B. Gutzeit Method

Although the sensitivity and convenience of the molybdenum blue method have resulted in it becoming the favourite for low arsenic determinations in the mining and metallurgical industries, on rare occasions the Gutzeit method can be employed. In this classical test, less than 0·1 mg of arsenic can be volatilized as arsine, AsH_3, by the action of zinc in hydrochloric or sulphuric acid solution, and the colour produced on strips of paper impregnated with mercuric chloride compared with standard stains similarly prepared [1]. When extreme care is taken to maintain conditions constant, the Gutzeit test is highly sensitive and reproducible for very minute amounts of arsenic.

The equipment required consists of sensitized mercuric chloride paper and a Gutzeit apparatus, both obtainable from any chemical supplier. Folded dry lead acetate paper is placed in the lowest tube of the apparatus to remove hydrogen sulphide from the arsine; it is renewed in each test. In the tube above, glass wool moistened with lead acetate solution is packed; this removes the last traces of hydrogen sulphide and moistens the arsine. In the upper constricted tube of the apparatus is carefully placed the sensitized mercuric chloride paper.

The sample is prepared in the usual way by dissolving in nitric acid, removing the latter with sulphuric acid, and reducing the arsenic to the trivalent state with ferrous sulphate or other reducing agents. Elements such as Hg, Ag, Pt, Pd, Ni, Co, large amounts of copper and more than 0·1 mg antimony should be separated prior to the test. Distillation of arsenic as the trichloride, outlined previously, is the most effective way of eliminating interferences; if mercury is present it can be removed initially as mercurous chloride before the distillation step. When traces of arsenic are collected in a ferric hydroxide precipitate, and a re-precipitation is made, nickel, cobalt, and copper are passed into the filtrate. Precipitation with hydrogen sulphide in acid solution will separate arsenic, with Sb, Hg, Ag, Pt, Pd, and Cu from Ni and Co; treatment of the sulphides with ammonia and ammonium sulphide will leave Hg, Ag, Pd, and Cu insoluble. Arsenic, like antimony, does not form a dithizonate; successive extraction in acid and alkaline solution with dithizone and chloroform or carbon tetrachloride will remove Hg, Ag, Cu, Pt, Pd, Ni, and Co.

Place 50 ml of the sample solution, 3 ml sulphuric acid, 2 ml ferric alum solution, and 0·5 ml stannous chloride solution in the bottle of the apparatus. In their respective tubes, place the lead acetate paper and the glass wool saturated with lead acetate solution, and connect with the upper tube containing the strip of sensitized paper. Add 35 g of the arsenic-free zinc to the solution in the bottle and insert the stopper carrying the tubes. Shake gently and allow to stand in a bath at 25 °C for one hour. Remove the test paper and compare with standards.

REAGENTS

Ferric alum solution. 8·4 g ferric ammonium alum are dissolved in water containing 1 g of sodium chloride and 2 ml sulphuric acid, and diluted to 100 ml. Two ml of the solution = 0·1 g Fe_2O_3.

Lead acetate solution. Dissolve 1 g lead acetate in water, add enough acetic acid to clear the solution, and dilute to 100 ml.

Zinc. Treat $\frac{1}{3}$–$\frac{1}{6}$ inch mesh arsenic-free zinc with hydrochloric acid until the surface is clean, and wash with water.

Stannous chloride. Dissolve 8 g stannous chloride in 9·5 ml water containing 0·5 ml hydrochloric acid.

Lead acetate paper. Soak qualitative filter paper in the lead acetate solution described above, dry, and cut into strips 7 × 0·5 cm.

Standard Stains. A set of standard stained papers ranging from 0·001 to 0·05 mg As_2O_3 can be prepared by volatilizing known amounts of arsenic as described above. The stains should be protected from the deleterious action of moisture and light by dipping in molten paraffin wax and storing in a well-stoppered test tube in the dark.

3. Optical Spectrography

Small amounts of arsenic may be determined in many materials by emission spectrography. Methods have been published for 0·001–0·03% arsenic in pig lead and 0·001 to 0·3% in tin alloys [2, 3], 1–100 ppm arsenic in nickel [10], and about 1–100 ppm in copper [4].

4. X-Ray Spectrography

Arsenic can be determined by X-ray fluorescence, usually by the K_α line 1·1758 or K_β 1·0572 [9, 11].

5. Atomic-Absorption Spectrophotometry

Arsenic from low to moderate amounts can be determined by atomic absorption, using the most sensitive line 1937·0 Å, or other absorption lines 1972·0 and 1889·9 Å [5]. The sensitivity is reported as 5 μg/ml.

REFERENCES

1. A.O.A.C., *Methods of Analysis of the Association of Official Analytical Chemists*, Washington, D.C., Association of Official Analytical Chemists, 1965.
2. A.S.T.M., *Chemical Analysis of Metals. Sampling and Analysis of Metal Bearing Ores*, Philadelphia, Pa., American Society for Testing and Materials, 1969.
3. A.S.T.M., *Methods for Emission Spectrochemical Analysis*, 5th ed., Philadelphia, Pa., American Society for Testing and Materials, 1968.
4. DOZINEL, C. M., *Modern Methods of Analysis of Copper and its Alloys*, 2nd ed., London, Elsevier, 1963.
5. ELWELL, W. T., and GIDLEY, J. A. F., *Atomic-Absorption Spectrophotometry*, 2nd ed., Oxford, Pergamon, 1966.
6. ELWELL, W. T., and SCHOLES, I. R., *Analysis of Copper and its Alloys*, Oxford, Pergamon, 1967.
7. FURMAN, N. H., ed., *Scott's Standard Methods of Chemical Analysis*, 6th ed., Vol. 1, Princeton, N.J., Van Nostrand, 1962.
8. HILLEBRAND, W. F., LUNDELL, G. E. F., BRIGHT, H. A., and HOFFMAN, J. I., *Applied Inorganic Analysis*, 2nd ed., New York, Wiley, 1953.
9. KOLTHOFF, I. M., and ELVING, P. J., *Treatise on Analytical Chemistry*, Part 11, Vol. 10, New York, Interscience Publishers, 1970.
10. LEWIS, C. L., OTT, W. L., and SINE, N. M., *The Analysis of Nickel*, Oxford, Pergamon, 1966.
11. LUKE, C. L., *Anal. Chim. Acta* **41**, 237–50 (1968).
12. LUKE, C. L., and CAMPBELL, M. E., *Anal. Chem.* **25**, 1588–93 (1953).
13. REED, J. F., *Anal. Chem.* **30**, 1122–4 (1958).
14. SANDELL, E. B., *Colorimetric Determination of Traces of Metals*, 3rd ed., New York, Interscience Publishers, 1959.
15. SCHOELLER, W. R., and POWELL, A. R., *The Analysis of Minerals and Ores of the Rarer Elements*, 3rd ed., London, Charles Griffin, 1955.
16. SNELL, F. D., and HILTON, C. L., *Encyclopedia of Industrial Chemical Analysis*, Vol. 6, New York, Interscience Publishers, 1968.
17. WILSON, C. L., and WILSON, D. W., *Comprehensive Analytical Chemistry*, Vol. 1C, 237–51, London, Elsevier, 1962.

4 Barium

In only a few mining laboratories is the determination of barium a frequent analytical procedure. The important barium minerals are the sulphate, barite, and the carbonate, witherite. Small quantities of barium are found in many carbonate and silicate rocks. Barium is used in paints, cosmetics, papers, electronics, drilling muds, and a few other applications.

ISOLATION OR SEPARATION OF BARIUM

Even in complex materials, the determination of barium does not present major difficulties, the insolubility of barium sulphate serving to isolate it from most elements. The chief interfering substances are silica, lead, calcium, and strontium. Silica can be eliminated at the outset by treatment of the sample with hydrofluoric and sulphuric acids in a platinum dish, or by dehydration in hydrochloric or perchloric acid. Lead may be removed from the combined sulphate precipitate by a hot concentrated solution of ammonium acetate, in which lead sulphate is soluble. Alternatively, lead may be eliminated initially by hydrogen sulphide in dilute hydrochloric acid solution. If calcium and strontium are present, barium is finally separated as the chromate [2, 5, 6, 7, 9, 11]. This method depends on the fact that barium is precipitated from a solution slightly acid with acetic by an excess of ammonium chromate, whereas calcium and strontium remain in solution.

The alkaline earth sulphates rank, in order of decreasing solubility in water or dilute hydrochloric acid: calcium, strontium, and barium. The last is so insoluble that both barium and sulphur are usually determined by precipitating and weighing barium sulphate. The oxalates of the alkaline earths, in order of decreasing solubility in water or dilute ammonium hydroxide, are barium, strontium, and calcium. When the usual double precipitation of calcium oxalate is carried out, 3–4 mg of barium can be present in the sample without contaminating the final calcium precipitate. The order of decreasing solubility of chromates in water or dilute acetic acid is calcium, strontium, and barium; when barium is precipitated as chromate, a re-precipitation is required if strontium is present, but calcium chromate is so soluble that a double precipitation of the barium is not necessary unless calcium is preponderant.

1. Gravimetric Method as $BaSO_4$ or $BaCrO_4$

In the mining and metallurgical industries, barium is usually determined by precipitating and weighing as barium sulphate or barium chromate.

Weigh 0·5–2 g sample into a platinum dish, add 10 ml 1 : 1 sulphuric acid and 30 ml hydrofluoric acid. Evaporate to strong fumes of sulphuric acid; for materials high in silica it may be necessary to cool, wash down the sides of the dish with a small quantity of water, add 20 ml hydrofluoric acid, and again evaporate to copious white fumes. Cool, add water, boil, and filter. If the sample contains lead, wash the combined sulphate precipitate thoroughly with a hot concentrated solution of ammonium acetate to remove all the lead, finally washing several times with hot water. If calcium and strontium are absent, place the precipitate in a platinum crucible, dry, ignite to constant weight at 900°C, and weigh as $BaSO_4$.

$$BaSO_4 \times 0{\cdot}5885 = Ba$$

If calcium and strontium are present, mix the precipitate of alkaline earth sulphates after ignition with about 8 g sodium carbonate and fuse. When cool, place the crucible and contents in a beaker of warm water until the cake is detached from the crucible and has disintegrated, filter the sample and wash thoroughly with hot 0·2% sodium carbonate solution. Discard the filtrate and washings. Dissolve the carbonates from the paper with hot dilute nitric acid, and wash the paper thoroughly with hot water. Cool the solution, neutralize with ammonium hydroxide, and make just acid with dilute nitric acid. Add 10 ml of 30% ammonium acetate solution, heat to boiling, and add with stirring 20 ml of 10% ammonium dichromate solution. Allow to stand on a warm plate for 3 hours, filter the supernatant liquid and wash by decantation with 0·6% ammonium acetate solution.

Dissolve the precipitate on the paper with warm dilute nitric acid, catching it in the original beaker, and wash the paper. Add more acid if necessary until the precipitate is dissolved, and then dilute ammonium hydroxide slowly with stirring until the precipitate forming again no longer dissolves. Add 10 ml of 30% ammonium acetate, and bring the liquid to boiling with occasional stirring. Let the sample stand in a warm place for 2 hours, filter, and wash with 0·6% ammonium acetate solution. Dry, ignite at dull red heat, or at 120°C for 2 hours, and weigh as $BaCrO_4$.

$$BaCrO_4 \times 0{\cdot}5422 = Ba$$

If silica is desired on the sample, decompose initially with hydrochloric and nitric acids, evaporate to dryness several times with hydrochloric acid, dissolve in dilute hydrochloric acid, filter, wash, and fuse any insoluble residue with sodium carbonate. Add the contents of the crucible to the filtrate, evaporate to dryness twice, moistening both times with hydrochloric acid, and filter off the dehydrated silica. Retain the filtrate for barium. Dry the precipitate in a platinum crucible, ignite to constant weight at about 1000°C, cool, add 20 ml hydrofluoric acid and 1 drop of 1 : 1 sulphuric acid, evaporate to dryness on the hot plate, ignite, and re-weigh. The loss in weight is silica. If a residue remains after the treatment of crude silica with hydrofluoric acid, it could contain a little barium. Fuse with 5 g sodium carbonate, dissolve in water, filter, and wash thoroughly with hot 0·2% sodium carbonate; discard the filtrate. Dissolve the barium carbonate from the paper with hot dilute hydrochloric acid and add it to the original filtrate from the silica precipitation. Proceed with the determination of barium as outlined above.

When silica is removed by filtration after dehydration in hydrochloric acid, if lead is present it can be precipitated by hydrogen sulphide in 5–10% hydrochloric acid solution, filtered, and washed with 1% hydrochloric acid saturated with hydrogen sulphide. Boil out all hydrogen sulphide from the filtrate, and precipitate barium as the chromate in the manner outlined previously. This is usually a shorter procedure than precipitating barium, with lead and strontium as sulphates, removing lead by ammonium acetate, fusing barium sulphate with sodium carbonate, leaching with water, dissolving barium carbonate in hydrochloric acid, and finally precipitating barium as the chromate. Removal of lead by hydrogen sulphide precipitation is particularly advantageous for barium carbonate ores.

It is often useful to remember that barium sulphate can be dissolved in hot concentrated sulphuric acid. When a small amount of barium is precipitated as sulphate in the presence of a large quantity of calcium, traces of the latter will contaminate the final barium sulphate precipitate. Dissolve the latter in hot concentrated sulphuric acid, cool, add a few drops of hydrochloric acid, and dilute to 200 ml with cold water. Allow to stand several hours or overnight, filter, and wash. The precipitate is now completely free of calcium.

Norwitz [8] found that there is no need to adjust the *p*H exactly in the precipitation of barium as chromate if 7 ml hydrochloric acid, or 6 ml perchloric acid + 2 ml hydrochloric acid are present in 225 ml, and the solution is boiled, until a precipitate appears in about 15 minutes, and then for an additional hour. Heat to 80°C a 225-ml solution of a barium salt containing 7 ml hydrochloric acid, add 10 ml of 40% ammonium acetate, 25 ml of 10% potassium dichromate, and 10 g urea. Boil for one hour, maintaining the volume approximately at 225 ml, filter, wash with 0·5% potassium dichromate and finally with water, dry at 120°C for 1 hour, and weigh as $BaCrO_4$.

2. Volumetric Method after Separation as $BaSO_4$ and $BaCrO_4$

For precise analyses, the following chromate procedure for the separation and determination of barium in the presence of strontium has been found most satisfactory [2, 12]. It is applicable to samples containing up to 0·14 g barium and to approximately 0·87 g strontium. Separate barium, strontium, etc. as sulphates, and leach out lead if present by means of ammonium acetate solution, as outlined previously. Fuse the alkaline earth sulphates with sodium carbonate, dissolve the soluble constituents of the melt in water, filter, and wash with 0·2% sodium carbonate. Discard the filtrate, transfer the precipitate of mixed carbonates to a 400 ml beaker, washing the paper with water and finally with several ml of 10% hydrochloric acid. Dissolve the precipitate by the dropwise addition of hydrochloric acid to the neutral point.

Add 3 ml of 2 M hydrochloric acid, dilute to 185 ml, and add 5 ml of 0·4 M sodium dichromate. With continuous stirring, introduce 10 ml of 1·31 M sodium acetate through a constricted funnel designed to deliver this quantity in 2 minutes. During this addition the *p*H is raised sufficiently to permit the quantitative precipitation of barium chromate. Heat the suspension to boiling over a burner in 3–6 minutes, stirring occasionally during the heating. When vigorous boiling is reached, remove the beaker from the flame, cool it rapidly to room temperature in a water bath, and allow it to digest for one hour at room temperature. For a quantity of barium below 0·05 g allow digestion to proceed overnight.

Pour the cold supernatant liquid through a medium-porosity sintered-glass filter crucible, such as Corning 30M, of about 30 ml capacity. Wash the fine granular precipitate in the beaker three times by decantation with about 20-ml portions of a buffered wash solution. The latter is prepared by adding 2·23 g $Na_2Cr_2O_7.2H_2O$, 9·63 g $NaC_2H_3O_2.3H_20$ and 3·1 ml of 12 M hydrochloric acid to enough water to make the volume 1 litre. Transfer the precipitate to the crucible by washing with more of the same solution. Wash the precipitate on the filter with more wash solution, so that the total volume of wash solution is 150–200 ml. Wash the precipitate and the crucible with 30 ml of the basic wash solution, i.e. water containing enough sodium hydroxide to raise its *p*H to 9 or 10.

Transfer the major part of the precipitate back to the beaker with water; retain the crucible. Dissolve the precipitate with 3 ml of 2 M hydrochloric acid, and heat to boiling. Cool, dilute to 185 ml and add 3·75 ml of 0·4 M sodium dichromate. Re-precipitate the barium chromate at room temperature by adding 8·55 ml of 1·31 M sodium acetate through the constricted funnel while stirring. Heat the suspension to boiling, cool, allow to digest one hour, or overnight for less than 0·05 g barium. Repeat the filtration and washing by decantation, and wash the precipitate into the original crucible containing the untreated residue from the first precipitation. Scrub the beaker thoroughly and wash into the crucible. Wash the crucible with 150–200 ml wash solution and finally with the alkaline water.

Transfer the major part of the precipitate with water into a 500-ml Erlenmeyer flask. Draw 50 ml of 0·5 M hydrochloric acid, followed by some water, through the filter crucible into the flask to dissolve the remaining residue. Dilute the solution in the flask, if necessary, to about 150 ml. Add about 2 g potassium iodide, and allow to react for about 5 minutes. Titrate the liberated iodine with 0·1 N sodium thiosulphate, adding 5 ml of 0·5% starch solution towards the end of the titration.

The titration depends on the following reactions:

$$BaCrO_4 + 2HCl \rightarrow BaCl_2 + CrO_3 + H_2O$$

$$2CrO_3 + 12HCl + 6KI \rightarrow 2CrCl_3 + 6KCl + 6H_2O + 3I_2$$

$$Na_2S_2O_3 + I_2 \rightarrow 2NaI + Na_2S_4O_6$$

1 ml of 0·1 N sodium thiosulphate = 0·004578 g barium.

3. Colorimetric Methods

A few colorimetric methods for barium have been proposed [7, 10] but are not in common use, probably because they require the separation of this element from nearly all others, and the determination of traces of barium is rarely encountered in industrial practice.

4. Optical Spectrography

Barium has been determined by emission spectrography, using lines 3071·59, 4554·04, or 4934·09 [7].

5. X-Ray Spectrography

Barium is occasionally determined by X-ray fluorescence, using the lines $K_{\alpha1}$ 0·3851 or $L_{\beta1}$ 2·5674 Å [7].

6. Atomic-Absorption Spectrophotometry

Small to moderate amounts of barium can be determined satisfactorily by atomic absorption, using the most sensitive line 5535·6 Å [4]. Aluminium and phosphorus interfere; the sensitivity is about 8 μg/ml.

7. Flame Photometry

Barium, like the other alkaline earths, has been determined by flame photometry [1, 3, 7], but separations are usually required and the methods have limited application.

General

In rock analyses, barium is expressed as BaO; Ba × 1·1165 = BaO. In nearly all ores or concentrates of barium the element occurs as the sulphate or carbonate. Ba × 1·6993 = $BaSO_4$, and Ba × 1·437 = $BaCO_3$.

REFERENCES

1. Andersen, N. R., and Hume, D. N., *Anal. Chim. Acta* **40**, 207–20 (1968).
2. Beyer, G. L., and Rieman, W., *Anal. Chem.* **19**, 35–7 (1947).
3. Dean, J. A., Burger, J. C., Rains, T. C., and Zittel, H. E., *Anal. Chem.* **33**, 1722–27 (1961).
4. Elwell, W. T., and Gidley, J. A. F., *Atomic-Absorption Spectrophotometry*, 2nd ed., Oxford, Pergamon, 1966.

5. FURMAN, N. H., ed., *Scott's Standard Methods of Chemical Analysis*, 6th ed., Vol. 1, Princeton, N.J., Van Nostrand, 1962.
6. HILLEBRAND, W. F., LUNDELL, G. E. F., BRIGHT, H. A., and HOFFMAN, J. I., *Applied Inorganic Analysis*, 2nd ed., New York, Wiley, 1953.
7. KOLTHOFF, I. M., and ELVING, P. J., *Treatise on Analytical Chemistry*, Part 11, Vol. 4, 153–217, New York, Interscience Publishers, 1966.
8. NORWITZ, G., *Anal. Chem.*, **33**, 312–13 (1961).
9. NORWITZ, G., *Analyst* **90**, 554–63 (1965).
10. SANDELL, E. B., *Colorimetric Determination of Traces of Metals*, 3rd ed., New York, Interscience Publishers, 1959.
11. SNELL, F. D., and HILTON, C. L., *Encyclopedia of Industrial Chemical Analysis*, Vol. 6, New York, Interscience Publishers, 1968.
12. WONSIDLER, G. J., and SPRAGUE, R. S., *Anal. Chim. Acta* **31**, 51–7 (1964).

5 Beryllium

Beryllium is a comparatively rare element; for many years its use was almost entirely confined to beryllium-copper. The lightness, substantial strength, high melting point, and low neutron-capture cross section of beryllium have attracted much interest in fields of atomic energy, aircraft, and missiles; active development work from geochemical prospecting to final fabrication has been in progress for some time.

The principal beryllium mineral is beryl, $3BeO.Al_2O_3.6H_2O$; other beryllium-containing minerals are found widely distributed but in small quantities. Pegmatite dykes, which are the chief source of the metal, contain less than 0·75% BeO, and commercial concentrates usually range 10–12% BeO.

Beryllium and its compounds can be toxic to humans, and laboratory workers should read one of the comprehensive reports on the toxicology of this element before commencing work in this sphere [2].

The analytical chemistry of beryllium has been reviewed in recent years in papers [12, 13, 19, 21] and a book [14].

ISOLATION OR SEPARATION OF BERYLLIUM

Beryllium closely resembles aluminium in its chemical properties, and in fact the chief difficulty in its determination for amounts greater than traces lies in complete separation from aluminium. The separation of beryllium from other members of the ammonium hydroxide and ammonium sulphide groups may take a number of forms:

(a) Beryllium is not precipitated by cupferron in acid solution and can thereby be separated from Fe, Ti, Zr, V, Nb, Ta, and W.

(b) Beryllium is not deposited on a mercury cathode in dilute sulphuric acid and is thus removed from Fe, Ni, Co, Cr, and Zn. This is a useful separation for many materials, because in 15 minutes 99·99% of the iron in 1 g of steel can be deposited, and in 1 hour 99·99% of the zinc in 5 g of a zinc-based alloy.

(c) Beryllium is not precipitated by hydrogen sulphide in ammoniacal tartrate solution, whereas Fe, Mn, Ni, Co, and Zn are.

(d) Sodium hydroxide treatment will leave beryllium with Al, V, Zn, and W in the filtrate; the precipitate contains Fe, Ti, Zr, Cr, U, Ni, Co, Mn, Nb, and Ta. If sodium peroxide is added, U and Cr will remain in solution.

(e) Iron is removed from beryllium by ether extraction of ferric chloride.

Unlike Fe, Al, etc., however, beryllium is not completely removed by a zinc oxide precipitation, even in the presence of iron.

(f) Chromium can be separated from beryllium by volatilizing chromyl chloride with boiling hydrochloric and perchloric acids; chromium may also be left in the filtrate from an ammonia separation of beryllium by prior oxidation with perchloric acid or sodium peroxide.

(g) In strong hydrochloric acid, Fe, Mn(VII), Cr(VI), V(V), U(VI), Zn, Zr, and W are adsorbed on anion exchange resins, whereas beryllium and Al, Ti, Ni, Mn(II), Cr(III), and V(IV) are not retained.

(h) Beryllium can be separated from aluminium by precipitating the latter with 8-hydroxyquinoline in acetic solution, or by passing hydrogen chloride gas into a cold solution of the chlorides in equal parts of hydrochloric acid and ether.

(i) A good separation is based on the fact that beryllium oxide is not dissolved by molten sodium carbonate, whereas aluminium oxide of course is decomposed into a water-soluble salt.

(j) Because a small quantity of aluminium does not interfere in the preferred colorimetric and fluorimetric methods for beryllium, traces of the latter can be collected with aluminium hydroxide and separated from Cu, Mo, Ni, Co, Mn, Zn, Ca, Mg, etc.

1. Gravimetric Determination

The following general procedure is applicable to beryllium in quantities above the trace level in a complex ore. In the absence of certain constituents, it can be considerably shortened. Following separation of all interferences, beryllium is precipitated by ammonium hydroxide, ignited, and weighed as BeO [8, 9, 10, 16, 20].

Weigh a quantity of sample which will yield a final weight of BeO not exceeding 0·2 g; otherwise the voluminous precipitate of beryllium hydroxide is difficult to wash, and to contain in the usual size of platinum crucible. Treat with 10 ml nitric acid, 10 ml hydrochloric acid, 10 ml 1 : 1 sulphuric acid and evaporate to dryness. Add 10 ml hydrochloric acid, 50 ml water, boil, filter, wash, dry, and ignite the precipitate; fuse with sodium carbonate, dissolve in hydrochloric acid, and evaporate to dryness twice to dehydrate silica. Dissolve soluble salts with dilute hydrochloric acid, and filter off silica, washing thoroughly with hot water. For referee work the small quantity of silica in the filtrate must be removed by evaporating to dryness and dehydrating as before.

Alternative methods of decomposition may be used. If the sample does not contain substances which might be injurious to a platinum crucible, it can be fused directly with sodium carbonate for 30 minutes at the full heat of a gas burner, followed by solution in hydrochloric acid and dehydration of silica. Another method is to remove silica initially by evaporating with hydrofluoric and sulphuric acids in a platinum dish; if an insoluble residue remains it may be fused with sodium carbonate.

It is well to remember that the silicate, beryl, is taken into solution with a carbonate fusion, using 2 parts of sodium carbonate to 1 of beryl, for 30 minutes at 800–1200°C, but not the oxide, beryllia. For the latter, treatment with aqua regia and sulphuric acid to strong fumes of sulphur trioxide is necessary; alternatively, beryllium oxide can be decomposed by a bisulphate fusion.

If a large amount of phosphorus is present, remove it by means of ammonium molybdate before the elimination of the hydrogen sulphide group. To the sample which has been evaporated to dryness in hydrochloric or sulphuric acid add 10 ml nitric acid, 50 ml water, and boil to solution of salts. Transfer to a 500 ml Erlenmeyer flask, dilute to 200 ml and bring the nitric acid concentration to 20 ml. Add 25 g ammonium nitrate, and 100 ml of ammonium molybdate reagent. The latter is made by adding 65 g of 85% molybdic acid to 142 ml of water and 143 ml of ammonium hydroxide, cooling, and pouring into 715 ml of 1 : 1 nitric acid. Stopper the Erlenmeyer flask, shake vigorously for 10 minutes, and allow to stand overnight. Filter, wash several times with 5% ammonium nitrate, and discard the precipitate. To the filtrate add 10 ml 1 : 1 sulphuric acid, and evaporate to dryness. Dilute with water, adjust acidity to 10% with sulphuric acid, and remove molybdenum together with the rest of the hydrogen sulphide group as described below.

If the sample contains members of the hydrogen sulphide group, remove these in 5–10% hydrochloric or sulphuric acid solution by hydrogen sulphide. Wash the precipitated sulphides thoroughly with hot acidulated hydrogen sulphide water and discard; boil the filtrate to remove all hydrogen sulphide, add hydrogen peroxide, and again boil.

To the cool, dilute hydrochloric acid solution add two drops of methyl red indicator, neutralize with ammonium hydroxide, and add several drops in excess. Beryllium is precipitated at *p*H 8·5, slightly higher than aluminium. Boil and filter, washing several times with hot, slightly ammoniacal 1% ammonium chloride. Dissolve the precipitated hydroxides of beryllium and associated elements with hot 1 : 1 hydrochloric acid, re-precipitate with ammonium hydroxide, and wash thoroughly with the 1% ammonium chloride solution.

Beryllium has now been isolated as hydroxide from members of preceding and succeeding groups, and from P, Mn, Ni, Co, and Zn; unfortunately it is nearly always accompanied by one or more of the following: Fe, Al, V, Ti, Zr, U, Cr, Nb, Ta, and W. The last three are largely removed with silica in the initial acid dehydration and filtration step. For the rare cases where niobium and tantalum are present, these can be removed with tannin in a slightly acid oxalate solution as described by Schoeller and Powell [16]; the infrequent occurrence of tungsten can be overcome by its conventional separation in acid solution with the aid of cinchonine. Uranium may be separated from beryllium by precipitation of the latter in ammonium carbonate solution, with recovery of the small quantities of beryllium passing into the filtrate. Chromium remains in the filtrate of an ammonium hydroxide separation if it has been previously oxidized to the chromate state with sodium peroxide or perchloric acid. Alternatively, both U(VI) and Cr(VI) can be removed from beryllium by absorption of the former on anion exchange resins in 9 M hydrochloric acid [10].

A convenient way to remove Fe, Ti, V, and Zr from the mixed hydroxide precipitate is by cupferron. Transfer the hydroxide precipitate to a beaker by a stream of hot water from a wash bottle, followed by hot dilute sulphuric acid and finally with hot water. Dissolve with more 1 : 1 sulphuric acid if necessary, adjust the volume to 10% sulphuric acid, cool to about 10°C, and transfer to a separatory funnel. Add a cold 6% aqueous solution of cupferron in successive small portions, shaking the funnel vigorously, until Fe, Ti, V, and Zr have been precipitated. Remove the precipitated cupferrates by extraction with several successive portions of chloroform, discarding the lower layer each time. Transfer the final upper aqueous layer, which contains the beryllium together with aluminium, to a beaker; place on the edge of a hot plate until the traces of chloroform have evaporated, add 10 ml nitric acid and evaporate to dryness. If necessary, add more nitric and sulphuric acids to destroy all organic matter.

The sample now contains only aluminium and beryllium. If aluminium is low it can be removed satisfactorily by precipitation with 8-hydroxyquinoline in feebly acid acetate solution. Neutralize the solution with ammonium hydroxide, make just acid with hydrochloric, heat to 60°C, and add an excess of 8-hydroxyquinoline. The latter is prepared by triturating 5 g of the solid reagent with 12 ml of acetic acid, and then diluting to 100 ml with hot water. One ml of this 8-hydroxyquinoline solution will precipitate 0·0031 g of aluminium. Add 2 N ammonium acetate until aluminium is precipitated, and then 25 ml excess. An excess of 8-hydroxyquinoline is indicated by a yellow colour in the supernatant liquid after the precipitate has settled. Filter and wash well with cold water. Discard the precipitate.

Evaporate the filtrate to fumes of sulphur trioxide after addition of 10 ml of nitric acid and 10 ml of 1 : 1 sulphuric acid. Cool, dissolve in water, filter if necessary, add two drops of methyl red and then ammonium hydroxide until just ammoniacal, finally adding 2 ml excess. Beryllium will be precipitated as a white voluminous hydroxide. Boil, filter through Whatman 41 paper, using paper pulp; wash thoroughly with hot 2% ammonium nitrate. Small quantities of the precipitate often adhere tenaciously to the walls of the beaker, but thorough scrubbing will remove these. Allow the filtrate and washings to stand for 30 minutes to be sure that no further precipitation occurs. Place paper and precipitate in a platinum crucible, dry, ignite cautiously, and finally for 2 hours at 1100°C. Cover, cool, and weigh quickly as BeO.

$$BeO \times 0{\cdot}3603 = Be$$

When aluminium is high, the best method of removing it is based on the fact that beryllium oxide is not attacked by molten sodium carbonate, whereas aluminium oxide is converted to sodium aluminate. Fusion of an impure beryllium oxide, followed by a water leach and filtration, will separate aluminium.

Precipitate the hydroxides of beryllium and aluminium, in the manner directed previously for beryllium alone, and ignite the mixed oxides. Cool, mix with 5 g sodium carbonate, and fuse for 20 minutes. Dissolve the cold melt in 100 ml of hot water, filter through Whatman No. 41 paper, using pulp, and wash very thoroughly with hot 2% ammonium nitrate. Dry the paper and precipitate in a tared platinum crucible, and ignite at a low temperature to burn off the carbon of the filter paper. If aluminium is very high, make a second fusion with sodium carbonate followed by leaching, filtering, etc. Finally, ignite for 2 hours at 1100°C, cover, cool, and weigh quickly as BeO.

A number of variations of the gravimetric method are practised, depending on the presence of interferences. In steels, iron can be conveniently removed by an ether extraction, but the presence of tungsten and niobium may necessitate several pyrosulphate fusions and a final elimination of silica and platinum before the final weighing of BeO [1].

For beryllium-copper, remove and determine copper by electrolysis, evaporate the solution to fumes of sulphuric acid, filter, and wash. Add 10 ml of ammonium ethylenediamine tetraacetate solution, made by dissolving 2·5 g of ethylenediamine tetraacetic acid in 100 ml of water made faintly ammoniacal. Add 1 : 1 ammonium hydroxide slowly with stirring until a permanent turbidity is formed, and then 10 ml in excess. Allow to stand several hours or overnight, filter and wash thoroughly with 1% ammonium nitrate which has been made faintly ammoniacal. Place paper and precipitate in a tared platinum crucible, dry, ignite slowly, and finally at 1100°C for 2 hours. Cover, cool, weigh quickly as BeO. The addition of EDTA permits beryllium to be determined in the presence of 0·30% of nickel, cobalt, and iron [1].

2. Volumetric Determination

Beryllium has only one state of oxidation, and a volumetric procedure must therefore be an indirect one. The element can be completely precipitated as hydroxide from a sulphuric or hydrochloric acid solution at *p*H 8·5. The addition of an excess of sodium fluoride liberates hydroxide ions, which can be determined by titration with a standard acid [8, 10]. For visual titration a mixed phenolphthalein–thymolphthalein–methyl orange indicator is used; the accuracy is improved when a potentiometric titration is employed.

The reaction is not stoichiometric, and the titrating conditions must be rigorously maintained. Aluminium, zirconium, uranium, thorium, and the rare earths interfere. The method is not as suitable as others for general work or occasional samples, but for routine analyses on certain materials it provides a procedure offering reasonable accuracy and speed. Details are given in the references cited above.

3. Colorimetric Determination

A number of colorimetric procedures are used for small quantities of beryllium. In addition to the usual advantages of colorimetry, this phase of beryllium analysis has received impetus from the toxicity of beryllium compounds. The occupational limits for beryllium in air, recommended by the U.S. Atomic Energy Commission, are 2 micrograms per cu. metre in an 8-hour day or 25 micrograms per cu. metre for 30 minutes.

A. 4-(*p*-Nitrophenylazoorcinol) Method

When an alkaline solution of 4-(*p*-nitrophenylazoorcinol), also called *p*-nitrobenzeneazo-orcinol or Zenia, is added to an alkaline solution containing beryllium, a reddish-brown lake is formed. This is the basis of the most popular colorimetric procedure for beryllium in a wide range of materials [3, 5, 6, 8, 10, 14, 15, 20]. It is sensitive, relatively few elements interfere, and aluminium, which is so difficult to remove from beryllium, can be tolerated in substantial amounts. Magnesium, zinc, copper, iron, nickel, cobalt, and calcium interfere, but small quantities can be rendered harmless by the addition of chelating agents such as EDTA or Versene T. Larger amounts of calcium, magnesium, copper, cobalt, nickel, and zinc can be eliminated initially by double precipitation of beryllium with ammonium hydroxide in the presence of ammonium chloride; iron in quantity can be removed by an ether extraction, cupferron, mercury cathode, sodium hydroxide, or other separations listed earlier.

Transfer the sample solution or a suitable aliquot containing from 0·06 to 1·1 mg beryllium, 3 ml of 1 : 1 sulphuric acid, and not more than 5 mg Mg, 10 mg Fe, 20 mg Ca, and 35 mg Al, to a 100-ml volumetric flask. Adjust the volume to about 35 ml, add 5·0 ml of 13·9% Versene T solution, mix, and allow to stand 5 minutes. Adjust the *p*H to 5–5·5 with 5 N sodium hydroxide or sulphuric acid solution. Add 10 ml of citrate–borate solution, mix, and let stand for 5 minutes. Add exactly 10 ml of 4-(*p*-nitrophenylazoorcinol) solution, dilute to the 100-ml mark, mix thoroughly, and allow to stand for 10 minutes. Determine the optical density in a spectrophotometer at 515 nm. Prepare a calibration curve by carrying through the above steps aliquots of a standard beryllium solution to cover a range of 0·05–1·5 mg Be.

Citrate–borate buffer solution. (a) To 1 litre of water in a 2-litre beaker add 144 g of boric acid and 256 g of citric acid monohydrate. Stir to dissolve as much of the two acids as possible. (b) Dissolve 480 g of sodium hydroxide in 1 litre of water. Cool solutions (a) and (b), add (b) to (a) slowly with constant stirring and cooling. Cool to room temperature, dilute to 2 litres, and filter.

4-(p-nitrophenylazoorcinol) solution. Dissolve 0·06 g of 4–(*p*-nitrophenylazoorcinol) powder in 200 ml of 0·1 N sodium hydroxide by stirring mechanically for several hours. Filter, and store in a brown bottle.

B. Aluminon Method

In a suitably buffered solution, aurintricarboxylic acid, or aluminon, forms a red lake with beryllium [1, 5, 10]. Aluminium also forms a red complex with this reagent, but the addition of EDTA will overcome the interference of small amounts of aluminium, together with small quantities of Cu, Fe, Ni, Co, Ti, Zn, and Zr. The recommended range for colour measurement is 0·005–0·1 mg of beryllium in 100 ml. The procedure is very suitable, for instance, for beryllium copper, following the removal and determination of copper by electro-deposition, and the evaporation of the solution to strong fumes of sulphuric acid.

Dissolve 0·5 g or other suitable weight of sample in mixed acids, or by fusion, and evaporate to dryness with sulphuric or hydrochloric acid. Dissolve in a little hydrochloric acid and hot water, filter, and wash. Make to volume in a flask, and transfer a convenient aliquot of 5–50 ml, which should contain 0·005–0·1 mg of beryllium, to a 100-ml volumetric flask. Dilute to about 75 ml, add 2 ml of EDTA solution, mix, add 15 ml of aluminon buffer composite solution, and mix. The *p*H must be between 6·2 and 6·6. Dilute to the mark, mix, transfer to an absorption cell, and allow to stand away from direct sunlight for exactly 20 minutes from the time of addition of the aluminon buffer composite solution.

Measure the optical density of the solution at 515 mμ. A standard beryllium solution can be prepared by dissolving 9·82 g of $BeSO_4.4H_2O$ in 100 ml of 1 : 3 hydrochloric acid, diluting to 500 ml, and standardizing by the gravimetric procedure described earlier. Transfer 10 ml of the above solution to a litre flask, add 10 ml of hydrochloric acid, dilute to the mark and mix. This solution contains 0·01 mg beryllium per ml.

EDTA solution. Dissolve 2·5 g ethylenediamine tetraacetic acid in 30 ml of water made faintly ammoniacal, and dilute to 100 ml.

Aluminon buffer composite solution. Add 500 g of ammonium acetate to 1 litre of water in a 2-litre beaker. Add 80 ml of glacial acetic acid, and stir until the salt has dissolved; filter if necessary. Dissolve 1 g of aluminon in 50 ml of water, and add to the buffer solution. Dissolve 3 g of benzoic acid in 20 ml of methanol and add, with stirring, to the buffer solution. Dilute the mixture to 2 litres. Add 10 g of gelatin to 250 ml of water in a 400 ml beaker. Place the beaker in a boiling water bath and allow to remain, with frequent stirring, until the gelatin has dissolved completely. Pour the warm gelatin solution into 500 ml of water, with stirring. Cool to room temperature, dilute to 1 litre, and mix. Transfer the aluminon and gelatin solutions to a 4-litre Pyrex bottle, mix well, and store in a cool, dark place.

4. Optical Spectrography

Small amounts of beryllium can be determined in many materials by emission spectrography. The most sensitive lines are 2348·61 and 3321·34 Å, but lines 2650·62, 3130·42, and 3131·07 are also frequently used. Spectrography is particularly valuable for determining low contents of beryllium in aluminium and its alloys [1, 4, 10]. The point-to-plane spark is used for 0·0002–0·5%, high voltage spark for 0·001–1·2%, and condensed arc for 0·001–0·05% beryllium in aluminium alloys.

5. Atomic-Absorption Spectrophotometry

Using a nitrous oxide–acetylene or fuel-rich oxy–acetylene flame, beryllium can be determined by atomic absorption using the most sensitive absorption line 2348·6 Å [7]. The toxicity of beryllium makes imperative an efficient fume extraction system. The sensitivity is good, 0·03–1 ppm.

6. Fluorimetric Method

The toxicity of beryllium has been responsible for the development of fluorimetric procedures for the determination of this element in extremely small quantities in air-borne dusts, bone, urine, etc.

In sodium hydroxide solution, morin forms a compound with beryllium which shows a strong yellow-green fluorescence. Less than 0·001 ppm of beryllium can be detected with a fluorophotometer. Relatively few elements interfere, and most of these can be rendered harmless by chelating agents. The morin fluorimetric method has been used chiefly in studies involving very low beryllium contents [10, 11, 15, 17, 18], but can also be employed conveniently for 0·01–0·025% beryllium in aluminium alloys [3]. Full details of this rapid and very sensitive procedure may be found in the references cited above.

REFERENCES

1. A.S.T.M., *Chemical Analysis of Metals. Sampling and Analysis of Metal Bearing Ores*, Philadelphia, Pa., American Society for Testing and Materials, 1969.
2. Breslin, A. J., and Harris, W. B., U.S. At. Energy Commission HASL–36 (1958).
3. British Aluminium Company, *Chemical Analysis of Aluminium and its Alloys*, London, British Aluminium Company, 1961.
4. British Aluminium Company, *Spectrochemical Analysis of Aluminium and its Alloys*, London, British Aluminium Company, 1961.
5. British Drug Houses, *The B.D.H. Book of Organic Reagents*, Poole, England, British Drug Houses, 1958.
6. Covington, L. C., and Miles, M. J., *Anal. Chem.* **28**, 1728–30 (1956).
7. Elwell, W. T., and Gidley, J. A. F., *Atomic-Absorption Spectrophotometry*, 2nd ed., Oxford, Pergamon, 1966.
8. Furman, N. H., ed., *Scott's Standard Methods of Chemical Analysis*, 6th ed., Vol. 1, Princeton, N.J., Van Nostrand, 1962.
9. Hillebrand, W. F., Lundell, G. E. F., Bright, H. A., and Hoffman, J. I., *Applied Inorganic Analysis*, 2nd ed., New York, Wiley, 1953.
10. Kolthoff, I. M., and Elving, P. J., *Treatise on Analytical Chemistry*, Part 11, Vol. 6, 1–67, New York, Interscience Publishers, 1964.
11. May, I., and Grimaldi, F. S., *Anal. Chem.* **33**, 1251–3 (1961).
12. Metcalfe, J., and Ryan, J. A., *The Analytical Chemistry of Beryllium. Proceedings of a Symposium, Blackpool, June*, 1960, London, H.M.S.O., U.K. At. Energy Authority Production Group, 1961.

13. NATIONAL CHEMICAL LABORATORY, D.S.I.R., *The Determination of Beryllium*, London, H.M.S.O., 1963.
14. NAVOSELOVA, A. V., and BATSANOVA, L. R., *Analytical Chemistry of Beryllium*, Hartford, Conn., Daniel Davey, 1968.
15. SANDELL, E. B., *Colorimetric Determination of Traces of Metals*, 3rd ed., New York, Interscience Publishers, 1959.
16. SCHOELLER, W. R., and POWELL, A. R., *The Analysis of Minerals and Ores of the Rarer Elements*, 3rd ed., London, Charles Griffin, 1955.
17. SILL, C. W., and WILLIS, C. P., *Anal. Chem.* **31**, 598–608 (1959).
18. SILL, C. W., WILLIS, C. P., and FLYGARE, J. K., *Anal. Chem.* **33**, 1671–84 (1961).
19. SMYTHE, L. E., and WHITTEM, R. N., *Analyst*, **86**, 83–94 (1961).
20. SNELL, F. D., and HILTON, C. L., *Encyclopedia of Industrial Chemical Analysis*, Vol. 7, New York, Interscience Publishers, 1968.
21. VINCI, F. A., *Anal. Chem.* **25**, 1580–5 (1953).

6 Bismuth

Bismuth is not an abundant element, but small quantities of its sulphide or oxides are found in many ores of lead, copper, cobalt, antimony, arsenic, nickel, molybdenum, tin, silver, and vanadium. Most of the world's bismuth is obtained as a by-product from the refining of lead and copper. It is an important constituent of numerous low-melting or "fusible" alloys, and much smaller quantities occur in bearing metals, solders, and other alloys. Bismuth is a detrimental impurity in many refined metals, such as copper, and its determination in the parts-per-million range is important in many mining and metallurgical laboratories.

ISOLATION OR SEPARATION OF BISMUTH

Bismuth, of course, is precipitated by hydrogen sulphide in 5–10% hydrochloric or sulphuric acid, together with As, Sb, Sn, Cu, Mo, Cd, Pb, Hg, Se, Te, Au, Ag, Ge, and Pt metals, and thereby separated from members of succeeding groups.

The separation of bismuth from other members of the hydrogen sulphide group may be carried out in various ways. Treatment with ammonium sulphide will dissolve As, Sb, Sn, Mo, Se, and Te from Bi, Cu, Cd, Pb, and Hg which remain insoluble. Mercury may be volatilized by ignition or fusion with sodium carbonate; mercury sulphide is insoluble in dilute nitric acid and may be separated in this manner. Copper and cadmium may be removed from bismuth by precipitating the latter as oxychloride, or as carbonate by adding a slight excess of ammonium carbonate to the solution nearly neutralized with ammonia. Copper and molybdenum can be separated from Bi, As, Sb, Sn, Pb, Se, and Te by collecting the latter elements in a ferric hydroxide precipitate with ammonia, filtering, dissolving in hydrochloric acid and re-precipitating. Traces of bismuth, together with Sb, Sn, Au, and Mo can be isolated from a large sample of copper or lead by means of hydrous manganese dioxide obtained by adding potassium permanganate to a hot, dilute nitric acid solution containing a manganous salt. Silver and mercurous mercury can be removed from bismuth with hydrochloric acid; selenium, tellurium, and gold can be precipitated with sulphur dioxide in hydrochloric acid. Arsenic, antimony, tin, and selenium can be volatilized from bismuth by repeated boiling with a mixture of 3 hydrobromic acid : 1 bromine, or hydrobromic and perchloric acids.

Bismuth can be completely precipitated from a hot nitrite solution of the platinum metals by neutralizing with sodium bicarbonate to the blue colour of thymol blue; the complex nitrite of palladium is unstable at this alkalinity, but this metal can be readily removed, subsequently, by precipitation with dimethylglyoxime in 3% hydrochloric acid solution [9].

Lead is usually separated from bismuth as lead sulphate; a little bismuth is always occluded and for accurate work it is necessary to dissolve the precipitate in ammonium acetate, add sulphuric acid, take to fumes, and again filter off lead sulphate. Bismuth can also be separated from lead by precipitation of the former in cold 0·1M nitric acid solution with cupferron and extraction of its cupferrate with chloroform.

The differing behaviour of dithizone towards metals can be utilized to separate traces of bismuth from small amounts of other members of the hydrogen sulphide group [14]. Antimony, arsenic, germanium, stannic tin, and selenium do not react with dithizone in carbon tetrachloride or chloroform. Copper, gold, silver, mercury, platinum, palladium, and tellurium can be extracted in dilute mineral acid. Bismuth and lead can be separated from cadmium by dithizone in ammoniacal cyanide solution; final removal of bismuth can be effected by extracting at *p*H 3·4.

Bismuth deposits on both platinum cathode and anode, and must be removed before the ordinary electrolysis of copper or lead.

In conventional copper electrolysis, small quantities of bismuth, for instance up to 10 mg in the presence of 0·1 g of copper, will usually co-deposit quantitatively on the cathode. Higher contents of bismuth will adversely affect the electrodeposit. With the internal electrolysis apparatus [1] where the lead anode is enclosed in an alundum sheath, the small quantities of bismuth in pig lead will deposit quantitatively on the platinum cathode, together with copper and silver.

1. Gravimetric Determination as Bi_2O_3

When more than 5 mg of bismuth are present, it is preferable to precipitate as basic carbonate in ammonium carbonate solution and weigh as Bi_2O_3 [8, 9, 17]. Weigh 0·5–1 g sample, add 25 ml water and 10 ml nitric acid. Evaporate to dryness on the hot plate, add 5 ml hydrochloric acid, 10 ml 1 : 1 sulphuric acid and evaporate to strong fumes of the latter. Cool, dilute with 50 ml water, boil, filter, and wash thoroughly with 2% sulphuric acid. Allow to stand an hour and re-filter if necessary to remove any lead which may have gone through the first filtration. If an insoluble residue of silicates or lead sulphate remains, it may contain a little bismuth. Transfer the paper and precipitate to a platinum crucible, dry, ignite at a low temperature, and fuse with sodium carbonate. Extract the melt with dilute hydrochloric acid, filter, wash, and combine this filtrate with the original solution.

To the clear filtrate, adjusted to 5–10% acid, pass in sufficient hydrogen sulphide to precipitate bismuth and all other members of the acid sulphide group. Filter and wash thoroughly with acidulated hydrogen sulphide water.

Leach the precipitated sulphides with hot 10% ammonium sulphide saturated with hydrogen sulphide until all As, Sb, Ge, Mo, Sn, Se, and Te are in solution. Filter and wash with dilute ammonium sulphide solution. Dissolve the insoluble sulphides of Bi, Cu, and Cd in dilute nitric acid and bromine water, and boil off the latter. Nearly neutralize with ammonia, dilute to 200–300 ml with water, and add a saturated solution of ammonium carbonate until a slight excess is present. Heat to boiling and digest warm for 2 hours. Filter on a tared Gooch crucible and wash thoroughly with a solution containing 30 ml ammonium hydroxide, 1 g ammonium nitrate, and 15 ml saturated ammonium carbonate per litre. If large quantities of copper or cadmium are present, dissolve the precipitate in nitric acid and re-precipitate the bismuth as basic carbonate. Dry, ignite at 900°C, and weigh as Bi_2O_3.

$$Bi_2O_3 \times 0{\cdot}8970 = Bi$$

The general procedure outlined above may be modified to suit the composition of the sample. Arsenic, antimony, tin, and selenium may often be conveniently eliminated at the outset by several evaporations with a mixture of hydrobromic acid and bromine, or of bromine and hydrochloric or sulphuric acids. Silica may be removed by initial decomposition with hydrofluoric and sulphuric acids in a platinum dish. Occlusion of bismuth in the lead sulphate precipitate may be overcome by solution of the latter in ammonium acetate, re-evaporation and re-precipitation of lead with sulphuric acid.

2. Gravimetric Determination as BiOCl

Intermediate quantities of bismuth can often be separated and weighed as the oxychloride. The method is suitable for a range of bismuth below 5 mg but above the trace amounts for which colorimetry is the preferred procedure [8, 9].

Take a weight of material such that the quantity of bismuth present does not greatly exceed 5 mg. Bring into solution with nitric and hydrochloric acids, and evaporate to dryness with hydrochloric acid; if an insoluble residue remains, fuse with sodium carbonate, dissolve in hydrochloric acid, and add to the main portion. Evaporate twice to dehydrate silica, moistening both times with hydrochloric acid; dissolve the soluble salts in dilute hydrochloric acid, and filter off silica and silver, washing thoroughly with 2% hydrochloric acid. Adjust the acidity to 5–10% hydrochloric acid and precipitate the acid sulphide group with hydrogen sulphide. Filter and wash thoroughly with acidulated hydrogen sulphide water.

Digest the precipitated sulphides with hot 10% ammonium sulphide solution until As, Sb, Sn, Mo, Ge, Se, and Te are in solution; filter and wash well with 1% ammonium sulphide solution. Dissolve the precipitated sulphides of Bi, Cu, Pb, Hg, and Cd in hot dilute nitric acid, boil, cool, and neutralize most of the acid with ammonium hydroxide. Add dilute ammonium hydroxide drop by drop until the nitric acid has been neutralized and a faint opalescence appears. Add 5 ml of 1 : 9 hydrochloric acid, dilute to 400 ml, with hot water, heat to boiling, and allow to stand warm for two hours. Filter on a small paper and wash several times with warm water. Dissolve the precipitate in 4 ml of hot 1 : 9 hydrochloric acid, dropping the acid around the paper and catching the solution in the original beaker. Wash the paper with hot water, 1 ml of hot 1 : 9 hydrochloric acid, and finally again with water. Dilute to 400 ml with hot water, and precipitate bismuth oxychloride as before. Filter through a weighed Gooch crucible, wash with hot water, alcohol, dry at 105°C and weigh as BiOCl.

$$BiOCl \times 0{\cdot}8024 = Bi$$

In routine work, if copper and cadmium are absent or low, one precipitation will suffice. It is often useful to collect small quantities of bismuth, along with As, Sb, Sn, Pb, Se, and Te, in a ferric hydroxide precipitate with an ammonia separation, before precipitating with hydrogen sulphide. If the iron is 10–20 times that of the bismuth + associated impurities, a quantitative separation is achieved from copper and molybdenum.

Depending on the sample, variations of this procedure may often be practised. Decomposition in a platinum dish with hydrofluoric and sulphuric acids, and volatilization of As, Sb, Sn, and Se by hydrobromic acid and bromine, are examples of useful modifications for some materials.

3. Volumetric Determination

Bismuth can be determined satisfactorily over a wide range by titration with EDTA in an acidic solution using thiourea indicator [10, 13]. At a *p*H of 1·5–2, and with 0·5–1 g thiourea per 50 ml, bismuth can be titrated in the presence of equal molar amounts of most cations, except tin, nickel, vanadium, and zirconium. Interference from iron can be prevented by adding ascorbic acid, and antimony can be tolerated if tartrate is added. Bismuth can be easily separated from interferences by sulphide precipitation in acid solution, or volatilization of As, Sb, Sn, and Se by hydrobromic acid and bromine. The EDTA method for bismuth is particularly useful for the many samples which contain lead; a ratio of 30 Pb : 1 Bi can be tolerated.

To a sample in dilute acid solution containing 30–80 mg of bismuth, none of the interferences listed above, and preferably only minor amounts of other cations, add 0·5 to 0·8 g of thiourea, and dilute to 10–15 ml. Warm gently to effect solution and add dilute ammonium hydroxide until a temporary precipitate forms, which requires a few seconds to dissolve. Dilute to 40–50 ml, and adjust to *p*H 1·5–2 with dilute perchloric acid. Titrate with 0·05 M EDTA solution to the complete disappearance of the yellow colour due to the bismuth-thiourea complex.

If iron is present, add 0·3–0·4 g of ascorbic acid after the addition of thiourea. In the presence of antimony, add enough tartrate to give a clear solution at *p*H 2, add thiourea, adjust the *p*H to 1·5–2, and continue as described above. If the material contains tin, or appreciable quantities of arsenic, antimony, or selenium, volatilize these initially by several evaporations with hydrobromic acid and bromine in a perchloric acid solution, heating until copious fumes of perchloric acid appear. If large amounts of other elements are present, isolate the bismuth with other members of its group by precipitation with hydrogen sulphide in 5–10% hydrochloric or sulphuric acid, filter, wash, dissolve the sulphides in nitric acid, and proceed as directed above.

1 ml of 0·05 M EDTA solution = 10·45 mg bismuth

4. Colorimetric Determination

Two colorimetric procedures for bismuth are in common use for the small quantities present in many ores, concentrates, mattes, blister copper, pig lead, and various metals and alloys. One depends on the yellow colour of bismuth iodide which is formed when potassium iodide is added to a bismuth salt in 1–2 N acid; the other is based on a yellow colour imparted to bismuth compounds by thiourea in strong acid solution. The latter is not as sensitive as the iodide method, but larger quantities of bismuth may be determined and fewer interferences are encountered.

A. Iodide Method

This sensitive and selective method has been in use for over 80 years, and is still one of the most reliable for traces of bismuth [5, 7, 8, 10, 14, 15]. Antimony, tin, platinum, and palladium form coloured compounds, and large amounts of lead and thallium occlude bismuth in their iodide precipitates.

Lead can be satisfactorily separated from bismuth by successive extractions of bismuth cupferrate with chloroform in 0·1 M nitric acid solution. The bismuth can be transferred back to the aqueous phase by shaking the chloroform with 2 M sulphuric acid [14]. A small amount of bismuth can also be separated from a large quantity of lead by internal electrolysis in nitric acid solution; this is a routine procedure to remove bismuth, copper, and silver in a single electrodeposit on a platinum cathode from a sample of pig lead [1]. Smaller amounts of lead can be separated from bismuth by precipitation of the former as sulphate, filtration, washing, solution of the precipitate in ammonium acetate, re-fuming with sulphuric acid, and a second filtration to recover occluded bismuth.

After elimination of lead, precipitation of the hydrogen sulphide group in 5–10% hydrochloric or sulphuric acid, followed by treatment with ammonium sulphide in ammoniacal solution, as already described, will leave bismuth as a precipitate with copper, cadmium, mercury, palladium, and most of the platinum. For the rare occasions when platinum or palladium are present, separate the bismuth and other base metals from the platinum by the hydrolytic precipitation in nitrite solution already described; remove palladium by precipitation with dimethylglyoxime in 3% hydrochloric acid. Antimony and tin may often be more conveniently eliminated at an initial stage by volatilization with hydrobromic acid and bromine than by subsequent treatment with ammonium sulphide and ammonia.

Use a weight of sample such that not more than 0·5 mg of bismuth is present in the solution for the final colorimetric determination. Obtain the filtered and washed precipitate of sulphides of Bi, Cu, and Cd as outlined in the previous sections, dissolve in hot nitric acid, add 4 ml 1 : 1 sulphuric acid and evaporate to fumes of the latter. If the sample contains copper, remove it before or after evaporation of the nitric acid solution by neutralizing with ammonium hydroxide and adding 15 ml excess, adding 5 ml of a saturated ammonium carbonate solution, boiling for 15 minutes and allowing to stand warm for one hour. Filter and wash ten times with hot ammonia wash solution containing 30 ml ammonium hydroxide, 1 g ammonium nitrate, and 15 ml saturated ammonium carbonate per litre, until all copper has been removed from the bismuth precipitate. Dissolve the latter in hot dilute sulphuric acid, washing the paper thoroughly with hot 1% sulphuric acid.

To the solution of bismuth in about 5% sulphuric acid, add 5 ml of 30% potassium iodide, and destroy the liberated iodine by careful addition of a few drops of 1 g per litre sulphur dioxide water. The latter is equivalent to 1 ml of saturated sulphur dioxide water at 20°C diluted to 100 ml with water. Too strong a sulphur dioxide solution in this method gives a yellow colour; too strong sulphuric acid does likewise. Dilute the sample to 100 ml in a volumetric flask and determine the concentration of bismuth photometrically at 460 nm from a calibration curve obtained with a series of standards. The standard solution is made by dissolving 0·0500 g bismuth in a litre of water containing 10 ml sulphuric acid; 1 ml = 0·05 mg Bi.

A convenient collection of traces of bismuth may sometimes be made in a small ferric hydroxide precipitate in ammoniacal solution. This isolation of bismuth, with similar traces of As, Sb, Sn, Pb, Se, and Te is particularly useful from large samples of the metals, oxides or salts of copper, cobalt, nickel, molybdenum, and zinc. The iron, which must be added if not present in the sample, should be 10–20 times the total of the above trace metals. Boil, allow to stand in a warm place for several hours or overnight, filter, wash thoroughly with hot dilute ammonium hydroxide containing ammonium salts. Discard the filtrate and washings. Dissolve the precipitate in hot dilute sulphuric or hydrochloric acid, and re-precipitate with ammonium hydroxide to liberate occluded copper etc. Filter, wash as before, and proceed with final isolation of bismuth, if necessary, by one of the methods described previously.

Another suitable collection of traces of bismuth from large samples of copper or lead can be carried out in dilute nitric acid solution by the manganese co-precipitation method [1, 7, 14]. To the sample in 1 : 4 nitric acid add 10 ml of 2% potassium permanganate, heat to boiling, add 20 ml of 10% manganese nitrate solution, and boil. Cool, filter, wash with hot water, reserve the precipitate, boil the filtrate with 10 ml of 10% ammonium persulphate solution, allow to stand an hour, filter, and wash with hot water. Discard the filtrate, and combine the two precipitates. Destroy the filter paper by evaporating with 35 ml of nitric acid and 15 ml of sulphuric acid, and finally isolate bismuth from accompanying As, Sb, Sn, Se, and Te by one of the procedures outlined earlier.

B. Thiourea Method

The addition of thiourea to acidified solutions containing bismuth produces a deep yellow colour [1, 3, 4, 8, 10, 14, 15]. When bismuth is separated with other members of the acid sulphide group by hydrogen sulphide, mercury, silver, and the platinum metals give a brown or red colour, and antimony gives a yellow colour similar to that of bismuth but much less intense. Tellurium also gives a yellow colour, whereas selenium is precipitated. Large amounts of copper, cadmium, lead, and tin give white precipitates. Remove silver and mercury as chlorides, platinum by hydrolytic precipitation of bismuth and other base metals in nitrite solution, and palladium by dimethylglyoxime in 3% hydrochloric or sulphuric acid. Volatilize arsenic, antimony, tin and selenium by evaporations with hydrobromic acid and bromine, or dissolve them, together with tellurium, in ammoniacal ammonium sulphide solution. Lead can be separated as described earlier by cupferron, internal electrolysis or with a re-precipitation of lead sulphate. Appreciable quantities of selenium and tellurium can be removed by precipitation with sulphur dioxide in strong and dilute hydrochloric acid solutions, respectively.

Select a sample weight such that not over 4 mg bismuth is present in the solution for final colour comparison. Obtain the precipitated sulphides of Bi, Cu, and Cd as outlined previously, dissolve in hot dilute nitric acid, and adjust the acid concentration to about 1 N. Cool, add 10–12 g thiourea for each 50 ml solution, and warm to dissolve. Cool, filter off metal complexes and excess thiourea, and wash with saturated thiourea solution. Make to volume in a 100 ml volumetric flask and determine bismuth photometrically at 420 nm. The same temperature, thiourea concentration, and acidity should be used for standardization and for determining the unknowns. A convenient standard solution contains 0·2000 g bismuth dissolved in 20 ml nitric acid diluted to 500 ml; 1 ml = 0·4 mg Bi.

The procedure outlined above, in which bismuth is finally isolated from nearly all other elements before the colorimetric determination, is a general one intended for any material. With many samples, the method can be greatly shortened for routine work. For example, 0·01–0·5% bismuth in solders and bearing metals can be determined by thiourea, after volatilization of arsenic, antimony, and tin with hydrobromic acid and bromine, in the presence of 0·1–99% lead, 0·01–10% copper, 0·005–0·5% iron, 0·001–0·05% zinc, and 0·001–0·01% aluminium, at 440 nm in perchloric acid solution [1]. Similarly, 0·01–0·6% bismuth in aluminium-base alloys can be determined directly in the presence of the following maximum percentages of other constituents: silicon 14, copper 12, magnesium 12, zinc 7, nickel 4, iron 2, manganese 1·5, tin, lead, and chromium 1, and titanium 0·5. Only copper interferes; each 1% of this element decreases the bismuth value by 0·001% [1].

5. Optical Spectrography

Emission spectrography has long been used for the determination of low bismuth contents, especially in blister and refined copper, copper alloys, nickel, pig lead, aluminium and its alloys, tin alloys, and other metals. Bismuth has a number of useful lines for quantitative spectrography, the principal ones being 2697, 2780, 2897·98, 2938, 3024·64, 3067·72, and 3397·21. Detailed procedures have been published for 0·03–0·7% bismuth in aluminium alloys [1, 2], 0·001–0·3% in tin alloys [1, 2], 0·0003–1% in lead [1, 2, 10], and 0·0001–0·01% in nickel [11].

Bismuth is one of the most detrimental impurities in copper, and very few electrolytic brands now contain more than 1 ppm. Routine spectrographic measurements of 0·2–5 ppm bismuth have been carried out in copper refineries for many years. The spectrographic determination of 0·5 ppm to 0·1% bismuth in various copper alloys has been discussed [5].

6. X-Ray Spectrography

Bismuth, having a high atomic weight, is very suitable for determination by X-ray fluorescence, but this method appears to be limited to specific applications. The most sensitive lines are $L_{\alpha 1}$ of 1·1438 and $L_{\beta 1}$ of 0·9519 Å.

7. Atomic-Absorption Spectrophotometry

Bismuth can be determined by atomic absorption, using an air–acetylene flame, and the most sensitive line 2230·6, or other lines 2061·7, 2268·2, and 3067·7 [6]. The sensitivity has been reported from 1 to 4 μg/ml.

8. Other Methods

A polarographic procedure suitable for 1 to 200 ppm of bismuth in copper and its alloys has been described [7]. It follows a chloroform extraction of bismuth with sodium diethyldithiocarbamate in potassium cyanide solution.

The minute quantities of bismuth in sea and natural waters have been determined by sorption on an anion exchanger, elution with nitric acid, and final photometric measurement with dithizone [12].

In 0·01 N nitric acid, n-propyl gallate forms a precipitate with bismuth and antimony, but not with copper, cadmium, lead, arsenic, mercury, or zinc [16].

REFERENCES

1. A.S.T.M., *Chemical Analysis of Metals. Sampling and Analysis of Metal Bearing Ores*, Philadelphia, Pa., American Society for Testing and Materials, 1969.
2. A.S.T.M., *Methods for Emission Spectrochemical Analysis*, 5th ed., Philadelphia, Pa., American Society for Testing and Materials, 1968.

3. BRITISH ALUMINIUM Co., *Chemical Analysis of Aluminium and its Alloys*, London, The British Aluminium Co., 1961.
4. BRITISH DRUG HOUSES, *The B.D.H. Book of Organic Reagents*, Poole, England, British Drug Houses, 1958.
5. DOZINEL, C. M., *Modern Methods of Analysis of Copper and its Alloys*, 2nd ed., London, Elsevier, 1963.
6. ELWELL, W. T., and GIDLEY, J. A. F., *Atomic-Absorption Spectrophotometry*, 2nd ed., Oxford, Pergamon, 1966.
7. ELWELL, W. T., and SCHOLES, I. R., *Analysis of Copper and its Alloys*, Oxford, Pergamon, 1967.
8. FURMAN, N. H., ed., *Scott's Standard Methods of Chemical Analysis*, 6th ed., Vol. 1, Princeton, N.J., Van Nostrand, 1962.
9. HILLEBRAND, W. F., LUNDELL, G. E. F., BRIGHT, H. A., and HOFFMAN, J. I., *Applied Inorganic Analysis*, 2nd ed., New York, Wiley, 1953.
10. KOLTHOFF, I. M., and ELVING, P. J., *Treatise on Analytical Chemistry*, Part 11, Vol. 8, 147–75, New York, Interscience Publishers, 1963.
11. LEWIS, C. L., OTT, W. L., and SINE, N. M., *The Analysis of Nickel*, Oxford, Pergamon, 1966.
12. PORTMAN, J. E., and RILEY, J. P., *Anal. Chim. Acta* **34**, 201–10 (1966).
13. REED, J. F., *Anal. Chem.* **32**, 662–4 (1960).
14. SANDELL, E. B., *Colorimetric Determination of Traces of Metals*, 3rd ed., New York, Interscience Publishers, 1959.
15. SNELL, F. D., and HILTON, C. L., *Encyclopedia of Industrial Chemical Analysis*, Vol. 7, New York, Interscience Publishers, 1968.
16. WILSON, A. D., and LEWIS, D. T., *Analyst* **88**, 585–9 (1963).
17. WILSON, C. L., and WILSON, D. W., *Comprehensive Analytical Chemistry*, Vol. 1C, 265–76, London, Elsevier, 1962.

7 Boron

Boron is not a common constituent of mining and metallurgical products; its occurrence, however, in some ores and minerals, fertilizers, soils, fluxes, ceramics, glasses, certain steels and alloys, abrasives, and electroplating solutions ensures that its determination may be required occasionally in any inorganic laboratory.

Boric acid is volatile in steam and may be lost if digestions are carried out at too high a temperature in open beakers; decomposition procedures must be carefully controlled. Because Pyrex glass contains boron, quartz or the virtually boron-free Corning No. 7280 glass should be used in work of high accuracy; in all cases blanks must be carried through the entire determination.

Unless removed, boron will contaminate a silica precipitate and will also be tenaciously retained by hydroxides of iron and aluminium when these are precipitated with ammonia. Boron can be entirely removed from a sample by evaporating to fumes of sulphuric acid with a mixture of sulphuric and hydrofluoric acids.

ISOLATION OR SEPARATION OF BORON

Boron is almost always determined volumetrically or colorimetrically, and a separation, if required, is effected by distillation or ion exchange. Occasionally, however, it may be useful to recall that boron is not extracted by a cupferron–chloroform solution in dilute acid, and can thereby be separated from Fe, V, Ti, Zr, Nb, Ta, Sn, Ga, Bi, Mo, Pd, and Sb(III). Likewise, boron is not deposited on a mercury cathode in dilute sulphuric acid, and can in this manner be removed from Cr, Mo, Fe, Co, Ni, Cu, Zn, Ga, Ge, Cd, In, Sn, Ag, Au, Pt, Pd, Hg, Tl, and Bi.

1. Volumetric Determination

A. General Procedure

When alkali or alkaline earth borates are acidified, treated with methyl alcohol, and boiled, all the boron is volatilized as an ester, methyl borate. This separates boron from nearly all other elements [13, 16, 19, 26, 29]. The methyl borate is caught and saponified in alkaline solution, the alcohol expelled and the boric acid titrated with a standard sodium hydroxide solution, or determined colorimetrically. Boric acid cannot be titrated directly with sodium hydroxide, because the neutralization point is too high for a sharp colour change with indicators. If a polyhydric alcohol such as mannitol is added, however, compounds of stronger acid properties are formed and the titration can be made with phenolphthalein or other indicators.

Weigh a 0·5–5 g sample, depending on the boron content, into a beaker; add 1:1 hydrochloric acid, cover, and heat quickly on a water bath until solution is complete. If the material is not wholly soluble in hydrochloric acid, fuse directly or fuse the insoluble residue with 6 parts of sodium carbonate; decompose the melt with hydrochloric acid but do not heat to boiling, because boric acid may be lost with the steam.

If necessary, use other methods to bring the sample into solution. Elemental boron in the amorphous form is soluble in a mixture of nitric and sulphuric acids, or in fused sodium carbonate. Crystalline boron is scarcely attacked by acid or alkaline solutions, but may be fused with potassium hydroxide. Boron carbide can be decomposed by a sodium carbonate fusion in a platinum crucible. Elemental boron or boron carbide may also be dissolved by prolonged fusion with potassium persulphate in a quartz flask. Other methods of dissolving elemental boron have been discussed [8, 9].

Most metal borides can be fused with a mixture of sodium carbonate and potassium nitrate; some, such as cobalt boride, should be dissolved in aqua regia under a reflux condenser and evaporated to dryness at a low heat. Ferroboron is fused with sodium peroxide in an iron crucible, and the fusion leached with hydrochloric acid under a reflux condenser [2]. Other materials may be dissolved in phosphoric acid [2], a mixture of phosphoric and perchloric acids [18], or sulphuric acid and hydrogen peroxide [10]. Pyrohydrolysis has also been used for the determination of boron in refractory borides; steam is passed over a sample heated in a furnace boat to 900–1400°C, depending on the boride [24].

Pour the solution, derived from one of the dissolution procedures given above, into a 500 ml decomposition flask; add about 1 g of anhydrous calcium chloride for each ml of solution. Into the decomposition flask pass methanol which has been vaporized by heating in another 500 ml flask. From the decomposition flask, the methyl borate is led through a condenser and the distillate collected in a 250 ml Erlenmeyer titration flask. The latter contains 10 ml of 0·5 N sodium hydroxide and is provided with a U-tube as a water trap to prevent escape of methyl borate.

The decomposition flask, partially immersed in a water bath, is not heated until about 25 ml of alcohol has condensed in it. Then the water bath is heated by a small flame or hot plate, and the flask kept hot enough to prevent further condensation of alcohol. The distillation must not be rapid enough to allow methyl borate to escape from the system.

When a distillate of about 100 ml has collected in the Erlenmeyer flask, this receiver is exchanged for a fresh one and the collection of a second distillate is continued. Add the contents of the trap tube to the first receiver and perform a preliminary titration as follows. Add a drop of a saturated water solution of *p*-nitrophenol and run in the standard sodium hydroxide solution until the indicator shows that the free mineral acid is neutralized. Then add 1 ml of 0·1% phenolphthalein in ethyl alcohol and continue the titration until the colour of the indicator commences to appear. The endpoint will not be sharp, but the alkali used between the two endpoints indicates approximately the amount of boric acid in the distillate. If the sample is likely to contain more than 3–4% B_2O_3, use 0·5N sodium hydroxide, the object being to keep the distillate as free from water as possible and thus facilitate the dehydration later.

After completing the titration, add to the distillate twice as much sodium hydroxide solution as was used between the two endpoints, transfer the distillate to the second distilling apparatus, and boil off the alcohol. The addition of an excess of sodium hydroxide prevents the loss of boron by converting unstable $NaBO_2$ to stable Na_3BO_3.

In the meantime the second distillate of 100 ml will have collected. Unless the alcohol contained water, or insufficient calcium chloride was used, the second distillate will contain the remainder of the boron. Stop the distillation before removing the receiver; treat the second distillate like the first and add it to the latter. The distillation can be considered complete if less than 1 ml of 0·1N sodium hydroxide was required for titrating the second distillate.

When the alcohol has been boiled off on the steam bath, transfer the liquid to a casserole; rinse the flask once or twice with water, and heat the liquid carefully over a hot plate until the remaining alcohol has been boiled out.

Return the solution to the flask and add 1:1 hydrochloric acid carefully until the colour of both indicators is just discharged. Put in a boiling tube, and heat the flask on the steam bath for a few minutes; attach to the filter pump and boil under reduced pressure until the liquid is nearly cold and only an occasional bubble appears. When all carbon dioxide is removed, break the vacuum, cool under the tap, and proceed to the final titration.

Neutralize the excess hydrochloric acid by adding 0·5N sodium hydroxide carefully until the yellow of *p*-nitrophenol appears. Adjust to acid reaction with 0·1N hydrochloric acid and then to the appearance of a faint yellow colour with 0·1N sodium hydroxide. The solution is now neutral, because the indicator itself shows this colour in a neutral solution.

Add 1 g mannitol, read the burette, and continue the titration with 0·1N sodium hydroxide to the endpoint for phenolphthalein. Add another gram of mannitol; if this causes a disappearance of the red colour add more sodium hydroxide until it reappears. The first red colour that is permanent in the presence of mannitol is the endpoint of the titration. About 0·6 g mannitol for each 10 ml of solution will give the smallest titration error and sharpest endpoint.

The blank is determined using the reagents and procedure outlined above. The equations representing the steps in this method are:

volatilization: $H_3BO_3 + 3CH_3OH \rightarrow B(OCH_3)_3 + 3H_2O$

saponification: $B(OCH_3)_3 + 3NaOH \rightarrow Na_3BO_3 + 3CH_3OH$

neutralization: $Na_3BO_3 + 3HCl \rightarrow H_3BO_3 + 3NaCl$

titration: $H_3BO_3 + 3NaOH \rightarrow Na_3BO_3 + 3H_2O$

$$1 \text{ ml } 0{\cdot}1 \text{ N NaOH} = 0{\cdot}003481 \text{ g } B_2O_3$$
$$= 0{\cdot}001081 \text{ g B}$$

B. Ores of Boron

Another approach to the determination of boron in its minerals and ores is exemplified by the procedure for Gerstley borate, kindly furnished by U.S. Borax Research Corporation of Anaheim, Cal. [28], and Borax Consolidated Limited, of London. Boil 1 g for 5 minutes with 75 ml of 1:15 hydrochloric acid in a 250 ml flask fitted with a reflux condenser. Transfer to a beaker, add methyl red, make alkaline with 10% sodium hydroxide solution, and add an excess of 2 drops of hydrochloric acid. Add 5 ml bromine water and boil for several minutes to remove the excess. Dilute to about 100 ml, add 2 g barium carbonate, and boil. Allow to stand four hours or preferably overnight, filter with vacuum, and wash well with hot water; discard the precipitate.

Add 2 drops methyl red, acidify with 2 drops hydrochloric acid, boil 5 minutes to remove carbon dioxide, and cool. Add 1 drop of 0·25% aqueous solution of methylene blue, and 3 drops of methyl red. With this combination of indicators, the solution is purple when acid, green when alkaline, and a muddy bluish-green when neutral. Neutralize with 0·5N sodium hydroxide, add a few drops of phenolphthalein, about 10 g mannitol, and titrate to a purple endpoint with 0·5N sodium hydroxide. When the end point is reached, add a small amount of mannitol, and if the colour changes to a green continue the titration with 0·5N sodium hydroxide. If the purple colour remains, the end point has been reached. Repeat the addition of small amounts of mannitol until the end point is permanent. 1 ml of 0·5N NaOH = 0·01741 g B_2O_3.

When mannitol is added to the filtered solution after it has been neutralized, its colour will change from green to purple. This colour is due to the acidity of the boric acid–mannitol complex, and must not be confused with the purple colour of the phenolphthalein end point. As sodium hydroxide is run into the solution, the purple colour will change to green, and then to purple again when the solution becomes alkaline to phenolphthalein. The end point is reached when all the boric acid in the solution has been converted to sodium metaborate:

$$2H_3BO_3 + 2NaOH + 4H_2O \rightarrow Na_2B_2O_4.8H_2O.$$

C. Ferroboron

In another variation of the volumetric procedure, boron is separated from interferences by ion exchange [2, 4, 20] and finally titrated potentiometrically to *p*H 6·9, after the addition of mannitol, with standard sodium hydroxide solution. Ferroboron, which contains 10–20% of boron and few interfering cations, is an example of a material which is usually treated in this way [2].

Fuse 1 g of the sample in an iron crucible with 10 g of sodium peroxide. Cover the crucible, cool, tap to loosen the cake, transfer the latter to a 250 ml Florence flask, and connect to a reflux condenser. Mix 30 ml of hydrochloric acid with 20 ml of water and rinse the crucible with successive portions, adding the rinsings to the flask through the condenser. If this does not decompose the cake and make the solution acid, add enough hydrochloric acid to make it just acid to litmus.

Boil gently for 5–10 minutes, cool, disconnect the flask from the condenser, add 1 g of ferrous sulphate, and agitate the flask. Blow a current of air on the surface of the solution to expel chlorine. Transfer the solution to a 250 ml volumetric flask, dilute to the mark, and mix thoroughly. Transfer 50 ml by means of a pipette to a 150 ml beaker, and dilute to 100 ml.

Prepare an ion-exchange column with a strong acid-type resin, such as Dowex-50, having 8% cross linkage, and $-50+100$ mesh size. Wash the resin with 100 ml of 1:2 hydrochloric acid and then with water, at a flow rate of about 20 ml per minute. Pass the 100 ml sample through the column, and wash with 200 ml of water in small portions. Transfer the eluate to a 500 ml flask of low-boron glass, add 5 drops of methyl orange indicator, and nearly neutralize by addition of sodium hydroxide pellets. Boil the solution under a reflux condenser for 5 minutes to remove carbon dioxide, cool, and transfer to a 600 ml beaker.

Using a *p*H meter, adjust the solution to *p*H 6·9 by addition of 0·1N sodium hydroxide solution from a burette. Add sufficient mannitol to make the concentration 0·6M, or 11 g per 100 ml of solution. Titrate immediately to *p*H 6·9 with the 0·1N sodium hydroxide solution. Make a blank determination following the same procedure and using the same quantities of all reagents.

Regenerate the resin before using again, by washing with 1:2 hydrochloric acid until all iron is removed, and finally with water to remove the acid.

2. Colorimetric Determination

For a small quantity of boron, after distillation a final photometric determination replaces the titrimetic step, using curcumin [3, 5, 10, 14, 15, 17, 22, 23], carminic acid [2, 7, 12, 17], dianthrimide [6], quinalizarin [1, 3], or tetrabromochrysazin [18]. The colour comparison may also follow isolation of boron by ion exchange [7, 12] rather than by volatilization. In a few materials, the colorimetric procedure can be applied directly, without any separation of boron [2, 3, 5, 10, 11, 14, 15, 22, 23].

A. Curcumin

Curcumin is probably the most important reagent for the colorimetric determination of boron [3, 5, 10, 14, 15, 17, 22, 23]. Usually boron is first separated by distillation as methyl borate, but for some materials a direct method can be used. The first illustrative procedure described below is intended for less than 0·008% boron in steels, and involves a distillation [2].

Transfer to a 100 ml quartz flask 0·5 g of steel if the boron content is 0·002%, add 10 ml of phosphoric acid, and assemble the reflux assembly, methanol vaporization flask, and trap as described earlier under Volumetric Determination. Alternatively, use the distillation apparatus recommended by A.S.T.M. [2]. Transfer the methanol solution containing the boron to a porcelain casserole, and evaporate to dryness on a steam bath. Cool, add 1 ml of 1:4 hydrochloric acid, and 5 ml of an oxalic acid solution made by dissolving 50 g of oxalic acid in 450 ml of acetone. Mix, and add 2 ml of 0·025% curcumin solution in ethanol.

When the residue in the casserole has dissolved, evaporate to dryness on a water bath at 55 $\pm$ 3°C, and bake for 30 minutes at the same temperature. Cool, add 25 ml of acetone, and filter through a fine fritted-glass crucible into a 100 ml volumetric flask. Wash with 25 ml of acetone, dilute to the mark with cold water and mix. Transfer a portion to an absorption cell and measure the optical density at 540 nm. Carry through all operations a blank containing 0·5 g of low-boron steel such as National Bureau of Standards sample No. 55 of ingot iron, together with 5 ml of a suspension containing 5·6 g calcium oxide per litre.

Prepare the calibration curve by carrying through all operations aliquots of boric acid containing 0·002–0·008 mg of boron, together with the low-boron steel and the suspension of calcium hydroxide.

This gives acid-soluble boron. If acid-insoluble boron is left in the decomposition flask, dilute to a volume of 90 ml with 1 : 8 hydrochloric acid. Filter, and wash thoroughly with hot 2% hydrochloric acid to remove the iron, and then with cold water to remove the hydrochloric acid. Transfer the paper and residue to a platinum crucible, add 5 ml of the calcium hydroxide suspension, and evaporate to dryness. Ignite at 600–700°C until all carbon is destroyed. Add 1 g of sodium carbonate and fuse. Cool, dissolve in small portions of phosphoric acid, and transfer to a 100 ml quartz flask. Assemble the apparatus as before and determine the acid-insoluble boron. Total boron is the sum of the acid-soluble and acid-insoluble.

An example of a direct photometric method for boron, using curcumin, is the determination of 0·01–0·16% boron in aluminium [5]. Dissolve 1 g with 25 ml of 15% sodium hydroxide solution in a covered nickel dish or large crucible. Evaporate to a pasty consistency, dilute with about 20 ml of water, and boil. Cool, pour into 30 ml of hydrochloric acid, stir, and warm until all the precipitated aluminium hydroxide has dissolved. Cool, filter if necessary, and transfer to a 100 ml volumetric flask; dilute to the mark and mix.

Transfer by pipette a 1·0 ml aliquot into a porcelain basin, add 10 ml curcumin reagent, and heat on a water bath at 55–60°C for at least 20 minutes after evaporation appears to be complete. Add to the residue 50 ml of ethanol, and with a small rubber policeman remove and break up the solids from the bottom of the basin. Filter the solution through a dry paper and measure the optical density at 550 nm. Carry through all operations a blank, and a series of standards for the calibration curve in the range 0–0·4 mg boron.

The curcumin reagent solution is prepared as follows. Add water to pure ethanol until the specific gravity is 0·8083–0·8093 at 15°C. To 300 ml of this diluted ethanol, add 15 g oxalic acid, 15 ml hydrochloric acid, and 85 ml of water. When solution is complete, add 1 g of curcumin. Dilute to 950 ml with the diluted ethanol, and dissolve the reagent by stirring and heating to 50°C. Cool to 20°C, filter, and dilute to 1 litre with diluted ethanol.

B. Carminic Acid

Boron, in the form of boric acid, reacts with carmine in concentrated sulphuric acid solution to produce a violet colour. This is the basis of a useful colorimetric method for many substances [2, 7, 12, 17]. Boron may be isolated by distillation or ion-exchange, as described earlier. For some materials, it may be even determined without a preliminary separation. The illustrative procedure outlined below is for 0·005–0·063% boron in aluminium-base alloys [2], which may contain up to the following percentages of other elements: Si 14, Cu 12, Mg 12, Zn 10, Ni 4, Fe 2, Mn 1·5, Sn, Pb, and Cr 1, Bi 0·6, Ti 0·5.

To a 1 g sample in a 250 ml Erlenmeyer flask add 15 ml of bromine water, followed by 10 ml of hydrochloric acid in small increments, with cooling, if necessary. When all metal has dissolved, heat just below boiling to remove bromine. Filter into a 50 ml volumetric flask and wash four times with a small quantity of water, keeping the filtrate below 40 ml.

Cover any residue with 0·25 g sodium carbonate and transfer to a small platinum crucible. Dry, char, fuse the contents until the melt is clear, and cool. Dissolve the melt with 5 ml of hot water, and add 1 : 1 sulphuric acid drop-wise until the effervescence ceases. Transfer the fusion solution to the main filtrate in the 50 ml flask, cool, dilute to volume, and mix.

Transfer a 2 ml aliquot to a dry 50 ml Erlenmeyer flask, add 10 ml of sulphuric acid, and cool to room temperature. Add 10 ml of carmine solution, stopper, mix, and allow to stand away from strong light for 45 minutes. The carmine solution is made fresh daily by dissolving 0·092 g carmine in 100 ml of sulphuric acid.

Measure the optical density at 585 nm. The only element normally present in aluminium-base alloys which might interfere is vanadium, if occurring in more than 0·3%.

C. Quinalizarin

Quinalizarin has long been a popular reagent for the colorimetric determination of boron, especially for soils, plant ash, and similar materials [1, 3]. It is often possible to dispense with a preliminary separation. Peroxide and oxidized manganese interfere, but the addition of ferrous sulphate and gentle heating overcomes the former and sodium sulphite the latter interference [3].

Pipette a 1-ml aliquot of the sample solution containing 1–8 μg of boron, to a boron-free test tube, and add 10 ml of quinalizarin solution. The latter is made by dissolving 45 mg of quinalizarin in 1 litre of sulphuric acid, and allowing to stand for a day with periodic agitation.

Stopper the tube, mix, and place in cold water. When cool, remove the tube and allow it to stand at room temperature for 20 minutes. Measure the optical density of the solution at 620 nm against a blank containing all reagents. Prepare the reference curve from a minimum of six known amounts of boron covering the above range.

D. Other colorimetric methods

Low-boron contents have also been determined using dianthrimide for 2–100 ppm in nickel-base alloys [6], and tetrabromochrysazin for 0·0002–0·025% in steels and high temperature alloys [18].

For geochemical work, the sample can be decomposed in sulphuric and hydrofluoric acids, and the blue complex formed between fluoborate ions and methylene blue extracted with 1,2-dichloroethane and measured visually or spectrophotometrically [27].

3. **Optical Spectrography**

The spectrographic determination of boron in agricultural materials in the parts-per-million range, using the lines 2496·8 or 2497·7, has been detailed [1, 25].

From 0·001 to 0·05% boron in aluminium, zirconium, and their alloys, is determined spectrographically, using the boron line 2497·7 [2]. The spectrographic determination of boron in nickel, in the range of 5–100 ppm, has been discussed [21].

REFERENCES

1. A.O.A.C., *Official Methods of Analysis of the Association of Official Analytical Chemists*, Washington, D.C., Association of Official Analytical Chemists, 1965.
2. A.S.T.M., *Chemical Analysis of Metals. Sampling and Analysis of Metal Bearing Ores*, Philadelphia, Pa., American Society for Testing and Materials, 1969.
3. Boltz, D. F., *Colorimetric Determination of Non-metals*, New York, Interscience Publishers, 1958.
4. Borland, H., Brownlie, I. A., and Godden, P. T., *Analyst* **92**, 47–53 (1967).
5. British Aluminium Co., *Analysis of Aluminium and its Alloys*, London, British Aluminium Co., 1961.

6. Burke, K. E., and Albright, C. H., *Talanta* **13**, 49–53 (1966).
7. Calkins, R. C., and Stenger, V. A., *Anal. Chem.* **28**, 399–402 (1956).
8. Donaldson, J. M., and Trowell, F., *Anal. Chem.* **36**, 2202–(1964).
9. Eberle, A. R., Pinto, L. J., and Lerner, M. W., *Anal. Chem.* **36**, 1282–5 (1964).
10. Elwell, W. T., and Scholes, I. R., *Analysis of Copper and its Alloys*, Oxford, Pergamon, 1967.
11. Elwell, W. T., and Wood, D. F., *Analyst* **88**, 475–6 (1963).
12. Fleet, M. E., *Anal. Chem.* **39**, 253–5 (1967).
13. Furman, N. H., ed., *Scott's Standard Methods of Chemical Analysis*, 6th ed., Vol. 1, Princeton, N.J., Van Nostrand, 1962.
14. Hayes, M. R., and Metcalfe, J., *Analyst* **87**, 956–69 (1962).
15. Hayes, M. R., and Metcalfe, J., *Analyst* **88**, 471–4 (1963).
16. Hillebrand, W. F., Lundell, G. E. F., Bright, H. A., and Hoffman, J. I., *Applied Inorganic Analysis*, 2nd ed., New York, Wiley, 1953.
17. Hopkin and Williams Ltd., *Organic Reagents for Metals*, Vol. 2, Chadwell Heath, England, Hopkin & Williams, 1964.
18. Karpen, W. L., *Anal. Chem.* **33**, 738–40 (1961).
19. Kolthoff, I. M., and Elving, P. J., *Treatise on Analytical Chemistry*, Part 11, Vol. 10, New York, Interscience Publishers, 1970.
20. Kramer, H., *Anal. Chem.* **27**, 144–5 (1955).
21. Lewis, C. L., Ott, W. L., and Sine, N. M., *The Analysis of Nickel*, Oxford, Pergamon, 1966.
22. Luke, C. L., *Anal. Chem.* **30**, 1405–6 (1958).
23. Luke, C. L., and Flaschen, S. S., *Anal. Chem.* **30**, 1406–9 (1958).
24. McKinley, G. J., and Wendt, H. F., *Anal. Chem.* **37**, 947–50 (1965).
25. Mitchell, R. L., *The Spectrochemical Analysis of Soils, Plants and Related Materials*, Farnham Royal, England, Commonwealth Agricultural Bureau, 1964.
26. Snell, F. D., and Hilton, C. L., *Encyclopedia of Industrial Chemical Analysis*, Vol. 7, New York, Interscience Publishers, 1968.
27. Stanton, R. E., and McDonald, A. J., *Analyst* **91**, 775–8 (1966).
28. U.S. Borax Research Corporation, Anaheim, Cal., personal communication, 1960.
29. Wilson, C. L., and Wilson, D. W., *Comprehensive Analytical Chemistry*, Vol. 1C, 89–102, London, Elsevier, 1962.

8 Cadmium

The principal natural occurrence of cadmium is in zinc ores, where small quantities are nearly always found, usually as the sulphide greenockite. Certain stages of zinc production give an enrichment of cadmium for eventual refining; the metal is used in various alloys, paints, electroplating solutions, amalgams, and other applications.

ISOLATION OR SEPARATION OF CADMIUM

Cadmium may be separated from other members of the hydrogen sulphide group in a variety of ways. Silver and mercurous mercury may be removed by prior treatment with hydrochloric acid; lead is separated with sulphuric acid. Mercury may also be volatilized by igniting its compounds or fusing them with sodium carbonate. Arsenic, antimony, selenium, and tin can be volatilized by boiling to a low volume several times with hydrobromic acid and bromine, or hydrobromic and sulphuric acids, or hydrobromic and perchloric acids. Selenium, tellurium, and gold can be precipitated by sulphur dioxide in hydrochloric acid solution. Copper can be separated by electrolysing in the usual solution of 2–3 ml nitric acid and 5 ml of 1 : 1 sulphuric acid per 100 ml.

Bismuth can be separated as $BiOCl$; with large quantities of cadmium a re-precipitation is required. Molybdenum can be isolated as soluble sodium molybdate by treatment with sodium hydroxide. Cadmium, with Cu and Bi, can be separated from As, Sb, Sn, Mo, Ge, Au, Ir, Se, and Te by treatment of the mixed sulphides with hot ammonium sulphide; the latter elements dissolve, leaving the sulphides of Cd, Cu, Pb, Ag, Pd, Rh, and Bi unattacked.

Cadmium can be separated from the platinum metals by hydrolytic precipitation from their complex nitrites [5]. If a hot nitrite solution is neutralized with sodium bicarbonate to the change of colour from yellow to blue of thymol blue, cadmium is precipitated from the platinum metals with the exception of palladium. The latter can be separated from cadmium by precipitation with dimethylglyoxime in 3–4% hydrochloric acid solution.

A small quantity of cadmium can often be separated by dithizone [13]. The following elements of the hydrogen sulphide group do not react with dithizone: As, Sb, Mo, Se, Rh, Ir, Ru, Os, and Ge. In strong mineral acid the following form dithizonates: Cu, Au, Hg, Ag, Te, Pt, and Pd. In a strongly alkaline solution containing tartrate and a low concentration of cyanide, Cd, Tl, and a little lead and zinc are extracted with a carbon tetrachloride solution of dithizone, whereas Bi, Co, Ni, and Sn remain in the aqueous phase.

Complete separation of cadmium from zinc, especially when the latter is present in large excess, by hydrogen sulphide precipitation in acid solution is a rather lengthy procedure. It is best carried out in 3N sulphuric acid solution with a long period of gassing, commencing hot and allowing to cool; the precipitate is dissolved in hot 1 : 1 hydrochloric acid, taken to fumes with sulphuric acid, and the precipitation repeated. Frequently a second or even a third precipitation is necessary.

Other methods of separating cadmium from zinc, the metal with which it is most frequently associated, can be employed. When zinc is preponderant, as in slab zinc, a sulphate solution can be treated with potassium iodide and β-naphthoquinoline; cadmium forms a precipitate whereas zinc remains in solution [1, 11]. A small amount of zinc accompanies the cadmium, but one subsequent separation with hydrogen sulphide will usually yield a pure cadmium precipitate. When zinc and cadmium are both found in substantial quantities, as in silver brazing alloys containing 15–30% zinc and 3–25% cadmium, sodium diethyldithiocarbamate can be used to precipitate cadmium in a solution containing tartaric acid, sodium hydroxide, and potassium cyanide [1].

1. Gravimetric Procedure

Weigh out 0·5–25 g, depending on the cadmium content, dissolve in hydrochloric and nitric acids, add 25 ml 1 : 1 sulphuric acid and evaporate to strong fumes of the latter. Cool, add water, boil, filter off silica and lead sulphate, and wash thoroughly. If the sample contains silver, evaporate the filtrate to dryness, dilute with a little water containing hydrochloric acid, allow to stand several hours and filter off the precipitated silver chloride. To the filtrate add 25 ml 1 : 1 sulphuric acid and evaporate to fumes.

Dilute to approximately 300 ml and pass in hydrogen sulphide for 30 minutes. If appreciable quantities of zinc are present, add ammonium hydroxide cautiously until a heavy precipitation of zinc sulphide occurs, keeping the solution sufficiently acid so that no iron precipitates. Allow the precipitate to settle, filter, and wash with cold water. Discard the filtrate. All the cadmium has now been isolated with the other members of the hydrogen sulphide group, together with a substantial part of any zinc present. Further separations may take two forms.

A. Separation as Sulphide

If zinc is present, as is generally the case, dissolve the mixed sulphides on the paper with dilute hydrochloric acid, add 15 ml 1 : 1 sulphuric acid and evaporate to fumes. Cool, dilute to 200 ml and pass in hydrogen sulphide as before. Filter off the precipitated sulphides of cadmium and other members of the group; the precipitate will contain zinc varying in quantity from traces to significant amounts depending on the original content of zinc. If necessary, make a third sulphide precipitation of cadmium to eliminate zinc.

For a small quantity of cadmium in the presence of a large amount of zinc, it may be more convenient to initially separate all the cadmium together with a small quantity of zinc, by β-napthoquinoline [1, 11]. To about 100 ml of a lead-free sulphate sample add 3–5 g of potassium iodide, and stir until dissolved. Add a 1% solution of β-naphthoquinoline in 5% sulphuric acid in an excess of 100% over that required for precipitation of cadmium; 10 ml will theoretically precipitate 0·025 g of cadmium. Allow to stand in ice water for 30–60 minutes, stirring occasionally. Filter without washing, and rinse the precipitate into a 400 ml beaker. Wash the original beaker and the filter paper with 50 ml of hot 1 : 9 sulphuric acid, and allow it to run into the precipitate in the 400 ml beaker. Wash with hot water and dilute to 200 ml. Heat to 60°C and pass hydrogen sulphide into the solution. Add ammonium hydroxide drop by drop until the yellow cadmium sulphide precipitates, and continue to pass in hydrogen sulphide for 20 minutes while the solution cools. Filter, dissolve the cadmium sulphide in hydrochloric acid, add sulphuric acid, evaporate to fumes, and again precipitate cadmium sulphide as before.

Digest the cadmium sulphide precipitate with hot 10% ammonium sulphide solution to dissolve any sulphides of As, Sb, Sn, Mo, Se, and Te. Filter and wash with hot 1% ammonium sulphide solution. If bismuth is present it can be removed as BiOCl by dissolving the sulphides of Cd, Bi, and Cu in hot dilute nitric acid, neutralizing with ammonium hydroxide, diluting to about 300 ml, adding 1–3 ml of 1 : 4 hydrochloric acid, and allowing BiOCl to settle an hour or two in a warm place. Filter, wash, and evaporate the filtrate to strong fumes after adding 5 ml 1 : 1 sulphuric acid. If copper is present, it can be removed at this stage by electrolysis in a solution containing 5 ml 1 : 1 sulphuric acid and 2–3 ml nitric acid per 100 ml. Cadmium is in the final solution; evaporate to fumes of sulphuric acid to eliminate nitric acid after the copper electrolysis.

Cadmium has now been isolated in sulphuric acid solution from all other elements; it may be determined in one of several ways, the choice depending on the quantity present.

(1) *Weighing as $CdSO_4$.* For substantial quantities of cadmium the best method is gravimetric, weighing as $CdSO_4$ [5, 11, 16]. Transfer the small amount of sulphate solution, after removal of all other elements, to a weighed platinum dish or large crucible, and evaporate carefully to dryness. Place in a muffle and slowly raise the temperature to approximately 500°C. Cool, dissolve the residue in a little water, carefully evaporate, and again heat to 500°C. This should expel all sulphuric acid. Cool, weigh, and repeat the operation until a constant weight is obtained.

$$CdSO_4 \times 0{\cdot}5292 = Cd$$

(2) *Weighing as CdS.* For routine determinations of small quantities of cadmium, a final weighing as sulphide may be used. Though frowned upon

by some authorities [5, 16], the procedure is advocated by others having long familiarity with the metal [1]. Because the composition of the sulphide does not quite correspond to CdS, the method should be confined to the lower quantities of cadmium in plant control practice; for exact work a final sulphate weighing or another method for cadmium should be used.

To a warm sulphuric acid solution of cadmium containing 6 ml of 1 : 1 sulphuric acid in 100 ml, pass in hydrogen sulphide for 30 minutes; if cadmium is slow in precipitating, add water in 5 ml portions until precipitation commences. Filter through a weighed Gooch or sintered glass crucible, wash 4–5 times with warm water; dry 1–2 hours at 110°C, cool and weigh as CdS.

$$\text{CdS} \times 0{\cdot}778 = \text{Cd}$$

B. Separation by Ion Exchange

The disadvantages of the lengthy separations required to isolate cadmium as sulphide from a complex sample, particularly when zinc is present, have led to the development of separations in a resin column. Both anion and cation exchange have been employed [7, 8, 9, 11, 14]. The anion exchange resin Dowex 1 will adsorb zinc and cadmium in 0·12N hydrochloric acid containing 100 g sodium chloride per litre; zinc is eluted with 2N sodium hydroxide containing 20 g sodium chloride per litre, and cadmium finally eluted with 1N nitric acid [9]. Cadmium forms a strong anion complex with iodides, which is retained on Dowex 1, while zinc and most other metals associated with cadmium are not adsorbed. Cadmium is finally eluted with 3N nitric acid [7].

A more rapid and convenient procedure is afforded by cation exchange, wherein zinc and most elements associated with cadmium are retained on a cation resin; the cadmium is eluted with 0·3N hydriodic acid–0·15N sulphuric acid [8], or with 0·5N hydrochloric acid. The following details for the isolation of cadmium by this method are suitable for a wide variety of materials [8]. Decompose the sample with hydrochloric and nitric acids, add 10 ml 1 : 1 sulphuric acid and evaporate to dryness. Add 90 ml of 2% sulphuric acid, boil, cool to around 80°C, add about 5 g iron chips, and place on a steam bath for 15 minutes. Cool, filter on a small filter paper containing a few iron chips, and wash with 2% sulphuric acid. Add to the filtrate, which should have a volume of 145–150 ml, 7·5 ml of concentrated hydriodic acid and pass the solution through a 1 × 10 inch column of Dowex 50, Amberlite IR–120, or similar strongly acidic cation exchange resin, 100–200 mesh with 8% cross linkage. Before use, prepare the column by passing 60 ml of the 0·3N hydriodic acid–0·15N sulphuric acid elutriant through the resin bed. Rinse the sample beaker with the hydriodic–sulphuric acid elutriant, and transfer the washings to the column. Wash the resin with 3 column volumes, or 180 ml, of hydriodic–sulphuric acid solution.

The iron treatment of the sulphate solution serves two purposes: it removes Bi, As, Cu, and Ag with the silica and lead sulphate, and it reduces ferric to ferrous ions and cupric ions to metallic copper. Ferric and cupric ions would otherwise oxidize hydriodic acid to iodine, which would have to to be reduced with sulphur dioxide or another reducing agent. The following ions are adsorbed by the resin: Zn, Fe, Al, Mn, Cr, Ca, Mo, Ni, Co, Sn, In, Ge, and Ga. The resin can be regenerated readily by passing 4–5 column volumes of 2N hydrochloric acid through it, followed by two volumes of water. All the cations which have been adsorbed in the iodide medium will be removed.

2. Volumetric Method

An EDTA titrimetic procedure is suitable for a wide range of cadmium concentrations [1, 7, 8, 11]. Unfortunately, about forty other cations form soluble complexes with EDTA, so that the cadmium must be isolated, or accompanied by only a few elements which can be inactivated or masked. In the example described below, for a silver brazing alloy containing 3–25% cadmium [1], the addition of formaldehyde to the cyanide solution decomposes the cyanide complex of cadmium but not the complexes of copper, cobalt, or nickel. Cadmium can thus be determined in the presence of Cu, Co, or Ni, but zinc must be absent. Decompose a 1 g sample in nitric acid, and remove and determine silver by precipitation as chloride. Remove and determine copper by electrolysis in the conventional sulphuric acid–nitric acid electrolyte. Evaporate the solution containing zinc and cadmium to fumes of sulphuric acid. Cool, dilute to 250 ml, add 15 ml of 30% tartaric acid solution, and 8 drops of methyl red indicator. Neutralize with 20% sodium hydroxide solution, add 10 ml of 10% potassium cyanide solution, and 75 ml of 2% sodium diethyldithiocarbamate solution. Stir and allow to stand about 30 minutes.

Filter with suction through a fritted-glass crucible. Wash several times with a solution containing 0·4% sodium hydroxide, 0·1% potassium cyanide, and 0·02% sodium diethyldithiocarbamate. Transfer the precipitate and crucible to a 250 ml beaker, add 25 ml of hydrochloric acid and 50 ml of water to completely cover the precipitate. Heat until the precipitate is dissolved, cool to room temperature, and dilute to 200 ml. To the final solution containing cadmium, add a few drops of methyl red and carefully neutralize with 20% sodium hydroxide. Add 30 ml of a buffer solution containing 54 g ammonium chloride and 350 ml ammonium hydroxide per litre. Add 10 ml of 10% potassium cyanide solution, and five drops of Eriochrome black T indicator. The latter is made by dissolving 0·4 g of Eriochrome black T, i.e. 1-hydroxy-2-naphthylazo-5-nitro-2-naphthol-6-sulphonic acid sodium salt, in 20 ml ethanol, adding 30 ml of triethanolamine, and storing in a polyethylene dropping bottle.

Add sufficient 10% formaldehyde to just give a red colour. Titrate slowly with 0·05M EDTA solution to a blue-green end point. Add 5 ml of 10% formaldehyde, and if the colour changes to red, titrate again with EDTA to a blue-green end point. Continue the formaldehyde additions, and, if necessary, the EDTA titrations until the blue-green end point is stable for at least two minutes after the last formaldehyde addition.

The EDTA solution is made by dissolving 18·6 g disodium ethylenediamine tetraacetate in 600 ml of warm water, cooling, adding 0·1 g $MgCl_2.6H_2O$, and diluting with water to 1 litre. One ml is approximately equivalent to 5·62 mg of cadmium. Standardize against a cadmium sulphate solution made by dissolving the pure metal or a high grade cadmium salt in sulphuric acid, evaporating to fumes, dissolving in water and making to a definite volume.

3. Electrolytic Procedure

It is sometimes convenient to determine cadmium, in quantities exceeding 10 mg, by electrolysis in alkaline cyanide solution [1, 11, 16]. Obtain a sulphate solution containing cadmium only, usually by solution of isolated cadmium sulphide in hot dilute sulphuric acid. Add a drop of phenolphthalein and then 10% sodium hydroxide until the solution is pink. Add carefully, with stirring, 10% sodium cyanide until the precipitate of cadmium hydroxide just dissolves, avoiding an excess. Electrolyse with a platinum gauze cathode for about 5 hours at a current density of 0·5 ampere per sq. dm, and then increase the current density to 1 ampere per sq. dm for an hour. Remove the cathode, rinse rapidly in water and ethanol, dry at 110°C, and weigh as metallic cadmium.

4. Polarographic Method

Because the cadmium wave precedes the zinc wave to an appreciable extent, cadmium can be readily determined in the presence of zinc.

The determination of cadmium in concentrates of lead and zinc, and in slab zinc, is usually carried out polarographically. The procedures outlined below are typical of those used in the mining industry; they have been kindly provided by The Zinc Corporation Limited [17], and Imperial Smelting Corporation Ltd. [6].

A. Lead Concentrate

Weigh 5 g, add 15 ml nitric acid and digest for 10 minutes. Add 10 ml 1 : 1 sulphuric acid and carefully evaporate to fumes of the latter. Cool, rinse the sides of the beaker with water, and again fume. Add 30 ml water and digest hot for 10 minutes; cool, add 10 ml of 21% hydroxylamine hydrochloride to the hot solution followed by 2 ml of 0·015% methyl violet. Add ammonium hydroxide until the purple colour just appears, transfer the solution to a 100 ml volumetric flask and cool to room temperature. Add 2 ml of 10% potassium thiocyanate solution and make to volume, mix, and transfer the solution back to the original beaker. The solution may require to be centrifuged. Decant the clear supernatant solution into a polarograph cell. Remove dissolved oxygen by bubbling O_2-free nitrogen through the cell for 3 minutes, and determine the polarogram. A standard solution is made from a chloride solution containing, per litre: 1 g Pb, 10 g Fe, 50 g Zn, and 0·2 g Cd. Ten ml of the standard solution, equivalent to 0·002 g cadmium, is carried through with the determination.

B. Zinc Concentrate

Weigh 1 g, add 20 ml hydrochloric acid, and digest for 10 minutes. Boil gently and finally evaporate to dryness. Take up with 10 ml of 37% by volume sulphuric acid and digest; add 50 ml water and warm for 15 minutes. Add 10 ml of 21% hydroxylamine hydrochloride, followed by 2 ml of 0·015% methyl violet. Add ammonium hydroxide until the purple colour just appears, transfer to a 100 ml volumetric flask and cool to room temperature. Make to volume, transfer back to the original beaker and mix well; allow the precipitate to settle. Decant the clear supernatant solution into a polarograph cell and continue as given for lead concentrate.

A slightly different procedure is employed for zinc concentrates containing traces of indium. Weigh 2 g, add 20 ml nitric acid and 7–10 ml bromine, and allow to stand one hour. Cautiously evaporate the bromine, take to dryness, cool, and digest with 20 ml hydrochloric acid. Dilute with 50 ml hot water, scraping the inside of the beaker with a rubber policeman, cool and transfer to a 500 ml volumetric flask. Add 100 ml of 25% ammonium chloride solution, neutralize the free acid with ammonium hydroxide and add 25 ml excess. Cool and add 30 ml of 40% sodium sulphite solution containing 1% glycerol; add 10 ml 1·25% gelatin solution containing 0·5% phenol. Dilute to 500 ml, mix, allow to stand 30 minutes and proceed with the polarographic procedure. Calculate the cadmium by reference to standard polarograms prepared with each batch of samples.

C. Zinc Metal

Weigh 25 g into a 600 ml beaker, cover, add 100 ml of hydrochloric acid and digest on the hot plate until reaction has ceased; a few drops of a dilute solution of ferric chloride may be added as catalyst initially. Add potassium chlorate, in small quantities at a time, and warm until the lead-cadmium sponge has been dissolved; avoid a large excess of chlorate. Boil the solution to a volume not over 50 ml to liberate all chlorine. Dilute to 250 ml, stir, and cool. Transfer to a 500 ml volumetric flask and dilute to about 450 ml, mix and cool; finally dilute to 500 ml and mix. Transfer 5 ml to a dry polarograph cell and add 1 ml of 0·09% gelatin solution; de-aerate with nitrogen and polarograph the solution. Calculate the percentage cadmium by reference to standard curves.

If indium exceeds 0·003%, the above procedure is modified as follows. From the final volume of 500 ml solution prepared according to the directions above, transfer an aliquot of 40 ml to another 500 ml flask containing 100 ml of 25% ammonium chloride solution. Dilute to about 250 ml, neutralize the free acid with ammonium hydroxide and add 25 ml excess. Cool, add 30 ml of 40% $Na_2SO_3.7H_2O$ containing 1% glycerol, 10 ml of 1·25% gelatin solution containing 0·5% phenol, dilute to 500 ml, and allow to stand for 30 minutes. Transfer 10 ml to a polarograph cell and record the polarogram. Calculate the percentage cadmium by reference to a standard curve.

D. Other Metals and Alloys

Directions have been outlined for 0·01–3% cadmium, and for 0·001–1% when copper is precipitated as oxalate, in copper and its alloys [4]. Polarographic procedures for cadmium in aluminium alloys have been described [2].

5. Colorimetric Procedure

The determination of traces of cadmium is important in toxicology. For example, the recommended limit in potable water is only 0·01 ppm, one of the lowest tolerances for impurities.

For quantities of cadmium less than 0·5 mg, extraction with dithizone and measurement of the resulting red colour gives a satisfactory method [7, 8, 10, 13]. To an aliquot in a separatory funnel, containing less than 0·01 mg cadmium, from the final cadmium sulphate solution obtained as described previously, add 25 ml of 2% tartaric acid, 0·25 ml of 20% hydroxylamine hydrochloride, 15 ml of dithizone solution made by dissolving 8 mg diphenylthiocarbazone in 1 litre of chloroform, and 5 ml of a 40% sodium hydroxide–0·05% potassium cyanide solution. Shake for 1 minute, filter the chloroform layer into a dry colorimeter tube, and measure the absorbance of the pink colour at 518 nm. Prepare a calibration curve by treating known quantities of cadmium up to 0·01 mg as described above.

6. Optical Spectrography

Small quantities of cadmium in many materials can be determined by emission spectrography, using the lines 2288·02 or 3261·06, or less frequently, 5085·82. Methods have been detailed for 0·001–2% cadmium in aluminium and its alloys [1], 0·0003–0·01% in pig lead [1], 0·0005–0·05% cadmium in zinc and zinc-base alloys [1], 1–100 ppm in nickel [12], and 0·2–5 ppm in zirconium and zirconium-base alloys [1].

7. X-Ray Spectrography

Cadmium can be determined by X-ray fluorescence, using the $K_{\alpha 1}$ line 0·5365 [11].

8. Atomic-absorption Spectrophotometry

Atomic absorption offers a very satisfactory method for cadmium, the line 2280·0 having a sensitivity of 0·03 mg/ml with the air–acetylene, or even air–coal gas flame [3]. The less sensitive 3261·1 line may also be used. The only interference so far reported is a slight one due to silicon, and this of course, can be eliminated by the usual dehydration step after acid dissolution, or by initial treatment with hydrofluoric acid.

REFERENCES

1. A.S.T.M., *Chemical Analysis of Metals. Sampling and Analysis of Metal Bearing Ores*, Philadelphia, Pa., American Society for Testing and Materials, 1969.
2. BRITISH ALUMINIUM CO., *Analysis of Aluminium and Its Alloys*, London, British Aluminium Co., 1961.
3. ELWELL, W. T., and GIDLEY, J. A. F., *Atomic-Absorption Spectrophotometry*, 2nd ed., Oxford, Pergamon, 1966.
4. ELWELL, W. T., and SCHOLES, I. R., *Analysis of Copper and its Alloys*, Oxford, Pergamon, 1967.
5. HILLEBRAND, W. F., LUNDELL, G. E. F., BRIGHT, H. A., and HOFFMAN, J. I., *Applied Inorganic Analysis*, 2nd ed., New York, Wiley, 1953.
6. IMPERIAL SMELTING CORPORATION LTD., personal communication, 1961.
7. KALLMANN, S., OBERTHIN, H., and LIU, R., *Anal. Chem.* **30**, 1846–8 (1958).
8. KALLMANN, S., OBERTHIN, H., and LIU, R., *Anal. Chem.* **32**, 58–60 (1960).
9. KALLMANN, S., STEELE, C. G., and CHU, N. Y., *Anal. Chem.* **28**, 230–3 (1956).
10. KNAPP, J. R., VAN AMAN, R. E., and KANZELMEYER, J. H., *Anal. Chem.* **34**, 1374–8 (1962).
11. KOLTHOFF, I. M , and ELVING, P. J., *Treatise on Analytical Chemistry*, Part II, Vol. 3, 171–229, New York, Interscience Publishers, 1961.
12. LEWIS, C. L., OTT, W. L., and SINE, N. M., *The Analysis of Nickel*, Oxford, Pergamon, 1966.
13. Sandell, E. B., *Colorimetric Determination of Traces of Metals*, 3rd ed., New York, Interscience Publishers, 1959.
14. STRELOW, F. W. E., *Anal. Chem.* **32**, 363–5 (1960).
15. WILSON, B., *Can. Mining Met. Bull.* **50**, 143–6 (1957).
16. WILSON, C. L., and WILSON, D. W., *Comprehensive Analytical Chemistry*, Vol. 1C, 402–8, London, Elsevier, 1962.
17. ZINC CORPORATION LIMITED, personal communication, 1960.

9 Calcium

Calcium is a common element in all types of mining, metallurgical, and other inorganic analyses. It occurs in rocks, minerals, ores, concentrates, slags, waters, fluxes, soils, fertilizers; in building materials such as cement, lime, plaster; and even in a few alloys.

In industrial analyses, Al, Ca, and Mg are frequently determined on the same sample; sometimes silica and iron are also determined on this sample. It must be borne in mind that calcium sulphate is fairly insoluble in water and some may be lost with silica if the sample is taken to fumes of sulphur trioxide and filtered in dilute sulphuric acid. If phosphates are present they must be removed as iron or aluminium phosphates with ammonium hydroxide or precipitated with ammonium molybdate. Acid-insoluble calcium compounds such as the sulphate may be fused with sodium carbonate and extracted with water before filtering and dissolving in hydrochloric or nitric acid. Calcium in calcium fluoride may be determined by removing fluorine with sulphuric acid, if necessary with the addition of a little silica, in the initial decomposition of the sample.

ISOLATION OR SEPARATION OF CALCIUM

After removal of silica, the hydrogen sulphide group, ammonium hydroxide and ammonium sulphide groups, calcium will be found in the filtrate with the other alkaline earths, magnesium, and the alkalies. Addition of ammonium oxalate to the ammoniacal filtrate will precipitate calcium, thereby separating it from the remaining elements. Nearly all the strontium, and some of the barium, will be in the oxalate precipitate with calcium, but the former two are so rarely encountered above trace levels that they are disregarded in most industrial determinations of calcium. In exact analyses, a double precipitation of calcium oxalate is required, and up to 3–4 mg of barium in the original sample will not report in the final oxalate precipitate.

1. Gravimetric Method

When calcium is precipitated by ammonium oxalate, and filtered, the determination may be completed by igniting the calcium oxalate to oxide, and weighing, or by dissolving the oxalate in sulphuric acid and titrating with potassium permanganate. The gravimetric finish is usually reserved for a few materials having a high calcium content, such as limestone; nearly all samples in the mining and metallurgical industries are titrated. Because the procedure, up to the final step, is the same for both, this will be described in detail under the present heading.

To a 0·5–5 g sample, depending on the calcium content, add 15 ml hydrochloric acid, 15 ml nitric acid, and boil. For sulphide ores and concentrates add a little bromine initially and allow to stand for ten minutes before adding nitric acid, followed by hydrochloric acid. Limestone or similar high-lime samples should be moistened with water first, then hydrochloric acid added carefully through the spout of the covered beaker to avoid loss from violent effervescence.

Evaporate to dryness, moisten with hydrochloric acid, and again evaporate to dryness. Take up in 10 ml hydrochloric acid, 50 ml water, boil, filter and wash. Place paper and precipitate in a platinum crucible, dry, ignite, fuse the residue with about 5 g of sodium carbonate, and add the contents of the crucible to the filtrate. Evaporate to dryness twice, moistening with hydrochloric acid both times to dehydrate silica. Dissolve in 10 ml hydrochloric acid, 75 ml hot water, boil, filter off silica, and wash thoroughly with hot water; discard the precipitate if silica is not required.

Adjust the acidity of the filtrate to 5–10% hydrochloric acid and pass in hydrogen sulphide to precipitate all members of the acid sulphide group. Filter and wash thoroughly with acidulated hydrogen sulphide water. Make the filtrate ammoniacal and pass in a little more hydrogen sulphide. Filter off all the sulphides of the ammonium sulphide group, washing well with 1% ammonium sulphide. Boil the filtrate thoroughly to remove all hydrogen sulphide; it now contains only Ca, Ba, Sr, Mg, and the alkali metals.

If the sample contains fluorine, calcium would be lost as calcium fluoride during the removal of iron and aluminium by ammonia. Fluorine must therefore be initially expelled as silicon fluoride by fuming with sulphuric acid, with the addition of a little silica if necessary. Calcium sulphate formed by this treatment must be kept in solution by hydrochloric acid and ammonium chloride to avoid removal with the silica precipitate.

When phosphorus is present to the extent of less than one-tenth the content of iron and aluminium, it will be completely removed in the precipitation of these by ammonium hydroxide. If phosphorus occurs in larger amounts, calcium phosphate will be removed with the ammonium precipitate. The remedy is either to add sufficient iron to remove the phosphorus, or to precipitate the latter by ammonium molybdate as described in the chapter on Phosphorus. Neither molybdenum nor phosphorus interferes in the precipitation of calcium oxalate in ammoniacal solution.

Gypsum and other calcium sulphates should be brought into solution by fusing with sodium carbonate, extracting the cooled melt with water, filtering, and washing with 1% sodium carbonate solution. Sodium sulphate and sodium silicate go into solution and pass into the filtrate, whereas calcium is precipitated as carbonate and after filtration can be dissolved in hydrochloric acid [6].

If, as frequently happens, iron and aluminium are virtually the only members of the ammonium sulphide group present, it is preferable to precipitate these with ammonium hydroxide in the presence of ammonium chloride rather than with ammonium sulphide. In this case, after removing hydrogen sulphide from the acid solution by boiling, add a few ml of hydrogen peroxide, boil, make slightly ammoniacal to precipitate all iron and aluminium, but avoid a large excess. Nickel and cobalt will then pass into the filtrate with calcium; small amounts can be tolerated, but large quantities tend to co-precipitate as oxalates with calcium and should be removed by electrolysis in ammoniacal solution after filtering off iron and aluminium. For instance, we have found that the presence of 0·04 g cobalt made no difference in the determination of 0·08 g CaO, but 0·8 g cobalt gave wholly erroneous results.

Heat the filtrate from the separation of either the ammonium sulphide or ammonium hydroxide group to boiling. To the hot, faintly ammoniacal solution add sufficient saturated ammonium oxalate solution to precipitate all the calcium, and leave an excess of about 1 g for each 100 ml of solution. Saturated ammonium oxalate contains about 0·5 g in 10 ml at room temperature, which is sufficient to precipitate approximately 0·2 g CaO; the latter quantity in a volume of 200 ml would require about 50 ml of saturated ammonium oxalate solution. Boil for a few minutes and allow to stand several hours, preferably overnight. Filter through a Whatman No. 40 paper, with pulp, wash thoroughly with 0·1% ammonium oxalate solution and finally once or twice with water.

Transfer the precipitate to a tared platinum crucible, dry, ignite carefully and finally for at least 5 minutes at 1200°C. Cover, cool, and weigh as CaO. In ore analysis, calcium is expressed as CaO, but if required, CaO $\times$ 0·7147 $=$ Ca.

Because magnesium oxalate is readily occluded by calcium oxalate, for accurate work it is essential to dissolve the precipitate of calcium oxalate in dilute hydrochloric acid, and re-precipitate in the same manner with ammonium hydroxide and ammonium oxalate. Combine this filtrate with the previous one and retain for magnesium.

2. Volumetric Procedure

A. Titration of Calcium Oxalate with Potassium Permanganate

Follow the same steps of decomposition, separations, precipitation, and filtration of calcium oxalate as recommended under **1** above. Remove all traces of ammonium oxalate from the paper and the precipitate by thorough washing.

Transfer the bulk of the precipitate, by a stream of water from a wash bottle, from the funnel to the original beaker and then place the latter under the funnel. Pour 50 ml of hot 20% sulphuric acid through the paper in several washes, and finally wash thoroughly with hot water. Heat the filtrate to 70–80°C, and titrate immediately with standard potassium permanganate to the first permanent pink. Now add the filter paper containing the last traces of the calcium oxalate precipitate to the titrated solution, together with a final rinsing of the funnel, and complete the titration if necessary to the first permanent pink.

The reactions for this determination are:

$$(NH_4)_2C_2O_4 + CaCl_2 \rightarrow CaC_2O_4 + 2NH_4Cl$$

$$5CaC_2O_4 + 2KMnO_4 + 8H_2SO_4 \rightarrow 5CaSO_4 + K_2SO_4 + 2MnSO_4 + \\ + 10CO_2 + 8H_2O$$

$$1\text{ ml } 0{\cdot}1\text{ N } KMnO_4 = 0{\cdot}002804\text{ g CaO} \\ = 0{\cdot}002004\text{ g Ca}$$

$$0{\cdot}1\text{ N } KMnO_4 = \frac{KMnO_4}{5} = 3{\cdot}16\text{ g/litre}$$

The potassium permanganate can be standardized against pure sodium oxalate dissolved in 5% sulphuric acid and heated to 70°C; 1 litre of 0·1N $KMnO_4$ = 6·700 g sodium oxalate. In technical work, a solution of potassium permanganate containing 2·85 g/litre is sometimes used; it can be adjusted to give a titre of 1 ml = 0·0025 g CaO. Permanganate solutions can also be standardized against calcium carbonate dissolved in acid and carried through all the steps of the procedure; 0·1000 g pure $CaCO_3$ = 0·05603 g CaO, giving a convenient titration of about 20–23 ml potassium permanganate.

The procedures detailed above may be appreciably shortened if the residue after acid solution is known to contain no calcium, and if few members of preceding groups are present in the sample. For many materials, solution in hydrochloric acid and removal of iron and aluminium with ammonium hydroxide will give a filtrate in which calcium can be precipitated immediately as the oxalate.

Separation from Barium and Strontium. In the rare cases when significant quantities of barium and strontium are encountered, the procedure given above will require slight modification. After isolation of the alkaline earths by a group precipitation as oxalates and ignition to oxides, nitric acid is added to convert calcium, barium, and strontium to nitrates; all moisture is expelled by heating. Calcium nitrate is then extracted by a mixture of equal parts anhydrous ether and absolute alcohol, in which the nitrates of barium and strontium are nearly insoluble [7, 9]. After filtration, the ether–alcohol mixture of the filtrate is evaporated at a low heat, and the organic matter destroyed by evaporating to dryness, with the addition of more nitric acid and a second evaporation if necessary. Calcium is then dissolved in dilute hydrochloric acid, the solution neutralized with ammonia, and oxalate added as described earlier.

Precipitation of calcium oxalate in acid solution. Calcium can be precipitated as oxalate in a weakly acid solution of *p*H 3·5–4·5, in the presence of iron, aluminium, titanium, zirconium, magnesium, and phosphorus.

Small amounts of barium do not appear in the precipitate, but strontium divides. The procedure described below is particularly valuable for phosphate rock, fertilizers, and similar materials [1, 7, 10, 12, 17].

To the solution containing about 10 ml of hydrochloric acid in 150 ml, add a few drops of methyl red. Heat to about 50°C, neutralize with ammonium hydroxide and add 1 ml in excess. Acidify the solution with a 10% solution of oxalic acid, and add 12 ml in excess. Stir and boil for 1–2 minutes, add approximately 50 ml of a saturated solution of ammonium oxalate, dilute to 250 ml and boil for 1–2 minutes. Allow to stand several hours or overnight, filter, and wash with a 0·2% ammonium oxalic–0·1% oxalic acid solution. Finally wash with 5–10 small portions of cold water, and determine calcium by the volumetric permanganate method described earlier.

Another approach is to complex the R_2O_3 elements with sodium gluconate before precipitating calcium with oxalate in ammoniacal solution [19].

B. Titration of Calcium with EDTA

Calcium may be determined in many products by titration with EDTA, using calcon as indicator. Many heavy metals can be masked by adding potassium cyanide. A large number of papers on EDTA titrations for calcium are a reflection of inherent difficulties with this procedure, not only with the colour to be taken as the end point but also to rather critical conditions of *p*H control, concentration of reagents, and amounts of impurities [5, 10, 15, 20, 21]. For certain materials, however, this titrimetric procedure is capable of providing rapid and excellent results; the method outlined below is typical.

EDTA determination of calcium and magnesium in carbonate rocks. Dissolve 0·5 g in hydrochloric acid and evaporate to dryness. If more than about 0·5% manganese is present, remove it initially in nitric acid and potassium chlorate, filter, evaporate to dryness, add 10 ml hydrochloric acid and again evaporate to dryness. If appreciable quantities of iron and aluminium are present, remove these with ammonium hydroxide, wash with 1% ammonium chloride, evaporate the filtrate to dryness, remove all traces of ammonium chloride by several evaporations with nitric acid, and finally evaporate to dryness with hydrochloric acid.

(a) *Titration of* Ca + Mg. To the evaporated chloride residue obtained above, free of ammonium salts, add 5 drops of hydrochloric acid, 25 ml of water, and heat gently to dissolve soluble salts. Cool, filter into a 250 ml volumetric flask, wash with water, dilute to volume, and mix. Measure 20 ml of this solution from a burette into a 250 ml Erlenmeyer flask and add 80 ml of water. Add a little solid potassium cyanide, shake to dissolve, and add 5 ml of the NH_4OH–NH_4Cl buffer. Add a small quantity of Eriochrome black T indicator, a few grains on the tip of a spatula usually sufficing. Titrate to a blue colour with 0·01 M di-sodium EDTA solution. This gives Ca + Mg.

(b) *Titration of* Ca *alone.* Measure out another 20 ml portion of the sample solution, add 80 ml of water and 5 ml of the NaOH–NaCN solution. Add a little calcon indicator, and titrate with the standard EDTA solution to a blue colour. Deduct the volume of di-sodium EDTA used in this titration, for calcium alone, from that of the previous titration for Ca + Mg to give the volume required for magnesium. In these titrations, vigorous shaking is essential.

Reagents. 0·01 M EDTA. Dissolve 3·7235 g of di-sodium EDTA in water and dilute to 1 litre. 1 ml = 0·4008 mg Ca, 0·5608 mg CaO, or 1·009 mg $CaCO_3$; 1 ml = 0·2432 mg Mg, 0·4032 mg MgO, or 0·8433 mg $MgCO_3$.

NH_4OH–NH_4Cl buffer. Dissolve 13·5 g of ammonium chloride in a little water, add 114 ml ammonium hydroxide, and dilute to 200 ml with water; store in a polyethylene bottle.

NaOH–NaCN solution. Dissolve 4 g sodium hydroxide in 1 g sodium cyanide in 50 ml of water. Cool to 20°C, dilute to 100 ml and store in a polyethylene bottle.

Eriochrome black T. Grind 0·25 g of the indicator with 50 g of sodium chloride.

Calcon. Grind 0·2 g of the indicator Eriochrome blue black R, or calcon, with 50 g of sodium chloride.

If desired, a few drops of 0·2% aqueous solution of the indicator may be used in place of the milligram additions of solid indicator. The aqueous solution will remain stable for many weeks if stored in a refrigerator.

3. Colorimetric Methods

Calcium is a major constituent in most of the materials where its determination is a routine procedure, such as ores, slags, cement, and waters. The permanganate or EDTA titrimetric methods can be applied to nearly all samples, even for a low calcium content. Consequently, the development of colorimetric procedures for calcium has been slow.

One of the best general reagents for traces of calcium is di-(*o*-hydroxyphenylimino) ethane, or glyoxal-bis-(2-hydroxyanil), which reacts with calcium in alkaline solution to produce a red complex which has its wavelength of maximum light absorption at 520 nm [8]. Of nearly all common metals, only Co, Ni, Cu, and Cd react with di-(*o*-hydroxyphenylimino) ethane. The complex of calcium can be extracted into chloroform to give a red colour, whereas cobalt, nickel, and copper are not extracted. Cadmium yields a blue solution in chloroform, but if alkaline potassium cyanide is added before extraction, cadmium will not interfere.

Transfer a 10 ml aliquot of sample solution containing not more than 50μg of calcium to a stoppered vessel. Add 1 ml of a buffer solution containing 1% sodium hydroxide and 1% sodium tetraborate, and 0·5 ml of a 0·5% di-(*o*-hydroxyphenylimino) ethane solution in methanol, and mix. Add 10 ml of a solvent mixture of equal volumes of ethanol and n-butanol, and mix. Allow to stand for 30 minutes and measure the optical density at 520 nm against a blank. Prepare a calibration curve in the range 0–50 μg of calcium.

Another spectrophotometric determination of 0·02–0·1% calcium in battery lead, using 8-quinolinol, has been described [13]. Other colorimetric procedures have been outlined [16].

4. Optical Spectrography

Small amounts of calcium can be determined in many materials by emission spectrography [2, 11]. A number of lines have been used for this purpose: 2398·6, 3009·2, 3158·87, 3179·33, 3180·5, 3350·4, 3624·1, 3933·67, 3968·47, 4226·73, 4249·0, 4425·44, 4434·96, 4436·0, and 4454·78.

Methods have been described for 0·0005–0·2% calcium in aluminium and its alloys [2], 0·7 to 80 ppm in nickel [11], and 20 to 200 ppm in zirconium and its alloys [2].

5. Atomic-absorption Spectrophotometry

Calcium can be determined by atomic absorption, using the most sensitive line 4226·7, which has a sensitivity of 0·08 μg/ml; the line 2398·6 can also be employed [4]. The absorption of calcium is suppressed by SO_4, PO_4, SiO_3, Al, and Zr, but the addition of strontium or lanthanum chloride minimizes this effect. Potassium and sodium enhance absorption, and these can be tolerated if the same quantities are added to the standard.

6. Flame Photometry

Flame photometry has been occasionally used to determine calcium [3, 14, 18].

SPECIAL PROCEDURES

1. Available Lime

In extractive metallurgy and many other fields of industry, large quantities of quicklime and hydrated lime are used in the form of dilute aqueous solutions for *p*H control or as a source of calcium ions. The term "available lime" is used to denote the actual content of CaO in these products; it gives a measure of the quality of the lime, because total lime would include unburned limestone which is insoluble in water and therefore of no value in milk-of-lime. There are several methods used to obtain the "available lime" in quicklime or hydrated lime.

A. Place 1 g of material in a 500 ml flask, add 30 g sugar and 300 ml water; shake vigorously until the sugar has dissolved. Make up to 500 ml and shake at 10-minute intervals for 1–2 hours. Allow the insoluble material to settle, filter part of the solution, and withdraw 50 ml of this filtrate for titration. Add 2 drops phenolphthalein and titrate with 0·1N oxalic acid to the disappearance of the pink colour. This determination is based on the fact that calcium in the form of CaO forms a water-soluble saccharate, $C_{12}H_{22}O_{11}.CaO.2H_2O$, whereas compounds such as $CaCO_3$, CaF_2, etc. do not form calcium saccharate. 1 ml of 0·1N oxalic acid = 0·0028 g CaO; 0·1N oxalic acid is made by dissolving 6·303 g $H_2C_2O_4.2H_2O$ in water and diluting to 1 litre.

B. Weigh 1·402 g sample into a 400 ml beaker, add 200 ml hot water, and boil for 3 minutes. Cool, add 2 drops phenolphthalein and titrate with 1·0N hydrochloric acid. When the pink colour disappears in streaks, reduce the rate of acid addition until the colour fades throughout the solution for 1–2 seconds. Note the reading and ignore the return of colour. Repeat the experiment, substituting for the 400 ml beaker a 1-litre graduated flask carrying a one-hole stopper fitted with a short glass tube drawn out to a point. Cool, and add dropwise with vigorous stirring about 4–5 ml less acid than before. Call this number of ml "A". Grind up any small lumps with a glass rod flattened at one end, dilute to the mark with freshly-boiled distilled water, close the flask with a solid stopper, mix well for 4–5 minutes and allow to settle for 30 minutes.

Pipette off a 200 ml portion, add phenolphthalein and titrate slowly with 0·5N hydrochloric acid until the solution remains colourless on standing 1 minute. Call this additional number of ml "B". Then the percentage of available CaO = 2A + 5B.

C. "Available lime" may also be obtained, in the absence of sulphates, silicates, or fluorides of calcium, by determining total calcium and the carbon dioxide evolved by treatment of the sample with hydrochloric acid. Assuming that carbonates other than $CaCO_3$ are absent, the difference between total calcium and that present as carbonate is a measure of the available lime.

2. Calcium Sulphate

The determination of calcium sulphate in sulphide ores by the solubility of the former in ammonium chloride has been described [22].

REFERENCES

1. A.O.A.C., *Official Methods of Analysis of the Association of Official Analytical Chemists*, Washington, D.C., Association of Official Analytical Chemists, 1965.
2. A.S.T.M., *Chemical Analysis of Metals. Sampling and Analysis of Metal Bearing Ores*, Philadelphia, Pa., American Society for Testing and Materials, 1969.
3. DEAN, J. A., *Flame Photometry*, New York, McGraw-Hill, 1960.
4. ELWELL, W. T., and GIDLEY, J. A. F., *Atomic-Absorption Spectrophotometry*, 2nd ed., Oxford, Pergamon, 1966.

5. Flaschka, H. A., *EDTA Titrations*, 2nd ed., Oxford, Pergamon, 1964.
6. Furman, N. H., ed., *Scott's Standard Methods of Chemical Analysis*, 6th ed., Vol. 1, Princeton, N. J., Van Nostrand, 1962.
7. Hillebrand, W. F., Lundell, G. E. F., Bright, H. A., and Hoffman, J. I., *Applied Inorganic Analysis*, 2nd ed., New York, Wiley, 1953.
8. Hopkin and Williams Ltd., *Organic Reagents for Metals*, Vol. 2, Chadwell Heath, England, Hopkin and Williams, 1964.
9. Kallmann, S., *Anal. Chem.* **20**, 449–51 (1948).
10. Kolthoff, I. M., and Elving, P. J., *Treatise on Analytical Chemistry*, Part 11, Vol. 4, 107–52, New York, Interscience Publishers, 1966.
11. Lewis, C. L., Ott, W. L., and Sine, N. M., *The Analysis of Nickel*, Oxford, Pergamon, 1966.
12. Low, A. H., Weinig, A. J., and Schoder, W. P., *Technical Methods of Ore Analysis*, New York, Wiley, 1939.
13. Luke, C. L., *Anal. Chim. Acta* **32**, 221–6 (1965).
14. Mavrodineanu, R., and Boiteux, H., *Flame Spectroscopy*, New York, Wiley, 1965.
15. Morris, A. G. C., *Anal. Chem.* **33**, 599–602 (1961).
16. Sandell, E. B., *Colorimetric Determination of Traces of Metals*, 3rd ed., New York, Interscience Publishers, 1959.
17. Snell, F. D., and Ettre, L. S., *Encyclopedia of Industrial Chemical Analysis*, Vol. 8, New York, Interscience Publishers, 1969.
18. Standen, G. W., and Tennant, C. B., *Anal. Chem.* **28**, 858–60 (1958).
19. Watts, H. L., *Anal. Chem.* **32**, 1189–90 (1960).
20. Welcher, F. J., *The Analytical Uses of Ethylenediaminetetraacetic Acid*, New York, Van Nostrand, 1958.
21. West, T. S., and Sykes, A. S., *Analytical Applications of Diaminoethane-tetra-acetic Acid*, 2nd ed., Poole, England, British Drug Houses, 1960.
22. Young, R. S., and Hall, A. J., *J. Soc. Chem. Industry* **66**, 375 (1947).

10 Carbon

Carbon may be encountered in a mining and metallurgical laboratory in many forms: as carbonates in minerals and ores, as oxides in gases, as carbon in iron, steel, and other metals and alloys, in coals, natural gas, petroleum, carbides, graphite, amorphous carbon, and industrial diamond.

1. Carbon in Carbonates

In these materials carbon is determined as CO_2 by treating with acid and determining the loss of weight of the sample, or the quantity of carbon dioxide evolved by passing it through a solid or liquid absorbent and measuring the increase in weight, or the excess liquid absorbent, respectively.

A. For rapid technical determinations required only occasionally, a Schroedter alkalimeter may be used to determine, by loss of weight, the carbonate in a sample.

Place 1 g of the sample in the bottom of the flask, and sulphuric and hydrochloric acids in the left and right bulbs respectively. Weigh the apparatus, open the stopper and allow hydrochloric acid to flow on the carbonate, then close the stopper. The liberated gas flows through sulphuric acid which absorbs moisture. After the vigorous action has ceased, place the apparatus on a hot plate and boil for three minutes. Aspirate CO_2-free air, purified by passing through soda lime, through the solution to expel the last traces of CO_2, by applying gentle suction at the outlet and opening the stopper. Weigh the apparatus again and take the loss of weight as the CO_2 of the sample. If sulphides are present, a little acidified copper sulphate or similar solution should be used in place of sulphuric acid.

B. A more accurate method for CO_2 in carbonates consists of decomposing the sample with hydrochloric acid and absorbing the evolved CO_2 in Ascarite in a standard absorption train similar to that used for carbon in iron and steel [2, 3, 8, 11]. The increase in weight of the absorption bulb gives the CO_2 in the sample.

Weigh 1–5 g into a decomposition flask, add hydrochloric acid and when initial effervescence has ceased, heat gently. A scrubbing unit of calcium chloride and Ascarite is placed ahead of the decomposition flask to remove CO_2 from the stream of air which is slowly aspirated through the train. Scrubbers containing $H_2SO_4 + Cr_2O_3$ or MnO_2, $CaCl_2$ and P_2O_5 to remove sulphur compounds, water vapour, and hydrochloric acid are inserted between the decomposition flask and the absorption bulb. A final $CaCl_2$ tube after the absorption bulb prevents water vapour reaching the latter from the aspirator.

C. For carbonate in materials which are decomposed by hot 10% hydrochloric acid, the CO_2 may be absorbed in an excess of standard barium hydroxide, giving a precipitate of $BaCO_3$ [12]. On titration with standard hydrochloric acid, using phenolphthalein indicator, none of the $BaCO_3$ is decomposed until all the excess barium hydroxide has been neutralized, indicated by the change from pink to colourless. Barium hydroxide is used as 0·2N solution, 31·6 g $Ba(OH)_2.8H_2O$ per litre. One ml of 0·2N HCl = 1 ml 0·2N $Ba(OH)_2$ = 0·0044 g CO_2.

D. Calcium carbonate may be determined in soil samples by a calcimeter; a convenient apparatus for routine use has been described [4]. It measures the volume of CO_2 evolved when dilute hydrochloric acid is allowed to react with the sample under standard conditions.

2. Carbon in Irons, Steels, and other Metals and Alloys

When a metal or alloy is heated to a sufficiently high temperature in a stream of oxygen the carbon in the sample is evolved as CO_2.

A. Volumetric determination of CO_2

When large numbers of carbon determinations must be made, as in the iron and steel industry, the measurement of the volume of CO_2 evolved when the sample is burned in oxygen, is common practice. Sulphur compounds are removed by MnO_2, the CO_2 and excess oxygen are collected and measured in a burette, then passed through an absorption vessel containing potassium hydroxide; the difference in volume before and after absorption in potassium hydroxide is the quantity of CO_2 evolved and consequently proportional to the carbon originally present in the metal. Determinations are very rapidly performed with these gasometric carbon analysers, and they are suitable for a wide range of products. Standard equipment is available from most laboratory supply houses.

B. Gravimetric determination of CO_2

The absorption of CO_2 in a bulb containing Ascarite or a similar soda-asbestos mixture and measurement of the increase in weight is a reliable and well-standardized procedure [3, 5, 8, 18, 19].

The apparatus usually consists of the following parts:

1. Mercury valve.
2. Bottle containing concentrated H_2SO_4.
3. Tower containing $CaCl_2$.
4. Tower containing stick NaOH.
5. Electric furnace with combustion tube.

6. Bottle containing 5% $KMnO_4$.
7. Tower containing MnO_2.
8. Bottle containing concentrated H_2SO_4.
9. Tower containing P_2O_5, or preferably Anhydrone.
10. Weighed absorption bulb filled with Ascarite and Anhydrone.
11. H_2SO_4 guard tube and gauge.

The assembly can be simplified, of course, where the oxygen is very pure and the steel or other metal sample does not contain significant quantities of sulphur or selenium. The absorption bulb may be any one of many standard types: Fleming–Martin, Miller, Stetson–Norton, Nesbitt, etc. Ascarite and similar soda–asbestos mixtures are now supplied with self-indicating granules, showing the degree of exhaustion of the absorbent.

In all combustion trains where gases must be dried it is advisable to remove the bulk of the moisture with stick NaOH, concentrated H_2SO_4 or $CaCl_2$; the expensive Anhydrone or similar magnesium perchlorate drying agents should be used to remove the last 2–3% of moisture.

Spread 1 5 g of the drilled or milled sample, sized between 14 and 60 mesh, in a groove of high quality alumina, such as RR Alundum, in a boat; insert carefully into the combustion tube in the furnace. The temperature should be not less than 1100°C, except for high alloy steels where 1200–1250° is necessary even in the presence of an accelerator.

As soon as the boat has been pushed into position in the hottest part of the tube, close the latter and turn on the oxygen slowly until the flow is about 300 ml per minute. When the sample has burned, usually in 1–2 minutes, reduce the flow of oxygen to 150 ml per minute for 5 minutes to sweep out all CO_2.

Close the oxygen valve, disconnect the CO_2 absorption bulb, place it in the balance case for 5 minutes, open momentarily, close and weigh. If desired, a similar bulb can be used as a counterpoise in weighing. Remove the boat from the tube and examine; if the drillings are not thoroughly fused in a solid pig, it indicates incomplete combustion and the determination must be repeated, using a higher temperature or an accelerator or both. A blank run must be made on carbon in reagents, boat, tube, tubing, oxygen, etc.; the blank should be less than 0·0005 g CO_2.

For high-alloy steels, and metals of high melting points, it is necessary not only to use a temperature of at least 1200–1250°C but to add an accelerator to the sample in the boat [3, 8, 9, 15, 18, 19, 20]. About 0·5 g of 20–40 mesh tin is commonly employed for this purpose.

For some metals and alloys, other modifications of the standard procedure are necessary. Nickel and cobalt are heated to 1200–1300°C, with an accelerator such as tin, for 30 minutes. Alloys of nickel, chromium, and iron require a temperature of 1300–1350°C. Ferrosilicon is mixed with 2 g of powdered cupric oxide, as well as with ingot iron to serve as an accelerator, before combustion at 1200–1350°C.

$$CO_2 \times 0{\cdot}2729 = C$$

C. Conductometric Method

Conductometric equipment is available commercially for determining low carbon contents in metals and alloys; measurement is made of the change in conductivity of barium or sodium hydroxide solution through which the evolved CO_2 is passed [1, 3, 5, 9, 15].

The conductometric method is highly precise with small samples, and is widely used for carbon contents in the range of 0·01 to 0·05%. Higher carbons would normally be determined gravimetrically, and lower ones by the low-pressure procedure. The conductometric method is not as rapid as the gravimetric procedure, but a single determination can be completed in about one hour.

Combustion of the sample is carried out in the manner described previously for the gravimetric method, and the evolved carbon dioxide is passed through a conductivity cell in a thermostatically controlled water bath. The equipment is standardized against a metal of known carbon content.

The conductometric method has been described in detail for carbon in copper [5], nickel [1, 3, 15], zirconium and its alloys [3], and many other metals and alloys [3, 6, 9].

D. Low-pressure Combustion Method

The low carbon content in some metals, for instance 0·0001–0·1% in electronic nickel, and below 0·01% in copper, may be determined by a low-pressure combustion method [3, 5, 20]. Combustion in oxygen is carried out at low pressure, the CO_2 is frozen and finally measured in the gas phase and in a known volume by means of a McLeod gauge.

E. Thermal Conductivity Procedure

Carbon in steel can be determined by burning the sample in oxygen as usual, and measuring the evolved carbon dioxide by the thermal conductivity of the gas mixture. Carbon dioxide is concentrated from the oxygen stream in a molecular sieve trap before measurement [16].

F. Infrared Method

Another procedure for carbon in steel utilizes the conventional resistance-combustion furnace with a constant flow of oxygen through the apparatus, and measurement of the carbon dioxide by an infrared analyser [22].

G. Graphite

Graphite in cast iron is determined by dissolving 1–3 g sample in 50 ml nitric acid on the steam bath [3]. Add a few drops of hydrofluoric acid to facilitate the filtration of any separated silicic acid. Filter through ignited asbestos on a small Gooch crucible which will fit in the combustion tube of the CO_2 apparatus. Wash thoroughly with hot water, hot 12% potassium hydroxide, hot water, 5% hydrochloric acid, and finally with hot water. Dry at a temperature not exceeding 150°C and determine the graphite by direct combustion at approximately 900°C in the apparatus used for the gravimetric determination of total carbon. The combustion tube must be closed immediately after inserting the sample.

H. Carbon by Wet Combustion

Carbon in organic materials may be determined by wet combustion. Oxidation is effected by chromic and sulphuric acids; the evolved CO_2 may be measured gravimetrically, or by absorption in standard barium hydroxide solution and back titration of the excess hydroxide with standard hydrochloric acid [2, 8, 14].

SPECIAL PROCEDURES

1. Methods for the evaluation of activated and industrial carbon have been included in reference books [10, 17].
2. The analysis for total [7, 21] and free [6, 21] carbon in cemented carbides has been described.
3. Determination of the diamond content of grinding wheels, impregnated tools, and other industrial products where a wide variety of matrices is found, has been discussed [23].
4. Another approach to the determination of carbon in iron and steel, following combustion in a stream of oxygen, has been proposed [13]. The carbon dioxide is absorbed in a solution of formdimethylamide and titrated with tetra-n-butylammonium hydroxide.

REFERENCES

1. Andrew, T. R., and Gentry, C. H. R., *Metallurgia* **60,** 29 (1959).
2. A.O.A.C., *Official Methods of Analysis of the Association of Official Analytical Chemists,* Washington, D.C., Association of Official Analytical Chemists, 1965.
3. A.S.T.M., *Chemical Analysis of Metals. Sampling and Analysis of Metal Bearing Ores,* Philadelphia, Pa., American Society for Testing and Materials, 1969.
4. Bascomb, C. L., *Chemistry and Industry,* 1826–7, 1961.
5. Elwell, W. T., and Scholes, I. R., *Analysis of Copper and its Alloys,* Oxford, Pergamon, 1967.
6. Feick, G., and Giustetti, W., *Anal. Chem.* **36,** 2198–9 (1964).
7. Furey, J. J., and Cunningham, T. R., *Anal. Chem.* **20,** 563–70 (1948).
8. Furman, N. H., ed., *Scott's Standard Methods of Chemical Analysis,* 6th ed., Vol. 1, Princeton, N.J., Van Nostrand, 1962.

9. Green, I. R., Still, J. E., and Chirnside, R. C., *Analyst 87*, 530–8 (1962)
10. Hassler, J. W., *Activated Carbon*, New York, Chemical Publ. Co., 1963.
11. Hillebrand, W. F., Lundell, G. E. F., Bright, H. A., and Hoffman, J. I., *Applied Inorganic Analysis*, 2nd ed., New York, Wiley, 1953.
12. Ingelson, H., and Bentley, J. A., *Analyst* **71**, 328–9 (1946).
13. Jones, R. F., Gale, P., Hopkins, P., and Powell, L. N., *Analyst* **90**, 623–9 (1965).
14. Kolthoff, I. M., and Elving, P. J., *Treatise on Analytical Chemistry*, Part II, Vol. 11, 297–403, New York, Interscience Publishers, 1965.
15. Lewis, C. L., Ott, W. L., and Sine, N. M., *The Analysis of Nickel*, Oxford, Pergamon, 1966.
16. Lewis, L. L., and Nardozzi, M. J., *Anal. Chem.* **38**, 1214–17 (1966).
17. Mantell, C. L., *Industrial Carbon*, New York, Van Nostrand, 1946.
18. Nunemaker, R. B., and Shrader, S. A., *Anal. Chem.* **28**, 1040–2 (1956).
19. Snell, F. D., and Ettre, L. S., *Encyclopedia of Industrial Chemical Analysis*, Vol. 8, New York, Interscience Publishers, 1969.
20. Torrisi, A. F., Kernahan, J. L., and Fryxell, R. E., *Anal. Chem.* **26**, 733–4 (1954).
21. Touhey, W. O., and Redmond, J. C., *Anal. Chem.* **20**, 202–6 (1948).
22. White, G., and Scholes, P. H., *Analyst* **91**, 482–9 (1966).
23. Young, R. S., Simpson, H. R., and Benfield, D. A., *Anal. Chim. Acta* **6**, 510–16 (1952).

11 Chlorine

Chlorine occurs in minerals, ores, leach solutions, waters, brines, salts, and many industrial products. Most substances containing chlorine can be decomposed by treatment with nitric acid; sulphuric acid should be avoided for fear of driving off hydrochloric acid. If chlorides of silver, lead, or mercurous mercury are present they may be digested or fused with sodium carbonate or sodium hydroxide, diluted with water, filtered, and washed.

Iodine may be expelled by adding 3 g sodium nitrite and 3 ml of 1 : 1 sulphuric acid for each 0·25 g of iodine, in a volume of about 700 ml in a large flask, and boiling for 45 minutes. Chlorine and bromine remain in the solution. Neutralize the latter with sodium hydroxide, evaporate to about 50 ml, cool, and add about 65 ml excess 1 : 2 acetic acid. Add 1·5 g potassium permanganate and pass a slow current of steam into the flask until all bromine has been expelled.

Chlorine may be determined, in the absence of the interfering elements mentioned above:

1. Gravimetrically as silver chloride in dilute nitric acid.
2. Volumetrically by:

(a) Volhard's method, wherein the chloride is precipitated in nitric acid by a slight excess of standard silver nitrate solution, and the latter determined by titrating with a thiocyanate solution.

(b) Mohr's method, in which the chloride is titrated with silver nitrate in neutral solution, using potassium chromate as indicator. The modification of Fajans, using dichlorofluorescein as indicator, may be substituted.

3. Colorimetrically with *o*-tolidine.

1. Gravimetric Method

Weigh 0·5–5 g, depending on the chlorine content, to give approximately 0·1 g Cl [4, 5]. Decompose with nitric acid and fuse any insoluble portion with sodium carbonate. If the interfering elements mentioned in the first two paragraphs are present, remove them as suggested.

Dilute to approximately 200 ml, and adjust the concentration of nitric acid to about 2 ml; add with stirring a 5% silver nitrate solution until the precipitate coagulates and no more is formed. If silicic acid separates out, remove it by solution of silver chloride in ammonia, filter, acidify the filtrate with nitric acid and precipitate silver chloride as indicated. One ml of 5% silver nitrate solution will precipitate approximately 0·01 g of chlorine. Heat the solution to boiling, remove from the hot plate, allow to settle, and add a few more drops of silver nitrate solution to be sure that all chloride has precipitated. Avoid a large excess of silver nitrate.

Set the beaker aside in the dark for several hours or preferably overnight. Filter through a weighed Gooch crucible which has been previously washed with nitric acid and dried in the oven. Wash by decantation with a solution containing 0·05 g silver nitrate per litre; transfer to the Gooch crucible, wash with 1% nitric acid to remove all silver nitrate, and finally several times with hot water. Dry at 130–150°C for an hour, cool in a desiccator, and weigh.

$$AgNO_3 + HCl \rightarrow AgCl + HNO_3$$

$$AgCl \times 0{\cdot}2474 = Cl$$

If tin or antimony are present, they may hydrolyse in the dilute acid solution. Precipitate them, together with silver, by initial treatment with hydrogen sulphide in 10% sulphuric acid, and separate the sulphides of antimony and tin by solution in alkali sulphides or polysulphides. Bismuth also tends to hydrolyse and contaminate silver chloride; remove it by repeated solution in ammonium hydroxide and precipitation with acid.

2. Volumetric Method

In the volumetric methods, if free chlorine and free hydrochloric acid are present they should be neutralized with ammonium hydroxide, and boiled to convert them to chlorides.

Cyanides and chlorides are usually not separated from each other because the combined chloride and cyanide can be determined by Volhard's method; the cyanide may then be titrated with standard silver nitrate, as described later in the section on Cyanides.

A. Precipitation with Silver Nitrate and Titration with Thiocyanate

In this method, often called Volhard's method, an alkali thiocyanate combines with the excess of silver nitrate which has precipitated the chlorine in nitric acid solution; the endpoint of the reaction is indicated by the appearance of reddish-brown ferric thiocyanate when an excess of thiocyanate reacts with an added ferric salt [3, 4, 5].

Decompose the sample in the manner outlined for the gravimetric method; adjust acidity to about 5 ml excess nitric acid for 150 ml. Add a measured excess of 0·1N silver nitrate, shake the mixture for a few minutes, or boil, to coagulate the silver chloride before filtering. All the adsorbed silver returns to solution during this ageing of the precipitate. Filter off the precipitated silver chloride, wash thoroughly with 1% nitric acid, and finally several times with hot water. Titrate the excess silver nitrate in the combined filtrate and washings with 0·1N NH_4CNS or KCNS, using 5 ml of a saturated solution of ferric ammonium sulphate as indicator. A permanent reddish-brown colour, best seen against a white base, marks the endpoint of the reaction. From this titration the quantity of silver nitrate which reacted with the chlorine can be obtained.

$$NH_4CNS + AgNO_3 \rightarrow AgCNS + NH_4NO_3$$

$$3NH_4CNS + NH_4.Fe(SO_4)_2.12H_2O \rightarrow$$
$$Fe(CNS)_3 + 2(NH_4)_2SO_4 + 12H_2O$$

$$1 \text{ ml } 0{\cdot}1 \text{ N } AgNO_3 = 0{\cdot}003545 \text{ g Cl}$$

In routine work it may be convenient to have 1 ml thiocyanate = 1 ml silver nitrate. Approximately 8 g NH_4CNS or 10 g KCNS, and 17·0 g $AgNO_3$ per litre, give 0·1N solutions. The silver nitrate may be standardized against 0·1N NaCl, which contains 5·846 g NaCl per litre.

B. Titration with Silver Nitrate in Neutral Solution

(1) *Using potassium chromate as indicator.* In this method, often termed Mohr's method, silver combines with chlorine in the presence of a chromate preferentially; so long as any soluble chloride is present the silver chromate which may be formed initially is decomposed. When all the chlorine has combined with silver, an excess of potassium chromate at once forms the red silver chromate, which indicates the endpoint of the reaction. Because this procedure must be carried out in a solution which is approximately neutral, its application is chiefly confined to waters, brines, and water-soluble chlorides [2, 3, 4, 5]. The following interfere: anions which form slightly soluble silver salts in neutral solution, such as bromide, iodide, or arsenate; reducing agents which reduce silver nitrate to metallic silver, such as ferrous iron; cations which form slightly soluble chromates such as barium and lead, or stable complexes with chloride such as Cd and Hg^{++}. Chlorates, perchlorates, and most organic compounds of chlorine do not react with silver nitrate under the conditions of this analysis. If the water contains hydrogen sulphide, add a drop or two of 1 : 1 sulphuric acid and boil thoroughly to expel all hydrogen sulphide before commencing the neutralization and titration. Ammonium salts of strong acids do not affect the sensitivity if the *p*H of the solution is maintained between 6·5 and 7·2.

The sample solution should have a *p*H between 6·5 and 10·5; if it lies outside these limits adjust with dilute sodium hydroxide or dilute nitric acid. To the nearly neutral solution in a beaker or porcelain casserole add 2–3 drops of a saturated solution of potassium chromate, and then the 0·1N silver nitrate solution slowly with vigorous stirring. When a permanent red tint is obtained, record the volume of silver nitrate and add an equivalent solution of sodium chloride carefully until the red colour is just destroyed. Deduct this from the volume of silver nitrate used.

$$AgNO_3 + KCl \rightarrow AgCl + KNO_3$$
$$K_2CrO_4 + 2AgNO_3 \rightarrow Ag_2CrO_4 + 2KNO_3$$
$$Ag_2CrO_4 + 2KCl \rightarrow 2AgCl + K_2CrO_4$$

$$1 \text{ ml } 0{\cdot}1 \text{ N } AgNO_3 = 0{\cdot}003545 \text{ g Cl}$$
$$0{\cdot}1\text{N } AgNO_3 = 17{\cdot}0 \text{ g/litre}; \ 0{\cdot}1\text{N NaCl} = 5{\cdot}846 \text{ g/litre}$$

(2) *Using dichlorofluorescein as indicator.* Dichlorofluorescein is a good indicator for the titration of chloride with silver. As soon as a slight excess of silver ion is present, the precipitate changes colour from white to pink. The addition of a protective colloid such as dextrin gives a sharper endpoint by keeping the precipitate in colloidal suspension, so that the pink colour is seen throughout the entire mixture at the endpoint. Dichlorofluorescein may be used with solutions as acid as *p*H 4.

Take a 10–100 ml sample which does not have a chloride content greater than 0·025N, i.e. about 0·09 g/100 ml. Adjust to approximate neutrality with 1M sodium hydroxide and 0·2M nitric acid, using phenolphthalein as indicator, until the red colour is just completely discharged. Add 5 drops of dichlorofluorescein, which is made by dissolving 0·1% in 60–70% alcohol, or by making a 0·1% aqueous solution with the sodium salt. Add about 0·1 g dextrin, and titrate with 0·1N silver nitrate to a faint permanent pink. 0·1N $AgNO_3$ = 17·0 g/litre; 1 ml = 0·003545 g Cl.

3. Colorimetric Determination

Small quantities of free chlorine in waterworks practice are determined by *o*-tolidine, the latter acquiring a yellow colour in the presence of this element [2]. Comparison is made with coloured discs ranging from 0·1 to 1 ppm chlorine. Owing to the widespread occurrence of traces of chlorides, it is necessary to run careful blanks when determining chlorine.

Mix 1 ml of the *o*-tolidine solution, prepared as described below, with 100 ml of the filtered water in a 100-ml Nessler tube. Allow to stand for 5 minutes or until maximum development of colour is attained. Match against coloured disc standards in a Nesslerizer or similar instrument. The following substances interfere and their presence must be considered in evaluating results: nitrites, ferric and manganic compounds, chlorates, and organic colouring matter.

Preparation of the *o*-tolidine solution may be done as follows. Weigh 1 g *o*-tolidine, transfer to a mortar, and add 5 ml of 1 : 4 hydrochloric acid prepared by adding 100 ml concentrated hydrochloric acid to 400 ml water. Grind to a thin paste and add 150–200 ml of water; the *o*-tolidine goes into solution at once. Transfer to a litre flask and make up to 505 ml with water; make up to the mark by adding the balance, 495 ml, of the 1 : 4 hydrochloric acid.

Other reagents have been proposed for free chlorine in water, such as

3,3′-dimethylnaphthidine, barbituric acid, and NN-diethyl-*p*-phenylenediamine; a critical survey of the suitability of these and other reagents has been published [7].

SPECIAL PROCEDURES

1. The procedure used for the determination of chloride such as $CaCl_2$ or $CoCl_2$, added to activated alumina [1], may be employed for some materials. This consists of adding to a 0·5 g sample in a flask, 75 ml of water, and 60 ml of 2 parts by volume sulphuric acid to 1 part phosphoric acid. Distil the hydrochloric acid into a receiving vessel containing dilute sodium hydroxide solution; evaporate the liquid in the distilling flask to strong fumes or until it is nearly dry. Neutralize the distillate with nitric acid, add a small excess, and determine chloride by precipitation with 0·1N silver nitrate and titration with thiocyanate, using ferric ammonium sulphate as indicator.

2. Small quantities of chlorine in selenium have been determined in a novel, indirect manner. Chlorine is distilled as hydrogen chloride, collected in a solution containing excess silver ions, the precipitated silver chloride filtered off, dissolved in ammoniacal solution, and silver determined by atomic absorption [8].

3. Chloride has been measured with a silver chloride type membrane electrode [9].

REFERENCES

1. ALCOA RESEARCH LABORATORIES, *Analysis of Activated Alumina, Tabular Alumina, Miscellaneous Materials*, New Kensington, Pa., Alcoa Research Laboratories, 1960.
2. AMERICAN PUBLIC HEALTH ASSOCIATION, *Standard Methods for the Examination of Water and Wastewater*, New York, American Public Health Association, 1965.
3. A.O.A.C., *Official Methods of Analysis of the Association of Official Analytical Chemists*, Washington, D.C., Association of Official Analytical Chemists 1965.
4. FURMAN, N. H., ed., *Scott's Standard Methods of Chemical Analysis*, 6th ed., Vol. 1, Princeton, N.J., Van Nostrand, 1962.
5. HILLEBRAND, W. F., LUNDELL, G. E. F., BRIGHT, H. A., and HOFFMAN, J. I., *Applied Inorganic Analysis*, 2nd ed., New York, Wiley, 1953.
6. KOLTHOFF, I. M., and ELVING, P. J., *Treatise on Analytical Chemistry*, Part 11, Vol. 7, 335–424, New York, Interscience Publishers, 1961.
7. NICOLSON, N. J., *Analyst* **90**, 187–98 (1965).
8. REICHEL, W., and ACS, L., *Anal. Chem.* **41**, 1886–8 (1969).
9. VAN LOON, J. C., *Analyst* **93**, 788–91 (1968).

12 Chromium

Chromium is a common element in many mining and metallurgical laboratories; it may be encountered in ores, iron and steels, alloys, electroplating solutions, refractories, timber preservatives, and pigments. In nature, chromium occurs as oxide in its principal mineral chromite, $FeO.Cr_2O_3$, and in those of much lesser importance such as crocoite, $PbCrO_4$, and picotite, a chrome spinel.

Chromium in most materials may be determined by oxidation to chromate, reduction by an excess of standard ferrous sulphate, and titration of this excess with a standard oxidant such as potassium permanganate. Small amounts of chromium are usually determined colorimetrically, using s-diphenylcarbazide.

ISOLATION OR SEPARATION OF CHROMIUM

Though few separations are usually required in an analysis for chromium, it is sometimes necessary or advisable to remove several elements. Some chromium alloys may contain Cu and Mo; these, with other members of the hydrogen sulphide group, are readily separated by hydrogen sulphide in 5–10% acid. A sodium peroxide fusion followed by a water leach will precipitate Fe, Ni, Co, Ti, and Zr, from Cr, Al, Mo, W, U, and V. A mercury cathode will deposit chromium with Fe, Mo, Ni, Co, Cu, Cd, Sn, Bi, Ga, Ge, and Zn; Al, Be, P, Th, Nb, Ta, W, B, Ca, Mg, Ti, Zr, V, and U are left in solution. A large deposit of chromium gives trouble, however, by making the mercury nearly solid. A precipitation with $(NH_4)_2S$ and tartrate gives Fe, Ni, Co, and Zn in the precipitate and chromium in the filtrate along with V and U. Hexavalent chromium can be extracted from pentavalent vanadium by methyl isobutyl ketone in 3N hydrochloric acid [15]; chromium can also be extracted with trioctylphosphine oxide [19]; Cupferron treatment in dilute sulphuric acid will precipitate Fe, Ti, Zr, V, Nb, Ta, W, Sn, Ga, Bi, and Mo from Cr, Al, Be, B, Pb, Zn, Cd, Ge, As, Mn, Ni, Co, and U(VI). Large quantities of iron may frequently be removed conveniently from chromium by an ether separation. Iron, aluminium, and uranium can be precipitated with ammonium hydroxide from chromium if the latter has been fully oxidized to the chromate state with reagents like silver nitrate–ammonium persulphate, nitric acid–sodium chlorate, or perchloric acid. Ion exchange resins have occasionally been used to separate chromium [16]. Dichromate can be removed in liquid secondary amines insoluble in water, and the chromium finally stripped with potassium hydroxide solution [3].

When decomposing samples with perchloric acid it must be borne in mind that the presence of hydrochloric or hydrofluoric acid may lead to partial volatilization of chromium. In fact, when its presence is undesirable, all chromium can be conveniently removed as chromyl chloride from a sample by repeated evaporation with a mixture of perchloric acid and hydrochloric acid.

1. Volumetric Procedure

A. Ores

For ores and similar materials which are not completely soluble in acids, fuse 1 g of the finely pulverized sample with about 8 g sodium peroxide in a zirconium or Armco iron crucible; transfer to a beaker and add about 150 ml warm water. When the melt has dissolved, remove the crucible, washing it thoroughly; boil the solution for five minutes, filter through asbestos, and wash with hot water.

To the filtrate add 60 ml 1 : 1 sulphuric acid and 5 ml nitric acid. Boil, add 1 ml of 1% silver nitrate solution for each 0·01 g of chromium. Heat to boiling and add 15–30 ml of 15% ammonium persulphate solution to oxidize the chromium completely. Boil for 10 minutes; if permanganic acid or oxides of manganese are present add 10 ml of 10% sodium chloride solution and resume boiling for 5 minutes after the manganese compounds are reduced, to ensure that all chlorine has been expelled.

Cool the solution to room temperature, dilute to 300 ml, add 3 ml phosphoric acid, and a measured excess of 0·1N ferrous ammonium sulphate. The colour will change from the orange of chromate to the green of reduced chromium. Before all the chromium solution has been reduced add 2 drops of ferroin. The latter is made by dissolving 0·2973 g *o*-phenanthroline and 0·1390 g $FeSO_4.7H_2O$ in 50 ml water. Ferroin has a red colour which is converted by strong oxidants into a faint blue ferric complex. In chromium solutions, especially with higher concentrations, the colour after adding an excess of ferrous ammonium sulphate in the presence of ferroin is a yellowish-purple green. When a slight excess of potassium permanganate is added, the colour changes from this off-shade green to a clear green or green with a faint blue tint. This colour change is more readily detected than the change to a darkening of the green colour when potassium permanganate is added to a reduced chromium solution in the absence of ferroin.

Titrate with 0·1N potassium permanganate to the clear green endpoint. Determine the permanganate required for the two drops of ferroin to change, in a solution of sulphuric and nitric acids similar to that used in the actual determination, from red to colourless. This blank is usually 0·05 ml; subtract it from the permanganate used in the back titration of excess ferrous ammonium sulphate.

1 ml of 0·1N ferrous ammonium sulphate $= \frac{Cr}{3} = 0{\cdot}001734$ g Cr.

$$2H_2CrO_4 + 6FeSO_4 + 6H_2SO_4 \rightarrow Cr_2(SO_4)_3 + 3Fe_2(SO_4)_3 + 8H_2O$$

$$2KMnO_4 + 10FeSO_4 + 8H_2SO_4 \rightarrow K_2SO_4 + 2MnSO_4 + 5Fe_2(SO_4)_3 + 8H_2O$$

The ferrous ammonium sulphate solution is prepared by dissolving 39·2 g in 1 litre of water containing 20 ml of 1 : 1 sulphuric acid; it is standardized against 0·1N potassium permanganate. The latter solution is made by dissolving 3·2 g $KMnO_4$ in water, and diluting to 1 litre; it is standardized against sodium oxalate, 6·7002 g of which is equivalent to 1 litre of 0·1N $KMnO_4$. Weigh out 0·2000 g pure anhydrous sodium oxalate, dissolve in 100 ml of 5% sulphuric acid, heat to 60°C, and titrate with $KMnO_4$; 0·2 g sodium oxalate = 29·85 ml of 0·1N $KMnO_4$.

Vanadium, if present, will also be titrated when ferroin or similar oxidation-reduction indicators are used. If the excess of ferrous ammonium sulphate is titrated with potassium permanganate alone, without ferroin, vanadium does not interfere; it is reduced by ferrous sulphate but again oxidized with an equivalent amount of permanganate.

The titration may also be carried out potentiometrically, as described below for chromium in metals and alloys [11, 12]. After the reduction of permanganic acid by hydrochloric acid, and cooling, chromium(VI) is titrated directly with standard ferrous ammonium sulphate solution. Vanadium is also titrated when the potentiometric method is used, but very few chrome ores contain more than a small amount of vanadium.

Chromium in rocks, and in most ores, is reported as Cr_2O_3;

$$Cr \times 1{\cdot}4614 = Cr_2O_3.$$

B. Metals and Alloys

For irons, steels, and other metals and alloys, a 0·5–2 g sample can usually be dissolved in about 60 ml of a sulphuric–phosphoric mixture containing 80 ml H_3PO_4 and 320 ml of 1 : 1 H_2SO_4 per litre [2, 11, 14, 16, 22]. Heat until action ceases, add 5–10 ml nitric acid, and evaporate to fumes; if any residue remains, filter, wash, ignite, fuse with sodium carbonate, and add the solution of the melt to the main filtrate. Some samples will require treatment with hydrochloric, nitric, sulphuric, and a few drops of hydrofluoric acid; evaporate to fumes of sulphur trioxide, and if a residue still persists, fuse it with sodium carbonate.

When the sample is completely dissolved, dilute to 300 ml with hot water, add 5 ml of 1% silver nitrate and 20 ml of 15% ammonium persulphate. Boil for 8–10 minutes, and if the colour of permanganic acid does not develop, add more silver nitrate and ammonium persulphate, and boil for a further 10 minutes. Add 5 ml of 1 : 3 hydrochloric acid and continue the boiling for 5 minutes after the pink colour has disappeared. If the permanganic acid colour is not destroyed by boiling for 10 minutes, or if a precipitate of manganese dioxide remains, add 2–3 ml additional 1 : 3 hydrochloric acid and boil again. The total period of boiling after the addition of ammonium persulphate should be not less than 15 minutes; 30 minutes will do no harm provided the volume is maintained at approximately 300 ml by the addition of hot water at intervals.

Cool the solution, dilute to 400 ml, add a measured excess of 0·1N ferrous ammonium sulphate, 2 drops of ferroin and titrate with 0·1N potassium permanganate to the clear green end point. Vanadium, if present, will be titrated; it may be determined in the same solution, after the initial end point is obtained, by reducing the acidity of the solution with 23 g of sodium acetate and titrating slowly at 50° C with 0·06N potassium permanganate to the green end point.

The titration of chromium may be done potentiometrically, in which case a direct titration with a standard ferrous ammonium sulphate solution is used. Vanadium is included in this titration; if present, it must be determined and subtracted from the apparent chromium value, using the relation $\%V \times 0{\cdot}34 = \%\ Cr$. With tungsten steels, the potentiometric method is carried out by omitting phosphoric acid and oxidizing tungsten to tungstic acid with nitric acid.

When the excess ferrous ammonium sulphate in chromium analyses is titrated visually with potassium permanganate without the aid of ferroin, a correction must be applied for dilution effect and colour interference. This is done most conveniently for occasional analyses by boiling the solution, which has just been titrated, for 10 minutes to destroy the slight excess of potassium permanganate, cooling to room temperature, and then titrating with potassium permanganate to the colour that was originally taken as the end point. The titrated solution may be reserved for the determination of vanadium.

2. Colorimetric Procedure

A. Chromate Method

Fuse the sample with approximately 10 times its weight of sodium peroxide in a zirconium or iron crucible. Dissolve the cooled melt in 100 ml of water, remove the crucible, add 1 g sodium peroxide and boil for 5 minutes. Filter through an asbestos pad on a large Gooch crucible, or small Buchner funnel; wash with a cold solution containing 20 g sodium hydroxide and 10 g sodium sulphate in 1 litre of water.

Dilute to a measured volume and compare the colour with standards containing approximately the same concentration of chromium and alkali [2, 11, 13, 14, 16, 21]. A solution containing 0·283 g $K_2Cr_2O_7$ per litre has a value of 1 ml = 0·0001 g Cr. Spectrophotometric determination may be made at 366 nm. Uranium is usually the only element which will interfere; it can be separated from chromium when the latter is in the hexavalent state by an ammonia precipitation. Alternatively, chromium can be precipitated by treatment with sodium hydroxide and sodium carbonate, uranium remaining in solution in the presence of carbonate.

For irons and steels, dissolve a 10-g sample in 110 ml of 1 : 9 H_2SO_4, boil until reaction is complete, and dilute with 100 ml of hot water [2]. Add a solution of 80 g/litre sodium bicarbonate from a burette until a permanent precipitate appears and then 4 ml in excess. Boil for 1 minute, allow to settle, filter, and wash three times with hot water. The filtrate will become cloudy in the funnel stem and receiving vessel, due to oxidation and hydrolysis.

Place the precipitate in a zirconium or iron crucible, dry, ignite, fuse with sodium peroxide and proceed as given above.

B. Diphenylcarbazide Method

Hexavalent chromium forms with s-diphenylcarbazide, also called 1,5-diphenylcarbohydrazide, a soluble red-violet complex. Few elements interfere when chromium is separated by a sodium peroxide fusion; molybdenum and vanadium are the only common ones. Molybdenum is readily removed with hydrogen sulphide in acid solution; vanadium, and molybdenum, are precipitated with cupferron in cold 10% sulphuric acid solution and can be extracted with chloroform.

The diphenylcarbazide procedure is more sensitive than the chromate method; various modifications to suit special materials have been published [1, 2, 4, 6, 7, 8, 10, 11, 16, 18, 20, 23].

Decompose the material with a sodium peroxide fusion, leach with water, and filter off Fe, Ni, Co, Ti, and Zr as described previously. A fusion in platinum with sodium carbonate and sodium borate has also been advocated [6, 20]. For irons and steels, remove the bulk of the iron with a bicarbonate separation as described for the chromate procedure. If vanadiun is present it will be in the filtrate with chromium and must be removed by acidifying with sulphuric acid and adding 10% excess, precipitating vanadium with cupferron in cold 10% sulphuric acid solution and extracting the cupferrate with chloroform. Transfer the upper aqueous layer containing the chromium from the separatory funnel to a beaker, add 10 ml 1 : 1 sulphuric acid, 10 ml nitric acid, and evaporate to fumes; if necessary, repeat the treatment to remove all traces of organic matter and chlorine. Convert the chromium to the chromate form by adding 10 ml perchloric acid, 10 ml 1 : 1 sulphuric acid, and evaporating to fumes of the latter. Adjust the sulphuric acid concentration to approximately 0·2N when diluted to 25 ml. Add 1 ml of 0·25% diphenylcarbazide in 50% acetone. Make up to 25 ml and determine the transmittance of the solution at 540 nm without delay. The calibration curve may be prepared by using a solution containing 0·1415 g $K_2Cr_2O_7$ per litre; 1 ml = 0·05 mg Cr.

3. Optical Spectrography

Chromium has a number of useful emission lines, and can be determined in a variety of materials by optical spectrography [2, 5, 16, 17]. The most sensitive lines are 4254·35, 4274·80 and 4289·72. Detailed procedures have been published for 0·03–2·20% chromium in plain carbon and low alloy steels [2], 0·01–0·5% in aluminium and its alloys [2, 5], 20–2000 ppm in zirconium and its alloys [2], 0·6–1000 ppm in nickel [17], and 0·01–0·5% in sapphire and ruby [6].

4. X-Ray Spectrography

Chromium has been determined in ores, steels, nickel, and other materials by X-ray fluorescence [16, 17]. The most sensitive line is $K_{\alpha 1}$ at 2·290 Å; other wavelengths are $K_{\alpha 2}$ 2·294 and K_{β} 2·085.

5. Atomic-Absorption Spectrophotometry

Atomic absorption offers another approach to the determination of chromium [9]. With a sensitivity of 0·05 μg/ml., line 3578·7 is the most sensitive chromium line, but 3593·5 and 4254·3 can be used. Iron reduces the absorption of chromium, and must be added to standards for the examination of high-iron materials. Interference is caused by molybdenum, tungsten, and nickel, but if these occur in quantities less than 1, 5, and 5% respectively, they may be tolerated by suitable adjustment of the height of the absorption path [9].

6. Polarographic Methods

When chromium is oxidized to the hexavalent state by boiling the alkaline sample solution with hydrogen peroxide, the element can be determined polarographically [16]. None of the common constituents of copper-base alloys interfere in this determination, and a detailed procedure for chromium above 0·05% has been published [10].

REFERENCES

1. Allen, T. L., *Anal. Chem.* **30**, 447–50 (1958).
2. A.S.T.M., *Chemical Analysis of Metals. Sampling and Analysis of Metal Bearing Ores*, Philadelphia, Pa., American Society for Testing and Materials, 1969.
3. Bennett, H., and Marshall, K., *Analyst* **88**, 877–81 (1963).
4. British Aluminium Co., *Chemical Analysis of Aluminium and its Alloys*, London, British Aluminium Co., 1961.
5. British Aluminium Co., *Spectrochemical Analysis of Aluminium and its Alloys*, London, British Aluminium Co. 1961.
6. Chirnside, R. C., Cluley, H. J., Powell, R. J., and Proffitt, P. M. C., *Analyst* **88**, 851–63 (1963).
7. Dean, J. A., and Beverly, M. L., *Anal. Chem.* **30**, 977–9 (1958).
8. Easton, A. J., *Anal. Chim. Acta* **31**, 189–91 (1964).
9. Elwell, W. T., and Gidley, J. A. F., *Atomic-Absorption Spectrophotometry*, 2nd ed., Oxford, Pergamon, 1966.

10. ELWELL, W. T., and SCHOLES, I. R., *Analysis of Copper and its Alloys*, Oxford, Pergamon, 1967.
11. FURMAN, N. H., ed., *Scott's Standard Methods of Chemical Analysis*, 6th ed., Vol. 1, Princeton, N.J., Van Nostrand, 1962.
12. HARTFORD, W. H., *Anal. Chem.* **25**, 290–6 (1953).
13. HAYWOOD, F. W., and WOOD, A. A. R., *Metallurgical Analysis by Means of the Spekker Absorptiometer*, 2nd ed., London, Hilger and Watts, 1956.
14. HILLEBRAND, W. F., LUNDELL, G. E. F., BRIGHT, H. A., and HOFFMAN, J. I., *Applied Inorganic Analysis*, 2nd ed., New York, Wiley, 1953.
15. KATZ, S. A., MCNABB, W. M., and HAZEL, J. F., *Anal. Chim. Acta* **25**, 193–9 (1961).
16. KOLTHOFF, I. M., and ELVING, P. J., *Treatise on Analytical Chemistry*, Part 11, Vol. 8, 273–377, New York, Interscience Publishers, 1963.
17. LEWIS, C. L., OTT, W. L., and SINE, N. M., *The Analysis of Nickel*, Oxford, Pergamon, 1966.
18. LUKE, C. L., *Anal. Chem.* **30**, 359–61 (1958).
19. MANN, C. K., and WHITE, J. C., *Anal. Chem.* **30**, 989–92 (1958).
20. RICHARDS, C. S., and BOYMAN, E. C., *Anal. Chem.* **36**, 1790–3 (1964).
21. SANDELL, E. B., *Colorimetric Determination of Traces of Metals*, 3rd ed., New York, Interscience Publishers, 1959.
22. SNELL, F. D., and ETTRE, L. S., *Encyclopedia of Industrial Chemical Analysis*, Vol. 9, New York, Wiley-Interscience, 1970.
23. WILLIAMS, A. I., *Analyst* **93**, 611–17 (1968).

13 Cobalt

Cobalt occurs in some minerals, ores, concentrates, mattes, speisses, and residues in association with arsenic, copper, iron, manganese, nickel, and sulphur; sometimes it is found with lead, silver, zinc, and a few other elements. Many cobalt minerals exist, but those of economic importance are few. The principal sulphides are carrollite $CuCo_2S_4$, and linnaeite or siegenite Co_3S_4; cobalt substitutes for nickel in many of the copper–nickel–iron sulphide orebodies. The chief cobalt arsenides are smaltite and safflorite, $CoAs_2$, skutterudite, $CoAs_3$, and the sulpho-arsenide, cobaltite, CoAsS. The main oxidized cobalt minerals are heterogenite, $CoO.2Co_2O_3.6H_2O$, and asbolite, $CoO.2MnO_2.4H_2O$; of lesser importance are sphaerocobaltite, $CoCO_3$, stainierite, $Co_2O_3.H_2O$, and erythrite $3CoO. As_2O_5.8H_2O$.

Cobalt is an important constituent in many high temperature alloys, magnets, cemented carbides, glass-to-metal seals, hard-facing alloys, springs, low-expansion alloys, enamels for steel, high speed steels, catalysts, electroplating solutions, dental and surgical alloys, driers, and colouring agents for the glass and ceramic industries. As a dietary essential for sheep and cattle, it is often added to livestock licks and feedstuffs, and to lime and fertilisers. Cobalt-60, produced by neutron bombardment of cobalt metal placed within a nuclear fission reactor, is widely used as a tracer, and as a source of ionizing radiation in teletherapy and other fields. The analytical chemistry of cobalt has been reviewed in journals and books [34, 39, 42]; two monographs on this subject have appeared [30, 44].

ISOLATION OR SEPARATION OF COBALT

Separations are important in the determination of cobalt; most analytical procedures for the majority of materials require the elimination of at least one element. Cobalt can be separated from members of the acid sulphide group by precipitation of the latter with hydrogen sulphide in 5–10% hydrochloric or sulphuric acid. If copper is the only member of this group present, it may often be advantageously removed and simultaneously determined by electrolysis. Silver may be separated as chloride and lead as sulphate, prior to the electrolysis of copper. Cobalt, with other members of the ammonium sulphide group, can be separated from the alkaline earths and alkalies by precipitation in ammonium sulphide solution. Practically all tungsten, niobium, and tantalum remain with silica when the latter is dehydrated with hydrochloric, sulphuric, or perchloric acids, and can thereby be removed from cobalt.

The principal separations of cobalt from other members of the ammonium sulphide and hydroxide groups are given below. The behaviour of a few other metals sometimes found in cobaltiferous materials is also included.

1. A zinc oxide separation removes Fe, Al, Ti, V, Cr, Zr, U, P, Nb, Ta, W, Tl, and Th as a precipitate from a filtrate containing Co, Ni, and Mn; most of the beryllium is precipitated but even in the presence of iron some passes into the filtrate. This is the most important separation for cobalt, because a subsequent precipitation of the latter with 1-nitroso-2-naphthol, formerly called α-nitroso-β-naphthol, can be made in the presence of Ni, Mn, Zn, Be, Ca, and Mg which are in the filtrate with cobalt after the zinc oxide precipitation.

2. Cupferron in cold 10% sulphuric or hydrochloric acid solution precipitates Fe, Ti, V, Zr, Nb, Ta, W, and Mo from Co, Ni, Al, Be, U(VI), Mn, Cr, and P. The latter are left in the upper aqueous layer if a chloroform extraction of the insoluble cupferrates is carried out.

3. Cobalt is deposited on a mercury cathode in dilute sulphuric acid solution, together with Ni, Fe, Cr, Mo, Cu, and Zn, and can thereby be removed from Al, Be, Ti, Zr, P, V, and U.

4. Fusion in sodium peroxide and later addition of water gives cobalt in the precipitate with Ni, Fe, Ti, Cu, Th, and Zr; the filtrate contains Cr, Be, Al, Mo, U, W, V, and Zn.

5. Treatment with ammonium sulphide in the presence of tartrate precipitates cobalt with Fe, Ni, and Zn, whereas Cr, U, and V remain in the filtrate.

6. Precipitation of the ammonium hydroxide group with sodium hydroxide gives cobalt in the precipitate, with Fe, Cr, Ti, Zr, Th, and U, whereas Al, Be, Zn, V, and W remain in the filtrate.

7. An ether separation of iron in hydrochloric acid of s.g. 1·10 is often useful to remove large quantities of this metal from cobalt.

8. Chromium can be volatilized from cobalt solutions by several evaporations with perchloric acid in the presence of hydrochloric acid or sodium chloride.

9. Cobalt can be isolated from uranium and manganese by addition of potassium ethyl xanthate to an acetic acid solution in a citrate buffer, and extraction of the cobalt complex into a chloroform layer.

10. For rapid technical work, small quantities of Fe, Al, and other members of the ammonium hydroxide group can be separated from cobalt by several ammonia precipitations; if appreciable quantities of these elements are present, however, this procedure cannot be employed.

11. Manganese can be removed from cobalt in acid solution by precipitation of MnO_2 in nitric acid and sodium chlorate. In ammoniacal solution, if only small quantities of manganese are present, they may be precipitated by bromine and ammonium hydroxide; for larger quantities, electrolysis in ammoniacal solution will deposit cobalt at the cathode, whereas manganese will be partly deposited on the anode, the rest remaining in solution. Cobalt

can also be isolated from manganese by precipitation of the former with phenylthiohydantoic acid in ammoniacal solution in the presence of citrate.

12. Nickel is removed from cobalt by precipitation of the former with dimethylglyoxime or sodium diethyldithiocarbamate, or of the cobalt with 1-nitroso-2-naphthol or potassium nitrite.

13. Cobalt may be separated from U, V, Zn, and W by precipitation of the former with sodium hydroxide and sodium carbonate.

14. At a *p*H of about 3·5, sodium phosphate will precipitate iron, aluminium, titanium, zirconium, and uranium; cobalt remains in solution with nickel, chromium, manganese, and vanadium.

15. Cobalt forms a complex with ammonium thiocyanate which can be extracted into a layer of amyl alcohol and ether, or of acetone. This serves to separate cobalt from nickel and a large number of other metals.

16. Following a preliminary extraction with diphenylthiocarbazone in carbon tetrachloride or chloroform in 0·1–1N hydrochloric or sulphuric acid, cobalt can be extracted at *p*H 8–9 with dithizone. The elements Bi, Cd, In, Pb, Mn, Ni, Tl(I), Sn(II), and Zn accompany cobalt.

17. Anion exchange resins such as Dowex 1, 1-X8, and 1-X10, and cation exchange resins like Dowex 50W-X8, 50W-X22, and Amberlite IRA 120 have been used to isolate cobalt from iron, nickel, and other common interfering metals [6, 16, 22, 44].

18. Small quantities of cobalt have been separated from a number of other elements by various chromatographic techniques [44].

Cobalt is tenaciously retained by the hydroxides of iron and aluminium; it is impossible to separate cobalt completely from large quantities of these hydroxides even by repeated ammonia precipitations. A large excess of ammonia and ammonium salts, hot solutions, and pouring the iron-cobalt solution into ammonia, all tend to minimize the retention of cobalt, but fail to reduce it to a satisfactory level. For example, with 0·15 g iron and 0·015 g cobalt, two careful ammonia precipitations yielded only 90% of the cobalt in the filtrate, and a further re-precipitation gave no significant increase in cobalt extraction. With 0·015 g of iron and the same quantity of cobalt, two ammonia precipitations gave 96% of the cobalt in the filtrate. Aluminium hydroxide retains even slightly more cobalt, probably because a large excess of ammonia cannot be used.

1. Gravimetric Determination as Co_3O_4 after Precipitation with 1-nitroso-2-naphthol

One of the first organic reagents to be used for the determination of a metal was α-nitroso-β-naphthol, now usually called 1-nitroso-2-naphthol. In 1885 it was applied to cobalt analysis, and it is still one of the most reliable and convenient methods for the determination of this element over a wide range of concentrations [2, 11, 17, 19, 30, 34, 40, 44]. The principle of this procedure is that, in the absence of Cu, Ag, Au, Pd, Sb, Sn, Bi, Mo, Fe, Cr, V, Ti, Zr, Nb, Ta, and W, cobalt is precipitated by 1-nitroso-2-naphthol in hydrochloric acid solution in the presence of nickel and all remaining metals. The list of interferences looks formidable; actually, precipitation with hydrogen sulphide in 5–10% hydrochloric acid followed by precipitation with zinc oxide gives a solution containing cobalt without any element which interferes in the final 1-nitroso-2-naphthol determination. Because iron is often the only interfering element, apart from members of the acid sulphide group, separation with phosphate or ether is sometimes substituted for the zinc oxide precipitation.

A. Separation of Fe, Al, Cr, V, Ti, Zr, U, P, Nb, Ta, W, Tl, and Th with Zinc Oxide

Weigh 0·5–10 g, depending on the cobalt content of the sample, and take into solution with 10–25 ml nitric acid, 10–25 ml hydrochloric acid, 10 ml 1 : 1 sulphuric acid, and a few drops of hydrofluoric acid if necessary. For high-sulphide products initial treatment with a few drops of bromine or several crystals of potassium chlorate is advisable, to avoid the inclusion of undissolved particles in a bead of sulphur. Certain high-cobalt alloys, such as Stellites, are best decomposed by prolonged treatment with perchloric acid after the addition of the other acids specified above. Evaporate to strong fumes of sulphuric acid, cool, dilute with water, and boil. If the siliceous residue is known to contain cobalt, filter, wash, dry, ignite, fuse with sodium carbonate, and add to the main filtrate.

Other methods of decomposition may be employed, if more convenient. High-silica materials may be treated in a platinum dish with hydrofluoric and sulphuric acids. Cemented carbides can be decomposed by nitric and hydrofluoric acids in a platinum dish, or by a sodium peroxide fusion in a nickel or zirconium crucible. Some refractory oxides and residues may be fused with sodium or potassium bisulphate in a Pyrex "copper" flask. In all cases, evaporate to strong fumes of sulphuric acid, cool, dilute with water, and boil.

Adjust the acidity of the solution to 5–10% with sulphuric or hydrochloric acid, and pass in hydrogen sulphide until all members of the acid sulphide group have been precipitated. Filter off the precipitated sulphides and silica, and wash thoroughly with acidulated hydrogen sulphide water Boil the filtrate for 10–15 minutes to remove all hydrogen sulphide, oxidize the iron with hydrogen peroxide, and boil to remove the excess of peroxide. Add a thick suspension of zinc oxide until all metals of the ammonium sulphide and hydroxide groups have been precipitated, and a slight excess of zinc oxide is visible at the bottom of the beaker. Because iron is nearly always present, the precipitate at this point will usually have the colour of coffee with cream. Boil and filter, washing several times with hot water. Transfer the precipitate back to the original beaker, dissolve with hydrochloric acid, and re-precipitate with the zinc oxide emulsion. Filter through the original paper and wash 7–8 times with hot water. The filtrate will contain only cobalt, with Ni, Mn, Ca, Mg, and a few other elements which do not affect the 1-nitroso-2-naphthol precipitation. A turbidity in the filtrate caused by colloidal zinc oxide may be disregarded; subsequent acid treatment will dissolve it. For work of the highest accuracy on small quantities of cobalt in the presence of large quantities of iron or other elements which are precipitated by zinc oxide. a third precipitation may be made; this, however, is rarely required.

Add 4 ml hydrochloric acid per 100 ml of solution to dissolve the colloidal zinc oxide and hold the nickel in solution. Heat to boiling and cautiously add sufficient 1-nitroso-2-naphthol in 1 : 1 acetic acid to precipitate all the cobalt. Use 0·5 g 1-nitroso-2-naphthol for each 0·01 g cobalt. Allow the red precipitate of cobaltinitroso-2-naphthol to boil for several minutes, and set the beaker aside in a warm place for several hours.

Filter on Whatman No. 40 paper, using pulp, wash several times with hot water, scrubbing the beaker thoroughly with a rubber policeman. Wash the precipitate 6–8 times with hot 4% hydrochloric acid, and finally with hot water until free of chlorides. Do not attempt to continue washing until the filtrate becomes colourless; a pale yellow colour from the excess reagent will persist long after all inorganic salts have been removed. Place paper and precipitate in a weighed porcelain or platinum crucible, dry, ignite slowly and carefully in a muffle, and finally at a temperature not exceeding 800°C. Cool in a desiccator and weigh as Co_3O_4.

$$Co_3O_4 \times 0{\cdot}7342 = Co$$

Results for cobalt are nearly always expressed as the metal, except in rock analysis where they are often given as CoO; Co $\times$ 1·2715 = CoO.

For routine work, the precipitate may be ignited in a fireclay crucible or annealing cup and brushed on to the counterpoised watch glass of the balance.

The efficiency of the zinc oxide precipitation as outlined above has been confirmed by tracer work with cobalt–60 [44]. Sometimes it is recommended that a suspension of zinc oxide be added to the sample in a volumetric flask, that the latter be shaken and the contents allowed to settle, and an aliquot portion of the supernatant liquid be withdrawn for a cobalt determination. Our results have indicated that this practice is unsatisfactory for accurate work [44].

It must be borne in mind that if the decomposition has been strongly oxidizing by means of perchloric acid or sodium peroxide, and a hydrogen sulphide precipitation has not been necessary, chromium will be in the hexavalent state and will not be precipitated by zinc oxide. If so, heat the solution briefly with a freshly-prepared, saturated solution of sulphur dioxide or a 10% solution of sodium sulphite before proceeding with the zinc oxide precipitation. A strongly oxidizing decomposition will put vanadium into the quinquevalent state; in the absence of a hydrogen sulphide separation, moderate amounts will be precipitated by zinc oxide, but if the quantity is large a little vanadium may pass into the filtrate. The same reduction with sulphur dioxide mentioned above, will put vanadium into the quadrivalent condition.

For some materials, particularly alloys, it may be advantageous to remove chromium by volatilization as chromyl chloride. Heat the sample in an Erlenmeyer flask with perchloric acid until dense white fumes are evolved.

Carefully add 2–3 ml of hydrochloric acid, and heat over a flame while constantly swirling the liquid in the flask until brown fumes of chromyl chloride cease to be evolved. Continue heating until white fumes of perchloric acid are given off, add another 2–3 ml of hydrochloric acid, and repeat the operation until brown fumes are no longer evolved when hydrochloric acid is added.

Uranium is precipitated with 1-nitroso-2-naphthol in acetic acid, but if hydrochloric acid is present, uranium remains in solution.

The precipitate of cobaltinitroso-2-naphthol is bulky; not more than 0·05 g of cobalt should be determined in this way when the precipitate is to be ignited and weighed as oxide in the usual 25 ml platinum crucible. It is sometimes convenient, however, to isolate up to 0·2 g cobalt by precipitating with 1-nitroso-2-naphthol, filtering in a large funnel, digesting the precipitate with nitric, hydrochloric, and sulphuric acids, evaporating to fumes of the latter, and finally determining cobalt by electrolysis in ammoniacal solution. Alternatively, the bulky precipitate may be ignited in a large platinum dish to Co_3O_4, dissolved in hydrochloric acid, evaporated to fumes with sulphuric acid and finally electrolysed as described later.

Nickel and manganese accompany cobalt in the filtrate from a zinc oxide separation. Though they are not precipitated by 1-nitroso-2-naphthol in hydrochloric acid solution, when large quantities of either occur in the presence of much cobalt the latter may occlude traces of nickel or manganese. For work of the highest accuracy it may be necessary to dissolve the final residue of Co_3O_4 in hydrochloric acid and re-precipitate the cobalt. If iron contamination is suspected, dissolve the cobalt oxide in hydrochloric acid, add sulphuric acid, evaporate to fumes of the latter and proceed with an electrolytic determination as described later.

If the ignited residue of Co_3O_4 must be taken into solution for re-treatment or further analysis, it is useful to remember that it will readily dissolve in hot concentrated hydrochloric acid; Co_3O_4 is also soluble in hot 1 : 1 sulphuric acid, but solution in hot nitric acid is slow.

Twenty ml of 1 : 1 acetic acid will dissolve approximately 1 g of 1-nitroso-2-naphthol. It is advisable to dissolve the latter first in warm concentrated acetic acid and then dilute with water to about 1 : 1.

The reactions for the zinc oxide and 1-nitroso-2-naphthol precipitations are:

$$3ZnO + 2FeCl_3 + 3H_2O \rightarrow 3ZnCl_2 + 2Fe(OH)_3$$

$$CoCl_2 + 3C_{10}H_6(NO)OH \rightarrow [C_{10}H_6(NO)O]_3Co + 2HCl + H$$

B. Separation of Iron with Sodium Phosphate

When iron is precipitated by sodium phosphate from a solution weakly acid with acetic, there is remarkably little occlusion of cobalt [44]. Even in the presence of large quantities of iron, one precipitation is sufficient for routine work. Unlike the zinc oxide separation, a filtrate is obtained which is suitable for immediate electrolysis of cobalt. The method has the disadvantage that chromium and vanadium are not wholly removed; in addition, if an electrolytic determination is contemplated, calcium remains with the solution of cobalt to give a precipitate of calcium phosphate in the ammoniacal electrolyte. The procedure is sometimes useful for certain classes of materials, however, and merits a brief discussion.

To a boiled and oxidized solution from which members of the hydrogen sulphide group have been previously removed, add sufficient trisodium phosphate to precipitate all the iron, aluminium, etc., plus a small excess. Particular care must be taken to avoid a large excess when cobalt is to be finally determined electrolytically, otherwise some cobalt will be precipitated as phosphate in the ammoniacal electrolyte. Ten ml of a solution containing 34 g $Na_3PO_4.12H_2O$ in a litre will precipitate approximately 0·05 g Fe; for materials high in iron, use a more concentrated solution. Add ammonium hydroxide with vigorous stirring until purple cobaltous phosphate is formed. When cobalt is low it may be difficult to see cobaltous phosphate, and red litmus should be used as indicator. When the litmus turns blue, a *p*H of approximately 5·6 has been reached; this is sufficient to ensure that cobaltous phosphate has been precipitated at *p*H 5·3.

Add 10 ml glacial acetic acid and stir vigorously to dissolve the cobaltous phosphate; this will give a *p*H of 3–3·5. Add 5–20 ml hydrogen peroxide and stir thoroughly to oxidize the solution. Iron will be precipitated as white ferric phosphate:

$$2\,Na_3PO_4 + Fe_2(SO_4)_3 \rightarrow 2FePO_4 + 3Na_2SO_4$$

Bring the sample to a boil, stirring to prevent bumping if a heavy iron precipitate is present. Filter through Whatman 31 or 41 paper on a fluted funnel, using paper pulp; a very heavy iron precipitate may be filtered on a Buchner funnel. Wash 8–10 times with hot 2–3% acetic acid. Proceed with the determination of cobalt by precipitation with 1-nitroso-2-naphthol; alternatively, add 10 ml 1 : 1 sulphuric acid, neutralize with ammonium hydroxide, add about 40 ml excess, and electrolyse. If nickel is present it will be deposited with cobalt on the cathode.

C. Separation of Iron with Ether

Removal of iron by an ether separation is often advantageously employed in cobalt analyses prior to precipitation with 1-nitroso-2-naphthol. It has the disadvantage that iron is the only member of Group 3 which is removed; if Cr, V, Ti, Zr, Nb, Ta, and W are present, the zinc oxide precipitation must be carried out. Many cobalt-containing materials, however, have large quantities of iron as their sole interference for the gravimetric cobalt determination with 1-nitroso-2-naphthol. It is often convenient to separate iron in this way from a large sample of ore, concentrate, matte, slag, iron, steel, or alloy; three extractions with ether on 5 g of iron will leave only about 0·05 mg of iron with the cobalt. With a little practice, the chemist will find the ether separation useful for removing iron, not only from cobalt, but also for many other analyses.

Obtain a solution of the sample in hydrochloric acid, from which members of the hydrogen sulphide group and silica have been removed by filtration. If iron is not already in the ferric state, boil for a few minutes with hydrogen peroxide. Evaporate the solution to a syrupy consistency. Transfer to a 250- or 500-ml separatory funnel; rinse the beaker with several small quantities of hydrochloric acid of s.g. 1·10. The latter is made by mixing 526 ml concentrated hydrochloric acid with 474 ml of water. Add diethyl ether to bring the concentration to 30 ml ether per g of iron in 20 ml of hydrochloric acid of s.g. 1·10. If desired, isopropyl ether may be used instead of the diethyl compound.

Shake the funnel vigorously under a cold water tap, allow to settle, and draw off the lower layer into a 400-ml beaker. This contains the cobalt, together with Al, Ni, Mn, Cr, V, Ti, Zr, Zn, Th, U, and Be; the upper layer contains the iron, with gallium and thallium if present. The lower layer must be shaken with one or two more portions of ether to remove traces of ferric chloride. The upper layers of ether and ferric chloride should be combined and shaken with one or more portions of hydrochloric acid of s.g. 1·10 that have been first saturated with ether. This will recover any traces of cobalt chloride that might be entrapped in the ether–ferric chloride layer; combine this small quantity with the main lower layer.

Place the beaker containing the cobalt and other elements from the lower layers of the ether separations at the side of the hot plate on thick asbestos until all ether has been driven off. Add 5 ml nitric acid, 5 ml 1 : 1 sulphuric acid, and evaporate to dryness. Add 8 ml hydrochloric acid, dilute to 200 ml with water, boil, and proceed with the 1-nitroso-2-naphthol precipitation.

D. Other Separations of Iron etc.

The separation of cobalt from iron and other members of the ammonium sulphide and hydroxide groups, which are enumerated at the beginning of this chapter, may be used in the gravimetric procedures. For example, precipitation with cupferron in cold acid solution in a separatory funnel and extraction of the insoluble cupferrates with chloroform will eliminate Fe, Ti, V, Zr, Nb, Ta, and W, all of which interfere in the 1-nitroso-2-naphthol precipitation; in the upper aqueous layer chromium is the only interfering element, and it can be volatilized initially as chromyl chloride.

Ion exchange and chromatographic separations have been proposed as alternative methods of removing iron and other interfering elements from cobalt [44]. For the gravimetric procedure, however, these are usually not as satisfactory as other separations.

2. Gravimetric Determination as Potassium Cobaltinitrite

Cobalt can be separated from nickel as a precipitate of potassium cobaltinitrite, now sometimes called potassium hexanitritocobaltate(III), by the addition of potassium nitrite in dilute acetic acid solution [11, 17, 19, 30,

34, 40, 44]. Free mineral acids, oxidizing agents, the hydrogen sulphide group, any appreciable quantity of the elements precipitated by ammonium hydroxide, and the alkaline earths, must be absent. For large amounts of nickel, sodium, and tungsten, or for large quantities of cobalt with a medium content of nickel, a re-precipitation is required.

Decompose the sample with nitric, hydrochloric, sulphuric, and perchloric acids; with nitric or sulphuric and hydrofluoric acids in a platinum dish; by fusion in a nickel or zirconium crucible with sodium peroxide; or by any other means to bring all cobalt into solution. Add 10 ml of 1 : 1 sulphuric acid and evaporate to dryness. Add 5 g tartaric acid, 25 ml water, and heat to solution of salts. If a precipitate remains, filter through Whatman No. 40 paper, and wash thoroughly with hot 1% tartaric acid solution. If arsenic is present, expel it at the outset by evaporating twice with 25 ml of hydrobromic acid added to the cold sulphuric acid solution. Calcium, barium, strontium, and lead, which interfere in this separation of cobalt from nickel, are removed as sulphates in the initial filtration.

To the sample in tartaric acid solution, from which arsenic, lead, calcium, barium, and strontium have been removed, add saturated potassium hydroxide until the solution is basic to litmus. Carefully add acetic acid until the sample is acid, and then add 4 ml excess. The solution should be clear, but a slight turbidity can be disregarded, because the re-precipitation step will remove the impurity. Add, with stirring, to the warm acetic acid solution a large excess of hot 50% potassium nitrite solution, usually 30–100 g potassium nitrite in 60–200 ml of water. Heat to boiling and keep near this point for 5–10 minutes. Allow to stand cold for four hours or preferably overnight. Filter on Whatman No. 40 paper, washing 6–8 times with cold 3% potassium nitrite solution and once with cold water. Discard the filtrate.

Return the paper and precipitate to the original beaker, add 15–30 ml nitric acid, 30 ml 1 : 1 sulphuric acid, and fume to dryness. If all organic matter has not been destroyed, add more nitric and sulphuric acids, and again evaporate to dryness. Add 50 ml water, 5 ml acetic acid, and heat to boiling. Carefully add 30 ml of 50% potassium nitrite which has been heated to boiling, stir occasionally over a period of 30 minutes at room temperature, and allow to stand for two hours. Filter through a tared Gooch crucible, or medium porosity sintered glass crucible; wash with 100 ml of 2% potassium nitrite, five 10-ml portions of 80% ethanol, and once with acetone. Dry for 1 hour at 110°C and weigh as $K_3Co(NO_2)_6$, which contains 13·03% cobalt.

The composition of potassium cobaltinitrite varies with the quantity of cobalt precipitated, but the deviation of the weight from $K_3Co(NO_2)_6$ is small enough to ignore in technical analyses. When arsenic is volatilized at the outset, and lead and alkaline earths removed as sulphates, precipitation in the presence of tartaric acid followed by a re-precipitation from a boiling solution of potassium nitrite gives accurate results. The following do not interfere: aluminium, antimony, bismuth, cadmium, chromium, copper, iron, manganese, molybdenum, nickel, niobium, tin, titanium, tungsten, vanadium, zinc, and zirconium.

When determining cobalt by the potassium nitrite method, always discard the filtrate from the precipitation of potassium cobaltinitrite; never attempt to determine any other element in this solution because boiling or evaporating may lead to a violent explosion.

3. Electrolytic Determination

For large quantities of cobalt an electrolytic determination is the most satisfactory procedure. Electrolysis is carried out in ammoniacal solution in the presence of ammonium sulphate, after removal of copper and nickel, which would be deposited, and of iron, aluminium, etc., which would impede the plating. Depending on the sample, electrolysis may be preceded by separation of iron and other elements with zinc oxide or phosphate, or of iron with ether, followed by precipitation of cobalt with 1-nitroso-2-naphthol. Alternatively, cobalt may be isolated, before electrolysis, by the potassium nitrite separation. In some instances, cobalt and nickel may be electrolysed together, with a subsequent separation.

In all cobalt electrolytic work it must be recalled that cobalt and nickel, unlike copper, are not easily dissolved off a platinum cathode in cold nitric acid. Cathodes should be boiled vigorously for five minutes in nitric acid to remove cobalt. If a black coating persists after adequate treatment of the cathode with hot nitric acid, it is almost certainly carbon and may be removed by heating for a few seconds in the hottest zone of a bunsen flame.

The familiar test for complete electrodeposition of copper, by simply diluting the solution and observing whether more copper is plated on the stem of the platinum cathode, cannot be used for cobalt; in the cloudy ammoniacal solution platinum and cobalt are nearly similar in appearance. Samples with which the analyst is not familiar should be checked for complete deposition by withdrawing a few drops with a pipette, transferring to a spot plate, and testing with one of the following solutions:

(1) *Nitroso-R-salt*, $C_{10}H_4OH.NO(SO_3Na)_2$. To a few drops of the cobalt solution add several drops of 50% sodium acetate, acidify with nitric acid, and add a few drops of a 0·5% solution of nitroso-*R*-salt in water. A permanent red colour indicates the presence of cobalt.

(2) *Phenylthiohydantoic acid*, $C_9H_{10}O_2NS$. When a few drops of a solution of 1 g phenylthiohydantoic acid in 100 ml ethyl alcohol are added to an ammoniacal cobalt solution, a pink colour is produced.

(3) *Potassium thiocarbonate*, K_2CS_3. This is prepared by saturating one half of a solution of 5% potassium hydroxide with hydrogen sulphide, adding the other half, and heating moderately with 1/25 of its volume of carbon disulphide. The dark red liquid is decanted from the undissolved carbon disulphide and kept in a well-stoppered flask. With this reagent, small quantities of cobalt in ammoniacal solution give a yellow colour, larger quantities a brown or black.

The sensitivity of some reagents, such as nitroso-R-salt, makes the interpretation of spot tests a little difficult for the inexperienced. It is a good plan to have on hand a stock cobalt solution of such concentration that a few tenths of a millilitre addition to the spot plate from a small capacity pipette represents about the limit of precision desired. For instance, when analysing a 1-g sample containing 20% cobalt, a quantity not exceeding 0·5 mg left in the electrolyte volume of 200 ml may be considered satisfactory in routine control work, and if so, 0·5 ml of this solution would contain 0·0012 mg cobalt. An equivalent quantity for checking the proper colour of the spot test would be given by 0·25 ml of a standard solution containing 0·0238 g $CoSO_4.7H_2O$ in 1 litre.

A. Electrodeposition after Isolation of Cobalt by 1-Nitroso-2-Naphthol or Potassium Nitrite

If the procedure outlined earlier in this chapter has been followed, the ignition of a cobaltinitroso-2-naphthol precipitate will yield Co_3O_4. With high-cobalt samples, this precipitate is bulky and consequently difficult to wash and to ignite in a platinum crucible; cobalt is preferably determined electrolytically rather than by weighing as oxide.

Follow the successive steps of decomposition, removal of the hydrogen sulphide group, separation with zinc oxide, and precipitation of cobalt by 1-nitroso-2-naphthol outlined earlier in this chapter. Filter the voluminous precipitate on a 12·5 or 15 cm Whatman No. 40 or 42 paper, with pulp, wash five times with hot water, five times with hot 4% hydrochloric acid, and five times with hot water. Place the paper and precipitate in a large platinum dish, dry, ignite slowly in a muffle and finally at a temperature of 800°C until all organic matter is destroyed.

Transfer the cobalt oxide from the platinum dish to a beaker, dissolve in hydrochloric acid or aqua regia, add 10 ml of 1 : 1 sulphuric acid, and evaporate to strong fumes of the latter. Cool, add water, warm until solution of salts is complete, neutralize with ammonium hydroxide and add 40 ml excess. If the sample has fumed to dryness or to a low volume, add more sulphuric acid, or add ammonium sulphate, to ensure the presence of the latter before making ammoniacal.

Electrolyse overnight on a stationary assembly at 0·5–1 ampere, or on a stirring type electrolytic apparatus at 2 amperes, and 6 volts, for 45–60 minutes. Check for complete deposition of cobalt by one of the spot tests discussed earlier. Without interrupting the current, wash the cathode quickly but thoroughly with water, switch off the current, immediately remove the cathode and immerse it successively in three beakers of alcohol. Dry for a few seconds over a hot plate, in an oven at 100°C, or in a stream of warm air from a hair-drier, and weigh the previously-tared cathode.

When cobalt is isolated from nickel and other elements by potassium nitrite, for high-cobalt samples the determination is also preferably completed electrolytically. Follow the procedures of decomposition, precipitation of potassium cobaltinitrite, dissolution of the latter in acids, and re-precipitation of potassium cobaltinitrite outlined earlier in this chapter. Collect the second precipitate on a Whatman No. 40 paper instead of a Gooch or sintered glass crucible, and wash with 100 ml of 2% potassium nitrite solution. Transfer paper and precipitate to a beaker, add 15–30 ml nitric acid, 30 ml 1 : 1 sulphuric acid, and evaporate to strong fumes of the latter. Destroy all organic matter by adding more nitric acid, and re-fuming. Cool, add water, boil to solution of salts, neutralize with ammonium hydroxide, and add 40 ml excess. Be sure the solution contains ammonium sulphate; if sulphuric acid has evaporated to a low volume, add a little more acid, or ammonium sulphate, before making ammoniacal. Proceed with the electrolysis of cobalt as outlined above.

B. Co-deposition of Cobalt and Nickel

For some high-cobalt materials it may be convenient to electrolytically determine cobalt and nickel together as a combined deposit. After dissolving the cathode deposit in nitric acid, nickel can be isolated by dimethylglyoxime and determined electrolytically, gravimetrically, or colorimetrically, depending on its content. Alternatively, cobalt may be isolated from the dissolved plating by 1-nitroso-2-naphthol or potassium nitrite, and determined electrolytically or gravimetrically, for high and medium contents, respectively.

Decompose the sample with hydrochloric, nitric, and perchloric acids in a beaker, or with nitric and hydrofluoric acids in a platinum dish, or by any other suitable technique to bring all cobalt into solution. Finally, evaporate the sample in a flask to fumes of perchloric acid, cool, add hydrochloric acid, and volatilize chromium by several evaporations with the mixture of perchloric and hydrochloric acids, swirling the flask over a flame until no more chromyl chloride is evolved. Add sulphuric acid and evaporate to dense fumes of this acid. Cool, add water, boil, filter, and wash. Adjust the acidity to 5–10% sulphuric acid, precipitate copper, molybdenum, and other Group 2 metals with hydrogen sulphide, filter, and wash with acidulated hydrogen sulphide water.

Remove hydrogen sulphide by boiling, oxidize iron by addition of hydrogen peroxide, evaporate to a volume of about 50 ml, cool to 10°C, and transfer to a separatory funnel. To the cold solution containing 10% by volume of sulphuric acid, add sufficient cold 6% aqueous solution of cupferron to precipitate all the iron, titanium, vanadium, and zirconium. An excess of cupferron is indicated by the formation of a temporary, fine, white precipitate which re-dissolves. Shake the funnel several times to mix the contents, add 10 ml chloroform, and shake for 1 minute. Allow to settle, draw off and discard the lower chloroform layer which contains the water-insoluble cupferrates of iron, titanium, vanadium, and zirconium. Add another 5–10 ml chloroform, and repeat the extraction until the lower chloroform layer remains colourless after shaking the funnel for 1 minute.

Transfer the upper aqueous layer to a beaker, and heat cautiously on the side of the hot plate until the traces of chloroform have evaporated. Add

10 ml nitric acid and evaporate to strong fumes of sulphuric acid to eliminate all organic matter; if necessary, add more nitric acid and re-fume. The sample will now contain only one element which interferes in the electrolytic determination — aluminium. If the latter is present, add to the sulphuric acid solution of the sample a slight excess of sodium hydroxide, boil, filter, wash several times with hot 1% sodium hydroxide, and discard the filtrate. Dissolve the precipitate in dilute sulphuric acid, reprecipitate with sodium hydroxide, and wash as before. Dissolve the precipitate in dilute sulphuric acid, neutralize the solution with ammonium hydroxide, and add 40 ml in excess. Electrolyse the solution as described earlier, and weigh as cobalt + nickel. Dissolve the cathode deposit in nitric acid and determine cobalt, or nickel, by procedures appropriate to the quantity present.

The method described above can be shortened when the sample contains only a few constituents in addition to cobalt and nickel. Some alloys, for instance, have only iron and chromium as interferences. Volatilize the chromium as chromyl chloride by several evaporations with a mixture of perchloric and hydrochloric acids, and remove the iron by an ether extraction in hydrochloric acid of s.g. 1·10. Transfer the lower aqueous layer to a beaker, place on the side of the hot plate, and carefully evaporate the traces of ether. Add nitric and sulphuric acids, and evaporate to strong fumes of the latter. Make ammoniacal and electrolyse as already described.

Other materials may have only iron and aluminium as interferences. These may be conveniently removed by a phosphate separation as described earlier. Evaporate the filtrate to a low volume, add 10 ml 1 : 1 sulphuric acid, neutralize with ammonium hydroxide and add 40 ml excess. Electrolyse at about 50°C at 3 amperes on a stirring assembly for 45 minutes or until deposition is complete. The warm electrolyte and higher current density appear to give better results when electrolysis is carried out directly on the filtrate from the phosphate separation of iron and aluminium. On other samples, electrolysis at room temperature, and about 2 amperes for stirring equipment, is preferable.

Removal of interfering elements, prior to electrodeposition, by ion exchange columns of Dowex 1 or 1-X10 has been described [44].

4. Volumetric Determination

Unlike many elements, cobalt is rarely determined in complex materials by a volumetric procedure. For the determination of cobalt in its compounds and salts, however, where impurities are often sufficiently low to render separations unnecessary in technical analyses, volumetric methods are very useful.

A. With Potassium Cyanide

In the absence of nickel, cobalt may be determined by cyanide titration in a manner similar to the procedure used for nickel [44]. When bivalent cobalt is converted to the cobaltic state by an oxidizing agent like potassium chlorate, one atom of cobalt reacts quantitatively with five molecules of potassium cyanide to form a complex cobalt cyanide. The endpoint of the reaction is determined by an indicator, silver iodide, which gives a turbidity to the solution until a slight excess of cyanide is present. Silver iodide is formed *in situ* by the action of silver nitrate on potassium iodide.

$$6Co(NH_3)_6Cl_2 + KClO_3 + 6NH_4Cl + 3H_2O \rightarrow$$
$$\rightarrow 6Co(NH_3)_6Cl_3 + KCl + 6NH_4OH$$
$$CoCl_3 + 5KCN \rightarrow K_2Co(CN)_5 + 3KCl$$

or, as sometimes represented:

$$Co_2(SO_4)_3 + 10KCN \rightarrow 2[2KCN.Co(CN)_3] + 3K_2SO_4$$
$$AgNO_3 + KI \rightarrow AgI + KNO_3$$
$$AgI + 2KCN \rightarrow KAg(CN)_2 + KI$$

Because it is generally easier to detect the appearance of a turbidity than its disappearance, a slight excess of potassium cyanide is usually added and back-titrated with silver nitrate until a faint opalescence is again visible.

Decompose the sample with nitric and hydrochloric acids, add 10 ml 1 : 1 sulphuric acid and fume strongly to dehydrate silica. If the insoluble residue could contain a small quantity of cobalt it should be fused with sodium carbonate and added to the main portion of the sample. Dissolve the soluble salts in water, remove the members of the acid sulphide group in 5–10 % acid with hydrogen sulphide, and filter off the silica with the precipitated sulphides. Copper is often the only member of the preceding groups present; if so, it may frequently be conveniently determined and eliminated by the usual electrolysis in nitric–sulphuric solution in lieu of a sulphide separation. Boil out all hydrogen sulphide; oxidize thoroughly by adding 10 ml nitric acid, 2 g potassium or sodium chlorate, and boiling gently for 10 minutes. Cool in a water bath, add ammonium hydroxide just to the neutral point, and for each 0·1 g of iron present add 10–20 ml ammonium citrate–sulphate buffer solution. The latter consists of 200 g citric acid, 270 g ammonium sulphate, and 200 ml ammonia per litre of solution.

Place a piece of red litmus paper against the side of the beaker in the solution and carefully add ammonium hydroxide until the paper turns blue. Add 2 ml ammonium hydroxide and boil gently for several minutes. Cool in a water bath, add 5 ml of 10% potassium iodide, and 2–3 ml standard silver nitrate solution from a burette. Add standard potassium cyanide from a burette until the cloudiness produced by silver iodide disappears. Finally, add standard silver nitrate solution, drop by drop, stirring constantly over a black background until a persistent cloudiness is again produced. The total number of ml of standard silver nitrate used is then subtracted from the number of ml of standard potassium cyanide to give the percentage of cobalt. After adding silver nitrate to the samples, keep them out of direct sunlight; the latter has a tendency to reduce silver iodide to metallic silver, giving erratic results.

Standardize the cyanide solution against the silver nitrate, and titrate the cyanide against a solution containing a known quantity of cobalt. It is useful to have two standard potassium cyanide solutions for routine work, one for low cobalt samples with a value approximately 1 ml = 0·001 g Co, and another for high cobalts possessing a titre of 1 ml = 0·005 g Co. These solutions are made up to contain 5·6 g KCN and 28 g KCN per litre, with the addition of 1 and 5 g, respectively, of potassium hydroxide as a stabilizer. The cyanide solutions are titrated against silver nitrate solutions containing 7·206 g and 36·03 g silver nitrate per litre, respectively. One gram of $CoSO_4.7H_2O$ theoretically contains 0·2097 g cobalt, and a 25 ml aliquot from a litre of solution containing 1 g of this salt gives a convenient quantity for titration against dilute potassium cyanide. These solutions should be standardized at least once a week, and the cobalt sulphate solution should be checked electrolytically. The following equivalents are useful:

$$\begin{aligned} 1\text{ g Co} &= 5{\cdot}5234\text{ g KCN} \\ &= 7{\cdot}2060\text{ g AgNO}_3 \\ 1\text{ g KCN} &= 1{\cdot}3046\text{ g AgNO}_3 \end{aligned}$$

With the exception of copper and, of course, nickel, very few elements interfere with the cyanide titration of cobalt. Mercury interferes, but is eliminated together with copper in the hydrogen sulphide separation. Zinc forms a compound with potassium cyanide, leading to high results; increasing the ammonium citrate decreases the interference of zinc for quantities of the latter below 5 mg. Accurate results for cobalt are only obtained in the presence of manganese if not more than 5 mg of the latter are present, if there is a fair amount of iron in the sample and if the manganese is not oxidized beyond the manganic state. The following are without effect: Fe, Al, Pb, Bi, Cd, As(ic), Sb, Sn, Se, Te, Mo, Cr(ate), Ti, U, V(ate), Zr, and W.

Sometimes cobalt is determined indirectly from a cyanide titration of combined nickel + cobalt by subtracting the quantity of nickel found in a separate analysis using dimethylglyoxime. In this calculation, because 1 atom of nickel reacts with 4 molecules of potassium cyanide and 1 atom of cobalt with 5 molecules of cyanide, percentage cobalt = 80% (% Ni + Co by KCN − % Ni by dimethylglyoxime).

B. With EDTA

The procedures for titrating cobalt with a solution of ethylene-diamine-tetra-acetate, EDTA, will be listed under the various indicators used to detect the end point.

(1) *Eriochrome Black T.* This indicator cannot be used for a direct titration of cobalt; an excess of EDTA must be added and the excess determined in ammoniacal solution by a standard zinc solution. For cobalt compounds or salts containing only small quantities of contaminants, the following procedure will prove satisfactory.

Transfer an aliquot containing not more than about 3 mg of cobalt to a small beaker, add an excess of 0·02 M EDTA, and adjust the *p*H to 10. This may be done by the dropwise addition of 1 : 1 ammonium hydroxide until red litmus paper in the solution turns blue, followed by the addition of 5 ml of an ammonium hydroxide–ammonium chloride buffer. The latter contains 13·5 g of ammonium chloride and 114 ml of ammonium hydroxide in a volume of 200 ml.

Add a small quantity of Eriochrome Black T which has been prepared by grinding 0·25 g of the reagent with 50 g of sodium chloride. A few drops of a 0·2% aqueous solution of Eriochrome Black T may be used in place of the solid indicator; the solution will remain stable for many weeks if stored in a refrigerator. The solution will now be blue. Titrate carefully with 0·02 M zinc solution to obtain the excess EDTA; the colour at the end point will change from blue to red. A standard 0·02 M zinc solution is made by dissolving 1·3074 g of "five-nines" zinc in hydrochloric acid and diluting to 1 litre; 1 ml contains 1·3074 mg of zinc. The standard 0·02 M EDTA solution is made by dissolving 7·556 g EDTA in water and diluting to 1 litre. This will give a solution which has the approximate value of 1 ml = 1·3074 mg Zn, but it must be standardized against the zinc solution described above. A solution of 0·02 M EDTA has the value of 1 ml = 1·1786 mg cobalt.

(2) *Xylenol Orange.* For fairly pure cobalt compounds and salts, the method given below has proved suitable. In the presence of copper, the addition of ascorbic acid and potassium iodide reduces and precipitates copper as the iodide.

To an aliquot containing not more than about 3 mg cobalt in a small beaker add 1 : 1 ammonium hydroxide until a piece of red litmus turns blue, followed by the addition of 10 ml of an acetate buffer. The latter consists of 6·8 g sodium acetate trihydrate and 3 ml acetic acid in a solution of 500 ml; the *p*H should be checked on a meter and adjusted to 5 if necessary.

Add 3 drops of 0·1% aqueous solution of xylenol orange, and titrate with standard 0·02 M EDTA solution until the colour changes to a lemon-yellow. The solution alters in colour from the original purple to an amber and finally to a lemon-yellow; the latter change is sharp and permanent. Standardize the EDTA solution against "five-nines" zinc as described in (1) above. One ml of 0·02 M EDTA = 1·1786 mg cobalt.

(3) *Zincon.* The method, using this indicator, is similar to (1) above.

(4) *PAN.* In a weak acetic acid solution, cobalt has been titrated, directly with EDTA, or indirectly to obtain the excess EDTA by a standard copper solution, using 1-(2-pyridylazo)-2-naphthol, PAN, as indicator [12, 29].

(5) *Naphthylazoxine.* In weakly acid solution this indicator has been used for the direct titration of cobalt with EDTA.

(6) *Pyrocatechol Violet.* In a buffered alkaline solution of cobalt, pyrocatechol violet has been used to indicate the end point in direct and indirect titrations with EDTA.

(7) *Calcein.* In strong acid, titration of excess EDTA with standard

calcium solution, using calcein indicator, has given nickel + cobalt. After oxidation of cobalt to form the Co(III)–EDTA complex, and masking with potassium cyanide, the liberated EDTA corresponding to the quantity of nickel present is titrated with more standard calcium solution. Cobalt is found by difference.

5. Colorimetric Determination

Many colorimetric procedures for small quantities of cobalt have been proposed, but for general application and reliability a choice can usually be made from ammonium thiocyanate, nitroso-R-salt, or 2-nitroso-1-naphthol.

A. Ammonium Thiocyanate

Cobalt forms a complex with ammonium thiocyanate which can be extracted with a mixture of amyl alcohol and ether to give a beautiful blue colour of ammonium cobaltothiocyanate, a sensitive test for cobalt long known by the designation "Vogel's reaction". The basic reaction is

$$CoCl_2 + 4NH_4CNS \rightarrow (NH_4)_2[Co(CNS)_4] + 2NH_4Cl$$

This test has been adapted for quantitative colorimetry, its reliability and sensitivity making it useful for a wide range of low-cobalt materials [44, 45].

Decompose a 0·5–2 g sample with nitric and hydrochloric acids, adding a few drops of bromine initially for sulphide materials, and hydrofluoric acid for silicates. If the residue which is insoluble in aqua regia is likely to contain cobalt, it must be fused with sodium carbonate and added to the main portion. Evaporate the sample to dryness but do not bake; traces of nitric acid do not affect the formation of the thiocyanate complex. Add 1–2 ml hydrochloric acid, 5–10 ml hot water, and heat to solution of salts. Keep the volume as low as possible in order that a concentration of 25% ammonium thiocyanate may be maintained subsequently in a small separatory funnel.

Cool, transfer the sample to a separatory funnel, with minimum water rinses. Add with swirling 5 ml of 10% hydroxylamine hydrochloride, 3 ml of 10% trisodium phosphate, 4 ml of 70% ammonium acetate, and a few drops of 50% tartaric acid solution. Add 10 ml of an ammonium thiocyanate solution containing 1 g per ml, and again swirl. Now add 10 ml of a solution containing 3 parts amyl alcohol to 1 part diethyl ether, and shake the separatory funnel vigorously for 15 seconds. Retain the upper blue layer in the original funnel, draw off the lower aqueous layer into a second separatory funnel, add 10 ml amyl alcohol–ether, and shake for 15 seconds.

Transfer the pale blue upper layer of the second extraction to the original upper layer; if necessary, make another extraction of cobalt with amyl alcohol–ether from the aqueous layer of the second extraction and combine this with the previous cobalt-containing layers.

Transfer the blue cobalt complex from the combined alcohol–ether extractions to a dry 50 ml volumetric flask, rinsing the separatory funnel twice with small quantities of alcohol–ether. Make up to the mark with alcohol–ether; mix thoroughly by inverting the flask a dozen times. Measure the absorbance of the solution in a spectrophotometer at 612 nm. The 50 ml volumetric flask may contain up to 2 mg cobalt, a straight-line relation between absorbance reading and mg Co per 50 ml being obtained from 0·05 to 2 mg of cobalt.

The standard graph can be prepared by carrying through all steps of the procedure successive increments from 0·05 to 2 mg of cobalt. A convenient solution contains 0·0200 g cobalt, or 0·0808 g $CoCl_2.6H_2O$ in 1 litre; one ml contains 0·02 mg Co.

Using the above procedure, the following elements are without interference in quantities at least 500 times that of the cobalt: Fe, Al, Cu, Ni, Cr, Mn, Be, Cd, Pb, Hg, Ag, Tl, Th, Sb, Se, Te, U, Zr, Mo, As, Bi, Sn, Ti, Zn, Nb, Ta, W, Mg, alkaline earths, and alkalies. Gold is precipitated as metal and can be removed by filtration; platinum and palladium impart a reddish-brown colour to the cobalt layer and their prior removal is necessary.

B. Nitroso-R-salt

This determination is based on the fact that the coloured complexes formed with nitroso-R-salt, $C_{10}H_4OH.NO(SO_3Na)_2$, by most of the common elements except cobalt are destroyed by hot nitric acid. The full development of the colour with cobalt is attained in the presence of sodium acetate, and for most samples, such variables as quantity of nitric acid and of nitroso-R-salt, boiling time, etc. exert an influence. This procedure is deservedly one of the most popular for small quantities of cobalt, being applicable to a wide variety of materials [1, 2, 4, 5, 6, 10, 11, 18, 19, 21, 30, 32, 34, 40, 44, 47].

Decompose the sample with hydrochloric and nitric acids, add 1 : 1 sulphuric acid and evaporate to strong fumes of the latter. If an insoluble residue remains, which might contain cobalt, fuse with sodium carbonate and add to the main sample. Other methods of decomposition may be used if convenient or necessary, such as aqua regia followed by perchloric acid, a sodium peroxide fusion, hydrofluoric and sulphuric or nitric acids in a platinum dish, etc.

Adjust the acidity to 5–10% sulphuric acid and remove members of the acid sulphide group with hydrogen sulphide. If molybdenum or the platinum metals are present, take the usual precautions of prolonged gassing in hot solutions, etc. to ensure their complete precipitation. The nitroso-R-salt procedure can tolerate a little molybdenum, but platinum and palladium must be absent. Filter, wash with acidulated hydrogen sulphide water, and discard the precipitate. Boil the filtrate to remove all hydrogen sulphide, and evaporate to a low volume.

Neutralize the solution with 20% sodium hydroxide and add 2 ml Spekker acid, which contains 150 ml sulphuric acid and 150 ml phosphoric acid per litre. Add 10 ml of a 0·2% aqueous solution of nitroso-R-salt, and 10 ml of a 50% sodium acetate solution. If cobalt is present the solution will turn from the yellow–green imparted by nitroso-R-salt to a red colour.

Place on the hot plate and boil for at least one minute, but avoid prolonged boiling. Add 5 ml nitric acid to destroy the colour which a few elements such as iron give with nitroso-R-salt. Boil for at least 1 minute but not more than 2 minutes. Cool, transfer to a 100-ml volumetric flask, and make to volume. Measure the absorbance in a spectrophotometer at 520 nm.

Prepare a standard graph covering the range 0·01–0·10 mg cobalt per 100 ml. For this, a solution containing 0·0200 g cobalt or 0·0808 g $CoCl_2.6H_2O$ per litre may be used; 1 ml = 0·02 mg cobalt. The maximum quantity of cobalt in the 100-ml flask should not exceed 0·5 mg; above this concentration a deviation from Beer's Law occurs.

For extremely low cobalt contents, the greater sensitivity at 420 nm may be utilized; interferences, however, are more pronounced at this wavelength. The standard graph at 420 nm can conveniently cover the range 0·002–0·02 mg cobalt per 100 ml. For most samples, measurement at 520 nm is preferable.

When small quantities of copper are present, measurements may be carried out at 520 nm if 30 ml sodium acetate and 10 ml nitric acid are used in place of the usual 10 and 5 ml quantities, respectively, of these reagents. By this means 0·1 mg of cobalt can be measured in the presence of 10 mg of copper without the need to separate the latter. The increased addition of sodium acetate and nitric acid is effective only for copper; other interferences discussed later are not overcome by this means.

If all acid sulphide group metals are removed by hydrogen sulphide, the only serious interference is thallium. The latter is rarely encountered, but can be removed by an ether separation or by potassium chromate in an ammoniacal solution containing sulphosalicylic acid [11, 17]. The following can be tolerated: iron in quantities at least 1000 times that of cobalt; at least 100 times as much Al, Zn, Ti, Th, U, Zr, W, Nb, Ta, Ba, Sr, Ca, and Mg; 50 times as much Mn; 25 times as much Ni, V, and Be; and 15 times as much Cr.

When the sample contains large quantities of chromium, manganese, nickel, vanadium, and other elements which can only be tolerated in moderate amounts, the following procedure, though lengthy, is thoroughly reliable. Remove all interfering elements except nickel, manganese, and zinc by hydrogen sulphide and zinc oxide separations, as described earlier in this chapter. Precipitate cobalt by addition of 1-nitroso-2-naphthol, adding a little iron as a collector if the cobalt is very low, and allow to stand several hours. Extract the cobaltinitrosonaphtholate and excess 1-nitroso-2-naphthol by shaking with successive portions of chloroform in a separatory funnel until the lower layer remains colourless. Discard the upper aqueous portions. Carefully evaporate the chloroform from the combined lower layers on the side of the hot plate, add nitric and sulphuric acids, and evaporate to fumes of the latter. Add more nitric acid, if necessary, and re-fume to destroy all organic matter. Proceed with the nitroso-R-salt procedure as described earlier.

Many other methods for isolating a small quantity of cobalt from large amounts of other elements, prior to final determination by nitroso-R-salt, can be employed [1, 2, 4, 5, 6, 27, 44]. These include ether separation of iron, volatilization of chromyl chloride, cupferron–chloroform extraction, ion exchange, initial extraction with dithizone, removal of nickel by dimethylglyoxime or sodium diethyldithiocarbamate, and other procedures listed earlier under Separation.

When nickel, manganese, and copper are present, the quantity of nitroso-R-salt must be increased at the rate of about 1·5 ml for each 5 mg of nickel or manganese, and 3 ml for each 5 mg of copper. Ten ml of 0·2% nitroso-R-salt are sufficient for only 0·6 mg of cobalt alone; in some stainless steels where a 0·5 g sample may contain 40 mg nickel, 6 mg manganese, and 1·25 mg copper in addition to 0·15 mg cobalt, the volume of nitroso-R-salt should be more than doubled.

If substantial quantities of titanium and thorium are present they may hydrolyse and precipitate in the solution after the colour has been developed; filtration through a dry filter paper at this stage will remove the turbidity.

Occasionally it is useful to know the behaviour of hydrogen sulphide group elements with nitroso-R-salt. Though usually removed as a group with hydrogen sulphide, some members in normal concentrations are without effect on the determination. The following do not interfere in concentrations 100 times that of the cobalt: As, Cd, Pb, Mo, Se, Te, Ag, and Sn. Antimony, gold, and mercury can be tolerated if they are present in quantities equal to cobalt; bismuth, palladium, and platinum must be absent.

C. 2-Nitroso-1-Naphthol

This reagent, known also as β-nitroso-α-naphthol, forms a complex with cobalt which can be extracted with carbon tetrachloride, chloroform, isoamyl acetate, benzene, or toluene. The cobalt complex is stable towards moderate concentrations of acids and alkalies, unlike most other metal derivatives of 2-nitroso-1-naphthol. Washing with dilute hydrochloric acid and sodium hydroxide removes the interferences and the excess reagent [23, 25, 44].

For rock or soil, decompose the sample with acids and fuse any insoluble portion with sodium carbonate. Transfer an aliquot representing 1–10 μg of cobalt to a separatory funnel. Add 0·5 ml bromine water, 10 ml of 20% ammonium citrate, and 1 ml of 10% sodium thiosulphate solution. Add 1 drop of phenolphthalein, and then ammonium hydroxide to a pink colour. Add 2 ml of 0·04% 2-nitroso-1-naphthol; the latter is made by dissolving in water faintly alkaline with sodium hydroxide. Add 5 ml isoamyl acetate, shake for 1 minute, and allow to stand for 1 hour. The upper acetate layer is usually amber to red; the lower aqueous phase is yellow. Discard the lower layer, wash the acetate layer three times with 5 ml of 1 N hydrochloric acid, shaking for 1 minute and allowing to stand for 2 minutes each time. Discard the aqueous layers, and wash twice in the same manner with 5 ml of 1 N sodium hydroxide. Finally wash twice with 1 N hydrochloric acid. Discard the aqueous layers, and measure the absorbance of the acetate layer at 530 nm. Determine the unknown by reference to a standard graph derived from known quantities of cobalt carried through all steps of the procedure.

For cobalt in stainless steel, dissolve 0·5 g in a platinum dish with perchloric and hydrofluoric acids. Evaporate to strong fumes of perchloric acid, cool, transfer to a flask, and evaporate to fumes of perchloric acid. Add hydrochloric acid and expel all chromium as chromyl chloride by several evaporations with the mixture of perchloric and hydrochloric acids, swirling the flask over a burner. Cool, add water, and heat to solution of salts. Add sufficient ammonium citrate to complex the iron, adjust the *p*H to 3–4, and add hydrogen peroxide to oxidize all iron to the ferric state. To the sample solution in a separatory funnel add an acetic acid solution of 2-nitroso-1-naphthol, and chloroform, and shake to extract the cobalt complex into the lower chloroform layer. Transfer the latter to another separatory funnel, and wash by shaking with a small quantity of 2N hydrochloric acid to remove interfering ions. Wash by shaking with another small quantity of 2N sodium hydroxide to remove excess reagent. Measure the absorbance of the cobalt complex at 530 nm.

D. Other Colorimetric Methods

Traces of cobalt may also be determined as tetraphenylarsonium cobaltothiocyanate, or by reaction with 2-nitroso-1-naphthol-4-sulphonic acid or 1-(2-pyridylazo)-2-naphthol [19, 44].

6. Potentiometric Determination

An excellent procedure for cobalt in a wide range of products is the potentiometric one, based on the oxidation of cobalt in ammoniacal solution by potassium ferricyanide. In the cold, and the presence of ammonium citrate, the only common element that interferes is manganese. The latter can be removed as dioxide by prior strongly oxidizing treatment, or cobalt can be separated as a precipitate by phenylthiohydantoic acid. The potentiometric procedure is applicable in the presence of large quantities of copper, nickel, and iron — the elements which interfere most frequently in other methods of determining cobalt. The details given below are applicable to most mining and metallurgical products [44], and can be modified slightly for irons and steels. The method has given good results for cobalt in cemented carbides, and in glass-to-metal seals [11, 19, 40, 44].

Decompose the sample with nitric and hydrochloric acids; if an insoluble residue remains, fuse it with sodium carbonate, and add to the main sample. Boil the acid solution thoroughly to drive off nitrous fumes, but it is not usually necessary to evaporate to low volume. If manganese is present, evaporate the mixed acid solution to dryness, add 40–50 ml nitric acid, and carefully add several successive small portions of sodium chlorate, boiling gently after each addition. Dilute slightly, and filter off the precipitated manganese dioxide through a Whatman 42 paper or a Gooch or sintered glass crucible. In the absence of manganese, insoluble matter need not be filtered off unless a large quantity is present.

Cool the sample, and carefully add ammonium hydroxide until iron commences to precipitate. Re-dissolve the iron by the addition of a drop or two of nitric acid, and again cool the sample. From a burette into a 600 ml beaker, measure out sufficient excess potassium ferricyanide solution such that it will react with all the cobalt and require a back-titration of about 10 ml of cobalt nitrate solution. Add to the ferricyanide solution 10–20 ml ammonium citrate solution for every 0·1 g iron present in the sample, followed by 80 ml ammonium hydroxide. Stirring constantly, slowly pour the sample into the beaker containing potassium ferricyanide, etc. There must be no heating effect produced at this stage, otherwise some of the cobalt will be oxidized, giving low results. If the sample has been almost neutralized and cooled beforehand, no heat will develop.

Place the beaker containing the sample on the potentiometric assembly and switch on the current and stirrer. Back-titrate the excess potassium ferricyanide with standard cobalt nitrate solution until the end point is shown by a large and permanent deflection of the galvanometer beam.

If chromium and vanadium are present, decompose the sample with nitric, hydrochloric, and perchloric acids, and take to a low volume of fuming perchloric acid. Pour the sample directly into the beaker containing excess ferricyanide, ammonium citrate, and 80 ml ammonia, without preliminary neutralization. Chromium and vanadium are in their highest valance states and will not interfere with the titration.

Cobalt can be isolated from manganese, chromium, and vanadium by precipitation with phenylthiohydantoic acid. The precipitate is not formula-pure, and cannot be weighed directly [3]. Dissolve the sample, containing about 25 mg of cobalt, in nitric and hydrochloric acids, dilute to 100 ml, and add 30 ml of 50% citric acid solution. Make ammoniacal, dilute with water to about 300 ml, and add 2 g phenylthiohydantoic acid dissolved in 60 ml of a hot 1 : 1 mixture of ethanol and water. Boil for 1 minute, digest hot for 5 minutes, filter through Whatman 541 paper, and wash 6 times with hot water. Transfer the paper and precipitate to the original beaker, add 20 ml nitric acid, 10 ml 1 : 1 sulphuric acid, and evaporate to fumes. Add more nitric acid if necessary until the solution is free of organic matter. Cool, add water, warm to solution of salts, neutralize with ammonium hydroxide, and make just acid with sulphuric. Cool, and proceed with the addition of ammonium citrate, potassium ferricyanide, etc. described above.

Reagents

Ammonium citrate solution. Dissolve 200 g citric acid in water and slowly add 270 ml ammonium hydroxide. Cool and make up to 1 litre. Tartaric acid may be substituted for citric acid.

Cobalt nitrate solution. Weigh out 5·00 g $Co(NO_3)_2.6H_2O$, dissolve in water and make up to 1 litre. One ml of this solution should contain 1 mg Co. Standardize by determining cobalt electrolytically.

Potassium ferricyanide solution. Weigh out 11·17 g $K_3Fe(CN)_6$, dissolve in water and make up to 1 litre. Keep the solution in a brown bottle; it is decomposed by light. One ml of this solution should be equivalent to 2 mg Co. Standardize by measuring 10 ml ferricyanide from a burette into a 600 ml beaker, adding 20 ml ammonium citrate, 80 ml ammonium hydroxide, diluting with water to about 400 ml and titrating potentiometrically with the standard cobalt nitrate solution.

It has been suggested that in the potentiometric determination of cobalt, the reactants be covered with a layer of light petroleum to exclude air and to shield the operator and equipment from the strongly ammoniacal vapours.

7. Polarographic Determination

The polarograph has furnished another useful tool for the determination of small concentrations of cobalt, particularly in the presence of large amounts of nickel and only minor quantities of most other metals. Polarographic determinations of cobalt have been the subject of a number of papers [7, 19, 44]. Iron can be separated from cobalt by hydrous oxide precipitation in pyridine. If cobalt is first oxidized to the trivalent state, it may be determined polarographically in the presence of a large quantity of nickel because the reduction wave of cobalt (III) to cobalt (II) precedes that of nickel (II). The oxidation of cobalt is usually carried out by three methods.

A. Sodium Perborate

In ammoniacal solution, cobalt (II) may be rapidly and quantitatively oxidized to an amine complex of cobalt (III) by boiling under a reflux condenser with sodium perborate. Manganese, which is also oxidized, co-precipitates cobalt; in its presence, cobalt should be separated by 1-nitroso-2-naphthol. Chromium may be separated by barium perchlorate, and a small amount of copper by zinc amalgam.

B. Lead Dioxide

Another procedure is based on the reduction of trioxalatocobaltate (III) to dioxalatocobaltate (II). Potassium oxalate is added to a cobalt solution, followed by lead dioxide, and filtration. Copper, iron, and nickel do not interfere, but calcium, cerium, chromium, manganese, and vanadium do.

C. Potassium Permanganate

In an ammoniacal ammonium chloride solution, cobalt (II) is oxidized by an excess of potassium permanganate. The latter is destroyed by addition of an excess of hydroxyammonium sulphate; this does not reduce the cobalt complex at room temperature. In this procedure, unlike A and B above, refluxing or filtering to eliminate the excess oxidizing agent is not required. Chromium, lead, tungsten, and vanadium interfere; for 1 mg cobalt the following can be tolerated, in mg: bismuth 100, copper 25, iron 50, manganese 30, and molybdenum 50.

8. Optical Spectrography

Cobalt is easily excited by conventional arc or spark sources, and its spectrum has a number of useful emission lines. The principal cobalt lines used in quantitative spectrographic work are 2286·16, 2378·6, 3044·01, 3061·82, 3333·4, 3453·51, 3465·8, and 3529·81. Small quantities can be determined in aluminium alloys, biological materials, nickel alloys and salts, irons and low-alloy steels, zirconium alloys, stainless steels, rocks, minerals, ores, and concentrates [1, 2, 19, 21, 24, 26, 28, 36, 44].

For some materials, the determination may be made directly on the sample, its solution, or evaporated salts. On other products, particularly if the cobalt is very low and the sample is complex, a concentration step is required. The latter may take one of many forms: collection of cobalt with 1-nitroso-2-naphthol or dithizone, removal of iron by an ether separation, elimination of several metals by ion exchange, etc.

9. X-Ray Spectrography

Cobalt can be determined by X-ray fluorescence from about 0·1 to 100%. The spectral lines used are 1·621 and 1·790 Å. These cobalt lines are close to nickel and iron lines; in addition, chromium and molybdenum have a strong absorption. In complex steels, therefore, it is difficult to determine cobalt by X-ray spectrography, but if the alloying elements are few this rapid procedure is very satisfactory over a wide range [19, 20, 41, 44].

10. Atomic-Absorption Spectrophotometry

Atomic absorption can be used to determine cobalt in a variety of materials. The most sensitive line, 2407·3, has a sensitivity of 0·2 μg/ml. Other lines may be used: 2424·9, 2521·4, 3453·5, and 3526·9 [8, 9, 10, 13, 15, 31, 33, 37]. For very low amounts of cobalt, a concentration step may precede the atomic absorption; examples are collection of cobalt by co-precipitation with manganese dioxide in strong sodium hydroxide solution [13], and concentration on chelating resins [31].

SPECIAL PROCEDURES

The diverse processes now employed in the extractive metallurgy of cobalt have made it necessary, at times, to determine metal, sulphide, or oxides when these forms occur together.

1. Metallic Cobalt

A. With Mercuric Chloride Solution

The following procedure has been very useful in the study of cobalt mattes and slags at Rhokana Corporation. The results of these determinations showed close agreement with the volume percentages of metallic cobalt or cobalt-iron alloy estimated by careful microscopic studies [35].

To 2 g of —200 mesh slag, add 50 ml of 7% mercuric chloride solution. Boil 1 minute, filter through Whatman 42 paper, and wash thoroughly with hot water. The following reaction occurs: $Co + HgCl_2 \rightarrow CoCl_2 + Hg$. Metallic cobalt passes into the filtrate as the chloride and can be determined by any suitable procedure. The 1-nitroso-2-naphthol and ammonium thiocyanate methods can be applied directly, as can the potentiometric procedure if an excess of ammonium chloride is used. Electrolytic or nitroso-R-salt determinations must be preceded by elimination of mercury with hydrogen sulphide.

The following, if present in metallic form, will be extracted along with the cobalt: Bi, Cr, Cu, Fe, Mn, and Ni. Cobalt sulphide, cobalt silicate, and cobaltosic oxide, Co_3O_4, are virtually insoluble in the mercuric chloride solution. Cobaltic oxide, Co_2O_3, is only soluble to the extent of 0·1–0·3%; cobaltous oxide, CoO, is slightly soluble.

B. With Chlorine-Alcohol Solution

A differentiation of metal and sulphide from oxide in roasted and reduced ores and concentrates, based on leaching with a chlorine–alcohol solution, is sometimes employed in the base metal industries. For cobalt, however, in the presence of sulphide, this procedure has a limited value because the time required to leach all the sulphide also results in considerable solution of the higher oxides [43]. For example, when the same particle size and experimental techniques are used with a 0·25 g sample, one ten-minute leach will dissolve 99% of the metallic cobalt, about 80% of the cobalt sulphide, 0·2% of the cobaltous oxide, CoO, and 1–3% of the higher oxides, Co_2O_3 and Co_3O_4. To attain 99% extraction of the sulphide, three ten-minute leaches are required; the solubility of the oxides then increases to 2–3% for CoO, 10–20% for Co_2O_3, and 8–11% for Co_3O_4.

In the absence of cobalt sulphide, the method separates metallic cobalt from its oxides; if the higher oxides of cobalt are not present, cobaltous oxide can be differentiated from the metal and sulphide by this means.

Weigh out a suitable quantity of —200 mesh sample, depending on the analytical procedure to be employed for cobalt; transfer to a dry 400 ml tall beaker. Add at least 10 times the sample weight of anhydrous methyl alcohol, stir, and place in a fume cupboard. Introduce a vigorous stream of gaseous chlorine from a cylinder into the dilute pulp for 10 minutes. Cobalt in metallic form, and most of the sulphide cobalt, will dissolve in the chlorine–alcohol solution, leaving the oxides virtually unattacked. Filter, and wash the residue with anhydrous methyl alcohol; carefully evaporate the filtrate to dryness and acidify with nitric acid. Make up to a definite volume and determine cobalt in a suitable aliquot.

If cobalt sulphide is present, a second or third ten-minute leach may be required to dissolve all of this compound; in this case, the procedure is suitable only when cobaltous oxide is the sole form of cobalt oxide present.

2. Oxide Cobalt in Ores and Concentrates

In the extractive metallurgy of cobalt, it is sometimes necessary to differentiate the oxidized from the sulphide state in natural forms of the element. For instance, it may be required to obtain a measure of the ratio of oxidized minerals, such as asbolite, heterogenite, sphaerocobaltite, and stainierite, to the cobalt sulphides, carrollite, linnaeite, and siegenite. The following procedure, based on the selective solvent action of dilute sulphuric or hydrochloric acids on oxidized cobalt minerals in the presence of the reducing agent sulphurous acid, is applicable to ores and concentrator products [46].

Depending on the quantity of oxide cobalt present in the material and the analytical method used, select a weight of —200 mesh sample that will enable an accurate determination of cobalt to be made on the leached portion. The ratio of leach solution to sample may vary considerably, depending on the quantity of oxide cobalt present and the other acid-consuming constituents of the ore. In general, add 15–25 ml of 10% by volume hydrochloric acid saturated with sulphur dioxide, or 5% by volume sulphuric acid saturated with sulphur dioxide, to each gram of material in a stoppered Erlenmeyer flask or covered beaker. When the initial effervescence and attack has subsided, add 0·1–0·3 ml hydrofluoric acid per gram of sample. For large samples, these quantities of hydrofluoric acid can be kept to a minimum.

Shake the flask or beaker for about 10 seconds every 10 minutes for 1 hour; allow to stand for another hour, and agitate again at intervals of 10 minutes for the third hour. Filter the sample through Whatman No. 40 paper, with pulp, and wash thoroughly with hot water. Add 10 ml 1 : 1 sulphuric acid to the filtrate, boil out sulphur dioxide, and evaporate to strong fumes of sulphuric acid. Cool, dilute, boil to solution of salts, and remove copper and other members of the acid sulphide group with hydrogen sulphide. Boil out hydrogen sulphide from the filtrate and proceed with the determination of cobalt by any standard method.

3. Differentiation of Cobalt Oxides

The proportions of cobaltous oxide and the higher oxides Co_2O_3 and Co_3O_4 in a mixture resulting from refining operations can be determined in the following manner [44]. Place 1 g of —200 mesh sample in a 250 ml Erlenmeyer flask with 20 ml water and shake gently until all particles are completely wetted. Add 30 ml glacial acetic acid and attach a reflux condenser to the flask. Boil quietly for one hour. Pour the contents of the flask on to a tared sintered glass crucible and wash well with hot water. Dry to constant weight in an oven at 105°C.

Under these conditions, cobaltous oxide is soluble in the dilute acetic acid, whereas the higher oxides are not. This treatment leaves cobalt sulphide practically intact [48], and consequently affords a procedure for separating sulphide from cobaltous oxide. When metallic cobalt is present, about 20% dissolves; its absence is essential for this differentiation.

4. Field Tests in Geochemical Prospecting

Several procedures have been described for rapid, approximate determinations under field conditions for cobalt in waters, soils, stream sediments, and vegetation, to aid the search for underground orebodies. Vegetation is ashed before solution in acid; soils and sediments are usually fused in a test tube with potassium bisulphate. Extraction of the cobalt complex of 2-nitroso-1-naphthol with carbon tetrachloride or toluene, and comparison with a set of standards, forms the basis of one procedure [38]. Another method utilizes the extraction of the blue cobalt thiocyanate complex into amyl alcohol [44].

With a potassium bisulphate fusion, only a small percentage of the cobalt is extracted from pyroxenes, amphibole, norite, and amphibolite. From micas, biotite schist and gneiss, and some soils, the cobalt extraction could be as low as 64, 42, 75, and 25% respectively [14].

Cobalt Information Bureaux

An international association of cobalt producers supports cobalt information offices at Centre d' Information du Cobalt, 35 rue des Colonies, Brussels 1, Belgium; Cobalt Information Center, Battelle Memorial Institute, 505 King Avenue, Columbus 1, Ohio, U.S.A.; Cobalt Information Centre, 7 Rolls Buildings, Fetter Lane, London, E.C.4A 1JA, Britain; Kobalt-Information, Dusseldorf, Elisabethstrasse 14, Germany.

Chemists who have special analytical problems with cobalt, as well as all other users, are invited to avail themselves of the facilities of these bureaux.

REFERENCES

1. A.O.A.C., *Official Methods of Analysis of the Association of Official Analytical Chemists*, Washington, D.C., Association of Official Analytical Chemists, 1965.
2. A.S.T.M., *Chemical Analysis of Metals. Sampling and Analysis of Metal Bearing Ores*, American Society for Testing and Materials, Philadelphia, Pa., 1969.
3. Bashar, A., and Townshend, A., *Analyst* **93**, 125–30 (1968).
4. Beeson, K. C., Kubota, J., and Lazar, V. A., *Agronomy* **9**, 1064–77 (1965).
5. Black, C. A., *Methods of Soil Analysis*, Part 2, American Society of Agronomy, Madison, Wisconsin, 1965.
6. Bradford, G. R., Pratt, P. F., Bair, F. L., and Goulben, B., *Soil Sci.* **100**, 309–18 (1965).
7. Bush, E. L., and Workman, E. J., *Analyst* **90**, 346–50 (1965).

8. Donaldson, E. M., and Rolko, V. H. E., Dept. Energy, Mines and Resources, *Tech. Bull.* TB 93, Ottawa, 1967.
9. Elwell, W. T., and Gidley, J. A. F., *Atomic-Absorption Spectrophotometry*, 2nd ed., Oxford, Pergamon, 1966.
10. Elwell, W. T., and Scholes, I. R., *Analysis of Copper and its Alloys*, Oxford, Pergamon, 1967.
11. Furman, N. H., ed., *Scott's Standard Methods of Chemical Analysis*, 6th ed., Vol. 1, Princeton, N.J., Van Nostrand, 1962.
12. Flaschka, H., and Garrett, J., *Talanta* **15**, 595–600 (1968).
13. Fleet, B., Liberty, K. V., and West, T. S., *Analyst* **93**, 701–8 (1968).
14. Harden, G., and Tooms, J. S., *Trans. Institution Mining Met.* **74**, Part 3, 129–41 (1964–5).
15. Harrison, W. W., *Anal. Chem.* **37**, 1168–9 (1965).
16. Hazan, I., and Korkisch, J., *Anal. Chim. Acta* **32**, 46–51 (1965).
17. Hillebrand, W. F., Lundell, G. E. F., Bright, H. A., and Hoffman, J. I., *Applied Inorganic Analysis*, 2nd ed., New York, Wiley, 1953.
18. Hopkin and Williams Ltd., *Organic Reagents for Metals*, Chadwell Heath, England, Vol. 1, 1958, Vol. 2, 1964.
19. Kolthoff, I. M., and Elving, P. J., *Treatise on Analytical Chemistry*, Part II, Vol. 2, 311–76, New York, Interscience Publishers, 1962.
20. Lassner, E., Puschel, R., and Schedle, H., *Talanta* **12**, 871–81 (1965).
21. Lewis, C. L., Ott, W. L., and Sine, N. M., *The Analysis of Nickel*, Oxford, Pergamon, 1966.
22. Maynes, A. D., *Anal. Chim. Acta* **12**, 211–20 (1965).
23. Marshall, T. K., Dahlby, J. W., and Waterbury, G. R., *U.S. At. Energy Comm.* LA-3124 (1964).
24. Mitchell, R. L., *The Spectrochemical Analysis of Soils, Plants and Related Materials*, Commonwealth Agricultural Bureau, Farnham Royal, England, 1964.
25. Needleman, M., *Anal. Chem.* **38**, 915–17 (1966).
26. Nichol, I., and Henderson-Hamilton, J. C., *Trans. Institution Mining Met.* **74**, 955–61 (1964–65).
27. Norwitz, G., and Gordon, H., *Anal. Chem.* **37**, 417–19 (1965).
28. Publicover, W. E., *Anal. Chem.* **37**, 1680–4 (1965).
29. Puschel, R., Lassner, E., and Illaszewicz, A., *Chemist-Analyst* **55**, 40–2 (1966).
30. Pyatnitskii, I. V., *Analytical Chemistry of Cobalt*, Moscow, Nauka, 1965.
31. Riley, J. P., and Taylor, D., *Anal. Chim. Acta* **40**, 479–85 (1968).
32. Sandell, E. B., *Colorimetric Determination of Traces of Metals*, 3rd ed., New York, Interscience Publishers, 1959.
33. Scholes, P. H., *Analyst* **93**, 197–210 (1968).
34. Snell, F. D., and Ettre, L. S., *Encyclopedia of Industrial Chemical Analysis*, Vol 9, New York, Interscience Publishers, 1970.
35. Talbot, H. L., and Hepker, H. N., *Trans. Institution Mining Met.* **59**, 147–79 (1949–50).
36. Tombu, C., *Cobalt*, No. 20, 103–110, No. 21, 185–9 (1963).
37. Ure, A. M., and Mitchell, R. L., *Spectrochimica Acta* **23B**, 79–96 (1967).
38. *U.S. Geological Survey Bull.* 1152, Washington, D.C., 1963.
39. Williams, W. J., *Talanta* **1**, 88–104 (1958).

40. WILSON, C. L., and WILSON, D. W., *Comprehensive Analytical Chemistry*, Vol. 1C, 656–71, London, Elsevier, 1962.
41. WYBENGA, F. T., *Appl. Spectry.* **19**, 193 (1965).
42. YOUNG, R. S., *Cobalt: Its Chemistry, Metallurgy and Uses*, A.C.S. Monograph 149, New York, Reinhold, 1960.
43. YOUNG, R. S., *Chemist-Analyst* **49**, 46 (1960).
44. YOUNG, R. S., *The Analytical Chemistry of Cobalt*, Oxford, Pergamon, 1966.
45. YOUNG, R. S., and HALL, A. J., *Ind. Eng. Chem. Anal. Ed.* **18**, 264–8 (1946).
46. YOUNG, R. S., HALL, A. J., and TALBOT, H. L., *Am. Inst. Mining Met. Eng. Tech. Publ.* 2050 (1946).
47. YOUNG, R. S., PINKNEY, E. T., and DICK, R., *Ind. Eng. Chem. Anal. Ed.* **18**, 474–6 (1946).
48. YOUNG, R. S., and SIMPSON, H. R., *Metallurgia* **45**, 51 (1952).

14 Copper

Copper is one of the most widely used metals, and in either the pure or alloyed form is among the commonest elements determined in most mining and metallurgical laboratories. It may be encountered in quantities varying from traces in many rocks and minerals to nearly 100% in refined copper; the range in number and amount of associated elements is also very wide.

Copper minerals of major economic importance exceed in number those of most other common metals. The sulphides include bornite Cu_5FeS_4, chalcopyrite $CuFeS_2$, chalcocite Cu_2S, covellite CuS, enargite Cu_3AsS_4, tetrahedrite $Cu_8Sb_2S_7$, and tennantite Cu_8AsS_7. The principal oxidized ores are cuprite Cu_2O, tenorite CuO, malachite $CuCO_3.Cu(OH)_2$, azurite $2CuCO_3.Cu(OH)_2$, chrysocolla $CuO.SiO_2.2H_2O$, brochantite $CuSO_4.3Cu(OH)_2$, atacamite $CuCl_2.3Cu(OH)_3$, and antlerite $3CuO.SO_3.H_2O$.

Most copper deposits are worked for their copper alone, but a substantial proportion of the world's output is derived from orebodies of copper–nickel, copper–cobalt, copper–zinc, copper–lead–zinc, copper–molybdenum, copper–gold, or copper–iron.

The analysis of copper has been developed in speed and accuracy to a high degree: with reasonable care in the choice of a separation procedure, even complex materials should not present difficulties.

Most copper-containing materials are readily decomposed by treatment with nitric, hydrochloric, and sulphuric acids; if necessary, any insoluble residue may be fused with sodium carbonate. Many other decomposition methods may be used if convenient: fusion with potassium bisulphate in a Pyrex "copper flask", fusion with sodium peroxide in an iron, nickel, or zirconium crucible, hydrofluoric and sulphuric or nitric acids in a platinum dish, 30% hydrogen peroxide and hydrochloric or 1 : 1 sulphuric acid, nitric and perchloric acids, or nitric acid and potassium chlorate. In technical work, some materials are decomposed in a beaker by aqua regia, with the addition of a few drops of hydrofluoric acid to break up silicates; sulphuric acid is added and the sample taken to fumes of this acid.

The determination of copper is sometimes required in organic material, and the latter must be destroyed before proceeding with the analysis. If the sample is carefully heated in a muffle at 510 ± 10°C for several hours until ignition is complete, no volatilization of copper compounds will occur [3]. When silica dishes are used, traces may be retained on the container, but with new dishes and thorough acid treatment the loss is insignificant. The alternative to dry combustion is wet oxidation by nitric and sulphuric acids, followed if necessary by perchloric acid, until all organic matter has been destroyed.

ISOLATION OR SEPARATION OF COPPER

In many products copper can be determined in the presence of numerous elements; in other samples it is necessary to isolate copper or remove many interferences. One of the most valuable separations is precipitation with hydrogen sulphide in 5–10% hydrochloric or sulphuric acid; after filtration, the precipitate of copper together with As, Sb, Sn, Ag, Pb, Cd, Hg, Bi, Se, Te, Au, Pt metals, Mo, and Ge is separated from all members of succeeding groups, which pass into the filtrate. If the precipitate is large it may occlude some members of the other groups; in this case, for accurate work a re-precipitation is required.

The separation of copper from other members of the hydrogen sulphide group, where this is necessary, may take many variations. Silver is usually removed from copper solutions as insoluble silver chloride in dilute nitric acid, and lead as the sulphate in dilute sulphuric acid. Lead may also be separated from copper by electrolysis in nitric acid solution, copper depositing on the cathode and lead on the anode as PbO_2. Mercury is rarely associated with copper but may be removed as the insoluble mercurous form in hydrochloric acid solution or by digestion of sulphides in hot dilute nitric acid, in which mercuric sulphide is insoluble; mercury may also be volatilized by igniting the sample or fusing with sodium carbonate.

Arsenic, antimony, tin and selenium can be volatilized from copper samples by boiling down to a low volume several times with a mixture of hydrobromic acid and bromine, or hydrobromic and perchloric or sulphuric acids. Arsenic may be separated from copper by distillation of arsenic trichloride at 108°C, and antimony by the same means at the higher temperature of 200°C. Small quantities of arsenic, antimony, and tin, with Bi, Pb, Se, and Te can be removed from acid copper solutions by adding iron, if this metal is not already present, to the extent of about 10–20 times the amount of As, Sb, etc., making ammoniacal, boiling, and filtering the precipitate of ferric hydroxide which contains all the As, Sb, Sn, Bi, Pb, Se, and Te. Traces of copper in the precipitate must be recovered by dissolving the latter in acid and re-precipitating with ammonia. Arsenic, antimony, tin and molybdenum can be separated from copper by digestion of the mixed sulphides in warm sodium sulphide or ammonium sulphide solution; As, Sb, Sn, and Mo are soluble whereas Cu, Bi, Cd and Pb are insoluble in alkaline sulphides.

Tin can be separated from copper by ammonium hydroxide in the presence of ammonium salts, copper passing into the filtrate; for small amounts of tin, ferric hydroxide should be used as a collector. For routine work, tin can be separated from copper by precipitation in nitric acid as metastannic acid. Copper may also be removed from tin by precipitation of the former on iron powder in a solution of hydrochloric acid.

Cadmium can be separated from copper by electrolysis in the usual sulphuric–nitric acid solution; small quantities of cadmium are not deposited. For larger quantities, separation is improved in nitric acid electrolyte alone. Cadmium can also be precipitated in alkaline cyanide solution, copper remaining soluble.

Bismuth can be separated from copper not only by collection of the former in a ferric hydroxide precipitate as already mentioned, but also by precipitation of bismuth as the oxychloride in faintly acid solution as described in the chapter on bismuth. Bismuth, with Sb, As, and Sn, can be separated from copper in weak sulphuric acid solution containing sodium sulphite, by potassium thiocyanate, which forms a white precipitate of cuprous thiocyanate.

Selenium and tellurium can be removed from copper by precipitation with sulphur dioxide in strong and dilute solutions, respectively, of hydrochloric acid. In addition to the separation by a ferric hydroxide collection mentioned above, selenium, but not tellurium, can be volatilized by several evaporations with a mixture of hydrobromic and sulphuric or perchloric acids, or by prolonged fusion with potassium bisulphate at a temperature above 720°C.

Molybdenum can be separated from copper by electrolysis; a little molybdenum is deposited the first time but a re-deposition will yield a complete separation. Molybdenum can also be removed by an ether separation; the latter is complete when a little iron is also present. Molybdenum accompanies arsenic, antimony, and tin when sulphides are treated with alkali or ammoniacal sulphide solutions, whereas copper remains insoluble. Molybdenum may be separated from copper by sodium hydroxide, the former remaining in solution. Copper can also be separated from molybdenum by precipitation of the former on aluminium or zinc in sulphuric acid, or lead in hydrochloric acid.

Gold can be removed from copper by precipitation of the former with sulphur dioxide, oxalic acid, or sodium nitrite; extraction of gold with ether in 5% hydrochloric acid can also be used. Gold can be removed in some samples by a combination of acid treatment and fire assay, as described in the chapter on gold. If gold is present in a cyanide solution, add sulphuric acid in a well-ventilated hood and evaporate to fumes of sulphuric acid; gold is precipitated as metal and can be removed from the water-soluble copper sulphate.

The platinum group can be separated from copper by hydrolytic precipitation with sodium nitrite. Convert the chloro compounds of the platinum group to soluble complex nitrites by digesting hot with sodium nitrite in a solution having an acidity not less than *p*H 1·5. Neutralize to *p*H 10 with sodium hydroxide to precipitate copper; the platinum metals, with the exception of palladium, remain in solution [21]. Palladium can be separated from copper by precipitation of the former with dimethylglyoxime in 4–5% hydrochloric acid. Gold accompanies copper in the nitrite hydrolysis, and must be previously removed with sulphur dioxide or oxalic acid. Platinum metals may also be isolated from copper by acid treatment and fire assay.

Indium is carried down with hydrogen sulphide in acid solution, especially if the acidity is low, but a re-precipitation in 10% sulphuric acid or hydrochloric acid will leave all indium in the filtrate. Nearly all germanium is volatilized as chloride if hydrochloric acid is used in the decomposition, but any remaining will accompany copper in the hydrogen sulphide precipitation; it may be removed by digestion in warm alkaline sulphide, in which germanium is soluble.

Tungsten and vanadium do not belong to the acid sulphide group, but have a tendency to co-precipitate with copper when the latter is separated by hydrogen sulphide. When appreciable quantities of tungsten or vanadium are present, the addition of 5 g of tartaric acid to the solution, before the introduction of hydrogen sulphide, will yield a clean copper sulphide precipitate.

Because copper is frequently a constituent of alloys which require for their analysis a separation by cupferron, mercury cathode, or 8-hydroxyquinoline, it is well to remember the behaviour of copper in such cases. In the cupferron separation in cold dilute hydrochloric or sulphuric acid, copper divides; with the mercury cathode electrolysis in dilute sulphuric acid all the copper is deposited, with Fe, Cr, Mo, Ni, Co, Zn, Sn, Cd, Ag, Au, Pt metals, Bi, Ga, Ge, In and Tl; it can thus be removed from Al, Be, Ti, V, Zr, U, and P, which are not deposited. With 8-hydroxyquinoline, in acetic acid–acetate solution copper is precipitated together with nearly every element except Be, Mg and a few others; in ammoniacal solution copper is precipitated with nearly all elements except B, P, Se, Te, F, As, and Pt.

1. Electrolytic Determination

The electrolytic procedure has the advantage that it is applicable to products ranging from about 0·1 to 100% copper [7, 10, 12, 15, 17, 18, 21, 25, 33, 35, 37]. Furthermore, the de-copperized solution can be used for the determination of other constituents. One weighing gives directly the quantity of copper, and personal factors are reduced to a minimum. Because large quantities can be deposited, accurate results on high-copper materials or heterogeneous samples may be readily obtained. With mechanical, electromagnetic, or air stirring, or with rotating anodes, electrolysis may be carried out rapidly; alternatively, it may often be conveniently done overnight with a stationary electrolytic apparatus.

The elements As, Sb, Sn, Bi, Mo, Se, and Te will partially deposit on the cathode with copper; some of them not only contaminate the copper but produce a dark, loosely adherent coating which tends to flake off the cathode when the latter is removed and weighed. Their removal, when they occur in more than traces, is therefore essential. Fortunately, methods of decomposition can eliminate all of the As, Sb, Sn, and Se; any small quantities remaining together with Bi and Te, can be easily removed by a ferric hydroxide separation. Molybdenum, if it occurs in quantities greater than about 15% of the copper, must be either removed beforehand, or its effect nullified by a re-deposition of the copper.

From a long experience with electrolytic copper procedures in the presence of a wide variety of impurities, the writer suggests, as a general rule, that a separation by ferric hydroxide, volatilization, or other means, is required if the impurity exceeds the following limits, expressed as a percentage of the copper present: As 0·5, Sb 0·5, Bi 0·2, Se 0·1, Te 0·1, Mo 15.

If selenium and tellurium are oxidized to the hexavalent state in the decomposition stage, by treatment with nitric acid and potassium chlorate for instance, they have far less tendency to co-deposit with copper in electrolysis. Small quantities of these oxidized impurities in blister or anode copper can often be tolerated. Similarly addition of 2% potassium permanganate solution to the appearance of a faint pink colour, followed by 5 ml of 2% manganese nitrate solution to a 2-g sample of a copper-tellurium alloy containing up to 0·6% tellurium, prevents the deposition of the latter. Too much reliance should not be placed, however, on non-interference from oxidized selenium and tellurium in the electrolytic determination of copper in complex samples. In work of high accuracy it is usually preferable to eliminate selenium and tellurium before a copper electrolysis if they constitute, individually, more than 0·1% of the copper.

When small quantities of copper in the presence of large amounts of iron must be determined electrolytically, it is advisable to make a prior separation of the copper by hydrogen sulphide or other means. Otherwise, difficulty is usually encountered in the deposition of copper, because the ferric ion resulting from the oxidizing action at the anode is reduced at the cathode in preference to the cupric ion. Sometimes, an initial deposition is followed by dissolution of copper from the cathode; this is noticeable when an excess of nitric acid is present in the electrolyte, and is attributed to the solvent action of ferric nitrate on copper. Thallium may similarly retard the deposition of copper; hydrogen sulphide precipitation of the latter, repeated if necessary, will separate the copper from thallium.

It may be desirable to determine copper electrolytically in high iron products such as steels or slags without recourse to the separation of copper by hydrogen sulphide. If so, remove iron by an ether separation in hydrochloric acid of s.g. 1·10. To the aqueous layer add sulphuric and nitric acids, evaporate to fumes of sulphuric, and proceed with the steps for the electrolytic determination. Molybdenum is also removed with iron in this separation.

Weigh out 0·5–5 g sample, depending on the copper content, and decompose with hydrochloric and nitric acids. Add 10 ml 1 : 1 sulphuric acid and evaporate to strong fumes of the latter. Cool, dilute with hot water, boil, filter, and wash; if the insoluble residue is likely to contain copper it must be fused with sodium carbonate and added to the main portion. Any other convenient method of decomposition, which will get all copper into solution, may be used.

Chromium forms an insoluble anhydrous sulphate which tends to hold some copper. If large quantities of chromium are present they may be volatilized initially by several evaporations with boiling hydrochloric and perchloric acids, or the sample may be decomposed with nitric and hydrochloric acids only, evaporating several times with the latter before proceeding with hydrogen sulphide precipitation. A convenient solvent for many samples is a nitric acid–potassium chlorate mixture containing 100 g potassium chlorate per litre of nitric acid.

If arsenic, antimony, or tin are present, decompose the sample with nitric acid, add 10–15 ml perchloric acid, and heat to white fumes. Cool, wash down the cover and sides of the beaker, and add 15–25 ml hydrobromic acid. Heat to strong fumes of perchloric acid; if the solution is not clear add a further 15–25 ml hydrobromic acid and evaporate again to near dryness. Add 10 ml of 1 : 1 sulphuric acid and evaporate to fumes of the latter.

An excellent method of decomposing products containing refractory oxides and silicates, such as many converter mattes and slags, is by direct fusion with sodium or potassium bisulphate in a pear-shaped Pyrex "copper" flask. The fusions are made over a battery of gas burners or electric heaters, and, if necessary, they are finally heated strongly for a few minutes over a blast burner. The melt is allowed to solidify in a thin layer around the sides of the flask by swirling the latter after withdrawing from the heat. After cooling, about 75 ml of warm water and 5 ml 1 : 1 sulphuric acid are added, the flask placed on the hot plate, and the salts brought into solution. The latter is filtered into a beaker, 2 ml of nitric acid is added, and the solution is ready for electrolysis.

Blister and refined copper are dissolved in sulphuric and nitric acids, the nitrous fumes boiled out and the solution made up to 500 or 1000 ml in a volumetric flask. Usually 10-g samples are taken and dissolved in a solution of 75 ml water, 50 ml 1 : 1 sulphuric acid, and 25 ml nitric acid. If an insoluble residue remains, which might contain copper, it must be filtered off, ignited, fused with sodium carbonate or bisulphate, and added to the filtrate. If appreciable quantities of silver are present in the blister, the addition of just sufficient sodium chloride to precipitate the silver is necessary. An aliquot of 1 g is usually taken from the 10-g portion of blister; aliquots for refined copper are generally 2 or 5 g.

Where small quantities of As, Sb, Sn, Bi, Se, and Te are present, they can be conveniently removed in a ferric hydroxide precipitate. To the sulphuric or nitric acid solution of copper add a little ferric sulphate or ferric nitrate, unless iron is already present, so that there is approximately 20 times as much iron as the combined impurities. If the iron is not already in the ferric state, add a few drops of nitric acid or hydrogen peroxide and boil. Precipitate the iron with ammonium hydroxide, adding a slight excess; all the impurities will be in the ferric hydroxide precipitate. Boil, filter, wash, dissolve the precipitate in 1 : 1 sulphuric acid or dilute nitric acid and re-precipitate with ammonium hydroxide to liberate traces of copper which are occluded in the ferric hydroxide. Boil, filter, and wash thoroughly. The filtrate contains all the copper, free of As, Sb, Sn, Bi, Se, and Te. Acidify with 1 : 1 sulphuric acid, adding 10 ml excess, add 5 ml nitric acid and proceed with the electrolysis. Place the beaker of sample solution, containing 5 ml sulphuric and 2–5 ml nitric acid in a volume of 200 ml, on the electrolytic assembly, and electrolyse until all copper has been deposited on the cathode. The universal test for completion of deposition is to dilute the solution and observe whether any more copper has plated on the stem of the cathode after an interval of several minutes or longer. For a rotating electrolytic apparatus 45 minutes at 1 ampere and 2–2·5 volts is usually sufficient for the complete deposition of 0·4–0·5 g of copper; some types of equipment are capable of a much faster rate of electrolysis.

If the electrolysed solution is required for the analysis of other elements, without stopping the current lower the beaker below the cathode and wash the latter simultaneously from a wash bottle, catching all the washings in the beaker. Remove the cathode quickly and rinse in three beakers of alcohol.

If the electrolysed solution is not required, carry out the washing as given above, or quickly replace the beaker containing the electrolyte with a beaker containing distilled water, release the cathode from its support and allow it to drop gently into the beaker of water. Quickly rinse the cathode in three beakers of water and finally in two or three beakers of alcohol.

After removing from the alcohol, the cathodes may be dried in several ways: on a small hot plate, in an oven at 110°C for several minutes, in a stream of warm air from a hair drier, or by igniting the alcohol. If the latter is practised it is essential to keep the cathode moving to avoid local over-heating, and oxidation; likewise, the cathodes must not be kept on a hot plate any longer than is necessary to drive off the alcohol.

If a platinum cathode remains dark after boiling in nitric acid, it is almost invariably due to carbon. Ignite the cathode over a burner for a minute and the original colour will be restored.

Refined copper and similar samples are usually electrolysed overnight on a stationary cabinet at 0·5–0·8 ampere and 2–2·5 volts. The slow electrolysis at a low amperage in the cold reduces any tendency for co-deposition of traces of other metals and is used for work of the highest accuracy.

Pure electrodeposited copper is salmon pink in colour; dark red, brown, or black indicates traces of impurities such as As, Sb, Sn, Bi, Se, Te, Mo, or Cd.

The standard sulphuric–nitric acid electrolyte described previously is used for copper on most materials. The metal can, however, be electrolysed satisfactorily in a solution containing 10–15% nitric acid alone. This modification is particularly important for some alloys containing lead, because it enables a simultaneous electrolytic determination to be made for copper, deposited on the cathode, and for lead, deposited as oxide on the anode.

Boil the sample solution containing about 10–15% nitric acid to expel nitrous fumes, cool, and electrolyse until all copper has been deposited. When this has occurred, it can be assumed that all the lead has also been deposited. Wash and remove the cathode and anode. Immerse the cathode in three beakers of alcohol, dry, and weigh as already described. Dry the anode at 110–120°C for 30 minutes, cool, and weigh as PbO_2. Use the factor 0·866 to convert PbO_2 to Pb. Handle the anode carefully, because the deposit is not as firmly adherent as a copper coating.

Electrolysis of copper in nitric acid alone may also be practised if the cadmium content is very high. For most samples containing copper and cadmium, however, the usual sulphuric–nitric electrolyte is perfectly satisfactory. Sulphuric–nitric solution is used, for instance, in the determination of copper in silver brazing alloys [7], where the ratio of cadmium to copper might occasionally be nearly 2 : 1.

Details have been published for the micro electrodeposition of copper, enabling a precise assay to be carried out on a small sample [30].

2. Volumetric Determination

Copper may be conveniently determined in a great variety of materials, and over a wide range of low and intermediate concentrations, by an iodometric titration [2, 5, 7, 15, 17, 18, 21, 25, 33, 35, 37].

This procedure depends on the liberation of iodine from potassium iodide by acidified solutions of cupric salts; the liberated iodine is titrated with standard sodium thiosulphate.

$$2Cu(C_2H_3O_2)_2 + 4KI \rightarrow Cu_2I_2 + 4KC_2H_3O_2 + I_2$$

$$I_2 + 2Na_2S_2O_3 \rightarrow Na_2S_4O_6 + 2NaI$$

There are two procedures used in the volumetric iodide determination of copper. In one, which is usually termed the "long iodide" method, copper is isolated from most other elements which react with potassium iodide. In the other, generally known as the "short iodide" procedure, the effect of iron, which is the commonest and often the sole interference, is overcome by complexing it with a fluoride.

A. Long Iodide Method

By hydrogen sulphide, sodium thiosulphate, or aluminium, copper can be separated from most of the other elements which react with acidified potassium iodide.

Decompose the sample in nitric and hydrochloric acids, add 10 ml of 1 : 1 sulphuric acid, and evaporate to strong fumes of the latter. Cool, add water, boil, filter, and wash. Unless any residue is known to be free of copper, fuse it with sodium carbonate and add to the main portion. Alternatively, use any other appropriate method of decomposition which will bring all copper into solution.

(1) *Hydrogen Sulphide.* Adjust the copper solution to 5–10% by volume with sulphuric or hydrochloric acid, and pass in a vigorous stream of hydrogen sulphide until all copper and other elements of its group have been precipitated. Filter, and wash quickly with 1% acid solution saturated with hydrogen sulphide. Place the paper and precipitate in a porcelain crucible, dry, and ignite at a temperature of 600–700°C until the residue is cupric oxide. Dissolve in warm 1 : 1 nitric acid and transfer to a beaker or small flask.

Boil the solution of copper nitrate until all nitrous fumes are eliminated. Add 5 ml of bromine water and again boil. Remove from the hot plate, cool, carefully add ammonium hydroxide until a slight excess is present, and boil to remove most of this excess. Remove from the hot plate, carefully add acetic acid until the dark blue colour of the copper-ammonia complex has disappeared, and then 5 ml excess. Boil for a minute or two, cool in a trough of running water, and wash down the inside of the beaker or flask. Add 5 ml of 80% potassium iodide solution, or 4 g of the solid salt, and titrate with standard sodium thiosulphate until the dark brown colour turns to a straw yellow. Add 5 ml of 0·5% starch solution, and continue the titration until the colour changes from purple or blue to the white or yellowish-white end point.

The standard sodium thiosulphate solution may be 0·1 N, equivalent to 0·006354 g of copper per ml. This is prepared by dissolving 24·82 g $Na_2S_2O_3.5H_2O$ in water, diluting to 1 litre, and standardizing against high purity copper foil. A few drops of amyl alcohol or chloroform, as a preservative, may be added to the thiosulphate solution before standardizing. Sometimes it is convenient to have a solution such that 1 ml = 0·005 g of copper; 19·5314 g $Na_2S_2O_3.5H_2O$ per litre will provide this, but the solution must of course be standardized against copper. Many copper mining companies, where the ore grade is low, use a sodium thiosulphate solution having a value of 1 ml = 0·0025 g of copper. This is made by dissolving 195 g $Na_2S_2O_3.5H_2O$ and about 25 g NaOH in water, keeping in 4 litres of solution for at least a week to stabilize, and making up to 20 litres.

The standard separation given above will require modification if the sample contains molybdenum, selenium, or a large quantity of tellurium. Molybdenum may be separated by sodium hydroxide or by digestion of the mixed sulphides in sodium or ammonium sulphide.

Selenium and tellurium may be removed by precipitation with sulphur dioxide in hydrochloric acid of 30–40% volume; selenium alone is not quantitatively precipitated in less than 28% hydrochloric acid and tellurium alone is not precipitated if the hydrochloric acid concentration is over 42%. Small quantities of selenium and tellurium may often be conveniently removed by co-precipitation with 15–20 times their weight of iron in ammonium hydroxide, with a re-precipitation to liberate occluded copper.

Selenium is volatilized by one or more evaporations with hydrobromic acid and bromine; hydrobromic and sulphuric, perchloric, or hydrochloric acids; evaporating to dryness two or more times with nitric and sulphuric acids, washing down the inside of the beaker and cover glass after the first evaporation; fusion with potassium bisulphate at a temperature above 720° C; or by careful ignition of the cupric sulphide precipitate above 700°C but below the temperature of 900°C at which fusion with the glaze of a porcelain crucible may occur.

For technical analyses, tellurium up to nearly four times the copper present, has no significant influence on the iodometric titration.

No other elements in normal concentrations will interfere in the long iodide procedure. Arsenic and antimony when oxidized to their pentavalent states by bromine water are without effect. Lead and silver are usually removed in the initial decomposition, but, if not, they, together with bismuth, merely consume potassium iodide; their presence only requires the addition of a little more of this reagent. Bismuth, and lead if not removed as sulphate, slightly change the colour of the final tint, but the end point remains sharp.

Gold, platinum, and palladium in quantities up to 10% of the copper present have no significant effect on the long iodide method [37]. For the rare samples where the concentration of noble metals is high, separate gold by precipitation with sulphur dioxide or oxalic acid, re-precipitating if necessary. Remove platinum by the sodium nitrite method, and separate palladium by precipitation with dimethylglyoxime in 4–5% hydrochloric acid.

If vanadium or tungsten are high, dissolve about 5 g of tartaric acid in the solution before precipitating copper with hydrogen sulphide. This will prevent the co-precipitation of these elements, and yield a clean copper sulphide.

(2) *Sodium Thiosulphate.* When a concentrated solution of sodium thiosulphate is added to a boiling 2–3% sulphuric acid solution of the copper sample, copper is gradually precipitated as Cu_2S after boiling for about 10 minutes.

$$2CuSO_4 + 2Na_2S_2O_3 + 2H_2O \rightarrow Cu_2S + 2Na_2SO_4 + 2H_2SO_4 + S$$

Cuprous sulphide obtained by precipitation with sodium thiosulphate has several advantages over cupric sulphide resulting from the ordinary hydrogen sulphide precipitation. It is less easily oxidized by the air and may be washed with water; it is more granular and thus more readily filtered, washed, dried, and ignited; finally, the precipitation can be carried out on a laboratory hot plate without the disagreeable aspects of handling hydrogen sulphide. When other elements must be determined in the filtrate from the precipitation of copper sulphide, however, the use of sodium thiosulphate has a serious drawback which makes its use impractical. Unlike free hydrogen sulphide which is readily removed by boiling, sodium thiosulphate requires for its elimination long boiling, repeated treatment with nitric acid, and filtration of the precipitated sulphur.

Though only about 1 ml of a 50% solution of $Na_2S_2O_3.5H_2O$ is theoretically required to precipitate 0·1 g copper, it is necessary to add a large excess to reduce ferric and other salts and to precipitate other members of the hydrogen sulphide group. Add 5–20 ml of a 50% solution, depending on the sample, boil until the copper sulphide has coagulated and the solution is clear, filter, wash, dry, and ignite to CuO at 600–700°C. Dissolve in 1 : 1 nitric acid, and continue with the method described earlier under (1).

(3) *Aluminium.* By boiling a 5–10% sulphuric acid solution of copper with a small square of pure sheet aluminium the copper is gradually reduced to the metallic state and can thus be separated from iron. Continue boiling until the copper detaches itself from the aluminium sheet, filter, and wash. Discard the filtrate. Dissolve the precipitated copper off the filter paper and

the aluminium sheet with warm dilute nitric acid, catching the copper solution and washings in the original beaker. The sample is now ready for the final steps of the long iodide procedure, outlined earlier under (1).

Contrary to statements in some texts, cobalt and nickel are not precipitated with the copper by aluminium [38].

If molybdenum is present, be sure that complete reduction of all iron and molybdenum has occurred, that the acidity of the solution is low at the end of reduction, and that aluminium is kept in the filter paper during filtration. Otherwise, slight co-precipitation of molybdenum will occur and the subsequent long iodide titration will be affected. Because molybdenum has such a pronounced effect on the usual long iodide procedure, in its presence it is generally preferable to use the short iodide method wherein even large quantities of molybdenum do not interfere.

Copper may also be isolated from iron by precipitation of the former on sheet zinc in 5% sulphuric acid; the procedure is similar to that given above for aluminium. Occasionally, copper may be removed from iron by precipitation of the former on stick cadmium in 10–15% hydrochloric acid [8], or on lead in 5% hydrochloric or perchloric acid.

Copper may also be separated from vanadium by means of aluminium or zinc in sulphuric acid, or of lead in hydrochloric acid.

B. Short Iodide Method

In this procedure, copper is not separated from iron; the interference of the latter is prevented by adding a simple fluoride, or ammonium bifluoride, to complex the iron. Molybdenum even in quantities of 0·5 g or more does not interfere, possibly through the formation of a complex such as $Na_2[MoO_2F_4]$. The short iodide method is excellent for ores and most samples in the mining and metallurgical industries where the copper content is not over about 15% and the amounts of other substances such as iron are not excessive. Vanadium interferes, and in its presence copper must be isolated and determined by the long iodide procedure. An alternative to the isolation of copper by hydrogen sulphide or aluminium is the removal of vanadium with iron by an ammonium hydroxide precipitation; a re-precipitation must be made to liberate occluded copper. Appreciable quantities of selenium and tellurium lead to a slow release of iodine in the solution after the titration appears to be complete. This is due to liberation of iodine from potassium iodide by selenic and telluric acids, or copper salts of these acids, and may be minimized by having only a very slight excess of acetic acid at the end point in the simple fluoride method. Substantial quantities of selenium and tellurium, particularly the former, should be removed as described previously for the long iodide procedure. If a large quantity of chromium is present, the sample is not treated with sulphuric acid, but all nitric acid is eliminated by evaporating twice with hydrochloric acid. Alternatively, chromium may be volatilized as chromyl chloride, as already described.

If the following precautions are taken, substantial quantities of manganese are without effect in either modification of the short iodide method. If the sample has been fumed to dryness, add a little sulphuric acid before neutralizing with ammonium hydroxide. Avoid a large excess of ammonia; add acetic acid immediately after the ammonia neutralization in the simple fluoride method, and add ammonium bifluoride similarly for this procedure, to dissolve any manganous hydroxide before it becomes oxidized to higher oxides of manganese. In the simple fluoride method, do not boil the solution after the addition of ammonium hydroxide unless it has been re-acidified with acetic acid.

(1) *Simple Fluoride.* Decompose the sample with nitric, hydrochloric, and sulphuric acids, and evaporate to strong fumes of the latter, but not to dryness. If the sample should fume dry, add a little 1 : 1 sulphuric acid. Cool, dilute with water, boil, filter, and wash; if an insoluble residue remains which might contain copper, fuse it with sodium carbonate and add to the main portion. Any other convenient method of decomposition, to get all copper into solution, may be employed. Add 5 ml bromine water and boil for several minutes, add a small excess of ammonium hydroxide and at once acidify with acetic acid, adding about 5 ml excess. Boil, remove from the hot plate and add 1–3 g of sodium, potassium, or ammonium fluoride depending on the iron content of the sample. The colour of the solution will change from the brown imparted by iron to the various tints of blue due to differing copper contents. Wash down the inside of the flask or beaker, and place in a cooling tray. Add potassium iodide and titrate with standard sodium thiosulphate, using starch indicator, as described for the long iodide method; a slightly larger excess of potassium iodide is required with the short iodide procedure.

(2) *Ammonium Bifluoride.* The procedure is the same as in (1) above, to the point where excess bromine has been removed by boiling. Remove from the hot plate, and add ammonium hydroxide until the odour of ammonia can be detected. Add 1 g ammonium bifluoride for each 0·1 g of iron or aluminium present. Stir, and if the precipitate of iron hydroxide does not dissolve completely, add more ammonium bifluoride. Be sure that all traces of ferric hydroxide on the walls of the beaker have been brought in contact with the solution and have dissolved. Cool to room temperature. Add 3 g potassium iodide, and titrate the liberated iodine with standard sodium thiosulphate, using starch indicator, as already described for the long iodide procedure.

C. Other Volumetric Methods

The accuracy and rapidity of the iodide methods for copper have made these the preferred volumetric procedures. Occasionally, others may be used for specialized applications. Examples of these are the cyanide titration of copper sulphate in leach plants or plating works [18, 37], and titrations with EDTA which are usually restricted to fairly pure copper salts [33].

3. Colorimetric Methods

The chemical literature abounds with colorimetric procedures for copper, but the five described below have been thoroughly tested in the mining and metallurgical industries, and are applicable to a wide variety of complex materials.

A. With Sodium Diethyldithiocarbamate

This reagent, $\begin{matrix} C_2H_5 \\ C_2H_5 \end{matrix}\!\!>\!N\!-\!\overset{\overset{\large S}{\|}}{C}\!-\!S\!-\!Na$, is used in the form of a 0·1% solution in water. It gives a distinct golden brown colour with 0·1 ppm copper in slightly acid, neutral, or alkaline solution; at 1 ppm a dark brown colour and a turbidity appear. In the absence of Fe, Al, Co, Ni, Mn, Zn, Cd, Sb, Bi, Sn, Pb and Hg, the colour may be measured directly [5, 9, 18, 22, 25, 32, 33, 37].

Decompose the sample by any appropriate means to bring all copper into oxidized acid solution, make to a convenient volume, and transfer an aliquot containing not more than 20 μg of copper to a 125 ml separatory funnel. The aliquot should not contain more than 0·5 mg of iron or manganese, or 10 μg of bismuth, cobalt, or nickel. If larger quantities are present, a separation of copper must be made by one of the methods such as dithizone, listed earlier under Separations.

Add 2 ml of 40% ammonium citrate, 1 drop of phenolphthalein, 5 ml of 0·1% sodium diethyldithiocarbamate solution in water, and 1 : 1 ammonium hydroxide until pink. Add exactly 10 ml carbon tetrachloride, shake for 5 minutes, dry the funnel stem, and withdraw the organic layer directly into an absorption cell. If the carbon tetrachloride layer is cloudy, centrifuge before transferring it to the absorption cell. Measure the transmittance of the solution at 440 nm. Prepare the standard graph from solutions containing 5–20 μg of copper. Weigh 0·3930 g $CuSO_4.5H_2O$, dissolve in water, add 5 ml sulphuric acid, and make to 1 litre. From the latter withdraw 10 ml, add 5 ml sulphuric acid, and dilute to 1 litre; one ml of this solution contains 1 μg of copper.

Carry out the colorimetric operations as rapidly as possible in subdued daylight to minimize fading of copper diethyldithiocarbamate.

A number of modifications of the basic diethyldithiocarbamate procedure to suit particular samples have been published [1, 4, 27, 34].

B. Dithizone

Diphenylthiocarbazone, or dithizone, $C_6H_5.N{:}N.C.S.NH.NH.C_6H_5$, which is green in a solution of carbon tetrachloride or chloroform, reacts with copper in dilute mineral acid to form the red–violet keto complex. In oxidized 0·05 – 1·0N solution of hydrochloric or sulphuric acid, only the following metals react to form dithizonates: Cu, Bi, Hg, Au, Ag, Te(IV), Pd(II), and Tl(III). In samples encountered in most mining and metallurgical laboratories when a colorimetric determination of copper is required, the last seven will be absent, or exceedingly low in relation to copper. If present, these elements must be separated in accordance with the recommendations of Sandell [32] or of similar colorimetric reference books. Silver and mercury may be initially precipitated as chloride. Bismuth and thallium react with dithizone quantitatively at *p*H values > 2 and 3–4, respectively; they will not interfere at the acid concentration specified unless present in large quantities. Gold can be separated from copper by precipitation of the former in 5% hydrochloric acid by sulphur dioxide, palladium can be removed by dimethylglyoxime in 4–5% hydrochloric acid, and tellurium is precipitated from a copper solution containing 10–20% hydrochloric acid by sulphur dioxide.

The dithizone colorimetric procedure for copper is very sensitive; about 5 μg of copper is a useful range in the final aliquot. It is generally better to isolate the copper initially with dithizone before commencing the colorimetric determination.

From a 0·1N hydrochloric or sulphuric acid solution of the sample, which has been oxidized with nitric acid and boiled to remove nitrous fumes, pipette into a 75 ml separatory funnel an aliquot containing about 5 μg of copper. Add several ml of 2% hydroxylamine hydrochloride solution. From a small burette add successive small quantities of a solution containing about 0·05 g dithizone per litre of carbon tetrachloride to the separatory funnel, shaking vigorously after each addition, until the lower green dithizone layer remains green, instead of turning red, indicating that all the copper has been extracted. Transfer the lower layer to a small beaker, evaporate carefully on the hot plate, add nitric and sulphuric acids, and evaporate to dryness to eliminate all organic matter.

Dissolve the sample in about 10 ml of 0·05 – 0·1N hydrochloric or sulphuric acid, and transfer to a small separatory funnel. Add 5 ml of 0·001% dithizone, w/v, in carbon tetrachloride, and shake for 2 minutes. If the colour of the lower layer is red-violet, an excess of dithizone is not present; add another 5 ml of dithizone and shake again. The colour of the final solution should deviate from that of pure copper dithizonate. Transfer the lower layer to the cell of a spectrophotometer and measure the transmittance at 535 nm. Construct the standard graph from known copper solutions, using the same volume, acidity, etc.

C. With 2,2′-biquinoline

The reagent 2,2′-biquinoline, also called 2,2′-diquinolyl or cuproine, forms with cuprous copper a coloured complex which can be isolated from most other elements by extraction in amyl alcohol and determined spectrophotometrically [20, 22, 26, 29, 32, 33].

Decompose the sample by any convenient means, add sulphuric acid, and evaporate to dryness. Dissolve the soluble salts in 1–5 ml hydrochloric acid, about 50 ml water, boil, filter, wash, and make up to 100 ml in a volumetric flask.

Place an aliquot containing about 30 μg of copper in a 100 ml beaker, add 5 ml of 10% hydroxylamine hydrochloride and 5 ml of 10% tartaric acid solution. Make the two latter additions to a blank determination which is carried along with each group of samples. With a *p*H meter, adjust the solution to 5·0–5·5 with 1 : 1 ammonium hydroxide and 1 : 1 hydrochloric acid. Transfer the sample to a 60-ml separatory funnel, and add exactly 10 ml of 0·02% 2,2′-biquinoline in n-amyl alcohol. Shake for 2 minutes and discard the aqueous layer. Draw off the organic layer into a 1·5 cm centrifuge tube and centrifuge the sample for 1 minute to remove the cloudiness of the organic extract.

Measure the transmittance of the sample in a spectrophotometer at 545 nm against a reagent blank. Prepare a standard calibration curve from increments of copper up to 80 μg. Make up a solution containing 1 g copper + 5 ml hydrochloric acid in a volume of 500 ml, or 2 mg copper per ml. From this solution withdraw 2 ml, place in a litre flask and make to volume; one ml of this latter solution then contains 4 micrograms of copper.

With this procedure, the following were found to be without interference in quantities greatly exceeding that of the copper: Fe, Al, Ti, Bi, Co, Ni, V, U, P, Sn, Na, Cl, B, SO_4, NO_3, As, Ca, Cd, Mg, Zn, Mo(VI), Mn(II), Sb(III), and W(VI).

The purple colour of the cuproine complex is stable for days, but *p*H adjustment and shaking time are important.

When chromium is present, it is necessary to reduce with sulphurous acid. The following procedure is recommended for alloy steels and ferro alloys. Decompose in aqua regia, evaporate to dryness, add hydrochloric acid, and again evaporate. Dissolve in about 25 ml water containing a few drops of hydrochloric acid, add 25 ml sulphurous acid, boil, cool, and add 15 ml of 50% citric acid. Transfer to a 100-ml flask, dilute to the mark, and mix. Withdraw a 10-ml aliquot to a small beaker and add 10% ammonium hydroxide until a *p*H of 5–6 is attained. Add 2 ml of 5% sodium acetate and transfer to a small separatory funnel. Adjust the volume to 17·5 ml, mix, add 10 ml of 0·05% 2,2′-biquinoline solution in n-amyl alcohol. Shake for 2 minutes, allow to stand for 5 minutes, and continue as outlined above.

For ferrotungsten or ferromolybdenum, dissolve in sulphuric and hydrofluoric acids; for ferromanganese, ferrotitanium, or ferrovanadium dissolve in sulphuric and nitric acids. When tungsten or manganese are present, a precipitate will be found after the initial decomposition, but it may be ignored. When the copper content is below 0·02%, use the entire sample instead of diluting to 100 ml and taking an aliquot.

D. Neocuproine

The reagent 2,9-dimethyl-1, 10-phenanthroline, or neocuproine, reacts with cuprous copper in the *p*H range 2–10 to give a yellow complex, which can be extracted by organic reagents such as chloroform, and measured photometrically at approximately 457 nm [6, 7, 19, 22, 26, 27, 33].

Decompose the sample by any suitable means to bring all copper into solution. Transfer the solution or an aliquot containing up to 80 μg of copper to a small separatory funnel, and add 10 ml of hydroxyammonium chloride–sodium citrate solution. The latter consists of 25 g hydroxyammonium chloride and 150 g of sodium citrate in 500 ml of water; for very accurate work, add 10 ml of neocuproine solution, mix, allow to stand for 15 minutes, and extract with chloroform to remove traces of copper.

Mix the sample and the hydroxyammonium chloride–sodium citrate solution, adjust the *p*H if necessary to 4–6 with dilute hydrochloric acid or ammonium hydroxide, using universal indicator paper. Add 10 ml of 0·1% neocuproine in ethanol or methanol free from pyridine, mix, and allow to stand for 15 minutes. Add 5 ml of chloroform, shake for 30 seconds, and allow to separate. Filter the chloroform layer through a small glass wool plug into a 25-ml volumetric flask containing 3–4 ml of alcohol. Care should be taken that none of the aqueous layer is transferred to the flask, to prevent the appearance of a turbidity. This can be facilitated by two techniques: release the pressure in the separatory funnel by removing the stopper rather than turning the stopcock, and fill the bore of the stopcock with chloroform before commencing the separation.

Extract again with 5 ml of chloroform by shaking for a few seconds; add this lower layer to the original in the volumetric flask. Make the volume up to 25 ml with alcohol, and mix. Measure the optical density of the solution at approximately 457 nm against the blank. Prepare the calibration graph to cover the range of 0–80 μg of copper. A standard copper solution having a value of 1 ml = 100 μg of copper can be made by dissolving 0·393 g $CuSO_4.5H_2O$ in 1 litre of water containing 5 ml of sulphuric acid. Dilute standard copper solutions can be made immediately before use by diluting the above standard to give values of 1 ml = 10 μg or 1 ml = 1 μg copper.

Traces of sulphide and cyanide interfere. In the presence of more than 2 mg of chromium, volatilize the latter by heating with a mixture of perchloric and hydrochloric acids over a burner until no more chromyl chloride is evolved. Alternatively, reduce the chromium by boiling the aliquot containing less than 0·15 mg of copper with 20 ml of sulphurous acid.

A related compound, 2,9-dimethyl-4,7-diphenyl-1, 10-phenanthroline, or bathocuproine, forms an orange chelate with cuprous ions at a *p*H between 4 and 10, which can be measured at about 480 nm. It is used for very small amounts of copper in water [6, 22], but for nearly all colorimetric determinations in the mining and metallurgical industries cuproine or neocuproine are preferable.

E. Hydrobromic Acid

Cupric copper in a mixture of hydrobromic and phosphoric acids forms a violet complex which is measured photometrically at approximately 600 or 660 nm. This is the basis of a method which is popular for low copper contents in a variety of metallurgical materials [7]. The procedure for solder and white metal, which usually contains 0·01–0·5% copper, is given below; appropriate modifications may be made for other samples.

Transfer up to 2 g of the sample to a wide-mouth 250-ml flask, and add 20 ml of a hydrobromic acid–bromine mixture made by pouring 20 ml of bromine into 180 ml of hydrobromic acid. Cover and heat gently until dissolution is complete. Add 10 ml of perchloric acid and heat with continuous swirling over a flame in a well-ventilated hood to volatilize antimony, arsenic, selenium, and tin. When copious white fumes of perchloric acid appear, heat gently and intermittently to decompose any lead bromide present. If necessary, repeat the volatilization treatment to remove all Sb, As, Sn, and Se. Evaporate the perchloric acid to about 2 ml, cool, and dilute to about 40 ml.

Add 1 g of test lead to the flask, cover, and boil for about 15 minutes to displace all the copper. Cool slightly, remove the solution by decantation, and wash once with water, decanting thoroughly. Heat the flask gently to remove moisture. Add 10 ml of hydrobromic acid–bromine mixture to the flask, cover, and heat gently to dissolve the metal. Boil to expel the excess bromine. Cool to room temperature. Transfer 10 ml of phosphoric acid plus one drop of hydrobromic acid–bromine mixture to a dry 25-ml volumetric flask. To the latter, transfer the sample solution, washing with a few ml of hydrobromic acid. Dilute to the mark with hydrobromic acid, mix, and transfer a portion to the absorption cell. Measure the transmittance at approximately 600 nm. Construct the standard graph by determining the photometric readings of copper solutions containing 0·05–0·8 mg of copper. A convenient standard copper solution is one containing 0·1000 g of copper per litre; 1 ml = 0·1 mg Cu.

The volatilization of arsenic, antimony, tin, and selenium by hydrobromic acid–bromine, followed by isolation of copper on lead, leaves very few elements which interfere. Gold and the platinum metals cannot be tolerated.

4. Gravimetric Method

Nearly all copper determinations are carried out by electrolytic, volumetric, colorimetric, or atomic absorption procedures. Occasionally, however, a gravimetric method will be convenient. This is based on the fact that in solutions of hydrochloric or sulphuric acid weaker than 0·5 N, copper in the cuprous state is quantitatively precipitated by ammonium thiocyanate. Lead, mercury, precious metals, selenium, and tellurium are likewise precipitated. Lead can be eliminated as sulphate, mercury can be removed as mercurous chloride or volatilized when a sample is fused with sodium carbonate; selenium and tellurium can be separated by sulphur dioxide in strong and weak hydrochloric acid solution, respectively. Gold can be separated from copper in 5% hydrochloric acid by sulphur dioxide. The platinum metals, except palladium, can be separated from copper by a sodium nitrite hydrolysis, and palladium can be precipitated by dimethylglyoxime in 4–5% hydrochloric acid.

To a warm solution of copper in hydrochloric or sulphuric acid, containing 2–3 g of tartaric acid, add sulphurous acid in small increments so that an excess remains after the sample has stood on a warm plate for about 30 minutes. Carefully add sodium hydroxide solution until a permanent precipitate is formed. Dissolve the precipitate by the careful addition of hydrochloric acid, and add 0·5 ml excess. Dilute to about 200 ml, boil, add 2% ammonium thiocyanate solution in a saturated solution of sulphur dioxide at the rate of 10 ml for every 50 mg of copper, with stirring. Remove from the hot plate and allow to stand for six hours or overnight.

Filter through a sintered glass or Gooch crucible, washing thoroughly with 0·01% ammonium thiocyanate solution containing 1% sulphurous acid. Wash with 20% ethanol, dry at 110°C to constant weight, and weigh as CuCNS, which contains 52·26% copper. The following do not interfere: Sb, As, Bi, Cd, Co, Fe, Ni, Mn, Sn, and Zn.

5. Polarographic Methods

Small amounts of copper may often be determined by polarography [25, 26]. The stepwise reduction of copper [11] occurs at the dropping mercury electrode to form fairly stable copper [1] in ammonia, concentrated chloride, pyridine, and thiocyanate media. The reduction of ferric iron interferes with the copper waves in many supporting electrolytes; it can be overcome by reducing ferric to ferrous iron by hydroxylamine, by complexing it with a fluoride, or by precipitation of iron from a pyridine supporting electrolyte.

A polarographic procedure for copper in pyrite is typical [13]. Dissolve the sample in aqua regia, evaporate to dryness, wash down the beaker and cover glass with water, and again evaporate to dryness. Add 2 ml hydrochloric acid, 15 ml water, and heat to solution of salts. Transfer to a 100-ml volumetric flask and bring the volume to about 75 ml. Add 2·9 ml of pyridine, and 2 ml of a 0·5% solution of sodium carboxymethylcellulose as maximum suppressor. Dilute to the mark and mix. Allow the precipitate of hydrous ferric oxide to settle, and decant the supernatant liquid through a filter. Transfer 40·0 ml of the filtrate to a polarographic cell. Add 0·8 ml of pyridine to raise the *p*H of the solution from about 3·6 to 5·2. De-aerate the solution and carry out the usual polarographic techniques from 0 to $-1 \cdot 3$ V. Wave heights corrected for dilution caused by further addition of pyridine equal 1·02 × measured wave heights.

Procedures have been outlined for copper in nickel [26], aluminium alloys, steels and other materials [25].

6. Optical Spectrography

Copper can be readily determined in many samples by arc or spark spectrography. Numerous spectral lines are used: 2242·61, 2247·00, 2276, 2294, 2824·37, 2961·16, 3247·54, 3273·96, and 5105·54 are most frequently employed. Detailed directions have been published for 0·03–0·5% copper in alloy steels, 0·001 to 30% in aluminium and its alloys, 0·0005–0·08% in lead,

0·001–0·3% in tin, 0·005–0·25% in zinc and zinc base alloys, and 10–400 ppm in zirconium and its alloys [7]. Procedures for determining 1–2600 ppm copper in nickel have been outlined [26].

Spectrographic methods have been extensively used for 5–20 ppm copper in plants [5], and for similar low concentrations of this element in various biological materials and soils [25].

7. X-Ray Spectrography

Copper can be determined by X-ray fluorescence over a range of about 0·1 to 100%, using the lines 1·392 and 1·542 Å. The determination of copper is now carried out by this means in process streams of the extractive industries [3, 14, 23, 25, 26, 36].

8. Atomic-Absorption Spectrophotometry

Copper can be determined rapidly and accurately by atomic-absorption spectrophotometry [11, 16, 24, 31]. The most sensitive line is 3247·5, with a sensitivity of 0·1 ppm, but other lines may be used. No interferences from the presence of other elements have been reported. Air–coal gas and air–acetylene flames are equally satisfactory.

SPECIAL PROCEDURES

Forms of Copper

1. 'Oxide' Copper in Mining and Extractive Metallurgy

In many ore deposits throughout the world copper occurs in both oxidized and sulphide forms, and because their recovery methods differ, it is necessary to distinguish these forms. This can be done chemically with a reasonable degree of accuracy, because dilute sulphuric acid saturated with sulphur dioxide dissolves oxidized forms of copper but does not attack sulphides. An exception is the mineral cuprite, Cu_2O, which liberates only approximately half of its copper in the dilute acid leaching solution. On the rare occasions when appreciable quantities of cuprite are present, therefore, this method gives a low figure for oxide copper [37].

Weigh 0·5–2 g of −100 mesh sample into a 250-ml Pyrex "copper" flask, or Erlenmeyer, and add 30 ml of 5% by volume sulphuric acid saturated with sulphur dioxide. Cover, and let stand for 1 hour, swirling occasionally. Filter into a 250-ml beaker and wash 6 times with hot water. The filtrate contains all the copper from malachite, azurite, tenorite, chrysocolla, and half the cuprite. Sulphides and native copper are not attacked and remain in the precipitate. Boil off excess sulphur dioxide from the filtrate, and determine copper by the electrolytic, long iodide, short iodide, or other appropriate procedure.

2. Chrysocolla

The determination of chrysocolla, a copper silicate, in copper ores, is of importance to a few producers because it is not amenable to normal flotation processes. Transfer a 1–5 g portion of −100 +200 mesh ore to a 150-ml separatory funnel, and add 20–40 ml tetrabromoethane. Stir thoroughly with a glass rod, and wash down any particles adhering to the walls of the funnel and to the stirring rod with a further 20–40 ml tetrabromoethane. If the mineral particles do not appear to be wetted readily, shake the funnel vigorously, and rinse the walls with additional tetrabromoethane.

Allow the material to remain in the liquid for an hour, occasionally tapping the funnel lightly to release any particles held mechanically in the upper or lower layers. Chrysocolla, together with other light minerals having a specific gravity below 2·96, will be on top of the tetrabromoethane; all other economically important copper minerals will sink to the bottom of the funnel. Draw off the bottom layer into a beaker, together with most of the tetrabromoethane; the latter can be recovered for re-use. Into another beaker transfer the float portion, washing any particles from this top layer, which may be adhering to the funnel, into the beaker.

Filter through a Whatman No. 40 paper, and wash thoroughly to remove tetrabromoethane. Dissolve the paper and precipitate in nitric and sulphuric acids, and evaporate to strong fumes of the latter. Determine copper by any appropriate method; copper $\times$ 2·76 = chrysocolla [39].

3. Metallic Copper

A. In the segregation process of the copper mining industry, where ores refractory towards conventional flotation are converted to segregated metallic copper, the latter is determined by the displacement of silver from a silver nitrate solution and measurement of dissolved copper by atomic absorption [28].

B. It is sometimes important to know the ratio of metallic copper to oxides of copper in materials such as marine paint pigments. Transfer a weighed portion to a 250-ml Phillips conical beaker with lip, add 25 ml of ethanol and a large spoonful of glass beads. Swirl the beaker vigorously to break up lumps of pigment. Add 100 ml of an extraction solution containing 40 ml of hydrochloric acid and 40 g of stannous chloride dihydrate per litre of ethanol. Cuprous oxide is dissolved; any cupric oxide which dissolves is reduced by stannous chloride to the cuprous state without any attack on metallic copper. Add lumps of dry ice to keep the temperature near 0°C, and continue to swirl the flask for 5 minutes. Filter the residue of copper + cupric oxide on asbestos, and wash with ethanol. Dissolve the copper + cupric oxide by warming in 25 ml of a solution containing 75 g $FeCl_3.6H_2O$, 150 ml hydrochloric acid, 400 ml of water, and 5 ml of 30% hydrogen peroxide boiled to remove excess of the latter. Maintain an atmosphere of carbon dioxide above the solution by the addition of dry ice. Add 50 ml of water, 3 drops of *o*-phenanthroline indicator, and titrate the ferrous iron with 0·1 N ceric sulphate to the change from orange to pale green.

Metallic copper + cuprous oxide can be determined by digesting an original sample directly with 25 ml of the ferric chloride–hydrochloric acid solution for 15 minutes in a carbon dioxide atmosphere, and titrating as described above. The total copper in the sample may be found by treatment with nitric and sulphuric acids, evaporation to fumes of the latter, and determination of copper electrolytically. The total copper minus the sum of metallic copper + cuprous oxide gives cupric oxide; total copper minus the sum of metallic copper + cupric oxide gives cuprous oxide. By subtracting either cupric or cuprous oxide, metallic copper is obtained. There may be a small difference between the values for metallic copper when calculated in these two ways, because cupric oxide is slightly soluble in hydrochloric acid–stannous chloride solution. The results, however, are comparable with those obtained in similar phase analyses, and are satisfactory for routine control work.

Copper Development Association

For information on special problems in the analysis of copper products, the chemist may find very helpful the services of the Copper Development Association, 55 South Audley Street, London W.1, and 405 Lexington Avenue, New York, N.Y. 10017.

REFERENCES

1. Abbott, D. C., and Harris, J. R., *Analyst* **87**, 497–9 (1962).
2. Agterdenbos, J., and Elberse, P. A., *Talanta* **13**, 523–5 (1966).
3. Alexander, G. V., *Anal. Chem.* **34**, 951–3 (1962).
4. Analytical Methods Committee, *Analyst* **88**, 253–8 (1963).
5. A.O.A.C., *Methods of Analysis of the Association of Official Analytical Chemists*, Washington, D.C., Association of Official Analytical Chemists, 1965.
6. A.P.H.A., *Standard Methods for the Examination of Water and Waste-water*, 11th ed., American Public Health Association, New York, 1965.
7. A.S.T.M., *Chemical Analysis of Metals. Sampling and Analysis of Metal Bearing Ores*, American Society for Testing and Materials, Philadelphia, Pa., 1969.
8. Borun, G. A., *Anal. Chem.* **34**, 720 (1962).
9. British Drug Houses. *The B.D.H. Book of Organic Reagents for Analytical Use*, Poole, England, British Drug Houses, 1958.
10. British Standards Institution, *British Standard* 1748, Parts 1–5, London, British Standards Institution, 1961.
11. Brooks, R. R., Presley, B. J., and Kaplan, I. R., *Anal. Chim. Acta* **38**, 321–6 (1967).
12. Butts, A., *Copper — The Metal, Its Alloys and Compounds*, New York, Reinhold, 1954.

13. Cooper, W. C., and Mattern, P. J., *Anal. Chem.* **24**, 572–6 (1952).
14. Cullen, T. J., *Anal. Chem.* **33**, 1342–4 (1961).
15. Dozinel, C. M., *Modern Methods of Analysis of Copper and its Alloys*, 2nd ed., London, Elsevier, 1963.
16. Elwell, W. T., and Gidley, J. A. F., *Atomic Absorption Spectrophotometry*, 2nd ed., Oxford, Pergamon, 1966.
17. Elwell, W. T., and Scholes, I. R., *Analysis of Copper and its Alloys*, Oxford, Pergamon, 1967.
18. Furman, N. H., ed., *Scott's Standard Methods of Chemical Analysis*, 6th ed., Vol. 1, Princeton, N.J., Van Nostrand, 1962.
19. Gahler, A. R., *Anal. Chem.* **26**, 577–9 (1954).
20. Guest, R. J., *Anal. Chem.* **25**, 1484–6 (1953).
21. Hillebrand, W. F., Lundell, G. E. F., Bright, H. A., and Hoffman, J. I., *Applied Inorganic Analysis*, 2nd ed., New York, Wiley, 1953.
22. Hopkin and Williams Ltd., *Organic Reagents for Metals*, Chadwell Heath, England, Vol. 1, 1955, Vol. 2, 1964.
23. Jenkins, R., Hurley, P. W., and Shorrocks, V. M., *Analyst* **91**, 395–7 (1966).
24. Kirkbright, G. F., Peters, M. K., and West, T. S., *Analyst* **91**, 411–17 (1966).
25. Kolthoff, I. M., and Elving, P. J., *Treatise on Analytical Chemistry*, Part II, Vol. 3, 1–41, New York, Interscience Publishers, 1961.
26. Lewis, C. L., Ott, W. L., and Sine, N. M., *The Analysis of Nickel*, Oxford, Pergamon, 1966.
27. Luke, C. L., *Anal. Chim. Acta* **32**, 286–7 (1965).
28. Mackay, K. E., and Gibson, N., *Trans. Institution Mining Met.* **77**, C 19–31 (1968).
29. Methods of Analysis Committee of B.I.S.R.A., *J. Iron Steel Institute* **182**, 301–3 (1956).
30. Murphy, T. J., and Taylor, J. K., *Anal. Chem.* **37**, 929–31 (1965).
31. Ramakrishna, T. V., Robinson, J. W., and West, P. W., *Anal. Chim. Acta* **37**, 20–6 (1967).
32. Sandell, E. B., *Colorimetric Determination of Traces of Metals*, 3rd ed., New York, Interscience Publishers, 1959.
33. Snell, F. D., and Ettre, L. S., *Encyclopedia of Industrial Chemical Analysis*, Vol. 9, New York, Interscience Publishers, 1970.
34. Wilson, A. L., *Analyst* **87**, 884–94 (1962).
35. Wilson, C. L., and Wilson, D. W., *Comprehensive Analytical Chemistry*, Vol. IC, 366–78, London, Elsevier, 1962.
36. Wood, R. E., and Bingham, E. R., *Anal. Chem.* **33**, 1344–7 (1961).
37. Young, R. S., *Industrial Inorganic Analysis*, London, Chapman and Hall, 1953.
38. Young, R. S., *Chemist-Analyst* **44**, 98–9 (1955).
39. Young, R. S., and Simpson, H. R., *Mining Mag.* **84**, 137–9 (1951).

15 Fluorine

Fluorine occurs in some minerals and ores, fluxes, insecticides, plastics, electrolytes, fertilizers, ceramics, wood preservatives, and waters; its determination is occasionally required in many industrial laboratories. The most important fluorine minerals are fluorite or fluorspar, CaF_2, and cryolite, Na_3AlF_6. Fluorapatite, $CaF_2.3Ca_3P_2O_8$, and apatite, $Ca_4.CaF(PO_4)_3$ or $Ca_5(Cl,F)(PO_4)_3$, also yield fluorine as a by-product from phosphate production. An accurate analysis for fluorine when it occurs in the presence of many interfering elements requires some experience and close attention to details of procedure.

Fluorine may be lost as gaseous silicon tetrafluoride or hydrogen fluoride during decomposition with acids, and as insoluble calcium fluoride when the the solution is made ammoniacal for the separation of iron, aluminium, etc. In the presence of fluorine, therefore, both silica and calcium may be lost unless precautions are taken.

The determination of silica and fluorine when these occur in the same sample is carried out as follows. Fuse with sodium and potassium carbonates as detailed below, digest for an hour with 200 ml of hot water, filter, and wash. Discard any insoluble residue, which is mainly calcium carbonate. To the filtrate, add about 10 g ammonium carbonate, boil 5 minutes, and allow to stand for several hours. Filter the precipitate of silicic acid, wash with dilute ammonium carbonate solution, and retain for silica. Recover the small amount of silica which passes into the filtrate by evaporating the solution almost to dryness, adding a little water, and neutralizing the ammonium carbonate with hydrochloric acid. Add 5 ml of an ammoniacal zinc oxide solution, and boil until all ammonia has been expelled. Filter the precipitate of zinc silicate and zinc oxide, and wash; retain the filtrate and washings. Add this precipitate to the main silica residue, and carry out the usual silica determination with acid dehydration. Fluoride is in the filtrate, and may be determined gravimetrically, volumetrically, or colorimetrically.

If fluorine is required in an organic material such as soil or grain, a compound must be present that will combine with the fluorine and keep the sample alkaline during the ashing. Fluoride-free calcium oxide is the usual fixative employed, and the ashing temperature should be 550–720°C.

When calcium is required in a sample which contains fluorine, evaporate to dryness several times with sulphuric or hydrochloric acids; to aid the elimination of fluorine, a little silica should be added if the material does not contain it.

1. Gravimetric Determination

A. As Calcium Fluoride

Fuse 0·5–2 g of material with ten times its weight of sodium or potassium carbonate, and digest for an hour with 200 ml of hot water. Break up all lumps and boil for 10 minutes, filter, wash, and discard the residue of calcium carbonate. Neutralize the filtrate with hydrochloric acid, and bring back to the basic side carefully with 2 N sodium carbonate. Add 1 ml of 2 N sodium carbonate in excess and sufficient 10% calcium chloride solution to completely precipitate the fluoride and the excess of carbonate. When no more precipitate forms, add 2–3 ml excess calcium chloride solution, allow to settle, filter, and wash with hot water; test the filtrate with additional calcium chloride solution.

Dry the combined precipitate of calcium carbonate and fluoride, and transfer to a platinum dish. Ignite the filter paper separately and add its ash to the dish; then ignite the sample, and cool. Add an excess of dilute acetic acid and evaporate to dryness on the water bath. The lime is converted to calcium acetate, whereas the fluoride remains unaffected. Add a little hot water, filter, and wash with hot water; calcium acetate is removed in the filtrate while calcium fluoride remains as a precipitate on the paper. Dry the residue, separate it from the filter paper, ignite the latter separately and combine its ash with the ignited residue.

$$CaF_2 \times 0{\cdot}4867 = F.$$

To confirm the result, treat the residue with a slight excess of sulphuric acid in a platinum dish, evaporate several times to fumes, and finally to dryness. Ignite at a dull red heat and weigh as calcium sulphate.

$$CaSO_4 \times 0{\cdot}2791 = F.$$

The reactions in this procedure are:

$$Na_2CO_3 + CaF_2 \rightarrow CaCO_3 + 2NaF$$
$$2NaF + CaCl_2 \rightarrow CaF_2 + 2NaCl$$
$$CaO + 2HC_2H_3O_2 \rightarrow Ca(C_2H_3O_2)_2 + H_2O$$

B. As Lead Chlorofluoride

There is no difference in accuracy between the calcium fluoride and lead chlorofluoride procedures; the latter may be more rapid and convenient for routine work [1, 6, 10].

Fuse the sample with ten times its weight of sodium carbonate, dissolve in water, and filter. Boil the precipitate with 50 ml of 2% sodium carbonate, filter, and add this filtrate to the previous one. Add a solution of 1 g zinc oxide in 20 ml of 10% nitric acid. Boil, filter, and wash thoroughly with hot water. Add 3 drops of methyl red to the filtrate, nearly neutralize with nitric acid, and evaporate to a volume of about 200 ml, keeping the solution alkaline throughout the evaporation. Cool, and add dilute nitric acid until a faint pink colour develops.

Add 1 g zinc oxide that has been dissolved in ammonium hydroxide with the aid of a little ammonium carbonate, and boil until the odour of ammonia can no longer be detected. This usually requires concentration to about 50 m. Dilute to 100 ml with hot water, allow to settle, and filter. Wash the residue with cold water.

The filtrate is now free of substances which interfere in the lead chlorofluoride precipitation. Add a few drops of bromophenol blue indicator to the solution, 3 ml of a 10% solution of sodium chloride, and dilute to 250 ml. Add dilute nitric acid until the colour changes to yellow, and then dilute sodium hydroxide until the colour just changes to blue. Add 2 ml of 1 : 1 hydrochloric acid, 5 g lead nitrate, and heat on the steam bath. Stir until the lead nitrate has dissolved; add 5 g sodium acetate. The reaction is $Pb(NO_3)_2 + HCl + NaF \rightarrow PbFCl + NaNO_3 + HNO_3$. Stir, and allow to remain on the steam bath for 30 minutes with occasional stirring. Allow to stand at room temperature overnight and filter through a weighed Gooch crucible. Wash the precipitate in the beaker and in the crucible once with cold water, 5 times with a saturated solution of lead chlorofluoride, and finally once with cold water. Dry the crucible and contents at 120°C for one hour, cool, and weigh. $PbFCl \times 0{\cdot}0726 = F$.

If desired, the precipitate can be filtered on a Whatman No. 42 paper, dissolved in dilute nitric acid, and titrated for chlorine by the Volhard method. From the amount of chlorine found, the fluorine content of the sample can be determined.

The lead chlorofluoride wash solution is prepared as follows. Dissolve 10 g lead nitrate in 200 ml water, and pour this solution into 100 ml of a solution containing 1 g sodium fluoride and 2 ml hydrochloric acid. Mix thoroughly, let settle, decant the supeanatant liquid, and wash the precipitate 5 times by decantation with 200 ml portions of water. Add 1 litre of water to the precipitate, stir, allow to stand 1 hour, and filter. More wash solution can be prepared as needed by treating the precipitate with fresh portions of water.

2. Volumetric Determination

A. Soluble Fluorides

(a) *Absence of interfering elements.* In the usual volumetric determination, fluoride is titrated with a standard solution of thorium nitrate; any ion such as Ba, Ca, Fe, Al, or PO_4 which forms a precipitate or a non-dissociated salt with fluorine or thorium interferes with the titration [1, 6, 7, 10]. If these interferences are present, use the treatment outlined under (b) below.

Dissolve a weighed quantity of the fluoride in water and make up to a definite volume. Transfer an aliquot to a small beaker, add water to make a volume of approximately 20 ml, and add 8 drops of 0·05% sodium alizarin sulphonate. Neutralize by adjusting with 2% sodium hydroxide and 1 : 200 hydrochloric acid until the solution is just acid, i.e. when the pink colour is discharged. Add 1 ml of chloroacetate buffer; this is made by dissolving

9·45 g monochloracetic acid and 2 g sodium hydroxide in 100 ml of water. Titrate with 0·1N thorium nitrate solution to the first permanent pink. Carry out a blank and subtract this from the titration of the sample. The reaction is slow near the end point.

$$Th(NO_3)_4.4H_2O + 4NaF \rightarrow ThF_4 + 4NaNO_3 + 4H_2O$$

One litre of 0·1N thorium nitrate contains 13·8055 g $Th(NO_3)_4.4H_2O$; one ml of this solution = 0·0019 g fluorine. If desired, it can be standardized against 10–25 ml taken from a 100-ml volumetric flask containing 0·221 g highest quality sodium fluoride; each ml of this contains 1 mg of fluorine.

If the volume of the solution to be titrated is sufficiently small, 4–8 ml, so that the colour can be distinguished, the accuracy of the titration is increased by using only several drops of indicator. For accurate results, thorium nitrate must be standardized in approximately the same volume and with the same number of drops of indicator as the unknown, and the two titrated to the same colour. The end point is not sharp in the presence of a large amount of thorium fluoride; not more than 10 ml of 0·1N thorium nitrate should be used for the titration. Carry out a blank and subtract this from the titration of the sample. The blank is usually 3 drops or about 0·10 ml.

(b) *Presence of interfering elements.* When elements such as Ca, Ba, Fe, Al and PO_4 are present, the fluorine must be isolated before the titration. This may be conveniently done by volatilizing it as hydrofluosilicic acid by adding perchloric acid, water and several pieces of glass to the sample in a distillation flask and distilling over the fluorine [1, 6, 7, 10]. When the quantity of the latter is less than 10 mg and the temperature is kept below 125°C the pieces of glass appear to supply the silica which is required to form hydrofluosilicic acid, and there is no significant etching of the flask. With more fluorine present and a higher temperature, the flask is noticeably etched, but fluorine is still recovered quantitatively.

Weigh 0·5–2 g sample and transfer, with minimum washing, to a 500 ml distilling flask which contains an inlet tube from a steam distillation flask, and a thermometer, both extending to the bottom of the flask. Add 15 ml of perchloric acid. Heat the steam distillation flask to boiling, and connect it with the sample distilling flask. The latter is provided with a water-jacketed condenser, and the condensate is delivered into an open 250-ml volumetric flask. Heat the sample flask to 135°C and maintain it as close to this temperature as possible until nearly 250 ml has been collected in the volumetric flask.

Cool the volumetric flask to 20°C, make up to the mark, mix, and withdraw 50 ml or any other appropriate aliquot for titration. Add 8 drops of 0·05% sodium alizarin sulphonate, and neutralize by adjusting with 2% sodium hydroxide and 1 : 200 hydrochloric acid until the solution is just acid, i.e. when the pink colour is discharged. Add 1 ml of chloracetate buffer; this is made by dissolving 9·45 g monochloracetic acid and 2 g sodium hydroxide in 100 ml of water. Titrate with 0·1N thorium nitrate solution to the first permanent pink. Carry out a blank and subtract this from the titration of the sample.

The separation of fluorine by distillation, as outlined above, is quite satisfactory for nearly all ores. Fluorine may also be isolated, prior to titration with thorium nitrate or to a colorimetric measurement, by ion exchange [9, 10, 16], or by pyrolytic separation in which moist oxygen is passed over the sample in a quartz tube heated to about 1000°C, in the presence of an accelerator [10, 14, 15].

B. Insoluble Fluorides

Water-insoluble fluorides which are decomposed by perchloric acid may be analysed by the procedure outlined in (b) above.

Fluorides which cannot be decomposed by perchloric acid, such as opal glass, must be fused with sodium carbonate and the silica separated with ammonium carbonate and zinc oxide because the presence of large quantities of silica retards the volatilization of fluorine. In rare cases the filtrate containing the fluoride may be directly titrated; usually the distillation procedure described above, to separate interferences, must precede the titration.

Fuse 0·5–1 g sample with 2·5–5 g sodium carbonate, leach with hot water, filter, and wash. Transfer the insoluble residue back to the dish in which it was leached by means of a fine jet with about 50 ml hot water, add sodium carbonate to make approximately a 2% solution, boil a few minutes, filter, and wash thoroughly with hot water. To the combined filtrates which should have a volume of 300 ml, add 0·5 g zinc oxide dissolved in perchloric acid, boil the alkaline solution for 1 minute, filter, and wash with hot water. Evaporate the filtrate to 200 ml, add a drop of methyl red, neutralize to a very faint pink with dilute perchloric acid, add a solution of 0·25 g zinc oxide and 0·5 g ammonium carbonate dissolved in 0·5 ml of ammonium hydroxide and 10 ml of water. Place on the steam bath until a clear solution is obtained. Boil until the odour of ammonia has entirely disappeared, evaporate to about 100 ml, filter, and wash with cold water. Evaporate the filtrate to 25 ml and transfer the whole or an aliquot to the distillation flask for the isolation of fluorine.

3. Colorimetric Determination

Obtain the sample, containing not more than 70 μg of fluoride in a volume of 50 ml, by distillation or other separation techniques, and adjust to the temperature of the standard curve [10]. If the sample contains free chlorine, add 2 drops of 0·1N sodium arsenite solution for each 1 ppm of chlorine present. Add 5 ml of Eriochrome Cyanine R solution, followed by 5 ml of zirconyl chloride solution, and mix thoroughly. After 5 minutes, measure the absorbance of the solution at 527·5 nm against a reference solution. The latter is made by adding 10 ml of Eriochrome Cyanine R reagent to 100 ml of water, and then 10 ml of a solution prepared by diluting

7 ml of concentrated hydrochloric acid to 10 ml with water. If the absorbance reading falls beyond the range of the standard curve, the combined sulphate and fluoride concentrations of the sample are too high, and a smaller aliquot must be used. If the presence of aluminium ions is suspected, allow the reaction to continue for another 15 minutes. If an appreciable drop in absorbance is recorded, interference by aluminium is indicated.

Eriochrome Cyanine R. Dissolve 1·8 g of the sodium salt of *o*-sulphohydroxydimethylfuchsondicarboxylic acid in water and dilute to 1 litre.

Zirconyl chloride octahydrate. Dissolve 0·265 g of $ZrOCl_2.8H_2O$ or 0·220 g of zirconyl nitrate dihydrate in 50 ml of water. Add 700 ml hydrochloric acid and dilute to 1 litre.

Other colorimetric procedures have been described [2, 4, 10, 11].

4. Optical Spectrography

Occasionally fluorine can be determined spectrographically, by measuring the band head of an alkaline earth metal fluoride molecule, usually calcium [10, 17].

5. X-Ray Spectrography

Traces of fluorine have been determined by precipitating as lanthanum fluoride in an ethanol–water mixture and measuring the lanthanum in the precipitate by X-ray fluorescence [12].

6. Atomic-Absorption Spectrophotometry

An indirect determination of fluorine by atomic absorption has been outlined; it depends on the depression of magnesium absorption, or enhancement of zirconium absorption, in the presence of fluorine [3].

7. Other Procedures

A simple, rapid field test for the measurement of hydrogen fluoride in air has been described. The gas is collected in an acidic solution of zirconium-Solochrome cyanine R reagent and the observed bleaching of the colour is compared with standards [13].

Fluorine is one of the elements which can be determined in solution by a selective ion electrode, and small quantities in water are satisfactorily measured in this manner [5].

A potentiometric determination for high concentrations of fluoride has been published. Fluoride is precipitated with lead nitrate in hydrochloric acid, and the excess chloride in the liquid portion is titrated with standard silver nitrate [8].

The determination of fluoride in preservative-treated wood has been described [18]. It is leached from thin sections with sodium hydroxide, and photometric measurement is made of the blue complex formed between fluoride ions and the red lanthanum chelate of alizarin fluorine blue.

REFERENCES

1. American Public Health Association, *Standard Methods for the Examination of Water and Wastewater*, New York, American Public Health Association, 1965.
2. A.O.A.C., *Official Methods of Analysis of the Association of Official Analytical Chemists*, Washington, D.C., Association of Official Analytical Chemists, 1965.
3. Bond, A. M., and O'Donnell, T. A., *Anal. Chem.* **40**, 560–3 (1968).
4. Chan, K. M., and Riley, J. P., *Anal. Chim. Acta* **35**, 365–9 (1966).
5. Crosby, N. T., Dennis, A. L., and Stevens, J. G., *Analyst* **93**, 643–52 (1968).
6. Furman, N. H., ed., *Scott's Standard Methods of Chemical Analysis*, 6th ed., Vol. 1, Princeton, N.J., Van Nostrand, 1962.
7. Gwirtsman, J., Mavrodineanu, R., and Coe, R. R., *Anal. Chem.* **29**, 887–92 (1957).
8. Hogan, J. M., and Tortorici, F., *Anal. Chem.* **39**, 221–3 (1967).
9. Kelso, F. S., Mathews, J. M., and Kramer, H. P., *Anal. Chem.* **36**, 577–9 (1964).
10. Kolthoff, I. M., and Elving, P. J., *Treatise on Analytical Chemistry*, Part II, Vol 7, 207–334, New York, Interscience Publishers, 1961.
11. Lim, C. K., *Analyst* **87**, 197–201 (1962).
12. Luke, C. L., *Anal. Chim. Acta* **43**, 245–52 (1968).
13. Marshall, B. S., and Wood, R. *Analyst* **93**, 821–6 (1968).
14. Nardozzi, M. J., and Lewis, L. L., *Anal. Chem.* **33**, 1261–4 (1961).
15. Newman, A. C. D., *Analyst* **93**, 827–31 (1968).
16. Shehyn, H., *Anal. Chem.* **29**, 1466–70 (1957).
17. Spindler, D. C., and Smith, M. F., *Anal. Chem.* **30**, 1330–2 (1958).
18. Williams, A. I., *Analyst* **94**, 300–5 (1969).

16 Gold

Gold is usually determined by fire assay methods. In the gold mining industry, routine fire assays have attained a remarkable level of accuracy and speed. Gold is frequently associated with base metals, silver, and the platinum metals; its determination may consequently be required in ores, concentrates, mattes, speisses, drosses, residues, blister copper, lead bullion, refinery solutions and anode slimes.

Gold usually occurs as metal, generally alloyed with silver. It also occurs with some tellurium minerals. The greater part of the world's gold is derived from underground or placer mining of gold ores, but a substantial portion is obtained as a by-product from base metal ores, principally copper, lead, and nickel.

It may be necessary to analyse high-grade materials such as refinery and electroplating products, or jewellery, by wet methods. In the base metal field, gold is determined on many samples by a combination of wet procedure and fire assay.

ISOLATION OR SEPARATION OF GOLD

Gold is a member of the hydrogen sulphide group, and can be separated from members of succeeding groups by precipitation with hydrogen sulphide in 5% hydrochloric acid. It will, of course, be accompanied by the following elements: Ag, Cu, Cd, Hg, Ge, Sn, Pb, As, Sb, Bi, Mo, Se, Te, and the platinum metals. Silver and mercurous mercury may be removed by prior isolation in hydrochloric acid as their insoluble chlorides. Mercury can be volatilized from gold by evaporation, repeated if necessary, of sulphuric and hydrobromic acids. Arsenic, antimony, and germanium can be separated by distillation of their volatile chlorides in hydrochloric acid. Gold, with molybdenum, can be extracted with ether in dilute hydrochloric acid, and removed from other members of the acid sulphide group except Ge, As, Sb, Sn, Te, and Ir. The most useful separation, however, is treatment of the combined acid sulphide group as chlorides with sulphur dioxide; gold chloride is reduced to metallic gold and, when a re-precipitation is made, only selenium and tellurium with traces of lead are found with the gold. Subsequent solution in nitric acid dissolves Se, Te, and Pb, leaving only the gold. When the content of gold is very low, it can be separated from base metals, by co-precipitation with added tellurium, in the presence of sulphur dioxide and hydrazine hydrochloride. Tellurium can be later separated by a 12-hour hydrogenation at 650–750°C [12]. Ion exchange [3, 4], chromatography [3, 5, 26], and solvent extraction [3, 4, 23, 28, 36] have also been used to isolate gold.

1. Wet Method

A. With Sulphur Dioxide and Oxalic Acid

This procedure, of course, is intended for appreciable quantities of gold; the small amounts which occur in ores and similar products are best determined by fire assay. The principle of this wet method is the reduction of a gold chloride solution to metallic gold with sulphur dioxide and oxalic acid [19, 22, 27, 32]. About the only elements which must be removed are selenium and tellurium; if desired, gold may be separated initially with hydrogen sulphide when large quantities of members of the succeeding groups are present.

Dissolve the sample, containing not more than 1 g of gold, in 10 ml nitric acid and 30 ml hydrochloric acid. If an insoluble residue remains, which might contain gold, it must be decomposed by hydrofluoric acid or fusion with sodium carbonate. Evaporate carefully to dryness, moisten with hydrochloric acid, and evaporate again. Dissolve the soluble salts in hydrochloric acid, boil, filter off AgCl, SiO_2, etc., and wash with 1% hydrochloric acid. With some materials, gold may tend to precipitate here and be lost with the discarded residue; if so, keep it in solution by addition of a little nitric acid before filtering off AgCl, etc.

To each 100 ml of the filtrate in 5% hydrochloric acid solution containing not over 0·5 g of gold, add 25 ml of a saturated sulphur dioxide solution; digest on the steam bath for 1 hour. If the filtrate contains nitric acid it must be evaporated to dryness several times with hydrochloric to get rid of the nitric acid before precipitation with sulphur dioxide. A little gold may tend to separate in hydrochloric acid solution but the precipitate is retained in the next step. Add another 10 ml of the sulphur dioxide solution and set aside to cool for another hour. Pour the supernatant liquid through a Whatman No. 42 paper, using pulp, and wash the precipitate thoroughly by decantation with hot 1% hydrochloric acid and finally with two small washes of water. Examine the filtrate carefully for fine particles of gold which might have passed through the paper.

If selenium or tellurium are present, they can be removed from the combined precipitate by dissolving in hot 1 : 1 nitric acid, filtering, and washing with 1% nitric acid. Then dissolve the gold in a hot solution of 8 ml hydrochloric acid, 2 ml nitric acid, and 10 ml water for each 0·5 g or less of gold. Filter the aqua regia solution from the disintegrated filter paper, washing the latter very thoroughly with hot 1% hydrochloric acid. Evaporate the filtrate to dryness on the steam bath, add 2–3 ml hydrochloric acid, evaporate to dryness again, and repeat several times to eliminate all nitric acid. Take up the residue with 3 ml hydrochloric acid, 5 drops sulphuric acid, and 75 ml water for each 0·5 g or less of gold. A little metallic gold may have separated but its presence can be disregarded, because it will be recovered in the subsequent precipitation.

For each 75 ml of this solution add 25 ml of a saturated solution of oxalic acid and boil for 10 minutes. Test for completeness of precipitation by adding another 5 ml of oxalic acid and boiling for another 5 minutes. Allow the solution to digest on a steam bath for 4 hours. Filter on a Whatman No. 42 paper, wash with 1% hydrochloric acid, wipe out the inside of the beaker carefully with pieces of filter paper and add these to the precipitate in a tared porcelain crucible. Dry, ignite at 800°C to constant weight, and weigh as metallic gold.

B. With Sodium Nitrite

Sodium nitrite is probably the most satisfactory precipitant for gold when the platinum metals are absent. In their presence it involves converting them to their nitrito complexes and then back to chloro complexes; it also means that palladium which is precipitated with the gold and most base metals at *p*H 10 must be recovered separately with dimethylglyoxime [20, 22].

Evaporate the chloride solution, which must be free from ammonium salts, to dryness with 1 g sodium chloride, moisten the residue with 4 drops of 1 : 1 hydrochloric acid and dissolve in 100 ml water. Add 20 ml of 25% sodium nitrite solution, boil until no further red fumes are evolved, and neutralize with 2% sodium hydroxide solution to *p*H 2, using indicator paper. Continue boiling and add 2% sodium bicarbonate solution to *p*H 8, indicated by the colour change of xylenol blue from yellow to blue. Filter, wash with hot 1% sodium chloride solution, transfer back to the original beaker, and dissolve in a little hydrochloric acid. Evaporate the solution to dryness and repeat the precipitation. If base metals are present, dissolve the second precipitate in bromine and 1 : 1 hydrochloric acid, boil to a low volume, and precipitate the gold with sulphur dioxide.

In the absence of base metals the gold may be precipitated directly with sodium nitrite by evaporating the original chloride solution to a paste, dissolving in 100 ml hot water and boiling with 7 ml of 25% sodium nitrite. Bring to *p*H 2 with dilute sodium hydroxide, add another 7 ml of 25% sodium nitrite, boil, and neutralize with dilute sodium hydroxide to *p*H 8. Filter the precipitated gold on a Whatman No. 42 paper, with pulp, wash with hot water, 1% nitric acid, and again with hot water. Dry and ignite at 800°C, and weigh as metallic gold.

2. Fire Assay

A. Ores, Concentrates, etc.

An assay fusion consists of heating a mixture of the finely-pulverized sample with about three parts of a flux until the product is molten. One of the ingredients of the flux is a lead compound which is reduced by other constituents of the flux or sample to metallic lead. The latter collects all the gold, together with silver, platinum metals, and small quantities of certain base metals in the sample, and falls to the bottom of the crucible to form a lead button. The gangue of the ore must be converted by fluxes into a slag sufficiently fluid so that all particles of lead may fall readily through the molten mass. The choice of a suitable flux depends on the character of the ore; the subject is fully covered in reference books on fire assaying [9, 14, 18, 33, 34].

In fire assay work, the term "assay ton" is used to express the weight of sample. This notation is employed because the weight of ore is calculated on the avoirdupois system, whereas gold is measured in Troy ounces. One assay ton is 29·1666 g, so that 1 ton avoirdupois : 1 ounce Troy = 1 assay ton : 1 milligram. In other words, with a sample weighing 1 assay ton, the final weight of the gold in milligrams gives directly the assay value in ounces per ton.

Weigh out 1–10 assay tons of -100 mesh sample; if the sulphur content is below 2%, mix with a suitable flux and transfer to the assay crucible. For low grade material, where it is necessary to use more than two assay tons, a number of separate fusions are made and the lead buttons afterwards are combined.

If the sulphur content of the material exceeds 2%, the sample must be roasted before mixing with the flux. Place in a flat fire-clay roasting dish and calcine in an open muffle or furnace. Apply the heat gradually to about 500°C and then increase it to 750°C until no more fumes are evolved. If appreciable arsenic and antimony are present, add at this stage a little fine charcoal and repeat the roasting. If chlorides are present, roasting must not be practised; for example, most of the gold will be volatilized from a sample that contains 10% sodium chloride if it is roasted at 750°C for 30 minutes. When the calcined sample has cooled, pulverize it in a porcelain mortar, mix with flux, and transfer to a fusion pot. The quantity of flux should be adjusted, depending on the ore, to obtain a lead button of about 50–60 g.

Allow the crucible to remain in the furnace until it has attained a temperature of between 900 and 1100°C. This usually requires 25–40 minutes. Pour into a cast iron mould and allow to cool. Hammer the lead button free of slag.

If base metals such as Cu, Ni, Co, Sb, etc. are present in the lead button, place the latter in a scorifier, add a little test lead, borax, and silica. Scorify in the muffle until the metals go into the slag and a lead button weighing about 25 g is obtained. Pour and hammer free of slag.

Place the lead button in a cupel and carry out the cupellation at approximately 850–900°C. If gold alone is to be determined, a temperature of 950–1000°C may be used. Remove the cupel from the muffle immediately the operation is completed, and cover with an inverted red-hot cupel. After cooling, remove the gold–silver bead from the cupel with forceps, flatten on the anvil carefully to remove any adhering cupel particles, and weigh on the assay balance.

Remove the silver by parting the bead in a small porcelain cup, first with 1 : 6 nitric acid and then with 1 : 1 nitric acid. Wash three times by decantation with hot water, dry, anneal by heating for a few seconds at a bright red heat. Weigh the gold bead on the assay balance; the difference in weights represents silver.

If there is less than three times as much silver as gold in the bead it will not part readily. The practice known as inquartation must be followed. After weighing the gold–silver bead, wrap it with about 10 times its weight of pure silver foil in 4 g of sheet lead, and cupel. Flatten the bead, part, and weigh the gold as described above. It is preferable to add the silver before the original cupelling, because inquarted gold tends to be low.

For some concentrates and other materials rich in base metals it is a good practice to separate most of the latter by wet treatment after roasting. Transfer the calcine to a 2-litre beaker, add 100 ml water, and 100 ml sulphuric acid. Evaporate to strong fumes, cool, dilute to a litre, boil until all soluble salts are dissolved, and cool. Add about 50 ml of 5% sodium chloride, stir; add 50 ml of 1% lead nitrate solution, stir, and dilute with cold water to about 1800 ml. The lead nitrate, which forms a precipitate of lead chloride, helps to collect the silver chloride precipitate.

Allow to settle overnight and filter through Whatman No. 40 paper, with paper pulp. Wash out the beaker thoroughly, wipe the inside carefully with another filter paper, and add this to the precipitate. Transfer the latter to an assay crucible, dry, and burn off the paper at a low temperature in the assay furnace. Remove from the furnace, cool, mix with a suitable flux in the crucible, and proceed with the fusion as described earlier. Better mixing of the flux and precipitate can be effected by placing the latter in a large porcelain crucible, burning off most of the paper at a low heat and transferring the residue to a large mortar. Add the flux to the mortar, grind thoroughly with a pestle, transfer to the fusion crucible; rinse the mortar by grinding a little dry litharge and adding this to the charge.

Though lead is the usual collector for gold in fire assaying, tin may also be used. After roasting one hour at 750°C, the ore sample is heated to 1250°C for 45 minutes with stannic oxide together with the usual flux constituents of sodium carbonate, borax, etc. When the tin button is parted in hydrochloric acid, gold remains in the insoluble residue [16].

B. Blister Copper, Refinery Slimes, etc.

Weigh out 1–3 assay tons, depending on the gold content, into a 2-litre beaker. Add 20 ml of a mercuric sulphate solution containing 37 g per litre, stir, and allow to stand for 1 hour [41]. The purpose of this is to accelerate the solution of copper by forming a galvanic couple. Add 75 ml water and 125 ml sulphuric acid; place on the hot plate and evaporate to fumes.

Cool, dissolve in water, and wash the sides and cover of the beaker. Boil and add 20 ml of 5% sodium chloride solution; fill the beaker to about 1 litre with water and allow to stand overnight. Test for complete precipitation of silver by adding a few more ml of 5% sodium chloride solution. Filter through Whatman No. 40 paper, with paper pulp, and wash with hot water. Clean out the beaker thoroughly with another filter paper, adding this to the precipitate. Transfer the latter to folded lead foil in a scorifier, and burn the paper off carefully in front of the muffle. Add 30 g test lead, 1 g borax, 0·5 g silica, scorify until the metals go into the slag and a lead button weighing about 25 g is left.

Pour into a scorifier mould, hammer free of slag, and place in a cupel. Carry out the cupellation at 850–900°C, the parting, and weighing as described under A.

For complex materials containing large quantities of acid-insoluble metallic compounds, it may be necessary to transfer the initial precipitate of Au, Ag, etc. to a large porcelain crucible, burn off most of the paper at a low heat, mix with fluxes in a large mortar, and transfer to a fusion pot. Heat in the assay furnace for 25–40 minutes at about 1000°C, pour into a crucible mould, and remove slag by hammering; transfer the lead button to a scorifier and scorify before cupelling as described under A.

In the presence of platinum metals it is advisable to dissolve the gold bead in $4HCl : 1HNO_3$ in a covered cup in a warm place. Filter into a 100 ml beaker through a small Whatman No. 40 paper, wash, and evaporate carefully to a syrup. Add dilute hydrochloric acid, and evaporate again, repeating this operation twice. Transfer the solution to double lead boats in a 3-inch scorifier and evaporate to dryness at low heat.

Add test lead until the total amount of lead is about 40 g. Add a sufficient quantity of silver so that there is at least 12 times as much silver as gold + platinum metals. Scorify, pour, and clean the button. Cupel, clean and part as described previously, using 1 : 6 and then 1 : 1 nitric acid. Wash, transfer the gold to a porcelain cup, dry, anneal, and weigh.

When the gold content of the bead is less than ten times the platinum content, much of the latter remains with the gold on parting. For beads obtained from platinum ores, therefore, the gold–platinum residues may be dissolved in aqua regia, the nitric acid removed in the customary way, and the gold precipitated by sulphur dioxide or hydroquinone. The recommendations of standard texts should be followed [3, 9, 32].

C. Cyanide Solutions

The determination of gold in various cyanide solutions of reduction works is frequently required. There are several procedures in common use [3, 9, 14, 24, 27]; the copper sulphate method described below is very satisfactory. It depends on the fact that gold and silver are quantitatively precipitated from a cyanide solution when copper sulphate, sodium sulphite, and sulphuric acid are added to it. An excess of cyanide and of sulphur dioxide must be present to give sufficient cuprous cyanide to act as a collector for gold and silver.

To 50–1000 ml of cyanide solution add, with stirring, 25 ml each of 15%

copper sulphate solution, 10% sodium sulphite solution, and 20% sulphuric acid solution. If sufficient cyanide is not present to give a fairly heavy white precipitate at this stage, add a few drops of a saturated solution of sodium cyanide and stir thoroughly. If the solution is not blue, add more copper sulphate solution.

Filter the solution through a Whatman No. 40 paper, containing paper pulp, and wash all the precipitate into the paper. Allow to drain, fold the paper, and place it in a $3\frac{1}{2}$ in (90 mm) scorifier. Dry it carefully in front of a muffle, away from draughts, char and heat until carbon is destroyed. Add 35 g test lead and 5 g borax. Place the scorifier in a muffle heated to about 1100°C, close the door, and after the lead has melted, open the door slightly and scorify for not longer than 3 minutes. Cupel the cleaned lead button and complete the assay in the usual manner.

Alternatively, ignite the precipitate in a fireclay crucible in an assay furnace, and fuse with a charge consisting of 30 g sodium carbonate, 10 g borax, 35 g litharge, and 1 g charcoal.

3. Volumetric Determination

Various titrimetric procedures for gold have been proposed [3], but nearly all have a restricted application. About the only one in common use in the metallurgical field is for works control of gold in plating solutions, which is outlined below [21].

To an appropriate volume of solution, add aqua regia, and evaporate to dryness. Moisten twice with hydrochloric acid, evaporating to dryness both times. Dilute with water, and add potassium iodide until the gold which is precipitated as aurous iodide is completely dissolved. Titrate with standard 0·01N sodium thiosulphate, using starch as indicator, to the disappearance of the blue colour. Deduct the amount of 0·01N iodine which is required to produce a perceptible rose tint in the solution. Theoretically 1 ml of 0·01N sodium thiosulphate = 0·000986 g of gold, but the titre should be determined on known amounts of gold carried through all steps of the analysis.

4. Colorimetric Methods

A large number of colorimetric procedures for small quantities of gold have been published [1, 2, 3, 30, 31]. On the whole, for routine purposes, they have not yet been able to equal the accuracy and speed of conventional fire assays. Some of them, however, possess fairly wide application and reasonable reliability, and may often be used in place of fire assaying. The subject has been critically reviewed several times [1, 2, 3].

Among the important colorimetric procedures are those utilizing *o*-tolidine [1, 2, 3, 16, 17, 25, 31], 5-(*p*-dimethylaminobenzylidene) rhodanine [3, 31], stannous chloride [3, 31], bromaurate [3, 13, 31], rhodamine B [3, 31], dithizone [3, 8, 31, 38, 42], phenyl α-pyridyl ketoxime [3], and brilliant green [35].

5. Optical Spectrography

Optical emission spectrography has been used for many years to determine very low concentrations of gold, generally with the lines 2427·95 and 2675·95. In many cases a prior concentration is made by fire assaying; the spectrographic determination is carried out on the gold–silver fire assay bead or a lead button from arrested cupellation [1, 3, 7, 13, 25].

6. X-Ray Spectrography

X-ray fluorescence has not become established as a routine procedure for gold [3, 7, 27].

7. Atomic-Absorption Spectrophotometry

Atomic absorption offers another useful technique for the determination of small quantities of gold [7, 10, 15, 29, 37, 39]. The selectivity of atomic absorption, following a concentration step of solvent extraction or fire assay, enables a very low content of gold to be measured. The extraction of gold chloride or bromide into methyl isobutyl ketone, and the aspiration of the latter into the flame, is one common technique [10, 37]. The most sensitive absorption line, with a sensitivity of 0·3 μg/ml, is 2428·0 Å; other absorption lines are 2676·0 and 6278·2 Å.

8. Other Procedures

Polarographic methods have been published [11], and neutron activation and tracer procedures have been reviewed [6]. Trimethylammonium iodide has been advocated as a quantitative precipitant for gold [40].

EXPRESSION OF RESULTS

In countries using the metric system, gold is usually reported in grams per metric ton, the latter unit being 1000 kilograms. Other countries express the results of gold analyses in Troy ounces per ton, where the latter is 2000 pounds, or in pennyweights (dwt). One ounce Troy = 20 dwt; 1 dwt = 0·05 oz; one oz/ton = 34·285 g/metric ton.

The term "inch-dwt" is sometimes used in mining practice. It is dwt of gold per ton × length in inches across the sample. The latter length is at least 36 to 40 inches; the value of ore mined, expressed in dwt/ton = 1/36 or 1/40 of the inch-dwt assays quoted.

REFERENCES

1. Beamish, F. E., *Anal. Chem.* **33**, 1059–66 (1961).
2. Beamish, F. E., *Talanta* **12**, 789–816 (1965).
3. Beamish, F. E., *The Analytical Chemistry of the Noble Metals*, Oxford, Pergamon, 1966.
4. Beamish, F. E., *Talanta* **14**, 991–1009 (1967).
5. Beamish, F. E., *Talanta* **14**, 1133–49 (1967).
6. Beamish, F. E., Chung, K. S., and Chow, A., *Talanta* **14**, 1–32 (1967).
7. Beamish, F. E., Lewis, C. L., and Van Loon, J. C., *Talanta* **16**, 1–25 (1969).
8. Beardsley, D. A., Briscoe, G. B., Ruzicka, J., and Williams, M., *Talanta* **13**, 328–31 (1966).
9. Bugbee, E. E., *A Textbook of Fire Assaying*, New York, Wiley, 1940.
10. Butler, L. R. P., Brink, J. A., and Englebrecht, S. A., *Trans. Institution Min. Metallurgy* **76**, C188–91 (1967).
11. Cathro, K. J., *Analyst* **86**, 657–64 (1961).
12. Chow, A., and Beamish, F. E., *Talanta* **14**, 219–31 (1967).
13. Chow, A., Lewis, C. L., Moddle, D. A., and Beamish, F. E., *Talanta* **12**, 277–80 (1965).
14. Dillon, V. S., *Assay Practice on the Witwatersrand*, Johannesburg, Transvaal and Orange Free State Chamber of Mines, 1955.
15. Elwell, W. T., and Gidley, J. A. F., *Atomic-Absorption Spectrophotometry*, 2nd ed., Oxford, Pergamon, 1966.
16. Faye, G. H., and Inman, W. R., *Anal. Chem.* **33**, 1914–16 (1961).
17. Frankel, E., *J.S. African Inst. Mining Met.* **60**, 367–70 (1960).
18. Fulton, C. H., and Sharwood, W. J., *A Manual of Fire Assaying*, 3rd ed., New York, McGraw-Hill, 1929.
19. Furman, N. H., ed., *Scott's Standard Methods of Chemical Analysis*, 6th ed., Vol. 1, Princeton, N.J., Van Nostrand, 1962.
20. Gilchrist, R., *Rev. Met.* **52**, 287–93 (1955).
21. Hanson-Van Winkle-Munning Co., *Simple Methods for Analyzing Plating Solutions*, Matawan, N.J., Hanson-Van Winkle-Munning Co., 1958.
22. Hillebrand, W. F., Lundell, G. E. F., Bright, H. A., and Hoffman, J. I., *Applied Inorganic Analysis*, 2nd ed., New York, Wiley, 1953.
23. Holbrook, W. B., and Rein, J. E., *Anal. Chem.* **36**, 2451–3 (1964).
24. Imperial Chemical Industries, *Analysis of Cyanidation Solutions and Gold Precipitate*, London, Imperial Chemical Industries, 1963.
25. Jordanov, N., Mareva, St., Krasnobaeva, N., and Nedyalkova, N., *Talanta* **15**, 963–8 (1968).
26. Kember, N. F., and Wells, R. A., *Analyst* **76**, 579–87 (1951).
27. Kolthoff, I. M., and Elving, P. J., *Treatise on Analytical Chemistry*, Part II, Vol. 4, 71–105, New York, Interscience Publishers, 1966.
28. Morris, D. F. C., and Khan, M. A., *Talanta* **15**, 1301–5 (1968).
29. Pollock, E. N., and Andersen, S. I., *Anal. Chim. Acta* **41**, 441–6 (1968).
30. Pugh, C. R., and Tucker, H. T., *J.S. African Inst. Mining Met.* **60**, 371–5 (1960).
31. Sandell, E. B., *Colorimetric Determination of Traces of Metals*, 3rd ed., New York, Interscience Publishers, 1959.

32. SCHOELLER, W. R., and POWELL, A. R., *The Analysis of Minerals and Ores of the Rarer Elements*, 3rd ed., London, Charles Griffin, 1955.
33. SHEPARD, O. C., and DIETRICH, W. F., *Fire Assaying*, New York, McGraw-Hill, 1940.
34. SMITH, E. A., *The Sampling and Assay of the Precious Metals*, London, Charles Griffin, 1947.
35. STANTON, R. E., and MCDONALD, A. J., *Analyst* **89**, 767–70 (1964).
36. STRELOW, F. W. E., FEAST, E. C., MATHEWS, P. M., BOTHA, C. J. C., and VAN ZYL, C. R., *Anal. Chem.* **38**, 115–7 (1966).
37. THOMPSON, C. E., NAKAGAWA, H. M., and VAN SICKLE, G. H., *U.S. Geol. Survey Prof.* Paper No. 600B, B130–2 (1968).
38. TITLEY, A. W., *Analyst* **87**, 349–55 (1962).
39. VAN LOON, J. C., *Z. Anal. Chem.* **246**, 122–4 (1969).
40. WHITE, W. W., and ZUBER, J. R., *Anal. Chem.* **36**, 2363–4 (1964).
41. YOUNG, R. S., *Anal. Chim. Acta* **4**, 366–85 (1950).
42. YOUNG, R. S., *Analyst* **76**, 49–52 (1951).

17 Iron

Iron is one of the commonest elements found in the earth's crust, and is our most important metal; its determination is consequently of frequent occurrence in all laboratories serving the mineral industries. Methods for iron are well standardized; the few interfering elements which are normally present can usually be eliminated by simple techniques.

The principal ores of iron are hematite, Fe_2O_3, magnetite Fe_3O_4, limonite or goethite $Fe_2O_3.nH_2O$, and siderite $FeCO_3$; pyrite FeS_2, and pyrrhotite Fe_7S_8, are important sources of sulphuric acid and of iron ore. Varying quantities of iron are found in virtually all rocks, minerals, ores, concentrates, tailings, slags, fluxes, refractories, flue dusts, refinery slimes, leach solutions, alloys, and other materials of the mining and metallurgical industries. Iron is frequently determined in soils, waters, and biological materials.

For referee iron determinations it should be borne in mind that most ordinary stainless spatulas used for weighing out samples are magnetic; a spatula made from austenitic stainless steel or similar non-magnetic material should always be employed for such work.

ISOLATION OR SEPARATION OF IRON

After removal of SiO_2, Nb, Ta, W, and Ag by evaporation in hydrochloric acid, and of members of the acid sulphide group with hydrogen sulphide in 5–10% hydrochloric acid, iron is precipitated by ammonium hydroxide in the presence of ammonium salts and is thereby separated from members of succeeding groups. It will be accompanied by Al, Be, Ti, Cr, Th, V, U, Zr, In, Ga, and the rare earths. Because iron is usually determined volumetrically by reducing the solution and titrating it with a standard oxidant, for most samples the only interference in the ammonium hydroxide group is vanadium if it is present in quantities exceeding about 1% of the iron.

It may frequently be more convenient to remove those members of the acid sulphide group which interfere in an iron determination by a technique other than hydrogen sulphide. Copper can be removed by electrodeposition, and copper with molybdenum by an ammonia separation. Selenium, tellurium, and gold can be separated by sulphur dioxide in hydrochloric acid solution. Arsenic, antimony, and selenium may be volatilized by several evaporations with a mixture of hydrobromic and sulphuric acids.

For some purposes it is desirable to separate iron from one or more elements of the ammonium hydroxide or sulphide groups. Electrolysis at a mercury cathode in dilute sulphuric acid deposits Fe, Cr, Zn, Ni, Co, In, and Ga, whereas Al, Be, Ti, Zr, P, V, Th, and U remain in solution. Cupferron in cold 10% sulphuric acid solution precipitates Fe, Ti, V, Ga, and Zr, leaving Al, Be, P, Cr, Ni, Co, U(VI), In, and Zn in the filtrate. Fusion with sodium peroxide and solution in water precipitates Fe, Ni, Co, Ti, Zr, Th, and Mn, while V, Al, P, Be, Cr, Zn, W, and U remain in solution. Precipitation with ammonium sulphide in a solution containing ammonium tartrate separates Fe, Mn, Ni, Co, and Zn from the filtrate containing Al, P, Ti, U, V, and Zr. Chromium may be separated from iron by volatilization of the former with boiling perchloric and hydrochloric acids. Manganese and zinc will not be precipitated with iron in the presence of ammonia and ammonium salts; manganese can also be separated from iron by a nitro-chlorate separation, and, together with nickel and cobalt, by zinc oxide. If a re-precipitation is made, an ammonia separation will remove nearly all nickel and cobalt from iron. Cobalt in particular is tenaciously held by ferric hydroxide; a large excess of ammonia and ammonium salts, and hot solutions, will minimize the retention. Fortunately, the volumetric procedure for iron does not require the complete elimination of nickel and cobalt; their interference is simply due to the colour imparted to the solution by large quantities.

An ether separation in dilute hydrochloric acid will remove iron into the organic layer, with Ga, Ge, Au, Mo, Tl, and small amounts of Te, As, Sb, Sn, P, and Ir, from Al, Cr, Co, Ni, Zn, Mn, Be, Ti, U, V, Zr, Ca, Mg, Ta, Nb, W, Pt, Pd, Cu, Pb, Bi, Ag, Cd, Se, S, and Ba, which are not extracted. Though usually employed to remove large quantities of iron when determining other constituents of a sample, an ether separation is occasionally useful for isolating iron prior to its determination. Details of the ether separation of iron are given in the section on Cobalt. Iron is separated from uranium by sodium bicarbonate, uranium remaining in solution as long as carbon dioxide is present. Iron can also be separated from a number of elements by ion exchange and chromatography [15]. Both anion and cation exchange resins have been used, usually with iron in the trivalent state and in concentrated hydrochloric acid. For instance, in concentrated hydrochloric acid, ferric iron is retained on Amberlite IRA 400 resin, while aluminium passes through; iron can then be eluted with N hydrochloric acid.

1. Volumetric Determination

For an appreciable quantity of iron, volumetric determinations are nearly always used; gravimetric methods are rarely employed.

A. With Potassium Dichromate

Decompose a 0·5–2 g sample with nitric and hydrochloric acids; for high sulphide products add a little bromine before the addition of nitric acid, and for slags or similar refractory products add a few drops of hydrofluoric acid after nitric and hydrochloric. Add 10 ml 1 : 1 sulphuric acid and evaporate to strong fumes of this acid. Dissolve the soluble salts in hot water, boil, filter, and wash. If an insoluble residue remains, which might contain iron, fuse it with sodium carbonate or sodium pyrosulphate and add to the main portion.

A number of other decomposition procedures may be used:

(1) Some substances such as silicates which are free of easily reducible metals, and of phosphides, arsenides, and sulphides, may be fused directly with sodium carbonate in a platinum crucible.

(2) Samples containing a mixture of sulphide, oxide, and silicate, such as converter slags, may be fused with sodium bisulphate in a Pyrex pear-shaped "copper flask".

(3) Treatment with hydrofluoric and sulphuric acids in a large platinum dish is often used for samples with a high silica content.

(4) In some laboratories it is customary to do iron and silica on the same sample; a sodium carbonate fusion, with prior acid treatment if necessary to eliminate substances which attack platinum, is followed by the usual silica dehydration steps, and iron is determined on the filtrate.

(5) A good dissolution procedure for many iron ores is to heat to boiling with perchloric and phosphoric acids. The solution of some iron ores is greatly assisted by the addition of a few drops of 5% stannous chloride solution.

(6) A rapid routine method for iron in ores consists in fusing the sample with sodium peroxide in a zirconium crucible, and adding water and hydrochloric acid; in the absence of Cu, Pt, Au, Mo, As, Sb, U, Se, Te, Ni, Co, W, and V, the solution may be titrated directly with potassium dichromate. If V and Mo are present, omit the addition of hydrochloric acid; filtration of the aqueous solution obtained after the sodium peroxide fusion yields iron in the precipitate, with V and Mo in the filtrate.

(7) If the material contains, in addition to iron, only the interferences Cu, Mo, Ni, or Co, the sample can be decomposed in aqua regia, and Cu, Mo, Ni, and Co can be removed in the filtrate with two ammonium hydroxide precipitations. This is a common separation in the metal extractive and fabricating industries; aluminium together with Be, Ti, Cr, V, U, Nb, Ta, Zr, and Th, will accompany iron, but none of these except vanadium affect the volumetric titration of iron.

(8) It is occasionally useful to remember that iron may be determined on the filtrate from a sulphur determination. The presence of excess barium chloride has no effect on the dichromate titration.

Adjust the acid content of the hydrochloric or sulphuric acid solution, obtained by complete decomposition of the sample with acids, fusion, or a combination of these, to 5–10% and precipitate all members of the hydrogen sulphide group. Filter, wash with acidulated hydrogen sulphide water; boil the filtrate for 15 minutes to remove all hydrogen sulphide, add sufficient hydrogen peroxide to oxidize all iron to the ferric condition, and boil until the excess peroxide is decomposed.

In the absence of Cu, As, Sb, Mo, Au, Se, Te, and Pt, this precipitation with hydrogen sulphide may be omitted; the other elements of the acid sulphide group do not interfere when iron is reduced with stannous chloride and titrated with $K_2Cr_2O_7$. If, as frequently occurs, copper is the only member of this group which occurs in significant amounts in the sample, it may be removed in the filtrate of the next step, together with Ni, Co, and Mo, when iron is precipitated with ammonium hydroxide. Copper may also be frequently separated and determined conveniently by electrolysis, prior to an iron titration. For some samples, As, Sb, and Se can be more readily removed by volatilization with several evaporations of a mixture of hydrobromic and sulphuric acids.

Remove the oxidized solution of iron from the hot plate, cool slightly, and cautiously add ammonium hydroxide until all iron is precipitated. If hydrochloric acid was not present initially, add 2–5 g ammonium chloride to keep Ca, Mg, Zn, etc. in solution. Boil for a few minutes, filter on Whatman No. 41 paper and wash thoroughly with hot water. If a heavy precipitate of ferric hydroxide is present, stir frequently to overcome bumping until boiling commences.

It is usually necessary to remove occluded impurities from the ferric hydroxide precipitate by a re-precipitation. Transfer the precipitate back to the original beaker, dissolve in hydrochloric acid, make ammoniacal, boil, filter, and wash as before.

Dissolve the precipitate in hot dilute hydrochloric acid, washing the paper thoroughly with hot water, and bring the solution nearly to boiling on the hot plate. Keep the volume of solution low; it should preferably not exceed about 100 ml. The concentration of hydrochloric acid should be about 1–2 N for the dichromate titration. Remove from the hot plate, and to the hot iron solution carefully add from a burette a solution of stannous chloride until the yellow colour of ferric chloride just disappears. Do not add more than a drop or two excess stannous chloride solution. A convenient solution of stannous chloride is made by dissolving 80 g $SnCl_2.2H_2O$ in 180 ml hydrochloric acid on a hot plate and diluting with 300 ml of water; place a stick of tin in the reagent bottle to maintain the solution in the reduced state. If more than a drop or two of excess stannous chloride has been inadvertently added, add a few drops of a dilute solution of potassium permanganate to restore the yellow colour; discharge the latter with dropwise addition of stannous chloride.

Immediately cool the sample in running water, and when cold add 10 ml saturated mercuric chloride solution rapidly and swirl to mix, followed by 15 ml of a sulphuric–phosphoric acid mixture containing 150 ml sulphuric acid and 150 ml phosphoric acid in 1 litre. The latter mixture serves two purposes: it prevents the yellow colour of ferric iron from producing a green at the blue endpoint, and it lowers the oxidation potential of the ferric-ferrous system below the value otherwise attained in either sulphuric or hydrochloric solutions.

If more than one or two drops of stannous chloride in excess have been added, especially if the solution is warm and mercuric chloride is added slowly, there is a tendency for the formation of dark grey or black mercury which vitiates results.

Add 3 drops of barium diphenylamine sulphonate; this indicator is prepared by dissolving 0·32 g in 100 ml of water containing 1 ml sulphuric acid and filtering. A 0·2% aqueous solution of sodium diphenylamine sulphonate may be substituted as an indicator. Titrate with standard potassium dichromate solution until the green colour of the chromic salt changes to a violet-blue. Add the oxidant slowly towards the endpoint. The reactions in this determination are:

$$2FeCl_3 + SnCl_2 \rightarrow 2FeCl_2 + SnCl_4$$

$$SnCl_2 + 2HgCl_2 \rightarrow 2HgCl + SnCl_4$$

$$6FeCl_2 + K_2Cr_2O_7 + 14HCl \rightarrow 6FeCl_3 + 2CrCl_3 + 2KCl + 7H_2O$$

$$1 \text{ ml of } 0{\cdot}1\text{N } K_2Cr_2O_7 = 0{\cdot}005585 \text{ g Fe}$$

$$0{\cdot}1\text{N } K_2Cr_2O_7 = 4{\cdot}9033 \text{ g/litre}$$

In industrial work, a solution having a value of 1 ml = 0·005 g Fe is frequently used; this contains 4·3899 g $K_2Cr_2O_7$ per litre. For lower iron contents, another $K_2Cr_2O_7$ solution having a value of 1 ml = 0·0025 g Fe is often convenient.

Titration with potassium dichromate after reduction with stannous chloride is the most satisfactory general-purpose method for iron [1, 8, 11, 15, 18, 31]. The only interferences are Pt, Cu, Au, Mo, As, Sb, Se, Te, W, and V. The first eight are eliminated by hydrogen sulphide precipitation. Most of the tungsten is separated initially with silica by acid digestion, and the remainder can be separated by ammonium hydroxide. If the vanadium content of the sample exceeds 1% of the iron, a sodium hydroxide separation should be made to pass vanadium into the filtrate; a re-precipitation is advisable to free the iron of occluded vanadium if the latter is high. Alternatively, iron can be precipitated from vanadium with ammonium sulphide in the presence of ammonium tartrate. Substantial quantities of nickel, cobalt, and uranium tend to mask the colour at the end point; the first two are removed with the ammonium hydroxide separation of the iron, while uranium can be separated from iron by the use of ammonium carbonate. If copper and molybdenum are the only members of the hydrogen sulphide group present, they can usually be more conveniently separated from iron by an ammonia precipitation.

The endpoint of the potassium dichromate titration is sharp and stable; it is not affected as much as permanganate or titanous solutions by small quantities of interfering elements or by organic substances. With it, iron can be readily determined in either chloride or sulphate solution. The standard solution can be made up simply by weighing pure potassium dichromate directly; standardization against iron or another compound is unnecessary. A further important advantage is the stability of the solution; no significant change in titre occurs when potassium dichromate solutions are stored for years.

The procedure outlined above, of course, assumes the presence of all interferences. For many materials it is not necessary to remove silica or to separate iron as the hydroxide. For instance, a determination of iron in Portland cement, which contains about 20% silica, entails only treatment with hydrochloric acid followed by reduction and titration. Again, for routine work on many ores it may be possible to fuse with sodium peroxide in a zirconium crucible, dissolve in hydrochloric acid, and proceed with reduction and titration.

It is sometimes recorded in the literature that, because stannous chloride and hydrogen sulphide reduce vanadium to the quadrivalent state in which it does not oxidize diphenylamine to give a blue colour as quinquevalent vanadium does, vanadium does not interfere in iron titrations. This is misleading; no significant interference is encountered when the vanadium is less than 1% of the iron, but above this figure interference becomes increasingly pronounced and a separation must be carried out.

The interference appears to be caused by the production of vanadium(III) during the reduction step [21]. The vanadium(III) is not re-oxidized on the addition of mercury (II) chloride, but is oxidized to vanadium(IV) when the iron(II) is titrated with potassium dichromate, yielding a positive error in the iron titration. Formation of vanadium(III) can be minimized by slow addition of stannous chloride and by rapid cooling of the solution after reduction. Reduction with hydrogen sulphide, in spite of certain disadvantages, may be preferable to stannous chloride when a small quantity of vanadium is present. Any appreciable amount of vanadium must, of course, be separated.

B. With Potassium Permanganate

The titration of iron with potassium permanganate follows the same steps of sample solution, elimination of interferences, and reduction of the iron with a solution of stannous chloride, which have been described in the previous section. To the cooled, reduced solution of iron add 10 ml saturated mercuric chloride, 25 ml of Zimmerman–Reinhardt solution, and titrate with standard potassium permanganate to a definite pink tint. The Zimmerman–Reinhardt solution is made by dissolving 200 g of $MnSO_4.4H_2O$ in 1 litre of water and adding a cooled mixture of 400 ml of sulphuric acid, 1200 ml water, and 400 ml of phosphoric acid. The reaction for the permanganate titration is:

$$5FeCl_2 + KMnO_4 + 8HCl \rightarrow 5FeCl_3 + KCl + MnCl_2 + 4H_2O$$

$$1 \text{ ml of } 0{\cdot}1\text{N } KMnO_4 = 0{\cdot}005585 \text{ g Fe}$$

0·1N $KMnO_4$ contains 3·16 g/litre. This solution must be standardized frequently against ferrous ammonium sulphate or sodium oxalate. If ferrous ammonium sulphate is used, it should be dissolved in 100 ml water containing 10 ml of 1 : 1 sulphuric acid and carried through all steps of the iron titration; 1 g of this salt contains 0·1424 g iron, equivalent to a convenient titration of about 25 ml of 0·1 N potassium permanganate. Potassium permanganate may also be standardized against sodium oxalate; 1 litre of 0·1N $KMnO_4$ = 6·700 g of $Na_2C_2O_4$. Dissolve 0·1500 g of the latter compound in 100 ml of 5% sulphuric acid, heat to 60°C, and titrate with potassium permanganate to a faint pink that persists for 30 seconds.

C. With Titanous Salts

If all iron is oxidized to the ferric state, it may be titrated with a standard solution of titanous sulphate or titanous chloride. Potassium thiocyanate is used as indicator, the endpoint being the disappearance of the red colour of ferric thiocyanate. The procedure suffers from the drawback that the standard solution must be protected from air; an arrangement of burette and storage bottle connected to a Kipp generator for carbon dioxide is a convenient assembly [11]. Storage under hydrogen or nitrogen may also be used. The 0·1 N titanous sulphate solution may be prepared by mixing 100 ml of a 20% solution with 80 ml of 1 : 1 sulphuric acid, and diluting to 1 litre with water. It is standardized by dissolving 1 g ferrous ammonium sulphate in 100 ml of 5% sulphuric acid, titrating with 0·1 N potassium permanganate to the usual pink end point, and boiling for 10 minutes to expel oxygen and the slight excess of permanganate. Cool, add 5 ml of 10% potassium thiocyanate solution, and titrate with the titanous salt solution to the complete disappearance of colour.

The interferences are practically the same as those for the dichromate titration: members of the hydrogen sulphide group, tungsten, and vanadium. For work of the highest accuracy, titration should be carried out in a flask containing an atmosphere of carbon dioxide or nitrogen; for many purposes standardization, and titration of the unknown, may be carried out in air. The reactions are:

$$2TiSO_4 + Fe_2(SO_4)_3 \rightarrow Ti_2(SO_4)_3 + 2FeSO_4$$

$$6KCNS + Fe_2(SO_4)_3 \rightarrow 2Fe(CNS)_3 + 3K_2SO_4$$

$$1 \text{ ml of } 0{\cdot}1 \text{ N } TiSO_4 = 0{\cdot}005585 \text{ g Fe}$$

Although the solution of titanous salt, if allowed to stand for a few days before standardization, remains fairly stable if stored under an inert gas, it is wise to frequently check such a sensitive solution.

2. Colorimetric Determination

A. With Potassium Thiocyanate

The red colour produced by the reaction of a thiocyanate with a ferric solution can be utilized to determine small quantities of iron [1, 8, 11, 15, 19, 25]. Substantial quantities of the following elements, which form coloured thiocyanates or complexes, must be absent: Ag, Hg, Cu, Co, Ni, Mo, Bi, Ti,

Cd, Zn, Sb, and U. Silver and mercury may be removed initially as insoluble chlorides in hydrochloric acid; mercury may also be volatilized by several evaporations with sulphuric and hydrobromic acids. Copper, nickel, cobalt, and molybdenum can be separated in ammonia from the small quantity of ferric hydroxide which is collected by the addition of a little manganese or aluminium. Bismuth, together with Ag, Hg, Cu, Cd, Sb, and Mo may be precipitated by hydrogen sulphide in acid solution. Iron can be isolated from titanium and uranium by an ether extraction, or by carbonate and tartrate separations. Zinc can be separated with ammonia and ammonium chloride.

Dissolve the sample in nitric and hydrochloric acid; if an insoluble residue remains, which might contain traces of iron, fuse it with sodium carbonate and add to the main portion. If the sample contained fluorides, or hydrofluoric acid was used in its decomposition, eliminate all fluorine by evaporating to dryness with sulphuric acid.

For high-purity nickel or cobalt, the rate of solution in hydrochloric acid can be greatly increased by adding a drop or two of 3% platinic chloride solution in hydrochloric acid.

Evaporate the acid solution nearly to dryness, dilute to a convenient volume so that 100 ml has at least 1 ml of 1 : 1 hydrochloric acid, and add to this volume 0·5 ml of 0·5% potassium permanganate solution. Make distinctly ammoniacal with ammonium hydroxide, boil, and allow to stand hot for an hour until the precipitate of hydrated manganese dioxide has coagulated. Filter, wash thoroughly with hot water, and discard the filtrate. Dissolve the precipitate in hot dilute hydrochloric acid, wash, and adjust the acidity to approximately 1 N. Add sufficient 1·5 molar potassium thiocyanate to make the final concentration about 0·3 M in thiocyanate, dilute to volume with water, and measure the transmittance at 480 nm without delay.

For some materials, the use of a final solution containing 50–60% by volume of acetone to increase the intensity and stability of the colour may be advantageous [25]. For other samples, extraction of the iron thiocyanate complex with isobutyl alcohol enables higher acidities to be used, thereby avoiding the difficulty of a separation of phosphates [25]. Iron can also be isolated from the bulk of the sample by solvent extraction with methyl isobutyl ketone from 7 N lithium chloride solution [19]. The ferric thiocyanate colour can be developed directly in the ketone extract and measured at 490 nm.

B. WITH 1,10-PHENANTHROLINE

With ferrous iron, 1,10-phenanthroline forms a coloured complex in weakly acid, neutral, and weakly basic solutions [1, 6, 8, 12, 15, 17, 25]. Iron is reduced to the ferrous condition with hydroxylamine hydrochloride or hydroquinone. Substantial amounts of many elements interfere by giving coloured or colourless complexes, or precipitates; among these may be noted Ag, Bi, Hg, Cd, Zn, Mo, W, Cu, Be, Ni, Co, and Sn. Separations in hydrochloric acid, with hydrogen sulphide in acid solution, and ferric hydroxide in ammonia, as outlined in the preceding section, may be necessary. Extraction of ferric iron by ether, or by isobutyl methyl ketone in 5·5–7M hydrochloric acid, can also be used to separate iron from nearly every interfering element. The list of interferences is not as formidable as it appears, because many elements can be tolerated in quantities greatly exceeding those of iron. For instance, beryllium and cadmium can be present up to 25 times the iron, and tin to 10 times the amount of iron in the sample.

Transfer an aliquot portion of the weakly acid sample solution containing 0·01–0·2 mg of iron to a 25 ml volumetric flask. On another portion determine the volume of 25% sodium citrate solution required to bring the *p*H to about 3·5. To the working aliquot add 1 ml of 1% hydroquinone, 2 ml of a 0·25% solution of 1,10-phenanthroline, and then the previously determined amount of sodium citrate solution. Allow to stand for an hour to complete the iron reduction, make up to volume, and determine the transmittance at 508 nm.

For a very low iron content, extract the iron contained in 15 ml of 7 N hydrochloric acid into 15 ml of a mixture of 2 volumes methyl isobutyl ketone and 1 volume of benzene by shaking for 1 minute and allowing the layers to separate for 1 minute [17]. Wash the organic phase with several small portions of 7 N hydrochloric acid, discarding the washings. Add 15 ml of water to the separatory funnel and transfer the iron back to the aqueous phase by shaking for 1 minute. Allow to stand 1 minute, draw off the aqueous layer to a small beaker, and repeat the extraction with another 5 ml of water. Proceed with the determination as outlined in the previous paragraph.

The reagent 4,7-diphenyl-1,10-phenanthroline, or bathophenanthroline, is sometimes used for extremely low iron contents. Its ferrous complex can be extracted in chloride solution at *p*H 4–5 into chloroform and measured at 533 nm [3, 12]. Bathophenanthroline has been used for iron in boiler feed water [30]. It has also been used to determine ferrous iron in the presence of a large amount of ferric iron, by complexing the latter with ammonium dihydrogen phosphate at *p*H 2 [22].

3. Optical Spectrography

Iron has a large number of spectral lines, and small quantities of this element are routinely determined by emission spectrography in many materials. Procedures for 0·02–2% iron in aluminium and its alloys, 0·005–0·25% in zinc and zinc-base alloys, 0·001–0·3% in tin, about 0·0005% in pig lead, and 100–5000 ppm in zirconium and its alloys have been described [1].

Iron has long been determined spectrographically in refined copper, and examples are given for copper and its alloys over a range of 1 ppm to about 1·5% [6]. Methods have been described for 10–5000 ppm of iron in nickel [17].

4. X-Ray Spectrography

The determination of iron by X-ray fluorescence for general application has not attained prominence. Various elements appear to enhance or diminish the iron radiation, other procedures for iron can determine wide ranges of this element rapidly and accurately, and it seems likely that X-ray techniques will be limited to specialized materials. The determination of 20 ppm to 1·2% iron in nickel has been described [17].

5. Atomic-Absorption Spectrophotometry

Small quantities of iron may be determined satisfactorily by atomic absorption. The most sensitive line, with a sensitivity of 0·1 μg/ml, is 2483·3. For maximum precision, the line 3719·9 is used. Other iron lines employed are 2488·1, 2522·8, 2527·4, 2719·0, 2720·9, 2966·9, 3020·6, 3440·6, 3719·9, and 3859·9 [7].

SPECIAL PROCEDURES

1. Simultaneous Reduction of Iron and Elimination of Copper

For some materials in extractive metallurgy, a rapid method for iron consists in the simultaneous reduction of iron and elimination of copper by boiling with test lead in 10% hydrochloric acid, or with aluminium or zinc in 10% sulphuric acid solution. Copper is precipitated in the metallic state and removed, with the Pb, Al, or Zn, by filtration. The reduced iron, if in hydrochloric or sulphuric acid solution, may be titrated with potassium dichromate; if in sulphuric acid solution, potassium permanganate may be used as oxidant.

2. Iron Ores by Hydrogen Sulphide Reduction

A favourite method for iron in iron ores utilizes hydrogen sulphide to reduce the iron and simultaneously precipitate Cu, Mo, and all other elements of the acid sulphide group [1]. Decompose with hydrochloric and nitric acids; if an iron-bearing residue remains, treat it with hydrofluoric and sulphuric acids in a platinum crucible, fuse with sodium pyrosulphate, and add to the main solution. Add sulphuric acid and evaporate to strong fumes of this acid. Cool, dilute, add a few drops of a solution of potassium permanganate until a pink colour is visible, and boil. Pass in hydrogen sulphide until all Group 2 elements are precipitated and all iron is reduced. Filter, wash, boil the filtrate until all hydrogen sulphide has been removed, and cool. Add the usual sulphuric–phosphoric acid mixture, 5 drops of 0·2% sodium diphenylamine sulphonate and titrate with standard potassium dichromate.

3. Ferric and Ferrous Iron

It is occasionally necessary in mining and metallurgical laboratories to determine the valence state of iron in samples. The results are usually only approximate, because the original state of the iron may change during analysis, and slight interferences are almost impossible to eliminate. Various methods have been critically reviewed [26].

A. FERROUS IRON

(1) A fairly close approximation of ferrous iron in silicates and similar samples may be made by the following procedure [11, 18, 23]. Place 0·5 g of −100 mesh material in a platinum crucible and moisten with freshly boiled and cooled water. Add 15 ml of 1 : 2 sulphuric acid, cover with a close-fitting platinum lid, and heat rapidly till almost boiling. Carefully move the lid to the side and quickly introduce 5 ml hydrofluoric acid, cover again immediately, and allow to boil gently for 10 minutes. In the meantime, to 200 ml of freshly boiled and cooled water in a 400-ml beaker, add 15 ml of 1 : 2 sulphuric acid and 10 ml of a saturated solution of boric acid. Immerse the platinum crucible and contents in the solution in the beaker, remove and rinse the crucible, and titrate the ferrous iron with standard potassium permanganate solution to the first permanent pink. Carbonaceous material, sulphides and vanadium interfere in the titration; iron-bearing minerals which dissolve with difficulty also introduce errors.

Potassium dichromate can be used in place of potassium permanganate; in this case the usual additions of sulphuric–phosphoric acid and diphenylamine indicator are made prior to titration [23]. Decomposition with hydrofluoric acid in a plastic apparatus has been described [27].

Ferrous iron may also be determined in silicates by an oxidizing decomposition [28]. A known excess of potassium iodate is added to the sample in a flask, together with hydrofluoric and sulphuric acids. Ferrous iron, which is released during the subsequent decomposition at boiling temperature, reacts with iodate to produce iodine. The latter is titrated with standard sodium thiosulphate, using starch indicator. It has been shown, however, that the presence of metallic iron from grinding operations can lead to appreciable errors with this procedure. Metallic iron consumes three times as much standard oxidant in the oxidizing decomposition method as it does in the non-oxidizing procedure outlined in the previous paragraphs [23].

(2) In solutions derived from leaching, refining or pickling operations, ferrous iron may be determined directly by adding to a measured volume 15 ml of 20% sulphuric acid, 3 drops of barium diphenylamine sulphonate indicator and titrating with standard potassium dichromate to the appearance of the violet-blue endpoint. One ml of 0·1N $K_2Cr_2O_7$ = 0·005585 g ferrous iron or 0·01519 g $FeSO_4$.

To the solution after the above titration add a slight excess of copper metal precipitated powder, stir for two minutes, filter and titrate again with standard 0·1N potassium dichromate. This titration gives the total iron, and the net titre represents ferric iron [32]. One ml of 0·1N $K_2Cr_2O_7$ = 0·01999 g $Fe_2(SO_4)_3$.

(3) Another approach to the determination of ferrous iron in the presence of a large quantity of ferric iron is to utilize the sensitivity of the ferrous-bathophenanthroline complex [22]. Ferric iron is complexed with ammonium dihydrogen phosphate at *p*H 2·0–2·1, the latter adjustment being made with sodium acetate.

B. Ferric Iron

(1) *Titration with titanous salt.* The iron solution, which may contain hydrochloric, sulphuric, or hydrofluoric acid if boric acid be present in excess, but not nitric acid, is titrated in the cold with a standard solution of titanous sulphate or chloride. The titration details are given under the discussion of titanous salts in the volumetric determination of iron. Decomposition of the sample may be effected by H_2SO_4 + HF in a closed platinum crucible, as described for ferrous iron.

When large numbers of ferric and ferrous irons must be determined, a more elaborate procedure will improve the speed and accuracy; a spectrophotometric titration method for ferric iron, using coulometric generation of titanous ion as the titrant, has been described [2]. Sulphides, organic matter, V, Mo, and Cu must be absent.

(2) *Titration with stannous chloride.* To the sample in a covered beaker add 50 ml of 1 : 1 hydrochloric acid, boil, remove from the hot plate, titrate while hot with standard stannous chloride solution to the colourless endpoint. A solution of 2 g stannous chloride dissolved in 10 ml hot hydrochloric acid and diluted to 1 litre has a value of approximately 1 ml = 0·001 g Fe. The stannous chloride solution should be protected from air. It is obvious that this procedure is applicable only to solutions or to substances which are decomposed in hydrochloric acid.

3. Magnetite in Copper Mattes and Slags

The determination of magnetite in copper reverberatory mattes and slags was originally developed many years ago [10, 24]; it is essentially a method for the determination of ferric iron, the magnetite content being calculated from the ferric iron value on the assumption that all the ferric iron present is in the form of magnetite. The procedure depends on the rapid dissolution of sulphides in a hot nitrochlorate solution and the relative insolubility of magnetite in the same conditions.

For mattes, weigh 0·5 g of −100 mesh sample into a 600-ml beaker, moisten with water, place in a cooling trough and add 25 ml of a nitrochlorate mixture. The latter is made by dissolving 30 g sodium chlorate in hot water, cooling, adding 30 ml nitric acid, cooling, and adding 30 ml of 85% sulphuric acid. When the action of the nitrochlorate solution on the sample has subsided, heat the beaker to boiling, add 50 ml cold water, filter, and wash. Determine ferric iron by titrating with a titanous salt, as described earlier in this chapter, and calculate to magnetite. Fe × 1·382 = Fe_3O_4.

A variation of this procedure for copper–nickel blast furnace mattes has been published [5]; mixtures of nitric acid with either potassium chlorate or bromine are allowed to stand in the cold for 16 hours.

For slags, the procedure is as follows [24]. To 0·5–1 g of −100 mesh sample add 20 ml of a hot mixture of equal parts nitric acid and 15% sodium chlorate, boil 1 minute, and add 100 ml cold water. Filter on a Gooch crucible and wash; transfer to a beaker containing a boiling solution of 7 ml water and 10 ml of a mixture of 3 parts hydrochloric acid and 1 part hydrofluoric acid. Cover and continue boiling until the residue has gone into solution, transfer to a larger beaker containing enough cold water to bring the temperature of the solution to 20°C, add 20 ml of hydrochloric acid saturated with boric acid, and titrate the ferric iron with standard titanous salt solution.

Another determination for magnetite in copper mattes and slags has been described [13]. The outline below contains additional information kindly supplied by the author in a personal communication. Weigh 1 g into a flask, pass in a stream of carbon dioxide, add an excess of standard stannous chloride solution and 50 ml hydrochloric acid. Boil for 20 minutes to remove all hydrogen sulphide, add sufficient hydrofluoric acid to complete the decomposition, add boric acid, dilute with air-free water, and cool. Add marble, 1 g potassium iodide, starch solution, and titrate the excess stannous chloride with 0·1 N iodine solution. One ml of the latter is equivalent to 11·577 mg Fe_3O_4. Results are satisfactory in the absence of metallic iron or copper, Fe_2S_3, CuS, Cu_2S, and CuO. The method can be used for nickel slags but not mattes.

Another method for magnetite in copper mattes and slags uses a bromine–methyl alcohol mixture to dissolve metallic iron and iron sulphide, leaving the oxides entirely in the residue [16]. To 1 g of −150 mesh material add 80 ml methyl alcohol and 20 ml bromine. Shake for 1 hour, filter, and wash. Dissolve the residue in hydrochloric and hydrofluoric acids under an atmosphere of carbon dioxide, and determine the magnetite from the ferric iron by titrating with a titanous salt.

4. Determination of Pyrite and Pyrrhotite

On mine and mill products a differentiation of pyrite from pyrrhotite is sometimes required. Most pyrites are nearly insoluble in dilute hydrochloric acid whereas some pyrrhotites are soluble. Where the solubility of these minerals falls into this category, the following procedure can be used [21]. Heat for 10 minutes 1–2 g of the −100 mesh sample with 50 ml of a solution containing 2 parts water to 1 part hydrochloric acid by volume, filter, and wash. For this particular orebody, pyrite was nearly insoluble and practically all the pyrrhotite soluble in this acid solution; determination of iron in the filtrate and conversion to Fe_7S_8 gave a good approximation of the pyrrhotite present. Treatment of the pyrite residue with mixed acids, determination of iron, and conversion to FeS_2 gave a reasonable estimate of the pyrite present. It must be emphasized that such a differentiation depends on the type of mineral; many pyrrhotites show a solubility little more than that of pyrite.

Another method for pyrrhotite is based on the evolution of hydrogen sulphide by acid treatment. Place 0·5–5 g sample in a 500 ml distilling flask fitted with a rubber stopper at the top, through which passes the stem of a separatory funnel [4]. The side tube is connected to two absorption bulbs containing about 3% cadmium sulphate. Pour 100 ml of 1:4 sulphuric acid solution into the funnel, and allow it to pass slowly into the flask. Heat the latter gently, and finally drive all the hydrogen sulphide out of the flask and into the absorption bulbs by a stream of air admitted through the separatory funnel. Transfer the contents of the absorption bulb to a beaker, rinse with water, add 25 ml of 0·1 N iodine solution, 10 ml hydrochloric acid, and titrate the excess iodine with 0·1 N sodium thiosulphate, using starch as indicator.

$$I_2 = H_2S = CdS = Fe_7S_8.$$

$$1 \text{ ml of } 0{\cdot}1\text{N } I_2 = 0{\cdot}0044 \text{ g pyrrhotite}$$

5. Determination of Metallic Iron

It is sometimes necessary to determine metallic iron in the presence of iron oxides in various metallurgical products. There are several simple methods [9, 14, 16, 23, 26, 29].

(a) Weigh a sample which is at least finer than −65 mesh into a 400-ml beaker, and add 15–20 ml of 10% copper sulphate solution containing 0·075% sulphuric acid [29]. Dilute with 25 ml hot water, cover the beaker and boil for 15 minutes, maintaining the volume of approximately 50 ml by addition of water. Filter, wash with hot water, dilute to 100 ml, add 14 ml 1 : 1 sulphuric acid and a strip of aluminium. Boil until all copper has precipitated, cool, filter, wash with cold water, dilute to about 300 ml with cold water, and titrate with standard potassium permanganate. This figure for iron represents the metallic iron in the sample. The presence of ferrous sulphide causes high results.

(b) In another procedure it is not so important to have the sample finely ground; a particle size up to 2–3 mm can be tolerated [9]. Shake 0·2–1 g of sample for 1 hour with 10 g copper sulphate, 20 ml water, and 1 g of mercury. The latter acts as a catalyst and keeps the surface of iron free from deposited copper. Filter the copper equivalent of the metallic iron, dissolve in nitric acid, and determine copper electrolytically or by any other standard procedure.

$$\text{The percentage of metallic iron} = \% \text{ of Cu} \times \frac{\text{at. wt. of Fe}}{\text{at. wt. of Cu}}$$

(c) In another method, introduce the finely ground sample to a 100-ml volumetric flask filled with carbon dioxide, and add 3 g of mercuric chloride and 50 ml of water [14]. Allow to remain on a steam bath until the iron has dissolved, or boil for 1 minute over a free flame. Close the flask with a Bunsen valve, cool, and fill to the mark with water. Mix, allow to settle, and filter into a vessel containing carbon dioxide. Acidify an aliquot with sulphuric acid and titrate the ferrous chloride with standard potassium permanganate. The amount of chloride present in the filtrate is so small that addition of manganous salt is unnecessary. Mercuric chloride acts upon the metallic iron to form ferrous chloride, mercurous chloride, and mercury; after filtration, the ferrous chloride alone is titrated.

(d) A fourth variation is carried out as follows. Place the −100 mesh sample in a 125-ml Erlenmeyer flask from which the air has been swept out by carbon dioxide. Pour 30 ml of a solution containing 53·35 g $CaCl_2.2H_2O$ and 46·65 g KCl per litre into a beaker, add 5 drops of glacial acetic acid, and saturate with carbon dioxide by bubbling the gas through for several minutes. Add this to the Erlenmeyer flask, and agitate for half-an-hour with a steady stream of carbon dioxide. Swirl occasionally to prevent caking. Using a platinum cone and a coarse paper, filter the solution under light suction into a 300-ml Erlenmeyer flask. Rinse the original flask and the filter with about 100 ml of water. Add 1 g of pure aluminium shavings and 6 ml of 1 : 1 hydrochloric acid. Boil gently under a carbon dioxide atmosphere until the solution becomes colourless, then cool under the same condition in an ice bath. Filter with gentle suction through a fine paper into a 500-ml Erlenmeyer flask, and wash the residue with cold water. Add 10 ml of Zimmerman–Reinhardt reagent and titrate with standard potassium permanganate solution to a faint pink [14, 23].

(e) Another approach depends on the fact that when a finely ground sample of iron oxides and metallic iron is shaken for an hour with an 80 methyl alcohol : 20 bromine mixture, filtered, and washed, all the metallic iron is in the filtrate and all iron oxides remain in the residue [16]. Iron sulphide accompanies the metallic iron. Another method uses bromine in ethanol [26]. These procedures are similar to the chlorine–alcohol leach described, in the chapter on Cobalt, to differentiate metallic and sulphide forms of the element from the oxides.

REFERENCES

1. A.S.T.M., *Chemical Analysis of Metals. Sampling and Analysis of Meta Bearing Ores*, Philadelphia, Pa., American Society for Testing and Materials, 1969.
2. Clemency, C. V., and Hagner, A. F., *Anal. Chem.* **33**, 888–92 (1961).
3. Cluley, H. J., and Newman, E. J., *Analyst* **88**, 3–17 (1963).
4. Dillon, V. S., *Assay Practice on the Witwatersrand*, Johannesburg, Transvaal and Orange Free State Chamber of Mines, 1955.
5. Drummond, P. R., *Can. Inst. Mining Met. Trans.* **XLIII**, 627–52 (1940).
6. Dozinel, C. M., *Modern Methods of Analysis of Copper and its Alloys*, 2nd ed., London, Elsevier, 1963.
7. Elwell, W. T., and Gidley, J. A. F., *Atomic-absorption Spectrophotometry*, 2nd ed., Oxford, Pergamon, 1966.
8. Furman, N. H., ed., *Scott's Standard Methods of Chemical Analysis*, 6th ed., Vol. 1, Princeton, N.J., Van Nostrand, 1962.
9. Habashy, M. G., *Anal. Chem.* **33**, 586–8 (1961).
10. Hawley, F. G., *Eng. Mining World* **2**, 270–2 (1931).
11. Hillebrand, W. F., Lundell, G. E. F., Bright, H. A., and Hoffman, J. I., *Applied Inorganic Analysis*, 2nd ed., New York, Wiley, 1953.
12. Hopkin and Williams Ltd., *Organic Reagents for Metals*, Chadwell Heath, England, Hopkin and Williams, Vol. 1, 1955, Vol. 2, 1964.

13. KINNUNEN, J., *Chemist-Analyst* **40**, 89–91 (1951).
14. KOLTHOFF, I. M., and BELCHER, R., *Volumetric Analysis*, Vol. III, New York, Interscience Publishers, 1957.
15. KOLTHOFF, I. M., and ELVING, P. J., *Treatise on Analytical Chemistry*, Part II, Vol. 2, 247–310, New York, Interscience Publishers, 1962.
16. KORAKAS, N., *Trans. Institution Mining Met.* **72**, Part I, 35–54 (1962–63).
17. LEWIS, C. L., OTT, W. L., and SINE, N. M., *The Analysis of Nickel*, Oxford, Pergamon, 1966.
18. LOW, A. H., WEINIG, A. J., and SCHODER, W. P., *Technical Methods of Ore Analysis*, New York, Wiley, 1939.
19. LUKE, C. L., *Anal. Chim. Acta* **36**, 122–9 (1966).
20. MCLACHLAN, C. G., *Am. Inst. Min. Met. Eng. Trans.* **112**, 593–6 (1934).
21. OTTAWAY, J. M., and FULLER, C. W., *Analyst* **94**, 313–20 (1969).
22. POLLOCK, E. N., and MIGUEL, A. N., *Anal. Chem.* **39**, 272 (1967).
23. RITCHIE, J. A., *Geochim. Cosmochim. Acta* **32**, 1363–6 (1968).
24. ROBERTS, L. E., and NUGENT, R. L., *U.S. Bur. Mines Rept. Invest.* 3120 (1931).
25. SANDELL, E. B., *Colorimetric Determination of Traces of Metals*, 3rd ed., New York, Interscience Publishers, 1959.
26. SANT, B. R., and PRASAD, T. P., *Talanta* **15**, 1483–6 (1968).
27. SCHAFER, H. N. S., *Analyst* **91**, 755–62, 763–70 (1966).
28. VAN LOON, J. C., *Talanta* **12**, 599–603 (1965).
29. WILLIAMS, C. E., BARRETT, E. P., and LARSEN, B. M., *U.S. Bur. Mines Bull.* 270 (1927).
30. WILSON, A. L., *Analyst* **89**, 389–401 (1964).
31. WILSON, C. L., and WILSON, D. W., *Comprehensive Analytical Chemistry*, Vol. IC, 635–55, London, Elsevier, 1962.
32. YOUNG, R. S., *Chemist-Analyst* **32**, 79 (1943).

18 Lead

The principal lead mineral is the sulphide, galena, but the element is also found as cerussite $PbCO_3$, anglesite $PbSO_4$, pyromorphite $PbCl_2.3Pb_3\text{-}(PO_4)_2$, and other minerals in association with molybdenum, vanadium, arsenic, copper, bismuth, antimony, and silver. Galena is frequently accompanied by sulphides of zinc and iron. Lead is an important metal and is a constituent of many alloys; the determination of lead may also be required in paints, pesticides, gasoline additives, foodstuffs, and other materials.

ISOLATION OR SEPARATION OF LEAD

The best separation of lead is based on the insolubility of its sulphate. It may also be separated from members of the ammonium hydroxide and succeeding groups by precipitation with hydrogen sulphide in acid solution. In this hydrogen sulphide group, lead accompanies Cu, Bi, and Cd in remaining insoluble in ammonium sulphide and can thus be removed from As, Sb, Sn, Ge, and Mo. It can be separated together with As, Sb, Sn, Bi, Se, and Te from copper and molybdenum by a ferric hydroxide precipitation.

When lead is precipitated as sulphate, it may be contaminated with SiO_2, W, Nb, Ta, and the alkaline earths. Silica may be eliminated with hydrofluoric acid, or lead sulphate may be dissolved in ammonium acetate. Tungsten can be separated by initial digestion in nitric acid, and any co-precipitated lead recovered by solution of the tungstic acid in ammonia + ammonium tartrate and precipitation of the lead with ammonium sulphide. If niobium and tantalum are present, lead can be precipitated with hydrogen sulphide in an acid solution containing tartaric acid. Lead can be separated from Ca, Sr, and Ba by precipitation of the former with hydrogen sulphide in 5–10% hydrochloric acid. It may also be separated from calcium and strontium by electrolysis in nitric acid solution, and, in technical work, from barium by extraction of the lead sulphate with ammonium acetate. For work of the highest accuracy, however, the latter procedure should be avoided; barium is not only slightly soluble in ammonium acetate but also tends to prevent solution of lead.

Lead may be separated from a number of elements by ion exchange or chromatography [8, 18, 29]. Lead may be collected by manganese dioxide in dilute nitric acid solution; the optimum acid concentration is 0.008 M [26]. The extraction of lead dithizonate in carbon tetrachloride or chloroform sometimes serves as a useful separation for small quantities before a final polarographic [14] or radiochemical [28] measurement.

1. Gravimetric Determination

A. Lead Sulphate

Dissolve 0·5–2 g in 20 ml 1 : 1 nitric acid, 10 ml hydrochloric acid, and a few drops of hydrofluoric acid if necessary [2, 3, 6, 13, 16, 18, 21, 31]. For pig lead and most lead alloys the best solvent is 1 : 3 nitric acid. Add 10 ml 1 : 1 sulphuric acid and evaporate to strong fumes of this acid. If hydrochloric acid has been used in the decomposition of the sample, wash down the sides of the beaker and evaporate a second time to fumes of sulphuric to eliminate all traces of hydrochloric acid.

Cool, add 100 ml of water for every 0·1 g of lead present. Boil, allow to stand one hour, and in the absence of silica, barium, and appreciable quantities of calcium and strontium, filter through a previously weighed Gooch crucible, wash thoroughly with 3% sulphuric acid, and finally with several small portions of hot water or with 50% alcohol solution, to remove the free acid. Dry and ignite to constant weight at 500–600°C; cool and weigh as $PbSO_4$. $PbSO_4 \times 0{\cdot}6833 = Pb$. Instead of filtering through a Gooch crucible, the precipitate may be filtered through Whatman No. 42 paper, washed as described, placed in a weighed platinum crucible, dried, and ignited at 500–600°C to constant weight.

In the presence of silica and barium, filter the crude lead sulphate on a Whatman No. 42 paper, and wash thoroughly with 3% sulphuric acid. Dissolve the lead sulphate by repeated washing with small portions of a strong ammonium acetate solution, leaving silica and barium sulphate on the paper. Evaporate the filtrate to dryness, add 10 ml of 1 : 1 sulphuric acid and evaporate to fumes. Cool, add 100 ml of water, boil, allow to stand for one hour, and filter off the precipitate of lead sulphate on a Gooch crucible as outlined previously.

Alternatively, precipitate lead as sulphide in ammonium acetate solution with hydrogen sulphide, filter, dissolve the lead sulphide in nitric acid and a few drops of bromine, add sulphuric acid and evaporate to fumes of this acid. Proceed with the precipitation of lead sulphate as described above.

When substantial quantities of calcium or strontium are present, lead should be precipitated initially as sulphide in 5–10% hydrochloric acid solution. Tungsten, niobium, and tantalum should be separated as described in the introductory paragraphs.

B. Lead Molybdate

It is sometimes advantageous to precipitate lead as the molybdate, $PbMoO_4$, a compound which is less soluble than the sulphate and has a more favourable weight factor. The principal interferences are the alkaline earths which also form insoluble molybdates, chromates, arsenates, or phosphates which form insoluble compounds with lead, and substances such as titanium or tin that are easily hydrolysed. An initial separation of lead as sulphide usually provides a medium suitable for the lead molybdate precipitation. If arsenic or tin are present, volatilize them in the initial decomposition step by evaporation with hydrobromic and perchloric acids. The example described below, for lead in irons and steels, is capable of wide application [3, 11, 13, 15].

Dissolve the sample in 1 : 1 hydrochloric acid and evaporate the solution just to dryness but do not bake. If a residue remains, which might contain lead, take appropriate steps to get it into solution by fusion or otherwise, and add to the main sample before taking to dryness. Add 400 ml of hot water, and heat until the iron salts have completely dissolved. Add 10 g of ammonium chloride, and pass a rapid stream of hydrogen sulphide through the solution for at least ten minutes. Filter through Whatman No. 541 paper and wash with cold 1% hydrochloric acid saturated with hydrogen sulphide.

Dissolve the lead sulphide in hot 1 : 1 nitric acid, catching the solution in a 250-ml beaker, and wash the paper thoroughly with hot water. Add 2 g of tartaric acid, and when this has dissolved, neutralize the solution with ammonium hydroxide and add 5 ml in excess for each 100 ml of solution. Heat to boiling, add 10 ml of 5% ammonium molybdate, and boil for several minutes. Filter through Whatman No. 40 paper containing paper pulp, and wash thoroughly with hot slightly ammoniacal 5% ammonium nitrate. Transfer paper and precipitate to a weighed porcelain, quartz, or platinum crucible, and ignite to constant weight at 600–650°C. Cool and weigh as $PbMoO_4$. $PbMoO_4 \times 0{\cdot}5644 = Pb$.

2. Volumetric Determination

A. Chromate Method

This procedure is suitable for a wide variety of ores and metallurgical products. Lead is separated initially as sulphate, dissolved in sodium acetate, precipitated as lead chromate, dissolved in hydrochloric acid, potassium iodide is added, and the liberated iodine is titrated with sodium thiosulphate [3, 11, 13, 15].

Dissolve the sample in hydrochloric and nitric acids, a few drops of hydrofluoric acid if necessary, add 10 ml 1 : 1 sulphuric acid and evaporate to strong fumes of the latter. Cool, dilute to 100 ml, allow to stand one hour, filter, wash with cold 10% sulphuric acid about five times, retaining the original beaker. Transfer as much of the lead sulphate precipitate as possible from the paper to the original beaker, using a small quantity of hot water. Place the filter paper in a small beaker containing 25 ml of an acetate extraction solution made by preparing a cold saturated solution of sodium acetate, filtering, adding 2 volumes of water, and then 25 ml acetic acid per litre. Heat the small beaker containing the macerated filter paper, filter into the original beaker containing the lead sulphate, and wash the beaker, paper, and funnel with hot acetate extraction solution and finally with hot water. Heat until all of the lead sulphate has dissolved.

Dilute, if necessary, to a volume of 150 ml, heat to boiling, and add carefully with a pipette 10 ml of a saturated solution of potassium dichromate. Boil for 10 minutes, filter, and wash ten times with hot 5% sodium acetate solution. Place the washed beaker beneath the funnel, and dissolve the precipitate of lead chromate with at least 50 ml of a hydrochloric acid–salt mixture; the latter is made by adding 300 ml of 1 : 1 hydrochloric acid to 1000 ml of a saturated solution of sodium chloride. Finally, wash the paper thoroughly with cold water.

Add 2 ml of 50% potassium iodide solution, stir gently, and titrate with standard sodium thiosulphate until the iodine is almost gone. Add 5 ml of 0·5% starch solution, and continue the titration until the colour of the solution changes to a clear green with no tinge of blue.

Standardize the sodium thiosulphate against 0·0500 g pure lead which has been dissolved in 1 : 3 nitric acid and carried through all steps of the procedure. Lead is one of the metals which can be obtained commercially in the 'five-nines' grade, 99·999%. A solution containing 17·985 g $Na_2S_2O_3.5H_2O$ per litre has an approximate value of 1 ml = 0·005 g Pb. If the thiosulphate solution has a value of 1 ml = 0·005 g Cu, it is equivalent to 1 ml = 0·005433 g Pb.

The reactions in this procedure are:

$$PbCrO_4 + 2HCl \rightarrow PbCl_2 + CrO_3 + H_2O$$

$$2CrO_3 + 12HCl + 6KI \rightarrow 2CrCl_3 + 6KCl + 6H_2O + 3I_2$$

$$2Na_2S_2O_3 + I_2 \rightarrow 2NaI + Na_2S_4O_6.$$

2 Pb = $3I_2$, and 1 ml of 0·1N sodium thiosulphate = 0·006906 g Pb.

If Ag, Sb, Bi, or Ba are present, they should be eliminated before or after the addition of sulphuric acid. Silver can be removed initially as the insoluble chloride; antimony can be volatilized by several evaporations with hydrobromic and sulphuric acids. If more than 1% bismuth is present, some of it will remain with the lead sulphate. Large quantities should be removed by the dissolution of lead sulphate in ammonium acetate, evaporation, and re-precipitation; small amounts can be removed before the lead chromate is filtered by adding 2 g citric acid dissolved in a little hot water, when bismuth chromate will go into solution.

If barium is present, rinse the crude lead sulphate precipitate back into the beaker with a small quantity of hot water. Add 10 ml hydrochloric acid and evaporate almost to dryness. Add 25 ml of the acetate extraction solution, boil, and continue the addition until all the lead sulphate has dissolved, using not over 75 ml. Filter through the original paper, and wash the residue of barium sulphate with hot water.

B. Molybdate Method

Lead is precipitated from an acetic acid solution by a standard solution of ammonium molybdate; the termination of the reaction is indicated by the yellow colour produced by the excess of reagent when a drop of the mixture comes in contact with a drop of tannin solution used as an outside indicator [13, 21]. A prior separation of lead from nearly all interfering elements is usually made by filtration of lead sulphate and extraction of the latter with a concentrated ammonium acetate solution. Though not as suitable for a wide range of materials as the chromate procedure, the molybdate method is used at some lead mines and reduction plants.

Decompose 0·5–2 g sample with 10 ml nitric acid, 10 ml hydrochloric acid, and a few drops of hydrofluoric acid if necessary. Add 10 ml 1 : 1 sulphuric acid and evaporate to strong fumes of this acid.

If calcium is present in appreciable quantities, precipitate lead with hydrogen sulphide from 10% hydrochloric acid solution. Filter and wash thoroughly with hot acidulated hydrogen sulphide water. Dissolve the precipitate off the filter paper with dilute nitric acid into the original beaker, wash, add 10 ml 1 : 1 sulphuric acid and evaporate to fumes of the latter.

Cool, add water, boil until all soluble salts are in solution, allow to stand one hour, and filter the precipitated lead sulphate on Whatman No. 42 paper. Wash thoroughly with 3% sulphuric acid and finally several times with hot water. For larger precipitates, wash by decantation, retaining the bulk of the precipitate in the beaker.

Dissolve the lead sulphate on the filter paper with hot saturated ammonium acetate, catching the solution in the beaker containing the bulk of the lead sulphate. Wash with hot water and add sufficient additional hot concentrated ammonium acetate to the filtrate to dissolve all the lead sulphate. If silica or barium sulphate are present in the crude precipitate of lead sulphate, they will remain insoluble and must be filtered off at this stage.

Heat the clear filtrate of lead acetate, diluted with water if necessary to about 150 ml, nearly to boiling and titrate with ammonium molybdate to the endpoint. The latter is indicated by the faint yellow colour imparted by a drop of the solution to a drop of 0·5% tannin used as an external indicator on a spot plate. The indicator should be prepared daily; or it may be kept for several weeks by adding 5 drops of a saturated mercuric chloride solution to 500 ml of indicator. The standard molybdate solution can be made by dissolving 4·4 g ammonium molybdate in 1 litre of water and allowing to stand for several days to stabilize. One ml of the above solution = approximately 0·005 g Pb; it can be standardized against pure lead dissolved in 1 : 3 nitric acid and taken through the procedure, or against pure lead sulphate dissolved in ammonium acetate and titrated. The reaction may be expressed:

$$Pb(C_2H_3O_2)_2 + (NH_4)_2MoO_4 \rightarrow PbMoO_4 + 2NH_4C_2H_3O_2$$

3. Electrolytic Determination

When an electric current is passed through a solution of lead containing sufficient free nitric acid, all the lead will be deposited on the anode as PbO_2 [2, 3, 11, 13, 15, 18]. The elements As, Sb, Sn, Bi, Ag, Hg, Mn, P, Se, Te, and Cl interfere, but a preliminary separation of lead as sulphate will overcome this.

Sand-blasted platinum gauze electrodes are preferable for this deposition, and about 0·1 g of lead is the maximum quantity that can be firmly deposited on a stationary anode. If the anode is rotated or the solution otherwise stirred, larger quantities can be satisfactorily deposited.

Dissolve a 0·5–2 g sample in 1 : 1 nitric acid, with the addition if necessary of hydrochloric acid and a few drops of hydrofluoric acid; add 10 ml 1 : 1 sulphuric acid and evaporate to strong fumes of the latter. Cool, dilute, allow to stand an hour, filter and wash the lead sulphate. Transfer the latter to the original beaker, add an excess of ammonium hydroxide, heat to boiling, and pour the mixture into enough nitric acid to leave an excess of 10 ml acid per 100 ml of solution.

Electrolyse overnight at 0·2–0·5 ampere and 2–2·5 volts, or by rapid deposition with a rotating cathode at 1·5–2·0 amperes and 2–6 volts. Test for the completion of deposition by adding a little water and after 15 minutes observing whether any further plating has occurred on the stem of the anode. Rinse the latter with water, alcohol, dry in an oven at 125°C for 30 minutes, cool, and weigh as PbO_2. Some analysts use an empirical factor of 0·8643 to convert PbO_2 to Pb, but the theoretical factor of 0·866 is now usually preferred. Remove the deposit of lead oxide from the anode by dissolving in warm 1 : 3 nitric acid containing a little hydrogen peroxide.

4. Polarographic Determination

Lead in the approximate range of 0·001–2% may be conveniently determined polarographically in many materials. Lead in aluminium [6], nickel [20], and zinc [3, 8, 17] for instance, is commonly determined by this means; another example is lead in blister and anode copper where a large weight may be taken, copper removed by electro-deposition in nitric acid, and lead dioxide dissolved from the anode [4].

The procedure outlined below is intended for lead in zinc [3, 17, 19, 23]. Dissolve 1–5 g in 10–30 ml hydrochloric acid; if dissolution is slow, warm gently. Take to incipient dryness but do not bake. Cool for not more than 5 minutes, dissolve the residue in 5 or 10 ml of a supporting electrolyte. The latter is made by dissolving 0·2 g gelatin in 50 ml of warm water, diluting to approximately 500 ml, adding 10 ml hydrochloric acid and 75 g potassium chloride, and diluting to 1 litre.

Transfer a suitable portion of the resulting solution to the polarograph cell, de-aerate for 10 minutes with nitrogen, and electrolyse through the range −0·2 to −0·9 volts. Adjust the sensitivity of the circuit to obtain a curve of optimum step heights. Carry through all steps of the method a blank, containing reagents only. Compare the wave height of the lead curve with that of a reference standard obtained by adding known quantities of lead to a zinc solution. It is essential that all conditions, including temperature, should be the same for both samples and standards.

Detailed directions have been published for the polarographic determination of lead in aluminium [6], copper [4, 11], nickel [20], and antimony sulphide concentrates [14].

5. Colorimetric Determination

Dithizone can be employed to isolate and determine small quantities of lead in a wide variety of materials [1, 2, 3, 6, 9, 12, 13, 18, 22, 27, 33]. The method depends on the inhibiting effect of cyanide ion in alkaline solution on the reaction between dithizone and all metals except lead, indium, stannous tin, thallous thallium, and bismuth. Interference from indium may be eliminated by extraction of lead at a *p*H above 10. Tin is usually converted to the stannic state which does not interfere, but a large quantity should be volatilized by two evaporations with hydrobromic and perchloric acids, and evaporated to dryness to expel the latter before proceeding with the determination in nitric acid solution. For the rare occasions when thallium is present, lead can be separated by extraction at *p*H 6·0–6·4 in chloroform, at which acidity thallium is not appreciably extracted. Thallium can also be removed by an ether separation in dilute hydrochloric acid. Bismuth frequently accompanies lead in mining and metallurgical products, and the following procedure assumes its presence; if bismuth is absent, lead can be extracted directly at *p*H 10·8 after rinsing with dilute ammonia.

Dissolve 0·5–5 g sample in sufficient nitric acid to bring all lead and bismuth into solution. If an insoluble residue remains, which could contain lead, fuse with sodium carbonate and add to the main portion. If substantial quantities of iron are present, evaporate to dryness with hydrochloric acid, and make an ether separation to remove all the iron. For lead in metallic copper, nickel, or cobalt, collect the lead with ferric hydroxide [33], make an ether separation and proceed with the isolation of bismuth and lead in cyanide solution.

To a suitable aliquot in a separatory funnel, preferably 10–25 ml containing 1–20 μg of lead, add 5 ml of 50% ammonium citrate, 1 ml of 20% hydroxylamine hydrochloride, and make just alkaline with ammonium hydroxide. Add 5 ml of 10% potassium cyanide solution and adjust the *p*H to 8, using test paper. Add successive small portions of 0·005% dithizone in carbon tetrachloride to the sample, shaking and withdrawing the lower coloured layer each time until all the lead and bismuth are extracted and retained. This point is indicated by the lower layer remaining green after shaking; a little excess dithizone does no harm.

Place the combined dithizonates of lead and bismuth in a separatory funnel. Add 10 ml of water containing a drop of 1 : 1 ammonia, shake, and draw off the carbon tetrachloride phase into another separatory funnel. Shake the aqueous phase with 1–2 ml of carbon tetrachloride and add this to the washed carbon tetrachloride extract. Remove lead and bismuth by shaking the carbon tetrachloride extract with two 5 ml portions of 1% nitric acid: discard the final lower green layer.

Adjust the combined extracts containing lead and bismuth to *p*H 2, remove the bismuth by successive small additions of 0·005% dithizone until the last portion remains an unchanged green after vigorous shaking. Remove any droplets of dithizone by shaking with a little carbon tetrachloride and withdrawing.

Add 10 ml of an ammonia–cyanide–sulphite solution; this is made by diluting 725 ml of 8 M ammonium hydroxide and 30 ml of 10% potassium cyanide to 1 litre and dissolving 1·5 g of sodium sulphite in the solution. Adjust the *p*H to 10·8 with 1 : 1 ammonium hydroxide. Add 10 ml of 0·001% dithizone solution; shake, and allow the phases to separate. Run the carbon tetrachloride solution, if clear, directly into a spectrophotometer cell and measure the absorbance at 520 nm against carbon tetrachloride in the reference cell. If the liquid is not clear, filter through a little glass wool. Establish the standard curve by taking 1–20 μg of lead, diluting to 10 ml with 1% nitric acid, adding 10 ml of ammonia–cyanide–sulphite solution, and shaking with 10 ml of 0·001% dithizone solution.

6. Optical Spectrography

Small quantities of lead in many materials can be determined by emission spectrography, usually with the lines 2663·17, 2833·07, or 4057·82. Detailed procedures have been described for 0·002–0·7% lead in aluminium and its alloys [3, 7], 0·001–0·3% in tin alloys [3], 0·001–0·05% in zinc and zinc-base alloys [3], and 20–200 ppm in zirconium and its alloys [3]. Methods have been published for 0·05–0·5% lead in steel [12, 25], and 4–80 ppm in nickel [20]. Electrolytic tough pitch copper normally contains less than 15 ppm lead, and spectrographic procedures have long been popular in copper refineries. From 0·0003 to 0·06% lead in glass has been determined spectrographically [5].

7. X-ray Spectrography

Apart from the determination of tetraethyl lead in gasoline, X-ray fluorescence has been little used in lead analyses. A description of an automatic X-ray analyser for 0·2–0·4% lead in the tailings of an ore dressing plant has been published [30].

8. Atomic-Absorption Spectrophotometry

Atomic absorption offers an excellent method for the lower ranges of lead in a variety of materials [10]. The most sensitive absorption line is 2170·0, with a sensitivity of 0·3 μg/ml. Other absorption lines used are 2833·1 and 2614·2. Procedures have been outlined for lead in copper and its alloys [9, 10, 11].

SPECIAL PROCEDURES

Forms of Lead

1. In Slags

A procedure for determining approximately the forms of lead in slag and similar metallurgical products can be useful [24].

To 0·5–5 g of −100 mesh sample, add 25 ml of a saturated solution of ammonium acetate, dilute to 100 ml, and boil for 10 minutes. Filter through an asbestos mat on a Gooch crucible, and wash with hot water. The filtrate contains the lead which was present as oxide, sulphate or basic sulphate; the residue contains lead sulphide, silicate, and metal. For most slags, omit the above treatment with ammonium acetate and proceed directly to the next step.

Transfer the residue or the weighed slag sample to a beaker; evaporate the residue nearly to dryness. Add 10–20 ml of 10% silver nitrate solution, and allow to stand for an hour with occasional stirring; filter through asbestos on a Gooch crucible and wash. The filtrate contains the lead that was originally present in the metallic form; the residue contains the lead present as sulphide and silicate.

Transfer the residue to a beaker, evaporate nearly to dryness, and add 25–50 ml of a saturated solution of sodium chloride containing 60 g $FeCl_3.6H_2O$ per litre. Allow to stand for 12 hours, with occasional agitation. Filter the solution through filter paper, wash the residue once with some of the original solvent, and then wash thoroughly with hot water. The filtrate contains the lead that was originally present as sulphide; the residue is lead silicate. The filtrates and residue are analysed for lead by conventional methods.

2. In Ores and Concentrator Products

A method has proved useful to differentiate various forms of lead in complex ores [32]. To 1 g of −200 mesh sample in a stoppered Erlenmeyer flask add 100 ml of 60% ammonium acetate solution. Allow to stand 1½ hours, shaking occasionally. Filter through Whatman No. 42 paper and wash thoroughly by decantation with cold water, retaining as much of the residue in the flask as possible. The filtrate contains lead which was originally present as sulphate, carbonate, or oxide; the precipitate represents lead present as sulphide, phosphate, and vanadate.

Replace the original flask under the funnel and pour through the paper 100 ml of cold 10% by volume perchloric acid. Lead sulphide is virtually insoluble in this reagent, whereas the phosphates and vanadates will dissolve. Reserve the lead sulphide residue on the paper for addition to the residue in the original flask. Stopper the flask and allow to stand for 2 hours, shaking occasionally. Filter through a Whatman No. 42 paper and wash thoroughly with cold 1% perchloric acid. The lead phosphate and vanadate are in the filtrate; lead sulphide remains on the paper and can be combined with the small quantity which was left on the paper from the first filtration.

The oxidized lead in the filtrate from the ammonium acetate and dilute perchloric acid treatments may be combined; they may also be determined separately, if desired, thus differentiating between lead present as sulphate or carbonate and that existing as phosphate or vanadate.

If the lead minerals are finely disseminated in a siliceous gangue, add a few drops of hydrofluoric acid to the sample during both leaches. Small quantities of hydrofluoric acid are without effect on lead sulphide; as high as 5% hydrofluoric acid by volume dissolves insignificant amounts with this procedure.

Finely divided lead sulphide has a tendency to oxidize on exposure to a combination of air and moisture; filtrations should not be unduly delayed.

Lead Associations

Lead producers in various parts of the world support Lead Associations, which are always ready to assist chemists who may have special problems with this metal. The addresses of these organizations are:

Lead Development Association,
34 Berkeley Square,
London, W1E 7QZ

Lead Industries Association,
292 Madison Avenue,
New York 17, N.Y.

REFERENCES

1. Abbott, D. C., and Harris, J. R., *Analyst* **87**, 387–9 (1962).
2. A.O.A.C., *Official Methods of Analysis of the Association of Official Analytical Chemists*, Washington, D.C., Association of Official Analytical Chemists, 1965.
3. A.S.T.M., *Chemical Analysis of Metals. Sampling and Analysis of Metal Bearing Ores*, Philadelphia, Pa., American Society for Testing and Materials, 1969.
4. Barabas, S., and Cooper, W. C., *Metallurgia* **LVI**, 101–3 (1957).
5. Bober, A., *Anal. Chem.* **37**, 1423–4 (1965).
6. British Aluminium Co., *Analysis of Aluminium and its Alloys*, London, British Aluminium Co., 1961.
7. British Aluminium Co., *Spectrochemical Analysis of Aluminium and its Alloys*, London, British Aluminium Co., 1961.
8. Carson, R., *Analyst* **86**, 198–200 (1961).
9. Dozinel, C. M., *Modern Methods of Analysis of Copper and its Alloys*, 2nd ed., London, Elsevier, 1963.
10. Elwell, W. T., and Gidley, J. A. F., *Atomic-Absorption Spectrophotometry*, 2nd ed., Oxford, Pergamon, 1966.
11. Elwell, W. T., and Scholes, I. R., *Analysis of Copper and its Alloys*, Oxford, Pergamon, 1967.
12. Flickinger, L. C., Polley, E. W., and Galletta, F. A., *Anal. Chem.* **29**, 1778–9 (1957).
13. Furman, N. H., ed., *Scott's Standard Methods of Chemical Analysis*, 6th ed., Vol. 1, Princeton, N.J., Van Nostrand, 1962.

14. GROENEWALD, I. D., *Analyst* **89**, 140–1 (1964).
15. HILLEBRAND, W. F., LUNDELL, G. E. F., BRIGHT, H. A , and HOFFMAN, J. I., *Applied Inorganic Analysis*, 2nd ed., New York, Wiley, 1953.
16. HILLS, F. G., *The Technical Analysis of Ores and Metallurgical Products*, New York, Chemical Publ. Co., 1939.
17. IMPERIAL SMELTING CORPORATION, personal communication, 1961.
18. KOLTHOFF, I. M., and ELVING, P. J., *Treatise on Analytical Chemistry*, Part II, Vol. 6, 69–175, New York, Interscience Publishers, 1964.
19. KOLTHOFF, I. M., and LINGANE, J. J., *Polarography*, New York, Interscience Publishers, 1952.
20. LEWIS, C. L., OTT, W. L., and SINE, N. M., *The Analysis of Nickel*, Oxford, Pergamon, 1966.
21. LOW, A. H., WEINIG, A. J., and SCHODER, W. P., *Technical Methods of Ore Analysis*, New York, Wiley, 1939.
22. MILNER, G. W. C., and TOWNEND, J., *Anal. Chim. Acta* **5**, 584–98 (1951).
23. MINO, H., *Eng. Mining J.* **156**, No. 9, 97–9 (1955).
24. OLDRIGHT, G. L., and MILLER, V., *U.S. Bur. Mines Rept. of Invest.* 2954 (1929).
25. PATERSON, J. E., *Anal. Chem.* **29**, 526–7 (1957).
26. REYNOLDS, G. F., and TYLER, F. S., *Analyst* **89**, 579–86 (1964).
27. SANDELL, E. B., *Colorimetric Determination of Traces of Metals*, 3rd ed., New York, Interscience Publishers, 1959.
28. SILL, C. W., and WILLIS, C. P., *Anal. Chem.* **37**, 1661–71 (1965).
29. STRELOW, F. W. E., and TOERIEN, F. Von S., *Anal. Chem.* **38**, 545–8 (1966).
30. SUNDKVIST, G. J., LUNDGREN, F. D., and LIDSTROM, L. J., *Anal. Chem.* **36**, 2091–4 (1964).
31. YOUNG, R. S., *Anal. Chim. Acta* **4**, 366–85 (1950).
32. YOUNG, R. S., GOLLEDGE, A., and TALBOT, H. L., *Am. Inst. Met. Eng. Tech. Publ.* 2303 (1948).
33. YOUNG, R. S., and LEIBOWITZ, A., *Analyst* **71**, 477–9 (1946).

19 Lithium

The use of lithium in nuclear developments has resulted in a remarkable increase in production of this alkali metal and its compounds in recent years. The important lithium minerals are the complex silicates: spodumene, lepidolite, and petalite. Lithium is now found occasionally in alloys, ceramics, greases, fluxes and many other materials.

Lithium is almost invariably determined by flame photometry at 670·8 nm or by atomic absorption [1, 5, 6, 7, 10, 13, 15, 16, 18]. For the rare occasions when traditional chemical methods must be employed, descriptions may be found in standard reference texts [11, 12, 17].

1. Flame Photometry

For flame photometry, lithium ores are usually decomposed with hydrofluoric and sulphuric acids in a platinum crucible or dish; they may also be fused with sodium peroxide in a nickel crucible. The procedure used for petalite and spodumene by Associated Lead Manufacturers Limited [3] is typical of modern practice. Weigh 1 g of the finely ground ore into a platinum crucible, moisten with water, add 15 ml hydrofluoric acid, several drops of sulphuric acid, and 8–10 drops nitric acid. Evaporate to dryness, cool, add 3 ml nitric acid, and again evaporate to dryness. Dissolve in water and make the solution to 500 ml in a volumetric flask. For ores containing 3·5–6·5% lithia, this solution will have about 30–60 ppm lithium. Make an approximate estimation of the concentration, and prepare a series of standards at intervals of 0·5 ppm lithium from a master lithium nitrate solution containing 100 ppm lithium. Determine the concentration of the unknown by comparing it with these standards in a flame photometer. With a few ores, it is necessary to add to the standards approximately the same quantities of Na, K, and Al which are present in the unknown.

The Foote Mineral Company uses the following procedure for spodumene and other lithium ores [4, 9]. Heat a 0·5 g sample of −200 mesh ore in a platinum dish at 1038°C for 30 minutes. Cool, add 1–2 ml sulphuric acid and about 10 ml hydrofluoric acid. Evaporate to dryness, add 5 ml sulphuric acid and evaporate just to fumes. Cool, transfer to a beaker, boil, filter into a 250 ml volumetric flask, and wash. Re-treat the precipitate by igniting the paper, adding hydrofluoric and sulphuric acids, evaporating to dryness, dissolving in water, and adding this portion to the solution in the volumetric flask. Compare the unknown in a flame photometer with a series of standards containing 2–70 ppm of lithium made from lithium sulphate monohydrate; the standards are 0·7N in sulphuric acid and contain 270 ppm aluminium using aluminium sulphate octadecahydrate.

Another procedure was developed by the same company for lepidolite [9]. Weigh 1 g of −200 mesh ore into a nickel crucible, add 7·33 g sodium peroxide, and mix. Fuse, holding at a dull red heat for about 5 minutes; cool, transfer to a beaker, add water, wash and remove the crucible. Add 30 ml hydrochloric acid and stir until complete solution occurs. Transfer to a 250 ml flask, make to volume, mix, withdraw a 25 ml aliquot to a second 250 ml flask, dilute to the mark and mix. The second flask, containing 0·1 g sample, 1·1 g sodium chloride, and 1 ml of free hydrochloric acid, is used for the flame comparisons. Three standard solutions which bracket the lithium sample are required; the standards must also match the composition of the sample solution with respect to all components which influence the intensity of the lithium radiation.

It has been shown that, in solutions containing about 50 ppm lithium, increasing the sulphuric acid content from 0·5 to 1·7N depressed the intensity of the lithium flame [14]. Variations in content of sodium and potassium from 5 to 35 ppm indicated no interference; aluminium was without effect from 0 to 600 ppm, and Ca, Mg, and Fe from 0 to 50 ppm likewise showed no interference [14].

2. Atomic-Absorption Spectrophotometry

Lithium can be determined by atomic absorption [2, 8]; the most sensitive line is 6707·8 having a sensitivity of 0·03 μg/ml, but line 3232·6 may also be used. Air–acetylene or air–coal gas flames are satisfactory, the latter being more sensitive. Other alkali metals cause an enhancement of the lithium absorption, but this can be overcome by addition of the same amount to the standards.

REFERENCES

1. Adams, P. B., *Anal. Chem.* **33**, 1602–5 (1961).
2. Angino, E. E., and Billings, G. K., *Atomic Absorption Spectrometry in Geology*, London, Elsevier, 1967.
3. Associated Lead Manufacturers, Liverpool 20, Britain, personal communication, 1960.
4. Brumbaugh, R. J., and Fanus, W. E., *Anal. Chem.* **26**, 463–5 (1954).
5. Burriel-Marti, F., and Ramirez-Munoz, J., *Flame Photometry*, Amsterdam, Elsevier, 1957.
6. Dean, J. A., *Flame Photometry*, New York, McGraw-Hill, 1960.
7. Ellestad, R. B., and Horstman, E. L., *Anal. Chem.* **27**, 1229–31 (1955).
8. Elwell, W. T., and Gidley, J. A. F., *Atomic-Absorption Spectrophotometry*, 2nd ed., Oxford, Pergamon, 1966.
9. Foote Mineral Co., Philadelphia, Pa., personal communication, 1961.

10. HERRMANN, R., and ALKEMADE, C. T. J., *Chemical Analysis by Flame Photometry*, New York, Wiley, 1963.
11. HILLEBRAND, W. F., LUNDELL, G. E. F., BRIGHT, H. A., and HOFFMAN, J. I., *Applied Inorganic Analysis*, 2nd ed., New York, Wiley, 1953.
12. KOLTHOFF, I. M., and ELVING, P. J., *Treatise on Analytical Chemistry*, Part II, Vol. 1, 301–460, New York, Interscience Publishers, 1961.
13. KUERNMEL, D. F., and KARL, H. L., *Anal. Chem.* **26**, 386–91 (1954).
14. LAIDLER, D. S., *Lithium and Its Compounds*, Royal Institute Chemistry Lectures, Monographs and Reports No. 6, 1–33 (1957).
15. MAVRODINEANU, R., and BOITEUX, H., *Flame Spectroscopy*, New York, Wiley, 1965.
16. PUNGOR, E., *Flame Photometry Theory*, London, Van Nostrand, 1967.
17. SCHOELLER, W. R., and POWELL, A. R., *The Analysis of Minerals and Ores of the Rarer Elements*, London, Charles Griffin, 1955.
18. SYKES, P. W., *Analyst* **81**, 283–91 (1956).

20 Magnesium

Magnesium is a common constituent of rocks, minerals, ores, slags, refractories, lime, cement, plaster, waters, soils, and many other materials. Magnesium and its alloys are important in light metal applications; the element is also added in varying quantities to some metals and alloys. The principal magnesium-bearing minerals are magnesite $MgCO_3$, dolomite $MgCO_3.CaCO_3$, brucite $Mg(OH)_2$, carnallite $MgCl_2.KCl.6H_2O$, kieserite $MgSO_4.H_2O$, epsomite $MgSO_4.7H_2O$, kainite $MgSO_4.KCl.3H_2O$, olivine $(Mg,Fe)_2SiO_4$, and serpentine $H_4Mg_3Si_2O_9$. Other important sources of magnesium are natural brines and sea water.

ISOLATION OR SEPARATION OF MAGNESIUM

The determination of magnesium nearly always requires the prior removal of silica, members of the hydrogen sulphide, ammonium hydroxide, ammonium sulphide, and ammonium carbonate groups. Magnesium is then precipitated as magnesium ammonium phosphate and ignited to the pyrophosphate, or precipitated with 8-hydroxyquinoline, dried and weighed as magnesium oxyquinolate. The separations for complex materials are accordingly time-consuming. Fortunately, many of the samples on which magnesium is determined on a routine basis contain only traces of metals of the hydrogen sulphide group, or of members of succeeding groups except Fe, Al, Ti, and Ca. A separation with ammonium hydroxide in the presence of ammonium chloride, followed by precipitation with ammonium oxalate, frequently leaves a filtrate which is free of interferences for the magnesium determination.

1. Gravimetric

A. Precipitation as Magnesium Ammonium Phosphate

Decompose 0·5–2 g sample with 10 ml hydrochloric acid, 10 ml nitric acid, 10 ml 1 : 1 sulphuric acid, and evaporate to strong fumes of the latter [3, 10, 11]. Cool, add 5 ml hydrochloric acid, dilute with hot water, and boil. If an insoluble residue remains, which might contain magnesium, filter, wash, dry, ignite, fuse with sodium carbonate, and add to the main portion; evaporate to dryness twice with hydrochloric acid to dehydrate silica, add 5 ml hydrochloric acid, and boil.

Filter off silica, silver, and other members of the acid group, and wash thoroughly with hot water. Any other convenient method of decomposition may be used, such as treatment with hydrofluoric and sulphuric acids in a platinum dish for highly siliceous material, or direct solution in nitric or hydrochloric acid for many alloys, or solution of aluminium and its alloys in sodium hydroxide. In all cases, unless silica occurs in traces only, it should be removed.

If members of the hydrogen sulphide group are present, adjust the acidity to 5–10% hydrochloric acid and pass in hydrogen sulphide until these are precipitated. Filter and wash with acidulated hydrogen sulphide water. Frequently, only one element of the hydrogen sulphide group will be present in a significant amount. In this case it may be separated by means other than hydrogen sulphide; for instance, copper can be removed and determined by electrolysis, and lead may be isolated as lead sulphate.

If elements of the ammonium hydroxide and ammonium sulphide groups occur in small quantities, precipitate them from the above filtrate by making the solution ammoniacal and passing in more hydrogen sulphide. Filter and wash with dilute ammonium sulphide solution; calcium and magnesium are in the filtrate. Many samples will contain substantial quantities of iron and aluminium, with minor amounts of titanium and manganese; if so, precipitation with ammonium hydroxide in the presence of ammonium chloride is the preferable procedure. Boil the filtrate from the precipitation of the hydrogen sulphide group for 10 minutes to remove all hydrogen sulphide, oxidize with hydrogen peroxide, boil to remove the excess of the latter, add a few grams of ammonium chloride, and make ammoniacal. If manganese is present, add a few drops of bromine and heat for several minutes before making ammoniacal, to throw manganese into the precipitate. Boil, filter, and wash several times with hot 1% ammonium chloride. Transfer the precipitate back to the original beaker, dissolve in hydrochloric acid, and reprecipitate to liberate occluded magnesium and calcium salts from the hydroxides. Nickel and cobalt will accompany calcium and magnesium in the filtrate. They will not interfere unless appreciable quantities are present; when this occurs, remove them at this stage by electrolysis in an ammonia–ammonium sulphate electrolyte as described in the sections on Cobalt and Nickel. This merely entails evaporating the ammoniacal chloride filtrate to fumes with sulphuric acid, cooling, making ammoniacal, and adding 40 ml excess ammonium hydroxide. After electrolysis, evaporate to dryness, add 10 ml hydrochloric acid, dilute with water, and boil. Alternatively, remove Ni, Co, and Mn by ammonium sulphide treatment.

From the faintly ammoniacal filtrate after precipitation of Fe, Al, etc, which has been boiled, if necessary, to expel hydrogen sulphide, remove calcium by adding sufficient saturated ammonium oxalate solution to precipitate the latter; 20 ml will precipitate about 0·2 g CaO in a volume of 100 ml. Boil for a few minutes and allow to stand for several hours. Filter, wash with hot 0·1% ammonium oxalate solution: if any appreciable quantity of calcium is present transfer to the original beaker, dissolve in hydrochloric acid and re-precipitate to set free the occluded magnesium salts.

Magnesium can now be precipitated by adding a five-fold excess of diammonium phosphate. Ten ml of 10% diammonium phosphate will precipitate about 0·3 g of MgO, so that a total of 50 ml will provide for this quantity a five-fold excess. Add ammonium hydroxide until the solution is ammoniacal, and then 10% excess; stir vigorously for several minutes to promote the formation of the precipitate, and allow to stand overnight. When magnesium is low, a stirring period of 5 minutes may be required before a faint opalescence indicates the commencement of precipitation.

Filter through Whatman No. 40 paper and wash with 5% ammonium hydroxide. For all accurate work, dissolve the precipitate in hydrochloric acid, and reprecipitate. Allow to settle overnight and filter on Whatman No. 40 paper, with pulp, washing thoroughly with 5% ammonium hydroxide. Place in platinum crucible, dry, and ignite slowly with free access of air to avoid any reduction of phosphate which might injure the crucible. The slow, low-temperature charring of paper is also necessary to prevent a form of phosphate "flameproofing" which leaves black specks of carbon in the ignited residue. Finally ignite for 30 minutes at 1000°C, and weigh as $Mg_2P_2O_7$.

$$Mg_2P_2O_7 \times 0{\cdot}2185 = Mg$$
$$\times\ 0{\cdot}3623 = MgO$$

The reactions are:

$$(NH_4)_2HPO_4 + Mg(NO_3)_2 + NH_4OH \rightarrow MgNH_4PO_4 + 2NH_4NO_3 + H_2O$$
$$2MgNH_4PO_4 \rightarrow Mg_2P_2O_7 + 2NH_3 + H_2O$$

The presence of large quantities of ammonium salts retards the precipitation of small quantities of magnesium. For most purposes, allowing the precipitate to stand overnight, followed by a reprecipitation, will overcome this interference. If magnesium is very low and the quantities of ammonium salts are extremely high, it is a good practice to evaporate to a low volume, add cautiously 50–75 ml nitric acid and evaporate to dryness. If necessary, add more nitric acid and evaporate again to get rid of all ammonium salts. Dilute to 200 ml and proceed with the precipitation of magnesium ammonium phosphate.

Other separations of interfering elements may be made, depending on the sample. For example, iron in cast iron can be nearly all eliminated by an ether separation [4]. Electrolysis at a mercury cathode in dilute sulphuric acid leaves magnesium in the solution with Ca, P, Al, Be, Ti, Zr, V, U, and W, whereas Fe, Ni, Co, Cr, Zn, Mo, Cu, Bi, Sn, Ag, Au, Pt and Pd are deposited. Precipitation with cupferron in 10% sulphuric acid and extraction of the cupferrates with chloroform puts magnesium in the upper acid layer with Ca, P, Al, Be, B, Mn, Ni, Co, Zn, Cr, Se, and Te; the lower chloroform layer contains Fe, V, Ti, Zr, Nb, Ta, W, Sn, Ga, Mo, and Bi. Magnesium can be separated from aluminium in metallic aluminium by dissolution of the latter in sodium hydroxide.

B. PRECIPITATION WITH 8-HYDROXYQUINONE

In a solution from which all interfering elements have been removed, as outlined in the preceding section, magnesium may be precipitated with 8-hydroxyquinoline instead of diammonium phosphate [4, 5, 7, 11]. To the solution which contains about 5 ml ammonium hydroxide in excess for each 0·03 g magnesium, add 10 ml of 8-hydroxyquinoline solution. The latter is made by dissolving 5 g of 8-hydroxyquinoline in 10 ml of acetic acid and diluting to 100 with water. Avoid a large excess of 8-hydroxyquinoline. Heat to boiling, remove, and allow to stand for 20 minutes. Filter on a weighed, medium porosity glass crucible and wash with hot 2% ammonium hydroxide solution. Dry in an oven for 2 hours at 160°C and weigh as magnesium oxyquinolate, $Mg(C_9H_6NO)_2$.

$$Mg(C_9H_6NO)_2 \times 0{\cdot}0778 = Mg$$
$$\times\ 0{\cdot}129 = MgO$$

If the precipitate is dried at 105°C, it may be weighed as the dihydrate $Mg(C_9H_6NO)_2.2H_2O$; the latter $\times$ 0·0697 = Mg, or $\times$ 0·116 = MgO.

2. Volumetric

A. OXYQUINOLATE-BROMATE

For some magnesium-containing materials which are determined on a routine basis, a more rapid procedure than the gravimetric ones is a volumetric method based on solution of magnesium oxyquinolate in hydrochloric acid, addition of an excess of $KBrO_3$–KBr solution to react with the quinolate, and measurement of this excess by addition of potassium iodide and titration with sodium thiosulphate [4, 5, 11].

Carry out the procedure as outlined in B above for the gravimetric separation of magnesium as oxyquinolate. Dissolve the latter from the filter paper with about 30 ml of 1 : 1 hydrochloric acid and wash with hot water. Dilute to about 150 ml and cool to room temperature. Add a measured amount of $KBrO_3$–KBr solution; this is made by dissolving 2 g KOH, 80 g KBr, and 9 g $KBrO_3$ in 700 ml of water and diluting to 1 litre. The quantity of $KBrO_3$–KBr solution needed depends on the magnesium content, and should be sufficient for a 5 ml back titration. Add 10 ml of 30% potassium iodide solution and titrate immediately with 0·33 N sodium thiosulphate, using 2 ml of 1% starch solution towards the end of the titration. Titrate a blank using the same volumes of $KBrO_3$–KBr and iodide solutions. Then

$$\%\ Mg = \frac{(A-B)C \times 0{\cdot}003 \times 100}{D}$$

where

A = ml of $Na_2S_2O_3$ required to titrate the blank

B = ml of $Na_2S_2O_3$ required to titrate the sample

C = normality of $Na_2S_2O_3$ solution

D = g of sample.

B. EDTA titration

In some samples, such as waters and carbonate rocks, magnesium and calcium can be determined together by titration in the presence of an NH_4OH–NH_4Cl buffer with EDTA, using Eriochrome black T as indicator; on another sample calcium alone may be titrated with EDTA in NaOH–NaCN solution using Murexide or Calcon as indicator. The difference in the titrations gives the magnesium content. An example of this procedure is outlined in the section on Calcium. Magnesium in aluminium-base alloys has been determined by EDTA titration [6].

3. Colorimetric

A. Titan Yellow

When a small quantity of magnesium is added to a solution containing Titan yellow, and the magnesium then precipitated by sodium hydroxide, a deep red lake is formed; the latter can be kept in suspension by stabilizers such as a mixture of glycerol and starch, and its intensity measured photometrically against a series of standards [7, 15]. Unfortunately, most metals interfere with the formation of the Titan yellow-magnesium lake. It is usually necessary, therefore, to remove members of preceding groups with hydrogen sulphide, ammonium hydroxide, ammonium oxalate, and finally evaporate to fumes with nitric acid to remove ammonium salts.

Transfer to a 50 ml flask a suitable aliquot so that the final volume contains 0·5 to 4 ppm magnesium. Add 1 ml of 5% hydroxylamine hydrochloride, 2 ml of 1% starch solution, 2 ml of 50% glycerol solution, 1 ml of 0·1% aqueous solution of Titan yellow, and 5 ml of 10% sodium hydroxide. Dilute to the mark, mix, and measure the absorbance in a spectrophotometer at 535 nm.

The effect of small quantities of certain elements such as Cu, Zn, Cd, and Hg can be masked by the addition of potassium cyanide; moreover, the interference of some elements is negligible until their concentration exceeds that of magnesium. For instance, 1–4 ppm of magnesium in titanium can be determined in the presence of up to about 6 ppm Ti, and 8 ppm of Mn, Ni, and V [7]. It is possible, then, with some materials to reduce the separations and markedly simplify the procedure.

B. 8-Hydroxyquinoline

Another colorimetric method for small quantities of magnesium depends on the formation of the oxyquinolate in ammoniacal solution with 8-hydroxyquinoline, extraction of this coloured complex with a mixture of chloroform and ethylene glycol monobutyl ether, and photometric measurement at 400 nm. The interference due to Cu, Ni, Co, Ag, Au, and Pt metals may be masked by the addition of sodium cyanide; the presence in more than traces of Ca, Zn, Cd, or Ga cannot be tolerated. The procedure is used for 0·01–0·1% magnesium in electronic nickel [4, 13], and for magnesium in calcium salts [14].

C. Magon

Small amounts of magnesium in rocks have been determined by means of an ethanolic solution of Magon or Magon sulphonate, triethanolamine, and a borax buffer at 510 nm [1].

4. Optical Spectrography

Magnesium has a number of useful spectral lines, and small quantities of this element are determined routinely on various materials. Detailed instructions have been published for 0·001–10% magnesium in aluminium and its alloys [4], 0·005–0·2% in zinc and zinc-base materials [4], 10–500 ppm in zirconium and its alloys [4], and about 1–2100 ppm in nickel [12].

5. Atomic-Absorption Spectrophotometry

Magnesium can be determined readily by atomic absorption, usually with line 2852·1 which has a sensitivity of 0·01 μg/ml; other sensitive lines are 2025·8 and 2795·5 Å [2, 8, 9]. Interferences are Si, Al, Ti, and P. Silicon can be removed by dehydration, and the addition of strontium or lanthanum overcomes the interference of aluminium. With the nitrous oxide–acetylene flame, even a large excess of aluminium does not interfere [9].

REFERENCES

1. Abbey, S., and Maxwell, J. A., *Anal. Chim. Acta* **27**, 233–40 (1962).
2. Allen, J. E., *Analyst* **83**, 466–71 (1958).
3. A.O.A.C., *Official Methods of Analysis of the Association of Official Analytical Chemists*, Washington, D.C., Association of Official Analytical Chemists, 1965.
4. A.S.T.M., *Chemical Analysis of Metals. Sampling and Analysis of Metal Bearing Ores*, Philadelphia, Pa., American Society for Testing and Materials, 1969.
5. British Drug Houses, *The B.D.H. Book of Organic Reagents for Analytical Use*, Poole, Dorset, British Drug Houses, 1958.
6. Burke, K. E., *Anal. Chim. Acta* **34**, 485–7 (1966).
7. Challis, H. J. G., and Wood, D. F., *Analyst* **79**, 762–70 (1954).
8. David, D. J., *Analyst* **85**, 495–503 (1960).
9. Elwell, W. T., and Gidley, J. A. F., *Atomic-Absorption Spectrophotometry*, 2nd ed., Oxford, Pergamon, 1966.
10. Furman, N. H., ed., *Scott's Standard Methods of Chemical Analysis*, 6th ed., Vol. 1, Princeton, N.J., Van Nostrand, 1962.
11. Hillebrand, W. F., Lundell, G. E. F., Bright, H. A., and Hoffman, J. I., *Applied Inorganic Analysis*, 2nd ed., New York, Wiley, 1953.
12. Lewis, C. L., Ott, W. L., and Sine, N. M., *The Analysis of Nickel*, Oxford, Pergamon, 1966.
13. Luke, C. L., *Anal. Chem.* **28**, 1443–5 (1956).
14. Newman, E. J., and Watson, C. A., *Analyst* **88**, 506–9 (1963).
15. Sandell, E. B., *Colorimetric Determination of Traces of Metals*, 3rd ed., New York, Interscience Publishers, 1959.

21 Manganese

Manganese is widely distributed in minerals, ores, and soils; it is a common constituent of irons and steels, many alloys, and a few industrial materials. The important manganese minerals are pyrolusite MnO_2, psilomelane colloidal MnO_2, manganite $Mn_2O_3.H_2O$, braunite 3 $Mn_2O_3.MnSiO_3$, hausmannite Mn_3O_4, rhodochrosite $MnCO_3$, rhodonite $MnSiO_3$, and bementite $2MnSiO_3.H_2O$.

Relatively few interfering elements must be removed in the volumetric methods for manganese, which are used most frequently in technical work; the same is true for potentiometric and colorimetric procedures. Metallic manganese dissolves in dilute acids; oxides and hydroxides are soluble in hot hydrochloric acid. Ferromanganese and silicomanganese are soluble in a mixture of nitric, perchloric, and hydrofluoric acids.

ISOLATION OR SEPARATION OF MANGANESE

Members of the hydrogen sulphide group can be separated from manganese by precipitation of the former with hydrogen sulphide in acid solution; manganese can then be isolated along with many other elements by precipitation with ammonium sulphide. Unless bromine is present, manganese is not precipitated by ammonia in the presence of ammonium salts, and can thus be separated from Fe, Al, Ti, Be, Cr, Zr, V, P, and U. Cupferron in cold 10% sulphuric acid will precipitate Fe, Ti, Zr, V, Nb, Ta, and W, leaving manganese in the filtrate with Al, Be, P, Cr, Ni, and Co. Treatment with sodium peroxide in alkaline solution will precipitate manganese, together with Fe, Ti, Zr, Th, Ni, and Co, from Al, V, Cr, Zn, Be, W, and U. A zinc oxide precipitation will leave Mn, Ni, and Co in the filtrate, free from Fe, Al, Ti, P, Cr, V, Zr, Nb, Ta, and W. An important separation of manganese from all members of the ammonium hydroxide and ammonium sulphide groups is precipitation by boiling with sodium or potassium chlorate in strong nitric acid. Manganese can also be separated from iron by an ether separation, and from chromium by volatilization of the latter with boiling perchloric and hydrochloric acids.

In the usual nitric–sulphuric acid electrolytic determination for copper, part of the manganese deposits on the anode as hydrated MnO_2, and the rest remains in solution: none is deposited on the cathode. The anode deposit may be dissolved in hydrochloric acid, or nitric acid + hydrogen peroxide, and added to the de-copperized solution. Nickel and cobalt can be separated from manganese by electrolytic deposition of the former in ammoniacal solution. Traces of manganese can be separated from nickel and cobalt by ammonia and bromine, with the addition of a little iron to collect the manganese.

1. Volumetric Methods

A. With Ammonium Persulphate and Sodium Arsenite

This method involves the oxidation of manganous salts to permanganate with ammonium persulphate in the presence of silver nitrate, and titration of the permanganic acid with sodium arsenite [2, 4, 5, 6]. The procedure is useful for routine work on some materials, but for most products of extractive metallurgy the bismuthate method is preferable.

To a 0·5–2 g sample add 30 ml nitric–sulphuric–phosphoric acid mixture, boil until oxides of nitrogen are expelled, and dilute with 100 ml hot water. If an insoluble residue remains, which might contain manganese, filter, wash, ignite, fuse with sodium carbonate, and add to the main portion. If hydrochloric, perchloric or hydrofluoric acids have been used in the decomposition, evaporate to strong fumes of sulphur trioxide to eliminate chloride and fluoride. Avoid a mixture of perchloric and hydrofluoric acids; small quantities of manganese can be volatilized at higher temperatures with this combination.

In the presence of over 0·5% tungsten or 1% chromium, separate manganese by zinc oxide or nitrochlorate. For the former, add a suspension of zinc oxide to precipitate Fe, Al, Cr, W, etc., boil, filter, wash, dissolve in nitric acid, reprecipitate with zinc oxide, filter and wash thoroughly; manganese will be in the filtrate, together with nickel, cobalt and zinc. For the nitrochlorate separation, dissolve the sample in aqua regia, filter off any residue of tungsten, add sulphuric acid and evaporate to strong fumes of the latter. Add 50 ml nitric acid, boil, carefully add a few crystals of sodium chlorate and continue boiling; repeat the latter operation until about 8 g of sodium chlorate have been added. Filter the precipitated manganese dioxide through a Gooch containing asbestos previously washed with hot colourless nitric acid. Wash with a little nitric acid and finally with cold water. Discard the filtrate and washings, dissolve the manganese dioxide by washing the beaker and the filter with 50 ml of 5% nitric acid to which 10 ml of 3% hydrogen peroxide has been added. Wash the beaker and filter three times with water, and boil the solution to decompose the peroxide. The reaction for the nitrochlorate separation is expressed—

$$4Mn(NO_3)_2 + 16NaClO_3 + 4HNO_3 \rightarrow 4MnO_2 + 12NaNO_3 + 4NaCl + 4Cl_2 + 4ClO_2 + 15O_2 + 2H_2O$$

To the solution obtained by one of the foregoing routes, free of chlorine and appreciable quantities of chromium and tungsten, add in succession 5 ml of 1% silver nitrate and 10 ml of a 30% solution of ammonium persulphate. Stir, heat rapidly to boiling, and boil for 1 minute after the foam disappears. Cool rapidly in a trough of running water, and titrate immediately with standard sodium arsenite until the solution assumes a yellowish colour or until one drop produces no further change in the colour of the solution.

Reagents

Nitric–sulphuric–phosphoric acid mixture. Add 185 ml of 1 : 1 H_2SO_4 to 440 ml water, then add 125 ml of H_3PO_4 and 250 ml of HNO_3.

Standard sodium arsenite solution. Prepare a stock solution by dissolving 22 g $NaAsO_2$ in water, dilute to 1 litre, and allow to age 3 weeks. Dilute 100 ml of the aged solution to 1 litre for use as needed; this gives a more stable solution than one freshly prepared when required.

Standardize the sodium arsenite solution against a standard steel of known manganese content, carrying the latter through all the steps of the determination. The sodium arsenite solution can be adjusted, if desired, so that 1 ml = approximately 0·0005 g Mn, or 0·1% on a 0·5 g sample. Sodium arsenite solutions may also be made by dissolving 0·674 g As_2O_3 and 2·2 g Na_2CO_3 in a small quantity of water and diluting to 1 litre; 1 ml = approximately 0·0002 g Mn. Alternatively, 0·1N arsenious acid solution can be made by dissolving 4·9455 g As_2O_3 in 40 ml of warm 10% NaOH, making just acid with 1 : 1 H_2SO_4, cooling, and diluting to 1 litre; suitable dilution of this primary standard may be used to provide the titrating solution for manganese.

The reactions of the ammonium persulphate titration may be expressed:

$$2Mn(NO_3)_2 + 5(NH_4)_2S_2O_8 + 8H_2O \rightarrow$$
$$5(NH_4)_2SO_4 + 5H_2SO_4 + 4HNO_3 + 2HMnO_4$$

$$As^{III} + Mn^{VI} \rightarrow As^{V} + \underset{\textit{Green}}{Mn^{IV}} + \underset{\textit{Yellow}}{Mn^{V}}$$

A variation of the ammonium persulphate method for irons and steels has been advocated [9]. Dissolve the sample in dilute sulphuric acid, and if any other acids are required for complete solution, evaporate to fumes of sulphuric. Make a zinc oxide separation, acidify the filtrate strongly to prevent interference from cobalt, and oxidize with ammonium persulphate. Titrate the permanganic acid with ferrous ammonium sulphate and potassium dichromate, using barium diphenylamine sulphonate as indicator.

B. With Sodium Bismuthate, Ferrous Ammonium Sulphate, and Permanganate

This procedure depends on the oxidation of manganese to permanganic acid with sodium bismuthate, its reduction with standard ferrous salt and final titration with standard potassium permanganate [2, 4, 5, 6, 7]. It is an excellent method, and can be employed with assurance over a wide range of manganese contents for all samples in extractive metallurgy.

In this method, chromium and vanadium must be eliminated in the precipitate of a zinc oxide separation or in the filtrate of a sodium peroxide precipitation. If alone, chromium can be volatilized by boiling with perchloric and hydrochloric acids. Cobalt must be removed because it is oxidized by the bismuthate and subsequently reduced with the permanganic acid; large quantities of nickel cause a colour interference at the visual end point. Manganese can be separated from nickel and cobalt by treatment with sodium chlorate and nitric acid as outlined in the previous section. Other interferences are cerium, which is rarely encountered, and chlorine, hydrofluoric acid, and organic matter; the last three can be eliminated by fuming with sulphuric acid.

If the sample is in nitric acid solution, boil out all oxides of nitrogen, cool, add a small quantity of sodium bismuthate until a pink colour develops or manganese dioxide precipitates; this will destroy any organic or reducing substances which may be present. Boil two minutes and add a freshly prepared solution of sulphur dioxide in water until the sample solution is clear. Boil until all sulphur dioxide and nitrogen oxides are expelled, and cool to 15°C.

If the sample is in sulphuric acid solution, heat to boiling, add ammonium persulphate first and finally a very small quantity of sodium bismuthate. Then reduce with sulphurous acid, boil, and cool as directed above. If the sample is in perchloric acid solution, add nitric acid to 15% of the solution, re-oxidize and reduce as indicated above for nitric acid solution.

To the cold solution containing 15–20 ml nitric acid or 5–10 ml sulphuric acid per 100 ml, add an excess of sodium bismuthate at least 25 times the weight of manganese present. The nitric acid solution for bismuthate precipitation should contain approximately 22% by weight nitric acid, and not much over 0·05 g manganese per 100 ml. Stir vigorously, and if all or nearly all of the bismuthate should dissolve, add more immediately. Let the solution stand one minute and add 50 ml of 3% nitric acid solution which is free of nitrous acid. The tendency of $HMnO_4$ to decompose is minimized by thus diluting the oxidized solution with an equal volume of water before filtering off the excess bismuthate. Filter without delay through a purified asbestos pad in a Gooch crucible, or through a sintered glass crucible. Wash the residue 4–5 times with the 3% nitric acid solution until the washes are colourless, and titrate the filtrate in the following manner.

If the sample is in sulphuric or nitric acid solution, add 2 ml of phosphoric acid and stir; if in perchloric acid omit the phosphoric acid. Add from a burette a measured quantity of a standard 0·1N ferrous ammonium sulphate solution until the pink colour is discharged, and 1–2 ml excess. Back titrate this excess with standard 0·1N potassium permanganate solution until one drop gives a distinguishable pink end point.

To ascertain the blank due to colour interference, if the solution is coloured boil for 10 minutes. Cool to the same temperature as before and titrate again to the same end point. If this titration requires more of the permanganate than is needed to give an end point in a colourless solution, subtract the difference from the ml of potassium permanganate used in the titration. Titrate the blank to the same end point with potassium permanganate as a check on reagents used. Then add the same volume of the ferrous ammonium sulphate solution as was used in the determination and titrate with permanganate, noting the ml of the latter added in both titrations. The difference in the volume of the potassium permanganate used in the two titrations represents the manganese in the sample.

REAGENTS

3% HNO_3. Boil 40 ml nitric acid, cool, and bubble air through it for 5 minutes. Dilute to 100 ml, add 1 g sodium bismuthate, and allow to settle. Keep in a cool, dark place.

Ferrous ammonium sulphate. Dissolve 39·2 g $Fe(NH_4)_2(SO_4)_2.6H_2O$ in a cold solution of 50 ml sulphuric acid in 950 ml water. This is approximately equivalent to 0·1N potassium permanganate.

$KMnO_4$, 0·1N. Dissolve 1 g potassium hydroxide in 1 litre of water and add 3·16 g potassium permanganate. Allow to age one week and filter through a purified asbestos pad or through a sintered glass crucible. Standardize the potassium permanganate against sodium oxalate; 1 litre of 0·1N $KMnO_4$ = 6·700 g sodium oxalate, or 0·1000 g of the latter is equivalent to 14·93 ml of 0·1N $KMnO_4$. Dissolve 0·1000 g sodium oxalate in 100 ml of boiled and cooled 5% sulphuric acid solution, heat to 60–70°C and titrate with potassium permanganate to the first permanent pink. Titrate a blank sulphuric acid solution in the same way and subtract this blank from the ml permanganate used to titrate the oxalate. Calculate the manganese value of 1 ml of potassium permanganate solution. Titrate the standard potassium permanganate daily against the ferrous ammonium sulphate to determine this ratio.

$$1 \text{ ml of } 0{\cdot}1 \text{ N } KMnO_4 = 0{\cdot}001099 \text{ g Mn}$$

For low manganese contents, it may be more convenient to use about 0·03N ferrous ammonium sulphate, containing 12 g per litre, and 0·03N potassium permanganate containing 1 g per litre. In this case, 1 ml of 0·03N $KMnO_4$ = 0·0003297 g Mn; it is conveniently standardized against 0·0500 g sodium oxalate which is equivalent to 24·88 ml of 0·03N potassium permanganate.

The reactions of the sodium bismuthate procedure may be expressed:

$$2Mn(NO_3)_2 + 5NaBiO_3 + 16HNO_3 \rightarrow 2HMnO_4 + 5Bi(NO_3)_3 + 5NaNO_3 + 7H_2O$$

$$2KMnO_4 + 10FeSO_4 + 16HNO_3 \rightarrow K_2SO_4 + 2Mn(NO_3)_2 + 3Fe_2(SO_4)_3 + 4Fe(NO_3)_3 + 8H_2O$$

2. Potentiometric Method

The oxidation of Mn(II) to Mn(III) by potassium permanganate in neutral sodium pyrophosphate solution is utilized in a potentiometric procedure which has few interferences [4, 6, 8, 11, 12]. Decompose 0·5–5 g sample with hydrochloric and nitric acids, add 10 ml 1 : 1 sulphuric acid, and evaporate to fumes of the latter. If an insoluble residue remains, which might contain manganese, fuse it with sodium carbonate and add to the main portion. If the material decomposes completely in hydrochloric or nitric acids, treatment with these alone is sufficient; when nitric acid is used, boil the sample thoroughly to remove oxides of nitrogen. If desired, add a small quantity of urea or sulphamic acid to remove the last traces of nitric oxide.

To a 25 or 50 ml sample of the faintly acidic manganese solution add 150–300 ml of saturated sodium pyrophosphate solution. It is important that the solution contain only a minimum quantity of free acid before the addition of sodium pyrophosphate; in the presence of Cu, Co, Fe, and Ni there can be air oxidation of manganese in the subsequent *p*H adjustment if the initial solution is strongly acid [11]. The pyrophosphate solution contains approximately 12 g $Na_4P_2O_7.10H_2O$ per 100 ml at room temperature; prepare fresh every two weeks. Adjust the *p*H to a value of 6–7 by the careful addition of 1 : 1 sulphuric acid or dilute sodium hydroxide, using indicator test papers or a *p*H meter.

Titrate with 0·02 molar potassium permanganate solution on a potentiometric assembly, using bright platinum and saturated calomel electrodes, to the usual end point where the sharp break in the curve appears. The 0·02 molar permanganate is made by dissolving 3·16 g $KMnO_4$ in 1 litre of water; 1 ml = 0·004394 g Mn. 1 $KMnO_4 \rightarrow 4$ Mn when the reaction is represented—

$$4Mn^{++} + MnO_4^- + 8H + 15\ H_2P_2O_7 \rightarrow 5Mn(H_2P_2O_7)_3 + 4H_2O.$$

The following, even in large quantities, are without effect on the titration: Li, K, Na, Rb, Cs, Cu, Hg, Rh, Ce(IV), La, Pt, Pd, Sn, Pb, Bi, Mo(VI), Fe, Al, Ti, Zr, Hf, Be, Cr(III), Co, Re(VII), Ni, U (VI), Zn, Th, Ca, Mg, Sc, Ga, In, Ge, Nb, W(VI), SiO_2, Y, Nd, Pr, Sm, Gd, B, PO_4, Cl, SO_4, Br, F, CO_3, NO_3, and ClO_3. Small quantities of Ag, Ba, Sr, Ta, and V can be tolerated; large amounts precipitate and occlude manganese. The following interfere: Sb, Ce(III), Se(IV), Te(IV), Au(III), Ir(IV), iodide, and large amounts of Tl(III); ruthenium causes a small positive error in the titration.

It is customary to eliminate chromium in ores and metals by volatilization with boiling hydrochloric and perchloric acids; vanadium is separated from manganese by sodium peroxide, nitrochlorate, zinc oxide, or cupferron.

3. Colorimetric Method

The pink colour obtained by oxidation of manganese solutions with potassium periodate in nitric or sulphuric acids furnishes the basis of a colorimetric procedure which is applicable to a wide variety of materials [1, 2, 4, 6, 10].

Decompose a weight of sample such that the final solution of the entire sample, or an aliquot, in a 100 ml flask prior to colour comparison, does not contain more than 2 mg manganese. If hydrochloric acid has been used in the decomposition, evaporate to strong fumes of sulphuric acid. If organic matter is present, ignite in a platinum dish at 800°C before proceeding with the decomposition. Decompose high-lime samples with nitric acid; if any residue remains which might contain manganese, fuse with sodium carbonate and add to the main portion. Avoid sulphuric acid for high-lime materials, because manganese is seriously occluded in heavy precipitates of calcium sulphate. Silica may be eliminated, in high-lime samples, by treatment with nitric and hydrofluoric acids in a platinum dish; for other samples sulphuric and hydrofluoric acids may be used.

To the solution containing 10–15 ml sulphuric or 20 ml nitric acid, add 5 ml phosphoric acid, and 0·3 g potassium periodate. There must be at least twice as much potassium periodate as theoretically required; the latter is 1 g potassium periodate per 0·1 g manganese. Boil for 1 minute, keep hot for 10 minutes and for very low manganese contents 30 minutes, cool, and dilute to 100 ml in a volumetric flask.

Measure the absorbance at 525 nm against standards. The latter can be made by dissolving 0·7192 g potassium permanganate in 500 ml of water, withdrawing a 50-ml aliquot into a beaker, adding 5 ml sulphuric acid and sufficient sodium sulphite to decolorize the solution, boiling out all sulphur dioxide, and making up to 500 ml in a volumetric flask. This second solution, containing 0·07192 g $KMnO_4$ in 500 ml or 0·00014384 g $KMnO_4$ per ml, has a manganese content of 1 ml = 0·05 mg Mn. From 1 to 40 ml portions of this solution can be carried through the steps of colour development to provide the calibration curve.

The oxidized solution containing an excess of periodate is stable and colour comparisons do not need to be carried out quickly. Moderate variations in acid concentration and excess of periodate do not affect the colour. High concentrations of coloured ions such as Cu, Ni, Co, and Cr interfere; reducing substances and chlorine must be destroyed by nitric and sulphuric acids. In sufficient acid, Ag, Pb, and Hg give no precipitate; Bi and Sn give a turbidity even in strongly acid solutions. In the presence of small amounts of chromium, measure the absorbance at 545 nm instead of 525. When chromium is high, volatilize it by boiling with perchloric and hydrochloric acids, or remove by a zinc oxide separation. The interference of coloured ions such as copper, nickel, and cobalt can often be overcome by use of a suitable reference solution. The alternative is a separation of copper by hydrogen sulphide in acid solution, or by electrolysis, and of nickel and cobalt by ammonia and bromine, with the addition of a little iron to collect the manganese in the precipitate. For high-iron samples it is preferable to use sulphuric rather than nitric acid in the final solution, because ferric periodate is not very soluble in nitric acid.

4. Optical Spectrography

Procedures have been published for 0·1–2% manganese in low alloy steel [2], 0·001–8% in aluminium and its alloys [2], 10–200 ppm in zirconium and its alloys [2], and 4–4000 ppm in nickel [7].

5. X-Ray Spectrography

Manganese has been determined in a number of materials by X-ray fluorescence [6, 7], but the latter method does not appear to be used widely for routine purposes in industry.

6. Atomic-Absorption Spectrophotometry

Manganese can be readily determined by atomic absorption. The most sensitive line at 2794·8 Å has a sensitivity of 0·06 μg/ml; other lines 2798·3, 2881·1, and 4030·8 can be used [3]. Only silicon interferes, and this of course can be easily removed by acid dehydration and filtration.

REFERENCES

1. A.O.A.C., *Official Methods of Analysis of the Association of Official Analytical Chemists*, Washington, D.C., Association of Official Analytical Chemists, 1965.
2. A.S.T.M., *Chemical Analysis of Metals. Sampling and Analysis of Metal Bearing Ores*, Philadelphia, Pa., American Society for Testing and Materials, 1969.
3. Elwell, W. T., and Gidley, J. A. F., *Atomic-Absorption Spectrophotometry*, 2nd ed., Oxford, Pergamon, 1966.
4. Furman, N. H., ed., *Scott's Standard Methods of Chemical Analysis*, 6th ed., Vol. 1, Princeton, N.J., Van Nostrand, 1962.
5. Hillebrand, W. F., Lundell, G. E. F., Bright, H. A., and Hoffman, J. I., *Applied Inorganic Analysis*, 2nd ed., New York, Wiley, 1953.
6. Kolthoff, I. M., and Elving, P. J., *Treatise on Analytical Chemistry*, Part II, Vol. 7, 425–502, New York, Interscience Publishers, 1961.
7. Lewis, C. L., Ott, W. L., and Sine, N. M., *The Analysis of Nickel*, Oxford, Pergamon, 1966.
8. Lingane, J. J., and Karplus, R., *Ind. Eng. Chem. Anal. Ed.* **18**, 191–4 (1946).
9. Methods of Analysis Committee of the Metallurgy (General) Division of the British Iron and Steel Research Association, *Metallurgia* **38**, 346–52 (1948).
10. Sandell, E. B., *Colorimetric Determination of Traces of Metals*, 3rd ed., New York, Interscience Publishers, 1959.
11. Scribner, W. G., *Anal. Chem.* **32**, 966–70 (1960).
12. Scribner, W. G., and Anduze, R. A., *Anal. Chem.* **33**, 770–3 (1961).

22 Mercury

Mercury is only occasionally determined in most mining laboratories, where it may be encountered in a few ores, concentrates, amalgams and compounds. The most important mercury mineral is cinnabar HgS; minor sources are the minerals calomel Hg_2Cl_2 and tiemannite HgSe, amalgams with silver, gold, or tellurium, and a little free mercury.

ISOLATION OR SEPARATION OF MERCURY

The boiling point of mercury is about 357°C, consequently great care must be taken in any decomposition procedure to avoid loss by volatilization. Mercury may be separated from elements of succeeding groups in the mercuric form with other members of the acid sulphide group by precipitation with hydrogen sulphide in 5–10% hydrochloric or sulphuric acid. Separations of mercury from some other members of the hydrogen sulphide group are not sharp; when these are required, it is necessary to pay careful attention to details which are given in standard references [2, 5, 7, 8]. Silver is separated from mercuric salts by precipitation of the former as chloride in 1% nitric acid; a re-precipitation is necessary. Mercury is separated from As, Sb, Sn, and Mo by digestion of sulphides in warm ammonium sulphide solution; mercuric sulphide is insoluble whereas the others dissolve. The sulphides of sodium or potassium dissolve mercury as well as As, Sb, and Sn, so cannot be used for separating mercury from these elements. From Cu, Cd, Pb, and Bi, mercury can be separated by digestion of sulphides in hot dilute nitric acid, in which mercuric sulphide is insoluble. A separation from selenium and tellurium can be made by adding to a cold dilute chloride solution a little phosphorous acid which precipitates mercurous chloride. Selenium can also be precipitated from mercury by ferrous sulphate in a strongly acid chloride solution. Mercury can be precipitated from tellurium by adding sodium hydroxide and sodium sulphide, boiling with an excess of ammonium chloride until the precipitated mercuric sulphide settles, filtering, and washing with colourless ammonium sulphide.

Mercury is completely volatilized when its compounds are ignited, or fused with sodium carbonate or potassium bisulphate. It is partially volatilized when solutions of mercuric salts, especially the chloride, are evaporated on the steam bath or boiled.

1. Gravimetric by Volatilization and Amalgamation

A simple and satisfactory method for determining mercury in ores is to volatilize the metal, reducing the material if necessary with iron filings or other fluxes, and collect the mercury on a cold, weighed sheet of silver or gold.

Transfer to an iron crucible a sample containing not over 0·1 g mercury. Mix with the sample 5–10 g of iron filings and sprinkle additional filings over the surface. The filings should be about 65 mesh. For high sulphide ores, or arsenic ores, mix with about twice their weight of a flux of 4 parts zinc oxide to 1 part sodium carbonate, and mix with this about 5 times its weight of iron filings. Insert the crucible in a circular opening in an asbestos board so that about half its height is above the board. Cover the crucible with a weighed sheet of silver or gold, and over the sheet to keep it cool place a cylindrical metal condenser through which cold water circulates. A small Erlenmeyer flask may be used as a condenser. Be sure that the silver sheet fits tightly over the crucible and is maintained in this position by slight pressure from the cooling condenser. Heat the crucible gradually, maintain heat for about 30 minutes, and allow to cool with water still flowing through the condenser. The silver sheet must never become hot; otherwise, mercury may be lost. Remove the sheet carefully, dip in alcohol, dry over warm air, and weigh. The increase in weight is metallic mercury. One sheet of silver will last for a number of determinations; it may be cleaned by rubbing with moist powdered calcium carbonate, and washing, before re-use.

A commercial assembly for this procedure, termed Whitton's apparatus [5], is obtainable from Braun-Knecht-Heimann Co., 1400 16th Street, San Francisco 19, Cal., U.S.A. It consists of a steel retort with a cover of sheet silver, and above these a flat-bottomed cooling dish, which is kept filled with water. The retort, silver sheet, and cooling dish are clamped together; a Bunsen burner is adjusted so that the blue cone of the flame strikes the bottom of the retort while the flame runs up the sides about $\frac{1}{2}$ inch. The water in the cooling dish commences to boil in about 7 minutes and is replaced as it boils away. At the end of 17 minutes withdraw the flame and allow the apparatus to cool for 5 minutes or until it can be dismantled. No water will remain on the silver sheet at the conclusion of the heating; drying is unnecessary. The deposit should be white; if the heating has been done at too high a temperature, or for too long a period, the deposit will be dark and an incorrect result will usually be obtained. If the ore contains much water, the latter will condense on the silver sheet, drop back into the crucible, and possibly splash the filings onto the sheet. Heat the charge slowly or cover with asbestos fibre.

2. Volumetric Determination with Thiocyanate

Mercuric nitrate forms an insoluble thiocyanate with a standard thiocyanate solution; the end point is indicated by a ferric salt [7, 8]. Mercury must be in the mercuric state; chlorides, palladium, and silver must be absent.

The procedure can be used in the presence of Pb, Cu, Bi, Cd, Sn, As, Sb, Tl, Fe, Zn, Mn, Ni, and Co.

To a solution of Hg(II) containing approximately 10 ml of nitric acid or 5 ml sulphuric acid per 100 ml, add 5% potassium permanganate solution dropwise with stirring until a red colour of manganese dioxide persists for 5 minutes. Destroy the excess by the careful addition of a dilute ferrous sulphate solution. Add 2 ml of a saturated ferric alum solution for every 100 ml of sample. This indicator solution is made by dissolving enough ferric ammonium sulphate or ferric sulphate to make a saturated solution, and then adding enough colourless nitric acid to clear the solution and produce a pale yellow colour.

Titrate with 0·1 or 0·01N ammonium thiocyanate until a distinct pink persists after vigorous stirring. The colour is removed rather slowly as the end point approaches, and care must be exercised at this stage. A 0·1N ammonium thiocyanate solution is made by dissolving about 9 g NH_4CNS in water and diluting to 1 litre; because thiocyanate is hygroscopic, it cannot be weighed directly as a standard. The thiocyanate is best standardized against a weighed portion of pure dry mercury which has been dissolved in 1 : 1 nitric acid and carried through the method. The thiocyanate may also be standardized against silver nitrate as described in the chapter on silver. The reactions are:

$$2NH_4CNS + Hg(NO_3)_2 \rightarrow Hg(CNS)_2 + 2NH_4NO_3$$
$$3NH_4CNS + Fe(NO_3)_3 \rightarrow Fe(CNS)_3 + 3NH_4NO_3$$

3. Colorimetric Determination with Dithizone

Divalent mercury reacts with dithizone in 1 N acid to give an orange colour in a solution of carbon tetrachloride or chloroform. The only other metals which will also react at this acidity are Ag, Cu, Au, Pt, and Pd. If a sufficient concentration of chloride is present, silver will not react; Au, Pt, and Pd are rarely encountered, and procedures have been developed to overcome the interference of small quantities of copper [1, 2, 4, 5, 8, 9].

A typical method for traces of mercury in organic matter is outlined below, based on the recommendations of an authoritative committee [1]. Oxidize the sample with nitric and sulphuric acids, collect the distillate from this operation and combine it with the residue. Dilute the combined digestate until it is approximately N in acid. Treat with 20% hydroxylammonium chloride to remove oxides of nitrogen. Extract the mercury with excess of 0·001% dithizone in carbon tetrachloride; any copper present will be extracted at this stage with the mercury. Destroy the dithizone with 1 ml of 5% sodium nitrite in 10 ml of 0·1N hydrochloric acid. Mercury reverts to the aqueous phase; discard the organic phase. Add 1 ml of 20% hydroxylammonium chloride, and set aside for 15 minutes, shaking occasionally. Add 1 ml of 10% urea solution and 1 ml of 2·5% EDTA solution. Extract the mercury with successive small portions of 0·001% dithizone in carbon tetrachloride; mercury is separated from copper at this stage. Combine the extracts and dilute to 4·0 ml with carbon tetrachloride. Measure the optical density of the solution at 485 nm. If more than 60 μg of copper is present, carry out the final extraction with a solution of dithizone in chloroform and determine the absorbance at 492 nm.

4. Optical Spectrography

Traces of mercury have been determined by emission spectrography, using lines 2536·52 and 4358·35 [8].

5. Atomic-Absorption Spectrophotometry

Small quantities of mercury can be determined by atomic absorption, using the line 2536·5 [3, 6]. The sensitivity of 10 μg/ml is not as great as that of many elements, but mercury can be initially concentrated by extraction from a large sample in N hydrochloric acid by a carbon tetrachloride solution of dithizone.

6. Mercury Vapour Detector

Because the toxic limit for continual breathing is 1 mg of mercury vapour in 10 cubic metres of air, special instruments are available commercially from the General Electric Company and other sources to determine 0·004 to 0·37 parts per million by volume, with an accuracy of about 5%. These detectors are based on the absorption of radiation at 2537 Å by mercury vapour.

REFERENCES

1. Analytical Methods Committee, *Analyst* **90**, 515–30 (1965).
2. A.O.A.C., *Official Methods of Analysis of the Association of Official Analytical Chemists*, Washington, D.C., Association of Official Analytical Chemists, 1965.
3. Elwell, W. T., and Gidley, J. A. F., *Atomic-Absorption Spectrophotometry*, 2nd ed., Oxford, Pergamon, 1966.
4. Elwell W. T. and Scholes, I. R., *Analysis of Copper and its Alloys*, Oxford, Pergamon, 1967.
5. Furman, N. H., ed., *Scott's Standard Methods of Chemical Analysis*, Princeton, N.J., Van Nostrand, 1962.
6. Hatch, W. R., and Ott, W. L., *Anal. Chem.* **40**, 2085–7 (1968).
7. Hillebrand, W. F., Lundell, G. E. F., Bright, H. A., and Hoffman, J. I., *Applied Inorganic Analysis*, 2nd ed., New York, Wiley, 1953.
8. Kolthoff, I. M., and Elving, P. J., *Treatise on Analytical Chemistry*, Part II, Vol. 3, 231–326, New York, Interscience Publishers, 1961.
9. Sandell, E. B., *Colorimetric Determination of Traces of Metals*, 3rd ed., New York, Interscience Publishers, 1959.

23 Molybdenum

Molybdenum is a common and important element in many ferrous and non-ferrous alloys. The commercially important minerals of molybdenum are molybdenite MoS_2, and wulfenite $PbMoO_4$. Small quantities of molybdenum are sometimes found in ores of copper and iron; in fact, a significant percentage of world molybdenum output is derived as a by-product of copper mining.

ISOLATION OR SEPARATION OF MOLYBDENUM

Molybdenum may be isolated, with other members of the acid sulphide group, by precipitation with hydrogen sulphide in dilute hydrochloric or sulphuric acid. Precipitation succeeds best in a solution containing not more than 5% acid, and hydrogen sulphide must be passed through the solution for a longer period of time than is customary for other metals of this group. It may even be necessary, if a substantial quantity of iron is not present in the sample, to filter off the molybdenum sulphide, oxidize the filtrate with ammonium persulphate, and again pass in hydrogen sulphide. This may be required because molybdenum in the reduced state is only slowly precipitated with hydrogen sulphide. Molybdenum sulphide may be also precipitated by thioacetamide [4, 17].

Arsenic, antimony, selenium, and tin may be separated from molybdenum by volatilization of the former with perchloric and hydrobromic acids. Mercury may be volatilized by fusion with sodium carbonate or potassium bisulphate. Molybdenum, with As, Sb, and Sn, may be separated from Cu, Cd, Bi, and Pb by digestion of the sulphides in warm alkaline or ammoniacal sulphide solution, in which Mo, As, Sb, and Sn dissolve. Molybdenum, with copper, may be removed from As, Sb, Sn, Bi, Se, Te, and P by a ferric hydroxide separation in ammonia; Mo and Cu are in the filtrate. Lead should be absent for this separation, as it tends to carry down molybdenum. Copper may be separated from molybdenum in several ways: a sodium hydroxide separation which puts copper in the precipitate and molybdenum in the filtrate, electrolysis in the usual sulphuric–nitric acid medium followed by solution of the copper deposit and a redeposition, precipitation of copper as cuprous thiocyanate, or digestion of the mixed sulphides in warm alkaline or ammoniacal sulphide solution. Silver may be removed as the insoluble chloride, and lead as the insoluble sulphate, from molybdenum. Selenium and tellurium may be isolated from molybdenum by precipitation of the former with sulphur dioxide in hydrochloric acid solution.

Because molybdenum is frequently associated in ores and alloys with elements of the ammonium hydroxide and ammonium sulphide groups, its behaviour in the common separations for these metals is of importance. Molybdenum, along with Fe, Cr, Ni, Co, and Zn, is deposited on a mercury cathode in dilute sulphuric acid, and can thereby be separated from Al, Be, Ti, Zr, P, V, U, Nb, Ta, Th, and W. Precipitation of molybdenum, with Fe, Ti, V, Zr, Nb, Ta, and W, by cupferron in cold 10% sulphuric acid and extraction of the cupferrate in chloroform will separate it from Al, Be, Ni, Co, Mn, and U(VI). Tracer studies have shown that all molybdenum is in the chloroform layer [9]. Molybdenum may be isolated by solvent extraction with α-benzoin oxime and chloroform [16]; only Si, W, Nb, Ta, and Pd accompany molybdenum. Precipitation of Fe, Mn, Cr, Co, Ni, Ti, Zr, Th, and U with sodium hydroxide will separate these from molybdenum which is accompanied by Al, Be, V, Zn, and W; if sodium peroxide is employed, Cr and U are thrown into the filtrate with molybdenum. Molybdenum accompanies iron in an ether extraction; if substantial quantities of iron are present the extraction of molybdenum is nearly complete [23]. Molybdenum may be separated from tungsten by precipitating the former with hydrogen sulphide in the presence of tartaric acid; it is preferable to first pass hydrogen sulphide into an acid solution, then make the latter alkaline and continue the stream of gas, and finally acidify and pass in more hydrogen sulphide. Separation of molybdenum from vanadium with hydrogen sulphide is also facilitated by the addition of tartaric acid.

Molybdenum ores are readily dissolved in nitric and hydrochloric acids; ferromolybdenum is soluble in a mixture of dilute nitric and a few drops of hydrofluoric acid. Metallic molybdenum is soluble in aqua regia, dilute nitric acid, or hot concentrated sulphuric acid; it is insoluble in hydrochloric, hydrofluoric, or dilute sulphuric acid. For a dithiol field test in geochemistry it has been stated that 2 ml of 5% sodium hypochlorite or ordinary household bleach to 0·25 g of −100 mesh sample will bring all molybdenum into solution in 1 hour [21].

1. Gravimetric Methods

A. Alpha Benzoin Oxime

Precipitation of molybdenum by α-benzoin oxime is suitable for the determination of the element in nearly all samples; it is particularly useful for irons and steels, and for occasional molybdenum determinations in almost any material [2, 8, 11]. The following elements are not precipitated: Ag, Pb, Hg, Bi, Cu, Cd, As, Sb, Sn, Se, Te, Al, Fe, Ti, Zr, Cr(III), V(IV), Ce, U, Ni, Co, Mn, Zn, and the platinum metals with the exception of Pd. Silica, with Nb, Ta, W, and Pd, will accompany the molybdenum in the precipitate. For the rare occasions when palladium is present, it may be removed from molybdenum by precipitation of the former with dimethylglyoxime in 3% hydrochloric acid. Nearly all Nb, Ta, and W will be removed with silica in an initial acid dehydration.

Decompose a weight of sample, which will give not more than 0·15 g of molybdenum, with nitric and hydrochloric acids, add 10 ml 1 : 1 sulphuric acid and evaporate to strong fumes of the latter. Cool, dilute with water, filter, and wash thoroughly with 1% sulphuric acid. If an undecomposed residue remains, which might contain molybdenum, it may be fused with sodium carbonate and added to the main portion. To make sure that all molybdenum is in the hexavalent state, add sufficient 0·1N potassium permanganate solution to impart a permanent pink colour to the sample solution. If chromium or vanadium is present, add sufficient freshly prepared sulphurous acid to reduce them, and boil until all sulphur dioxide has been removed. Cool the solution to 10°C; precipitation must be done in a cold solution to avoid reduction of the hexavalent molybdenum before precipitation is complete.

To the solution containing 5–10% by volume of sulphuric acid slowly add, with stirring, 10 ml of a solution of 2 g α-benzoin oxime in 100 ml of alcohol, and 5 ml additional for each 0·01 g of molybdenum present. Continue to stir the solution, add just enough bromine water to tint the solution a pale yellow, and finally add a few more ml of the benzoin oxime solution. Allow the sample to remain in the cooling trough at 10°C for 10–15 minutes with occasional stirring. Filter through Whatman No. 40 paper, using pulp; wash with about 200 ml of a cold freshly prepared solution containing 20–25 ml of the α-benzoin oxime precipitating solution and 10 ml sulphuric acid per litre. If sufficient reagent has been added, the filtrate on standing will deposit needle-like crystals. Filtration should be completed within about 30 minutes after precipitation; longer standing leads to low results.

Transfer the precipitate to a weighed platinum crucible, dry, char without flaming, and ignite to constant weight at 500–525°C. $MoO_3 \times 0{\cdot}6667 = Mo$. Molybdenum trioxide commences to volatilize slowly at temperatures above 500°C, but the rate is very slow below 600°C. If the molybdic oxide contains no impurities except tungsten it should dissolve completely in warm ammonium hydroxide. To the ignited residue add 5 ml ammonium hydroxide, digest on the hot plate, and filter through a small paper. Wash with 1% ammonium hydroxide, dry, ignite the paper and contents in the original crucible at 500–525°C, cool and weigh. The difference in weight represents MoO_3.

When tungsten is present, unless removed initially, it is precipitated and weighed with MoO_3. When the latter is dissolved with ammonium hydroxide, part of the tungsten remains as a precipitate of WO_3 and part passes into the ammoniacal filtrate. To the latter add hydrochloric acid until just acid to litmus paper, add 1–2 ml cinchonine solution containing 125 g cinchonine dissolved in 1000 ml of 1 : 1 hydrochloric acid, and allow to stand overnight in a warm place. Filter through Whatman No. 40 paper with pulp, wash with a solution containing 30 ml of the cinchonine precipitating solution in a litre, and ignite in a platinum crucible at 750–850°C. Cool, weigh the WO_3, and subtract from the weight of MoO_3 obtained previously by solution of the precipitate in ammonia. For work of extreme accuracy the WO_3 precipitate obtained from the ammoniacal filtrate must be corrected for its molybdenum contamination.

The precipitating reaction may be represented as:

$$Mo(SO_4)_3 + 3C_{14}H_{13}O_2N \rightarrow Mo(C_{14}H_{11}O_2N)_3 + 3H_2SO_4.$$

Precipitation is best performed in a sulphuric acid solution containing 5–10% by volume; solutions of 5% by volume hydrochloric, nitric, or phosphoric acids may also be used but precipitations in perchloric acid tend to give low results. Tartaric and hydrofluoric acids must be absent.

B. Lead Molybdate

Precipitation of molybdenum with lead acetate, with final weighing as lead molybdate, $PbMoO_4$, is a rapid method and, for an experienced analyst, an accurate one. It is used, for example, by Climax Molybdenum Company for mine samples, mill samples ranging from tailings to concentrates, ferro-molybdenum in the absence of tungsten, technical molybdic oxide, and calcium molybdate [6]. Interfering elements include those which form insoluble molybdates such as Pb, Tl, Ag, Ba, and large amounts of Ca, and Sr, and those which form insoluble lead salts in ammonium acetate–acetic acid solution, such as arsenates, phosphates, chromates, tungstates, vanadates, and fluorides. In the presence of iron, an ammonia separation will remove small quantities of As, P, V, W, and CaF_2. The following are without interference: Hg, B, Cu, Zn, Cd, Mn, Fe, Al, Mg, Co, and Ni. For samples of unknown composition, it may be preferable to separate molybdenum as the sulphide. To an alkaline solution containing tartrate pass in hydrogen sulphide and acidify with 1 : 1 sulphuric acid. Continue the hydrogen sulphide treatment until all molybdenum has precipitated, filter, and wash. Make the filtrate ammoniacal, boil, add a few drops of hydrogen peroxide, boil, and again saturate with hydrogen sulphide, acidify, pass in hydrogen sulphide, and filter as before to recover the small quantity of molybdenum left after the first precipitation.

Decompose the sample with nitric and hydrochloric acid, or with nitric acid saturated with potassium chlorate. Evaporate carefully to dryness, add 10 ml hydrochloric acid, dilute with water, and boil to solution of all soluble salts. The sample may be decomposed, if desired, with nitric acid followed by perchloric acid. Neutralize with ammonia, add an excess, boil, filter, wash, and retain the filtrate which contains the ammonium molybdate. Dissolve the precipitate in hydrochloric acid and re-precipitate with ammonia to recover the molybdenum which was occluded in the ferric hydroxide precipitate. If the latter was very bulky a second re-precipitation may even be necessary.

To the filtrate add a drop or two of methyl orange and neutralize the ammonia with hydrochloric acid, adding a few drops excess. Add 30 ml of a 25% solution of ammonium acetate and 10 ml of acetic acid, and heat nearly to boiling. In the presence of large quantities of sulphates, the amount of ammonium acetate–acetic acid may have to be increased considerably.

Precipitate the molybdenum by the slow addition, with stirring, of a lead acetate solution containing 20 g lead acetate and 20 ml acetic acid per litre. Add the lead acetate solution at the rate of about 6 ml per minute [6]. Test for complete precipitation by adding a drop of solution to a drop of 0·5% tannic acid on a spot plate; the endpoint is reached when no brown colour is produced. Then add 3–5 ml excess lead acetate solution, and allow to settle for an hour at a temperature just below boiling.

Filter by decantation through Whatman No. 40 paper with pulp, transfer the precipitate to the filter, scrubbing the beaker thoroughly. Wash with hot 0·3% ammonium acetate solution and finally with hot water. Place the precipitate in a tared porcelain crucible, dry, ignite at 500–525°C and weigh as $PbMoO_4$. Alternatively, the precipitate may be ignited in a No. 1 fireclay crucible and the contents brushed on to the weighing glass.

$$PbMoO_4 \times 0{\cdot}2613 = Mo$$

Because only the sulphide molybdenum is recoverable in extractive metallurgy, it may be desirable to determine only this, and not total molybdenum, in mine and mill products. If so, the ore sample is first boiled for 20 minutes with 30% hydrochloric acid, filtered and washed. The filtrate contains the oxide molybdenum and is rejected; the residue on the paper is analysed for molybdenum as outlined above [6].

2. Volumetric Method

Molybdenum can be reduced with zinc in a Jones reductor and titrated with standard potassium permanganate solution [2, 8, 11]. The sample must be treated to remove elements like Fe, As, Sb, Ti, V, U, W, Cr, and Nb which are also reduced by zinc and oxidized with permanganate. An ammonia separation in the presence of excess iron will remove all the above elements except tungsten; the latter is separated from molybdenum by hydrogen sulphide in the presence of tartrate. This procedure is used for accurate work with substantial quantities of molybdenum, such as ferro-molybdenum, molybdenum metal, pure molybdic oxide, and other molybdenum compounds [6].

Decompose a sample containing 0·2–0·3 g molybdenum with hydrochloric and nitric acid, add 20 ml 1 : 1 sulphuric acid and evaporate to strong fumes of the latter. Cool, dilute, filter off silica etc., and wash with hot water. Make the filtrate ammoniacal, boil, filter, wash, dissolve the precipitate in dilute sulphuric acid and re-precipitate with ammonia. Combine the filtrates which contain practically all the molybdenum; for extremely accurate work a second re-precipitation may be required or the traces of molybdenum occluded in the second ferric hydroxide precipitate may be determined colorimetrically.

Add to the ammoniacal filtrate 2 g tartaric acid and saturate the warm solution with hydrogen sulphide. Filter off the precipitated molybdenum sulphide and wash thoroughly with acidulated hydrogen sulphide water.

Dissolve the molybdenum sulphide with hot nitric acid, add sulphuric acid and evaporate to strong fumes of the latter. Cool, add water, and adjust the acidity to about 5% sulphuric acid. Add a few grams of granulated zinc and heat gently until the molybdenum is reduced. This preliminary reduction in an open beaker decreases the risk of incomplete reduction later, and also precipitates any copper which would otherwise coat the zinc in the reductor.

Place in the receiving flask of the reductor a 5-fold excess of a standard solution of ferric sulphate. The latter is made by dissolving 60 g ferric sulphate and 25 ml sulphuric acid in water, adding 150 ml of phosphoric acid, and diluting with water to 1 litre. One ml of this solution is equivalent to approximately 0·01 g molybdenum, and it is standardized by passing a measured volume through the reductor and titrating with 0·1 N potassium permanganate.

Connect the receiving flask to the reductor and pass the molybdenum solution through the latter at the rate of 20–30 ml per minute, receiving the reduced solution beneath the surface of the ferric sulphate solution. The rate of passage of the solution may be regulated by a filter pump attached to the filtering flask which is used as a receiver. When the funnel which forms the inlet of the reductor is nearly but not entirely empty, put through about 150 ml of cold water in successive 25 ml portions. Draw the liquid down almost to the constriction each time before adding the next 25 ml portion; in this way no air is drawn through the reductor.

The molybdenum solution is green as it passes through the lower part of the reductor; on coming in contact with the ferric phosphate it is changed to a bright red due to its immediate partial oxidation, which is accompanied by the reduction of a corresponding amount of the ferric salt to the ferrous condition. Titrate the solution in the receiver at once with 0·1N potassium permanganate to a faint permanent pink. Run a blank on the reductor by passing through dilute sulphuric acid of the same volume and concentration as the molybdenum solution, and titrating with standard permanganate. The number of ml of potassium permanganate in the titration, less the blank, gives a measure of the quantity of molybdenum present. 1 ml of 0·1N $KMnO_4$ = 0·003198 g Mo. Standardize the 0·1N $KMnO_4$ against 0·2 g sodium oxalate in 5% sulphuric acid at 60°C.

$$\text{Normality of } KMnO_4 = \frac{0{\cdot}2}{0{\cdot}067} \times \frac{1}{\text{ml of } KMnO_4 \text{ used in titration}}$$

$$2H_2MoO_4 + 3Zn + 6H_2SO_4 \rightarrow Mo_2(SO_4)_3 + 3ZnSO_4 + 8H_2O$$

$$3Fe_2(SO_4)_3 + Mo_2(SO_4)_3 + 8H_2O \rightarrow 6FeSO_4 + 2H_2MoO_4 + 6H_2SO_4$$

$$10FeSO_4 + 2KMnO_4 + 8H_2SO_4 \rightarrow 5Fe_2(SO_4)_3 + K_2SO_4 + 2MnSO_4 + 8H_2O$$

3. Colorimetric Method

A. Thiocyanate — Stannous Chloride Method

The coloured complex formed when hexavalent molybdenum is reduced by stannous chloride in the presence of a thiocyanate is frequently used for the determination of small quantities in ores and biological materials [1, 2, 3, 8, 10, 12, 15, 16, 18, 22]. The stability and selectivity are increased by extracting the coloured solution into an organic layer. Colorimetric procedures are of importance in biology, because molybdenum is an essential micronutrient for plants, but an excess in vegetation may lead to nutritional disorders in livestock [1, 20].

The following procedure is applicable to nearly all ores, particularly those containing tungsten [12]. Depending on the molybdenum content of the sample, and the subsequent aliquot taken, weigh out 0·5–2 g of ore into a zirconium crucible. Add about ten times its weight of sodium peroxide and fuse for 2–3 minutes at 600–700°C after the sample becomes molten. Cool, place crucible in a 400-ml beaker, cover, and cautiously add water. When no further reaction occurs, remove and rinse the crucible. Dilute to about 150 ml, filter through Whatman No. 4 paper, and wash with hot 0·5% sodium hydroxide solution. Molybdenum is in the filtrate, together with Al, P, V, Cr, W, Zn, U, and most of the copper, whereas the precipitate contains Fe, Ti, Zr, Th, most of the nickel and cobalt, and part of the niobium and tantalum.

If manganese is present, add a few drops of alcohol and boil the diluted sodium hydroxide solution before filtering off the insoluble hydroxides; manganese will be precipitated and removed at the same time. If more than traces of Cr, V, or U are present, make a preliminary separation of molybdenum by means of hydrogen sulphide in 5% acid solution, or by extracting with ether in 6N hydrochloric acid. Large quantities of nickel or cobalt also interfere, because they are not completely removed by a sodium peroxide fusion, and in their presence molybdenum should also be separated by hydrogen sulphide precipitation or ether extraction.

Make up the solution to a definite volume so that a convenient aliquot for colour development, usually 10 ml, contains 0·05 to 0·4 mg molybdenum. If tungsten is present in the sample solution in excess of 7 mg per 5 ml, use a 5 ml aliquot in 5 ml of 30% ammonium citrate solution to prevent precipitation of tungstic acid during colour development. Otherwise, a suitable portion up to 10 ml may be used, diluting to 10 ml with water.

Pipette the aliquot into a 50 ml volumetric flask, add a drop of phenolphthalein indicator, neutralize with 1 : 1 hydrochloric acid, and immediately add 25 ml of a solution made by diluting 333 ml concentrated hydrochloric acid to 1 litre with water. Then add 3 ml of 25% ammonium thiocyanate solution, 3 ml of 50% potassium iodide solution, and 2 ml of freshly prepared 1% sodium sulphite solution. Dilute to the mark, mix, allow to stand for 30 minutes, and read the optical density at 460 nm.

The final solution is 2M hydrochloric acid for maximum colour intensity after 20 minutes; the colour is stable for an hour before fading commences. The blank will have a faint yellow tinge which may be mistaken for traces of iodine not decolorized by sodium sulphite. The temperature should be held within the same narrow range for unknowns and for standards used to prepare the calibration curve.

Copper is removed as cuprous iodide by the addition of potassium iodide, but a large quantity of copper is preferably separated by preliminary treatment of the sample with sodium sulphide in alkaline solutions. Up to 5 mg antimony can be tolerated during colour development.

The calibration curve can be prepared by dissolving 0·1508 g of 99·5% MoO_3 in a little sodium hydroxide solution and diluting to 1 litre; 1 ml = 0·1 mg Mo.

When molybdenum is very low, the determination can be improved by extracting all the molybdenum compound, after colour formation, into a small volume of isopropyl ether or isoamyl alcohol and comparing with standards similarly prepared [18]. The procedure outlined below is typical.

Neutralize the cooled alkaline filtrate from the sodium peroxide fusion, obtained as described above, with 1 : 1 sulphuric acid. Add an excess of 2 ml for each 8 ml of sample solution, which will give a solution containing 10% of sulphuric acid by volume. Cool to room temperature and transfer to a 250 ml separatory funnel, rinsing the flask with 10% sulphuric acid.

Add 10 ml of 5% sodium thiocyanate solution, shake for 30 seconds, and add 10 ml of a stannous chloride solution. The latter is made by dissolving 350 g $SnCl_2.2H_2O$ in 200 ml of hot 1 : 1 hydrochloric acid and diluting to 1 litre. Shake the sample vigorously for 1 minute, and add 50 ml of isopropyl ether. Prepare the latter by shaking with 25 ml of 8% $Fe(SO_4)_3.9H_2O$, 10 ml of 5% sodium thiocyanate, and 10 ml of 35% stannous chloride, drawing off and discarding the lower or acid layer.

Shake the sample containing the purified isopropyl ether for several minutes, allow the layers to separate and draw off and retain the lower acid layer. Transfer the upper ether layer to a 100-ml volumetric flask, shake the lower layer with another 40 ml of isopropyl ether, discard the lower layer and add the upper ether layer to the first one in the 100-ml flask. Dilute to the mark with isopropyl ether, mix, allow to stand 2–3 minutes and measure the absorbance in a spectrophotometer at 540 nm. Obtain the percentage of molybdenum from a calibration curve prepared from a series of known quantities of the element. A solution made by dissolving 0·5 g of $Na_2MoO_4.2H_2O$ in 1 litre of water containing 5 ml of sulphuric acid has a theoretical value of 1 ml = 0·0002 g Mo; it can be standardized volumetrically or gravimetrically. With the above procedure, relatively large quantities of the following are without appreciable effect: Fe, Al, Ti, Mn, Ni, Co, U, Ta, Cr, V, W, Cu, Pb, and F.

B. Dithiol Method

Toluene-3, 4-dithiol, usually called simply dithiol, reacts with molybdenum (VI) in mineral acid solution to give a green complex which can be extracted into many solvents. This is the basis of many procedures for the determination of small quantities of molybdenum [2, 5, 18, 19, 21]. The method for 50–1000 ppm of molybdenum in niobium [2], summarized below, is typical.

Weigh 0·5 g sample into a small Erlenmeyer flask, add 10 ml sulphuric acid and 3 g ammonium bisulphate and heat strongly until the sample has dissolved. Cool, transfer to a 50 ml volumetric flask, using 1 : 2 sulphuric acid, and dilute nearly to volume with the latter. Cool, and dilute to volume with 1 : 2 sulphuric acid. Transfer a 15 ml aliquot to a small Erlenmeyer flask, evaporate to fumes of sulphuric acid, and cool. Add 15 ml of 1 : 2 hydrochloric acid, 10 drops of hydrofluoric acid, and 12 drops of 10% hydroxylamine hydrochloride solution.

Transfer to a 125-ml separatory funnel, using a minimum quantity of water, and cool to room temperature. Add 10 ml of 0·5% dithiol solution, and shake thoroughly six times at 3-minute intervals. Add 20 ml carbon tetrachloride, shake for 2 minutes, and allow to separate. Transfer the carbon tetrachloride layer to a dry 50-ml volumetric flask. Wash the aqueous phase twice with 10-ml portions of carbon tetrachloride, shaking for 1 minute and adding the washings to the volumetric flask. The aqueous phase can be retained, if required, for a tungsten determination. Dilute the organic layer to volume with carbon tetrachloride and mix. Measure the optical density of the solution at 680 nm.

There are many variations of the dithiol method. Traces of molybdenum have been collected at *p*H 2 by co-precipitation with oxides of manganese, followed by ion exchange extraction with n-butyl acetate, and addition of ascorbic acid, citric acid, or thiourea to overcome the interference of certain elements [5]. Molybdenum in soils and sediments has been determined, using zinc dithiol and extraction in light petroleum, after reducing iron by citric-ascorbic solution, inactivating copper by potassium iodide, and suppressing tungsten interference by control of the time allowed for complex formation [19].

4. Optical Spectrography

Small quantities of molybdenum can be frequently determined by emission spectrography, usually with the lines 2775·40, 2816·15, 3132·59, 3152·82, and 3170·35. Detailed instructions have been published for 0·10–0·60% molybdenum in plain carbon and low alloy steels [2], 20–200 ppm in zirconium and its alloys [2], and 8–80 ppm in nickel [13].

5. X-Ray Spectrography

A wide range of molybdenum contents can be determined satisfactorily by X-ray fluorescence.

6. Atomic-Absorption Spectrophotometry

Molybdenum in the lower ranges is conveniently determined by atomic absorption, generally with the line 3132·6, which has the maximum sensitivity of 3 μg/ml [7, 14]. Other lines may be used: 3112·1, 3158·2, 3170·3, 3194·0, 3208·8, 3798·3, 3864·1, and 3903·0. The absorption is dependent on the air–acetylene ratio and on the height of the hollow cathode beam above the top of the burner. The detrimental effect of Fe, Mn, Mg, Ca, and Sr can be overcome by dissolving the evaporated sample in a few drops of hydrochloric acid and 25 ml of water containing 2 g ammonium chloride and 1 g potassium chloride.

REFERENCES

1. A.O.A.C., *Official Methods of Analysis of the Association of Official Analytical Chemists*, Washington, D.C., Association of Official Analytical Chemists, 1965.
2. A.S.T.M., *Chemical Analysis of Metals. Sampling and Analysis of Metal Bearing Ores*, Philadelphia, Pa., American Society for Testing and Materials, 1969.
3. Bauer, G. A., *Anal. Chem.* **37,** 155 (1965).
4. Burriel-Marti, F., and Vidan, A. M., *Anal. Chim. Acta* **26**, 163–7 (1962).
5. Chan, K. M., and Riley, J. P., *Anal. Chim.Acta* **36,** 220–9 (1966).
6. Climax Molybdenum Co., personal communication, 1960.
7. Elwell, W. T., and Gidley, J. A. F., *Atomic-Absorption Spectrophotometry*, 2nd ed., Oxford, Pergamon, 1966.
8. Furman, N. H., ed., *Scott's Standard Methods of Chemical Analysis*, 6th ed., Vol. 1, Princeton, N.J., Van Nostrand, 1962.
9. Healy, W. B., and McCabe, W. J., *Anal. Chem.* **35,** 2117–9 (1963).
10. Hibbits, J. O., and Williams, R. T., *Anal. Chim. Acta* **26**, 363–70 (1962).
11. Hillebrand, W. F., Lundell, G. E. F., Bright, H. A., and Hoffman, J. I., *Applied Inorganic Analysis*, 2nd ed., New York, Wiley, 1953.
12. Hope, R. P., *Anal. Chem.* **29,** 1053–55 (1957).
13. Lewis, C. L., Ott, W. L., and Sine, N. M., *The Analysis of Nickel*, Oxford, Pergamon, 1966.
14. Julietti, R. J., and Wilkinson, J. A. E., *Analyst* **93,** 797–8 (1968).
15. Lounamaa, N., *Anal. Chim. Acta* **33,** 21–35 (1965).
16. Luke, C. L., *Anal. Chim. Acta* **34,** 302–7 (1966).
17. McNerney, W. N., and Wagner, W. F., *Anal. Chem.* **29,** 1177–8 (1957).
18. Sandell, E. B., *Colorimetric Determination of Traces of Metals*, 3rd ed., New York, Interscience Publishers, 1959.
19. Stanton, R. E., and Hardwick, A. J., *Analyst* **92,** 387–90 (1967).
20. Stiles, W., *Trace Elements in Plants*, 3rd ed., Cambridge, Cambridge University Press, 1961.
21. Stubbs, M. F., *Analyst* **93,** 59–60 (1968).
22. Waterbury, G. R., and Bricker, C. E., *Anal. Chem.* **29**, 129–35 (1957).
23. Young, R. S., and Leibowitz, A., *Iron Age* **164**, No. 21, 75–6 (1949).

24 Nickel

Apart from the extensive sulphide and oxidized ore deposits of nickel, varying quantities of the element may be found in mining operations which are devoted primarily to the production of copper, cobalt, lead, zinc, gold, the platinum group, and several other metals. Nickel minerals occur as sulphides, arsenides, and oxides or silicates. The sulphides have long been the principal source of nickel, the important minerals being pentlandite $(Ni,Fe)_9S_8$, millerite NiS, and less frequently one or more of the linnaeite series $(Fe, Co, Ni)_3S_4$, exemplified by polydymite Ni_3S_4, violarite Ni_2FeS_4, and siegenite $(Co,Ni)_3S_4$. Some pyrrhotites are nickeliferrous; in addition to pentlandite intergrown with it or between its grains, a small amount of nickel substitutes for iron in these pyrrhotites. Nickel arsenides are found in some sulphide orebodies, but occur more frequently in nickel-containing cobalt ores; the chief minerals are niccolite NiAs, maucherite $Ni_{11}As_8$, rammelsbergite $NiAs_2$, and gersdorffite NiAsS. Annabergite $Ni_3(AsO_4)_2.8H_2O$ is found near the surface of orebodies containing nickel arsenides and sulphides. The antimonide, breithauptite NiSb, occurs with nickel arsenides. Nickel oxides and silicates, which have to date made only a modest contribution to the world's nickel output, are found in enormous quantities; examples are garnierite $(Ni,Mg)_6Si_4O_{10}(OH)_8$, and nickeliferrous limonite $(Fe,Ni)O(OH).nH_2O$. The widespread use of nickel in its pure and alloyed forms makes its determination a common one in most metallurgical laboratories.

Nearly all nickel-containing materials, whether from extractive or fabricating metallurgy, respond readily to conventional decomposition procedures. For ores, the usual treatment with nitric, hydrochloric, and sulphuric acids, followed if necessary by a sodium carbonate fusion of any insoluble material, will bring all nickel into solution. For oxide and silicate ores, it is common practice in industrial work to add a few drops of hydrofluoric acid to the beaker containing the sample in aqua regia, in place of a fusion.

Concentrates and high-sulphide ores are often dissolved by adding a nitro-chlorate solution and allowing to stand for 10 minutes before adding sulphuric acid and placing on the hot plate; the nitro-chlorate solution is made by adding 100 g of potassium chlorate to 1 litre of nitric acid. Arsenic, antimony, tin, and selenium can be volatilized by one or two evaporations with a mixture of hydrobromic acid and sulphuric acid. Many other decomposition techniques can be used: hydrofluoric and sulphuric acids in a platinum dish, sodium peroxide fusion in an iron or zirconium crucible, and sodium bisulphate fusion in a pear-shaped Pyrex flask.

Metallic nickel and many of its alloys are soluble in hot 1 : 1 nitric acid. Other alloys may require, in addition, hydrochloric acid and sometimes prolonged treatment with perchloric acid. Occasionally, a residue will require solution in hydrofluoric and nitric acids. Tungsten occurs in some nickel-containing alloys and the formation of yellow tungstic acid on aqua regia treatment tends to hinder dissolution. For these alloys, filter the solution, wash the precipitate in dilute hydrochloric acid, dissolve the precipitate of impure tungstic acid in strong sodium hydroxide solution, and then treat the undissolved alloy particles on the paper with acids, combining this with the original filtrate.

The metals niobium, tantalum, and zirconium can be dissolved in a mixture of nitric and hydrofluoric acids. The same mixture will dissolve the carbides and nitrides of niobium, tantalum, titanium, and zirconium; it will also decompose the carbide of tungsten and the boride of zirconium. Ferro-nickel will usually dissolve completely in a hot mixture of hydrochloric, nitric, and sulphuric acids; if a residue persists, it can be treated with hydrofluoric and sulphuric acids in a platinum crucible, or fused with sodium carbonate, and added to the main portion. Ferrochrome and ferrosilicon can be decomposed by a sodium peroxide fusion. It is often convenient to dissolve aluminium-base alloys in a strong solution of sodium hydroxide.

Organic materials may be ignited at a temperature not exceeding 500°C without loss of nickel. Alternatively, wet oxidation with nitric and sulphuric acids, followed if necessary by perchloric acid, may be employed.

ISOLATION OR SEPARATION OF NICKEL

Nickel is separated from members of preceding groups by digestion in hydrochloric acid for silver and mercury [1], by precipitation of the acid sulphide group with hydrogen sulphide in 5–10% hydrochloric or sulphuric acid, and removal with ammonia of Fe, Al, Be, Cr, Ti, Zr, Th, U, Nb, and Ta. Together with Co, Mn, Zn, and Tl, nickel can then be precipitated with ammonium sulphide and thereby separated from the alkaline earths, magnesium, and the alkalies. While a precipitation of the acid sulphide group with hydrogen sulphide is a common and effective procedure, separation of nickel by an ammonia precipitation is an infrequent operation due to the tenacious absorption of nickel by the hydroxides of iron and aluminium.

If it is necessary to isolate nickel after the removal of the hydrogen sulphide group, this may be achieved by a variety of routes. Electrolysis at a mercury cathode in 1% sulphuric acid gives a deposition of Ni, with Fe, Co, Cr, and Zn, while Al, Be, Ti, Zr, V, Th, W, Nb, Ta, and U remain in solution. Nickel, with Al, Be, Mn, Cr, Zn, Co, P, and U(VI) is not extracted by cupferron-chloroform in cold 10% sulphuric acid, and is thereby separated from Fe, Ti, V, Zr, Nb, Ta, and W. Nickel, with Mn and Co, can be separated in the filtrate from Fe, Al, Cr, Ti, V, Zr, Th, U, W, Nb, and Ta by a zinc oxide precipitation. A sodium hydroxide precipitation leaves Al, Be, V, W, and Zn in solution, while nickel and Fe, Co, Cr, Ti, Zr, U, Th, Mn, and rare earths are precipitated; if sodium peroxide is used, Cr and U are thrown into the filtrate. Iron may be removed from nickel by an ether separation, manganese by the nitro-chlorate precipitation or by bromine and ammonium hydroxide, and chromium by volatilization with boiling perchloric and hydrochloric acids.

In pre-eminent position among separations, of course, is the precipitation of nickel with dimethylglyoxime, which is fully outlined later; unless cobalt and iron are both present in high concentrations, practically nothing interferes in ammoniacal solution in the presence of tartrate except palladium and gold. Cobalt can be separated from nickel by 1-nitroso-2-naphthol in 4-5% hydrochloric acid. When the 1-nitroso-2-naphthol precipitation follows a zinc oxide separation, nickel is accompanied in the final filtrate by only manganese, the alkalies, alkaline earths, and of course zinc Zinc can be removed from nickel by hydrogen sulphide in 0·01N sulphuric acid.

Nickel may also be separated from many elements by solvent extraction. The most important method is based on dimethylglyoxime as the extraction agent and chloroform as the solvent. Details are given later in the discussion of colorimetric procedures. Sodium diethyldithiocarbamate forms a complex with nickel over a wide *p*H range, which can be extracted with chloroform. Unfortunately, a number of other metals form similar complexes, but many can be removed initially by treatment with hydrogen sulphide or ammonium hydroxide. In ammoniacal solution containing hydrogen peroxide, cobalt does not form a complex with sodium diethyldithiocarbamate, and its separation from nickel is excellent.

Both anion and cation exchange resins have been utilized to separate nickel from interfering elements, particularly for complex alloys [16, 17, 22]. Iron and cobalt are strongly adsorbed on resins such as Dowex 1 or Amberlite IRA–410 in about 8 N hydrochloric acid, while nickel is not [17]. As a general rule, however, the ease of isolation by dimethylglyoxime for most materials has limited the use of ion exchange for nickel determinations. The same holds true for chromatographic separations [25].

As mentioned earlier, members of the hydrogen sulphide group can be removed from nickel by precipitation with hydrogen sulphide in dilute acid; the ones frequently associated with nickel are, however, often separated by other means. For instance, lead is conveniently removed as sulphate; arsenic, antimony, tin, and selenium can be volatilized by several evaporations with sulphuric and hydrobromic acids; copper is separated and determined by electrolysis in sulphuric–nitric acid solution, and molybdenum can be removed from nickel by cupferron, sodium hydroxide or ether.

1. Gravimetric with Dimethylglyoxime

The use of dimethylglyoxime to precipitate nickel is one of the oldest applications of an organic reagent for metal analysis; for over sixty years it has proved its selectivity and reproducibility on countless occasions for nearly every type of material [2, 3, 4, 5, 6, 9, 11, 14, 16, 19, 21, 29). The precipitate is voluminous, and collection and washing of a precipitate larger than 0·04 g nickel on a Gooch or sintered glass crucible is not advisable. It is often convenient, however, to precipitate amounts as large as 0·25 g nickel, filter through paper, dissolve with hot dilute nitric acid and re-precipitate in ammonium hydroxide to liberate any occluded impurities, dissolve the final precipitate in dilute hydrochloric or nitric acid, and complete the determination by the volumetric potassium cyanide method or by electrolysis.

In a solution containing a slight excess of ammonium hydroxide and ammonium citrate or tartrate, nickel is the only element which is completely precipitated by dimethylglyoxime. Iron and cobalt when together, palladium, and gold may be partially precipitated; copper, if in appreciable quantity, tends to co-precipitate. In most nickel materials, more than traces of palladium and gold are unlikely to be present, but they can be removed, with copper, by hydrogen sulphide in dilute sulphuric or hydrochloric acid solution. Copper can also be removed, and conveniently determined, by electrolysis in nitric–sulphuric acid solution. Bismuth is precipitated by dimethylglyoxime, but this reaction is entirely inhibited in the presence of tartrate.

Iron or cobalt, singly, even in substantial quantities, usually do not require a re-precipitation for routine industrial work. Small quantities of both, together, can be tolerated in the dimethylglyoxime method. When iron and cobalt together are present in high concentrations, however, a dark-brown, colloidal solution results; this retards nickel precipitation, is extremely difficult to filter, and yields a dark-coloured precipitate containing a large quantity of iron and cobalt. Methods are described later for overcoming this difficulty.

Dissolve a quantity of sample containing less than 0·04 g nickel in nitric and hydrochloric acids, add 10 ml 1 : 1 sulphuric acid and evaporate to strong fumes of the latter to dehydrate silica. If an insoluble residue remains, which might contain nickel, it can be fused with sodium carbonate and added to the main portion. Any other convenient method of decomposition, of course, may be employed. For high-chromium samples and high-temperature alloys it is usually preferable to add nitric and hydrochloric acids initially, then 20 ml of perchloric acid, and evaporate to dense fumes of the latter.

Add hot water, boil, filter, and wash. If copper is the only Group 2 metal it may frequently be removed and conveniently determined by electrolysis after adding 2 ml nitric acid to the sulphate solution. Otherwise, it may be precipitated with other members of the group by hydrogen sulphide; adjust the volume of the filtrate so that 5–10% sulphuric acid is present and pass in a vigorous stream of hydrogen sulphide until precipitation is complete. Filter, wash thoroughly with acidulated hydrogen sulphide water, boil the filtrate until all hydrogen sulphide has been expelled, and oxidize the solution with hydrogen peroxide, followed by vigorous boiling, to transform all iron to the ferric state. Small quantities of copper, below 1–2%, do not materially affect the dimethylglyoxime precipitation; larger amounts give rise to occlusion, which necessitates a re-precipitation, and a better practice is to remove this element, with others of its group, at the start.

To the oxidized solution containing nickel, free of SiO_2, W, $PbSO_4$, Pd, Au, and preferably free of copper and other members of the acid sulphide group, add sufficient 25% tartaric or citric acid solution to prevent the precipitation of hydroxides of iron, aluminium, etc. when the solution is made ammoniacal. It requires about 3 g, or 12 ml of a 25% solution of tartaric acid to hold 1 g of iron in solution. Add 10 ml of a 10% ammonium chloride solution, or more if necessary, to keep manganese and zinc in solution. Make the solution ammoniacal; if it does not remain clear, re-acidify with hydrochloric acid, add additional tartaric acid and ammonium chloride, and make ammoniacal. If a precipitate still persists, filter through Whatman No. 42 paper and wash with 1% ammonium hydroxide.

Make the solution faintly acid with acetic acid, and to the clear solution add, with stirring, sufficient 1% alcoholic solution of dimethylglyoxime to precipitate all the nickel. It requires 5 ml for each 10 mg of nickel, plus an additional 5 ml for every 10 mg of cobalt. A slight excess of dimethylglyoxime should be added to ensure complete precipitation, but the quantity of alcoholic dimethylglyoxime should not exceed 35% of the sample solution; nickel dimethylglyoximate is slightly soluble at higher alcohol concentrations. Make the solution distinctly ammoniacal, stir thoroughly, and allow to stand in a warm place for at least an hour; for small quantities of nickel allow to stand overnight.

Filter through a weighed Gooch or sintered glass crucible and wash with cold water. Small quantities of the precipitated nickel dimethylglyoxime sometimes adhere tightly to the side of the beaker and must be removed carefully with a rubber policeman. Test the filtrate, by adding more dimethylglyoxime solution and a little ammonium hydroxide, before discarding. Dissolve the precipitate by pouring through the crucible, in successive small portions into a clean suction flask, a total of about 50 ml of a hot mixture containing 40 ml of 1 : 1 hydrochloric acid and 10 ml of nitric acid, followed by about 50 ml of hot 2% tartaric acid. Neutralize the clear solution with ammonium hydroxide, acidify with acetic acid, and precipitate nickel with dimethylglyoxime and ammonium hydroxide as described above. Dry the precipitate at 150°C to constant weight, cool, and weigh as nickel dimethylglyoxime which contains 20·32% nickel.

$$NiCl_2 + 2C_4H_8N_2O_2 \rightarrow C_8H_{14}N_4O_4Ni + 2HCl$$

An important modification when the nickel is low, and iron or iron + cobalt high, is the elimination of iron before the dimethylglyoxime addition. For a low nickel content in irons and steels, the sample weight will be increased to a point where the high ratio of iron to nickel not only impedes the precipitation of nickel, but gives a dark-brown colloidal solution which is difficult to filter, and yields a dark-red precipitate containing considerable co-precipitated iron. When cobalt, in addition to iron, is high, nickel cannot be determined by the conventional dimethylglyoxime method outlined above, because a cobalt-iron complex is formed, which is practically impossible to filter, and the nickel dimethylglyoxime precipitate is grossly contaminated.

When these conditions exist, i.e. a large quantity of both iron and cobalt in any nickel-containing sample, or a large quantity of iron and a very small amount of nickel, separate the iron in the following manner. Evaporate the oxidized solution, after the removal of Group 2 elements, to dryness. Dissolve in 20 ml of hydrochloric acid of s.g. 1·10 for each gram of iron; this is made by mixing 526 ml of concentrated hydrochloric acid with 474 ml of water. Cool, transfer to a separatory funnel, rinsing with more hydrochloric acid of s.g. 1·10. Add 30 ml of diethyl ether or isopropyl ether for each gram of iron, and shake the funnel under a stream of cold water for 1 minute. Allow to settle, and draw off the lower aqueous layer into another separatory funnel. Shake the upper ether layer with another 10 ml of 1·10 s.g. hydrochloric acid and 15 ml of ether for each gram of iron present, and combine the lower layer with the reserved aqueous layer in the second funnel; discard the ether layer, which contains the iron. Shake the nickel-containing aqueous solution with another 10 ml of 1·10 s.g. hydrochloric acid and 15 ml of ether. Withdraw the lower aqueous phase to a beaker, and evaporate carefully on the side of the hot plate until all traces of ether have been expelled. Add 10 ml nitric acid, 5 ml 1 : 1 sulphuric acid, and evaporate to fumes of the latter to destroy all organic matter. Proceed with the dimethylglyoxime precipitation as described earlier.

For many materials, especially in routine analyses, one precipitation of nickel dimethylglyoxime suffices. The precipitate should be scarlet; a dark red or brownish-red indicates the presence of co-precipitated impurities, principally iron and cobalt, and a re-precipitation should be carried out. If the content of iron, aluminium, and similar elements is low, they can sometimes be removed by ammonium hydroxide, with a re-precipitation, in place of the conventional complexing with tartaric or citric acid, before the dimethylglyoxime precipitation of nickel. In routine work, the nickel precipitate is frequently dried to constant weight at the ordinary laboratory oven temperature of 110°C, but for work of high accuracy 150°C should be used.

If calcium is very high, citric acid is preferable to tartaric for complexing iron, because calcium citrate is much more soluble than calcium tartrate. For high-calcium samples, dehydrate silica with hydrochloric or perchloric acids rather than sulphuric; although most of the calcium sulphate is removed with silica, the solubility of the former allows some to pass into the filtrate where traces may accompany nickel in the final dimethylglyoxime precipitate.

2. Volumetric Method with KCN

In slightly ammoniacal solution, nickel reacts with potassium cyanide to form a complex anion. If the nickel solution contains a precipitate of silver iodide, which has been formed by adding a known amount of silver nitrate and a few drops of potassium iodide solution, the turbidity will not disappear until all the nickel has entered into reaction with the standard potassium cyanide. The reaction may be expressed:

$$Ni(NH_3)_6Cl_2 + 4KCN \rightarrow K_2[Ni(CN)_4] + 6NH_3 + 2KCl$$
$$AgI + 2KCN \rightarrow K[Ag(CN)_2] + KI$$

This is an excellent routine procedure for nickel in the absence of cobalt, or where small quantities of the latter are counted as nickel [2, 11, 14, 16, 18, 21]. Copper must be removed, zinc and tungsten in more than small quantities should be absent, but elements like iron and aluminium can be held in solution by addition of citric or tartaric acid prior to making the solution ammoniacal. The volumetric cyanide method is, perhaps, not as well suited for occasional analysis as the dimethylglyoxime or electrolytic procedures, but in the hands of an experienced chemist, very satisfactory results are obtained on products ranging from mine samples of 0·2% to mattes of 50% nickel. Preceded by a dimethylglyoxime separation, it is frequently used for the higher ranges of nickel where the gravimetric dimethylglyoxime method cannot be employed; in this case, of course, no elements interfere in the titration.

Decompose 0·5–5 g sample with nitric and hydrochloric acids, add 10 ml 1 : 1 sulphuric acid, and evaporate to strong fumes of the latter. Filter off the dehydrated silica and wash thoroughly with hot water. If an insoluble residue might contain nickel, fuse it with sodium carbonate and add to the main portion. Alternatively, use any other convenient method to bring the entire sample into solution.

If copper is to be removed and determined by electrolysis, add 1–2 ml of nitric acid per 100 ml of solution, and electrolyse; to the de-copperized solution add 2–3 ml nitric acid and a spoonful of potassium chlorate, and boil on the hot plate for 15 minutes. If copper is to be removed by hydrogen sulphide, adjust the acidity to 5–10% sulphuric acid, warm, and pass in hydrogen sulphide until all copper has been precipitated. Filter, wash thoroughly with acidulated hydrogen sulphide water, boil out all hydrogen sulphide from the filtrate, add 2–3 ml nitric acid and a spoonful of potassium chlorate and boil for 15 minutes.

To the nickel solution, free of copper, add enough citric or tartaric acid, or ammonium or sodium citrate, to hold all the iron, aluminium etc. in solution after making ammoniacal. It requires about 12 ml of a 25% tartaric acid solution to hold 1 gram of iron in solution. Where large numbers of nickel samples are determined, make a sodium citrate solution containing about 570 g per litre by cautiously mixing 1300 g citric acid with 1250 g sodium carbonate in 5 litres of water. Add from 10 to 60 ml of this solution to each sample, depending on the quantity of iron present.

The presence of zinc in quantities greater than 10% of the nickel leads to high results. Substitution of sodium pyrophosphate for sodium citrate decreases the interference of zinc; the results, though still slightly high, are within acceptable limits for most routine purposes when, for example, as much as 100 mg of zinc is present with 20 mg of nickel [30].

Add exactly 5 ml of a silver nitrate solution containing 1 g per litre, and if a silver chloride turbidity derived from the potassium chlorate added previously, does not form, add a drop of hydrochloric acid. Clear the turbidity by the cautious addition of ammonium hydroxide, stirring constantly, and add 2–4 drops excess. A larger excess of ammonium hydroxide leads to low results. Cool below 20°C, and add exactly 5 ml of a potassium iodide solution containing 80 g per litre. Titrate with standard potassium cyanide solution over a black background, with vigorous stirring, to the disappearance of turbidity.

A standard solution of potassium cyanide having an approximate value of 1 ml = 0·005 g nickel can be made by weighing out 23 g, dissolving in water containing a few grams of potassium hydroxide, and diluting to 1 litre. It is standardized against 1 g of nickel ammonium sulphate, which theoretically contains 0·1486 g nickel. Check the nickel content of the nickel ammonium sulphate by the electrolytic method. Dissolve the nickel ammonium sulphate in a small quantity of water and carry it through all the steps of the determination, together with a blank. For a potassium cyanide solution having a value of 1 ml = 0·005 g nickel, the quantity required for 1 gram of nickel ammonium sulphate should be 29·70 ml + 0·20 ml for the blank, or a total of 29·90 ml. The standard potassium cyanide solution changes slowly and must be standardized at weekly intervals.

For high nickel samples it is convenient to have a stronger potassium cyanide solution; the latter can have a value of 1 ml = 0·01 g nickel.

The general procedure outlined above must be modified if the sample contains a significant quantity of cobalt, zinc in a concentration much greater than 10% of the nickel, more than 1% tungsten, or a large amount of iron, aluminium, or other elements which are precipitated by ammonium hydroxide. To the oxidized solution, from which copper has been removed by hydrogen sulphide or electrolysis, add sufficient tartaric acid to hold iron, aluminium, etc. in solution on subsequent neutralization. Add 10 ml of 10% ammonium chloride solution, make the sample faintly ammoniacal, and then acidify with acetic acid. To the clear solution add sufficient 1% alcoholic solution of dimethylglyoxime to precipitate all the nickel, plus an equivalent amount to react with the cobalt. About 5 ml of the dimethylglyoxime solution are required for each 10 mg of nickel, plus an additional 5 ml for every 10 mg of cobalt. Make the solution distinctly ammoniacal, stir vigorously, and set aside in a warm place for an hour, or overnight when only a small quantity of nickel is present.

Filter on a Whatman No. 42 paper, wash with cold water, and discard the filtrate after testing it for nickel. Transfer the precipitate by washing to the original beaker, and dissolve traces of nickel on the paper by washing with successive small portions of a hot mixture of 20 ml 1 : 1 hydrochloric acid and 5 ml nitric acid. If the sample contains more than 2% cobalt, re-precipitate with dimethylglyoxime in ammoniacal solution in the manner described above.

Evaporate the solution to low volume until free of dimethylglyoxime and oxidizing gases. Cool, add a little tartaric acid or sodium citrate, and proceed with the addition of silver nitrate and succeeding steps of the cyanide titration method outlined above.

When the dimethylglyoxime separation is made to eliminate interferences, before the volumetric cyanide titration, the same precaution of eliminating iron as described previously must be taken if both iron and cobalt are high, or if iron is high and nickel very low. Iron can be conveniently removed by the ether extraction outlined earlier, or by a cupferron–chloroform extraction. For the latter, which separates iron, titanium, vanadium, zirconium, niobium, tantalum, and tungsten from nickel, add sufficient cold 6% cupferron solution to a cold 10% sulphuric acid solution of the sample in a separatory funnel to precipitate all iron, etc. Shake thoroughly, add successive small portions of cold chloroform, and shake to pass the iron cupferrate into the lower layers, discarding the latter each time. Continue until the chloroform layer remains colourless after shaking. Transfer the upper acid layer containing the nickel to a beaker, and evaporate carefully on the side of the hot plate until the traces of chloroform have volatilized. Add nitric acid and evaporate to strong fumes of sulphuric acid to eliminate all organic matter. Proceed with the determination of nickel by the cyanide volumetric method as described above, with prior dimethylglyoxime separation if required.

3. Electrolytic Method

For a large quantity of nickel, the preferred procedure is electrolytic deposition from an ammoniacal solution free of copper and nitrates, containing ammonium sulphate and about 40 ml excess ammonium hydroxide [2, 11, 14, 16, 21]. It is not as rapid as the volumetric method, but is superior for work of high accuracy or for occasional analyses.

When cobalt is present, it is deposited with the nickel and can be afterwards determined by any convenient method in the solution resulting from boiling the platinum cathode in nitric acid. Alternatively, nickel may be separated from cobalt by dimethylglyoxime before electrolysis, filtered, dissolved in dilute nitric acid, and evaporated to fumes with sulphuric acid to give a cobalt-free electrolyte.

Copper, molybdenum, and other members of Group 2 must be removed by prior separation with hydrogen sulphide in dilute acid solution, or by other means. Cobalt, zinc, vanadium, and tungsten cannot be present. Any appreciable quantity of other elements of the ammonium hydroxide and ammonium sulphide groups even though they are not theoretically deposited on the cathode, must be removed because they impede electrolysis and are mechanically co-deposited on the cathode. For complex samples, several separations must precede electrodeposition.

Fortunately, many high-nickel materials contain only one or more of the following interfering elements: copper, chromium, iron, and cobalt. Copper can be removed, and determined, by electrolysis in the usual nitric–sulphuric acid electrolyte, and the solution evaporated to fumes of sulphuric to eliminate nitric acid, before proceeding to the nickel electrolysis. Chromium can be volatilized as chromyl chloride by boiling with perchloric and hydrochloric acids, or nickel can be precipitated with dimethylglyoxime and the complex decomposed with nitric and sulphuric acids. Iron can be removed with an ether separation in hydrochloric acid, or nickel can be precipitated with dimethylglyoxime; iron can also be separated by a cupferron–chloroform extraction. Nickel can be separated from cobalt by a dimethylglyoxime precipitation of the former.

The procedure outlined below assumes the presence of all interfering elements; for most materials it can be appreciably shortened. Decompose the sample in nitric and hydrochloric acids, add 10 ml of perchloric acid, 10 ml of 1 : 1 sulphuric acid, and evaporate to dense fumes of the latter. Cool, dilute, filter, and wash. If the precipitate is believed to contain nickel, ignite, fuse with sodium carbonate, and add to the original filtrate. Alternatively, follow any other convenient technique to dissolve the sample completely. High-tungsten samples may give a precipitate or turbidity after the above treatments, but filtration and washing will pass all nickel into the filtrate.

If the copper content is low, and no other elements of Group 2 are present in significant amounts, proceed with the precipitation of nickel by dimethylglyoxime. If the copper content is high, and other elements of this group are absent, remove and determine copper by electrolysis in the usual nitric–sulphuric acid electrolyte. If copper is to be determined by a method other than electrolytic, or if other members of the copper group are present in high concentration, remove copper etc. by passing in hydrogen sulphide for 10–30 minutes to the solution adjusted to about 5% sulphuric acid. Filter, wash with 1% sulphuric acid saturated with hydrogen sulphide, and boil out all of the latter from the filtrate. Oxidize with a few ml of 30% hydrogen peroxide, and boil to remove the excess of the latter.

Precipitate nickel with dimethylglyoxime as described earlier under the gravimetric method. For very high nickel samples, the use of 1% alcoholic dimethylglyoxime may lead to inconveniently large volumes, because the volume of alcohol must not exceed 35% of the final solution; it is preferable to use dimethylglyoxime sodium salt which is soluble in water. Filter the nickel precipitate on a large Whatman No. 40 paper as described previously in the volumetric method. Carry out the appropriate modifications when iron and cobalt are high. Dissolve the precipitate in nitric and sulphuric acids, and evaporate to fumes of the latter to destroy all organic matter. Cool, dilute, neutralize with ammonium hydroxide, and add 40 ml excess. Ammonium sulphate must be present in the electrolyte; if the sample has accidently evaporated to dryness, add about 10 ml of 1 : 1 sulphuric acid before neutralizing with ammonium hydroxide.

Electrolyse at 2 amperes on a rotating-anode type of apparatus for 45 minutes or until all nickel has been deposited, or at 0·3–0·5 amperes overnight on stationary electrolytic equipment. To test for complete deposition, withdraw 1–2 ml by means of a pipette and place on a spot plate. Mix with a drop or two of dimethylglyoxime solution; if no detectable pink develops, electrolysis is complete. Without interrupting the current, lower the beaker and wash the cathode simultaneously. Switch off the current, quickly transfer the cathode to three successive beakers of alcohol, place in an oven at 105°C or on a small hot plate for several minutes, cool in a desiccator, and weigh. In routine work, the cathode is sometimes dried in a blast of hot air from a hair drier, or by ignition of alcohol, keeping the electrode in motion to prevent localized high temperature. In control laboratories the cathodes are not usually placed in a desiccator, following the evaporation of alcohol, but are taken on an electrode holder to the balance room. To remove nickel from the platinum cathode, boil for several minutes in 1 : 1 nitric acid, and rinse thoroughly in water.

It is sometimes convenient to determine nickel + cobalt electrolytically. After weighing the combined electrodeposit, strip it from the platinum cathode by treatment with hot nitric acid. Determine cobalt in this solution, depending on the amount present, potentiometrically by potassium ferricyanide, gravimetrically by 1-nitroso-2-naphthol or by potassium nitrite, colorimetrically, or by atomic absorption.

4. Colorimetric Procedure

When dimethylglyoxime is added to a basic solution of a nickel salt which has been treated with bromine water, a brown to wine-red colour is obtained, which forms the basis of the most important procedure for small quantities of nickel [2, 3, 6, 9, 10, 11, 14, 16, 21, 23]. Usually it is best to extract the nickel initially with dimethylglyoxime and chloroform, thereby isolating it from all but traces of interfering elements. With this preliminary separation, the colour produced by dimethylglyoxime and the nickelic ion furnishes the most convenient and reliable photometric procedure for nickel in all types of materials.

To a slightly acid solution containing less than 0·5 mg nickel in a separatory funnel, add 5 ml of 10% sodium citrate, or enough to keep in solution all iron, aluminium, etc. after making ammoniacal. Add ammonium hydroxide in slight excess, and 2 ml or more of 1% dimethylglyoxime in ethanol. Shake, extract three times with 2–3 ml chloroform, and combine the chloroform layers containing the nickelous dimethylglyoxime in another separatory funnel. Shake the combined chloroform extracts with 5 ml of 2% ammonium hydroxide and withdraw the chloroform layer into another separatory funnel. Shake the water phase with 1–2 ml chloroform, and combine the latter with the previously-washed chloroform extract.

Shake the chloroform solution vigorously for 1 minute with two 5-ml portions of 0·5N hydrochloric acid, to transfer the nickel back to the aqueous phase. Discard the chloroform layer, and transfer the combined hydrochloric solutions to a 100-ml volumetric flask. Add 5 drops of bromine water, and then ammonium hydroxide drop by drop until the colour of bromine is destroyed, followed by an excess of 3–4 drops. Add 0·5–0·7 ml of 1%

dimethylglyoxime, cool, and dilute to volume with water. Measure the transmittance at 530 nm against a blank carried through all steps of the procedure. Calculate percentage nickel by reference to a standard curve. The latter can be prepared by taking 2–10 ml aliquots from a standard nickel solution containing 0·0250 g nickel, or 0·1013 g $NiCl_2.6H_2O$, in 500 ml; 1 ml = 0·05 mg nickel. If desired, colour measurement may be made at the most sensitive wavelength, 450 nm; it is, however, more susceptible to traces of impurities than is 530 nm, which is preferable for most samples.

With the extraction procedure carried out before the colorimetric determination, as outlined above, scarcely any elements interfere. Large quantities of manganese interfere in the extraction of nickel, probably by oxidizing it to the nickelic state, in which condition it is not extracted by chloroform. The effect of manganese can be overcome by adding hydroxylamine hydrochloride to keep manganese in the reduced state. Cobalt consumes dimethylglyoxime, but the cobalt compound is not extracted by chloroform to a significant extent; any cobalt extracted is transferred completely to the aqueous phase by washing with dilute ammonium hydroxide. Copper tends to partially accompany nickel in the extraction, but most of it is removed from the chloroform layer when the latter is shaken with dilute ammonium hydroxide, whereas nickel remains in the chloroform phase.

The hue and colour intensity of the nickelic dimethylglyoxime solution depend on the order of reagent additions. If bromine water is added to an acid solution, followed by ammonium hydroxide and dimethylglyoxime, the solution is brownish. If bromine is added to a faintly ammoniacal solution, the final solution has a wine-red colour. The intensity of colour is stronger in solutions which are initially ammoniacal. It is essential, therefore, to follow the same order of reagent addition for unknowns and standards.

The procedure described above is suitable for nearly all materials, but many other colorimetric methods have been advocated for nickel. Directions have been given for the use of furil α-dioxime in geochemistry [24] and in boiler waters [28].

5. Polarographic

Small quantities of nickel can be determined polarographically in the presence of cobalt in a supporting electrolyte containing pyridine [3, 13]. At a *p*H of 4–5·5 ferric iron and small quantities of chromium are precipitated; manganese and small amounts of copper do not interfere. The procedure is particularly useful for the low quantities of nickel in cobalt compounds and in some iron ores.

6. Optical Spectrography

Procedures have been published for the spectrographic determination of nickel in various metals and alloys; 0·04–30% in copper and its principal alloys [6], 0·001–10% in aluminium and its alloys, 0·001–0·01% in pig lead,

0·03–1·5% in plain carbon and low alloy steel, 0·001–0·3% in tin alloys, and 20–1000 ppm in zirconium and its alloys [2]. Spectrographic methods for various materials have been detailed in other references [1, 12, 20].

7. X-Ray Spectrography

Nickel can be determined over a wide range of concentration in many materials by X-ray spectrography, using the line 1·6578 Å [16].

8. Atomic-Absorption Spectrophotometry

Small or intermediate quantities of nickel can be determined in many materials by atomic absorption [8, 26]. The most sensitive line, 2320·0 Å, has a sensitivity of 0·13 μg/ml, and no significant interferences have been reported when the air–acetylene flame is used.

SPECIAL PROCEDURES

Forms of Nickel

1. A procedure for the determination of metallic nickel in the presence of nickel oxides has been published [7]. For 0·1 g total nickel, ranging from 0·02 to 0·07 g metallic nickel, boil under a reflux for two hours with 100 ml of 18·5 g/litre mercuric chloride. Filter and wash; metallic nickel passes into the filtrate as the chloride and can be determined by any appropriate method.

2. In the nickel industry, metallic and sulphide nickel are sometimes differentiated from the oxide form of the element by leaching with a chlorine–alcohol solution. Metallic and sulphide nickel dissolve, whereas the oxide remains insoluble. As with all phase analyses, the method is empirical and the result is only approximate. For process control in such fields of extractive metallurgy as the roasting or reducing of ores, concentrates, and various other materials, it does have value. The procedure below is typical, but of course may be modified to suit the individual product.

Weigh into a dry, tall 400-ml beaker a suitable quantity of −200 mesh material, depending on the nickel content and the analytical method to be employed. Add at least ten times the sample weight of anhydrous methanol, stir, and place in a fume cupboard. Pass a vigorous stream of gaseous chlorine from a cylinder into the dilute pulp for 10 minutes. Nickel in metallic form, and nearly all the sulphide nickel will dissolve in the chlorine–alcohol solution, leaving the oxides virtually unattacked. Filter, and wash the residue with anhydrous methanol; carefully evaporate the filtrate to dryness and acidify with nitric acid. Make up to a definite volume and determine nickel in a suitable aliquot.

3. A similar procedure to 2 above, which involves boiling under reflux

with 5% bromine in methanol, has been described for dissolving metallic nickel in the presence of oxides [15]. A method has been published for the selective dissolution in alcoholic bromine of the metallic nickel binder in sintered titanium carbide [27].

REFERENCES

1. AHRENS, L. H., and TAYLOR, S. R., *Spectrochemical Analysis*, 2nd ed., Reading, Mass., Addison-Wesley, 1964.
2. A.S.T.M., *Chemical Analysis of Metals. Sampling and Analysis of Metal Bearing Ores*, Philadelphia, Pa., American Society for Testing and Materials, 1969.
3. BRITISH ALUMINIUM CO., *Analysis of Aluminium and its Alloys*, London, British Aluminium Co., 1961.
4. BRITISH IRON AND STEEL RESEARCH ASSOCIATION, METHODS OF ANALYSIS COMMITTEE, *J. Iron and Steel Institute* **180**, 262–9 (1955).
5. CLAASSEN, A., and BASTINGS, L., *Analyst* **91**, 725–31 (1966).
6. DOZINEL, C. M., *Modern Methods of Analysis of Copper and its Alloys*, London, Elsevier, 1963.
7. EK, C., *Anal. Chim. Acta* **14**, 311–17 (1956).
8. ELWELL, W. T., and GIDLEY, J. A. F., *Atomic-Absorption Spectrophotometry*, 2nd ed., Oxford, Pergamon, 1966.
9. ELWELL, W. T., and SCHOLES, I. R., *Analysis of Copper and its Alloys*, Oxford, Pergamon, 1967.
10. FORSTER, W., and ZEITLIN, H., *Anal. Chim. Acta* **35,** 42–53 (1966).
11. FURMAN, N. H., ed., *Scott's Standard Methods of Chemical Analysis*, 6th ed., Vol. 1, Princeton, N.J., Van Nostrand, 1962.
12. HEADRIDGE, J. B., and LAMBERT, A. K., *Analyst* **93**, 211–13 (1968).
13. HEYROVSKY, J., and KUTA, J., *Principles of Polarography*, New York, Academic Press, 1966.
14. HILLEBRAND, W. F., LUNDELL, G. E. F., BRIGHT, H. A., and HOFFMAN, J. I., *Applied Inorganic Analysis*, 2nd ed., New York, Wiley, 1953.
15. KINSON, K., DICKESON, J. E., and BELCHER, C. B., *Anal. Chim. Acta* **41**, 107–12 (1968).
16. KOLTHOFF, I. M., and ELVING, P. J., *Treatise on Analytical Chemistry*, Part II, Vol. 2, 377–440, New York, Interscience Publishers, 1962.
17. LIBERMAN, A., *Analyst* **80**, 595–8 (1955).
18. LUKE, C. L., *Anal. Chem.* **33**, 96–8 (1961).
19. LYLE, S. J., and MAGHZIAN, R., *Talanta*, **14**, 1021–8 (1967).
20. MITCHELL, R. L., *The Spectrochemical Analysis of Soils, Plants, and Related Materials*, Farnham Royal, England, Commonwealth Agricultural Bureau, 1964.
21. PESHKOVA, V. M., and SAVOSTINA, V. M., *Analytical Chemistry of Nickel*, New York, Daniel Davey, 1967.
22. SAMUELSON, O., *Ion Exchange Separations in Analytical Chemistry*, New York, Wiley, 1963.
23. SANDELL, E. B., *Colorimetric Determination of Traces of Metals*, 3rd ed., New York, Interscience Publishers, 1959.

24. STANTON, R. E., *Rapid Methods of Trace Analysis*, London, Edward Arnold, 1966.
25. STOCK, R., and RICE, C. B. F., *Chromatographic Methods*, 2nd ed., London, Chapman and Hall, 1967.
26. UNY, G., BRULE, M., and SPITZ, J., *Anal. Chim. Acta* **44**, 29–33 (1969).
27. VIOLANTE, E. J., *Anal. Chem.* **33**, 1600–1 (1961).
28. WILSON, A. L., *Analyst* **93,** 83–92 (1968).
29. WILSON, C. L., and WILSON, D. W., *Comprehensive Analytical Chemistry*, Vol. 1 C, London, Elsevier, 1962.
30. YOUNG, R. S., *Chemist-Analyst* **46**, 69, 75 (1957).

25 Niobium and Tantalum

Because the rare elements niobium and tantalum are almost invariably found together in minerals, it will be convenient to discuss them in the same chapter. The most important source of these metals is the iron manganese tantaloniobate $(Fe,Mn)O(Nb,Ta)_2O_5$, which forms an isomorphous series in which niobium and tantalum replace each other in all proportions. The minerals in which niobium predominates are called "niobites", while the term "tantalite" is used for those containing over 50 per cent of Ta_2O_5. Niobic and tantalic oxides also occur in other minerals, associated with titanium, calcium, uranium, tin, tungsten, and rare earths. Niobium is used as an alloying element in certain high-temperature alloys and in stainless steels; tantalum is employed in specialized applications which demand great resistance to corrosion and high temperatures.

ISOLATION OR SEPARATION OF NIOBIUM AND TANTALUM

These elements hydrolyse very easily, causing many difficulties in analysis. Most of the niobium and tantalum will separate with silica when a sample is dehydrated with hydrochloric, sulphuric, or perchloric acids; they will be found in the non-volatile residue left after the volatilization of silica with hydrofluoric and sulphuric acids. Members of the hydrogen sulphide group can be separated from niobium and tantalum by saturating with hydrogen sulphide a solution of the fused potassium bisulphate sample in 20% tartaric acid. Niobium and tantalum, unless removed beforehand, will be found in the ammonia precipitate. Separations from members of the latter group can be made as follows: a mercury cathode electrolysis in dilute sulphuric acid will leave Nb and Ta in the solution with Al, Be, Ti, Zr, V, Th, U, and W, whereas Fe, Ni, Co, and Zn are deposited; cupferron in cold 10% sulphuric or hydrochloric acid will precipitate Nb and Ta, with Ti, Zr, V, W, and Fe from Al, Ni, Co, Zn, Cr, Mn, U(VI), and Be. Niobium and tantalum can be separated from Ni, Co, and Mn by zinc oxide, from iron by ether extraction of the latter, and from Zr, Th, U, Be, and Al by precipitation of Nb, Ta, and Ti with tannin in oxalate solution.

Separation in a cellulose column can be used not only for isolating niobium and tantalum, with tungsten, from all other metals, but also to separate niobium from tantalum [6, 7, 8, 17, 18, 20, 24, 25, 27]. When a concentrated solution of the ammonium fluoro-salts is absorbed in a pad of cellulose fibre which is placed on top of a short column of the same material, and eluted with an 85 : 15 volume mixture of methyl ethyl ketone and 40% hydrofluoric acid, niobium and tantalum pass into the eluate. Other metals remain on the column except tungsten which moves slowly downward and is partly eluted with the earth acids. When a fluoride solution of niobium and tantalum containing a small excess of ammonium fluoride and very little free hydrofluoric acid is placed on the cellulose column and eluted with water-saturated methyl ethyl ketone, niobium moves only very slowly, whereas tantalum is readily eluted. Again, other metals remain on the column, except tungsten which moves downward slowly to contaminate the tantalum eluate.

Ion exchange can also be used to separate niobium and tantalum from all other elements, and to separate niobium from tantalum [1, 3, 5, 6, 13, 19, 20, 25, 30]. When a sample solution containing 5 hydrochloric acid : 4 hydrofluoric acid : 11 water is passed through a column of Dowex 1 or a similar resin, niobium and tantalum are retained whereas virtually all other elements pass through the column. Niobium can be eluted with 14% ammonium chloride : 4% hydrofluoric acid solution. Tantalum is finally eluted with 14% ammonium chloride : 4% hydrofluoric acid solution, which has been adjusted to *p*H 5·5 with ammonium hydroxide.

In view of the complexity of separations, a chemist undertaking occasional analyses would be well advised to include in each series a known quantity of niobium and tantalum. The analytical chemistry of niobium and tantalum has been reviewed extensively in recent years [10, 11, 20, 23, 25].

The behaviour of the metals in decomposition procedures is as follows: niobium metal is insoluble in hydrochloric acid, nitric acid, and aqua regia, but is slowly soluble in hot concentrated sulphuric acid; tantalum metal is insoluble in hydrochloric, nitric, or sulphuric acids, and in aqua regia, but dissolves slowly in hydrofluoric acid. Both metals are readily soluble in a mixture of nitric and hydrofluoric acids, or of hydrochloric and hydrofluoric acids. Minerals can be decomposed by digestion with hydrofluoric acid, fusion with potassium bisulphate, or fusion with potassium hydroxide or carbonate [14, 16, 20].

1. Gravimetric Procedures

A. Ion Exchange and Cupferron Precipitation

(1) *Ores and Concentrates.* The following procedure is applicable to all ores and concentrates [20]. To 0·5–1 g sample in a polyethylene beaker add 25 ml hydrochloric acid and 20 ml of hydrofluoric acid. Cover with a polyethylene lid, and heat for several hours on a steam bath until the sample is decomposed. Disregard any undissolved cassiterite or zircon. Add 55 ml of water and transfer to the top of a column of Dowex 1 resin. The latter is preferably 100–200 mesh, 8–10% divinylbenzene cross-linkage, contained in a polystyrene tube 15 inches long × 1 inch in diameter 380 × 25mm.

Wash the beaker with a 5 : 4 : 11 mixture, made by mixing 250 ml of 12M hydrochloric acid with 300 ml of water, adding 200 ml of 24M hydrofluoric acid, and diluting to 1 litre with water. Allow 300 ml of this eluate to pass through the column at the rate of about 100 ml per hour. Discard the eluate.

Replace the receiver with a 600 ml polyethylene beaker and pass through the column 350 ml of ammonium chloride–hydrofluoric acid solution, which is made by dissolving 140 g of ammonium chloride in 500 ml of water, adding 40 ml of hydrofluoric acid, and diluting to 1 litre. This eluate contains the niobium, contaminated possibly by traces of tin and antimony. If the sample contained considerable cassiterite, a small amount of tin would be rendered soluble by hydrochloric–hydrofluoric attack. Add to the niobium eluate 15 g of tartaric acid, 15 g boric acid, and stir until these are dissolved. Pass in hydrogen sulphide to precipitate tin, and antimony if present, filter, and wash with cold 5% sulphuric acid containing a little tartaric acid and hydrogen sulphide. Discard the precipitate, and expel hydrogen sulphide by boiling.

Cool the solution to below 10°C, add 60 ml of cold hydrochloric acid, 60 ml of cold 6% cupferron, some paper pulp, and filter. Wash thoroughly with 5% hydrochloric acid containing a little cupferron. Ignite carefully in a tared platinum crucible to constant weight at 1100°C and weigh as Nb_2O_5. $Nb_2O_5 \times 0{\cdot}699 = Nb$.

Elute the tantalum in the same manner with 350 ml of ammonium chloride–hydrofluoric acid, which has been made by dissolving 140 g of ammonium chloride in 500 ml of water, adding 40 ml hydrofluoric acid, adjusting the *p*H to 5·5 with ammonium hydroxide, and diluting to 1 litre with water. To the tantalum eluate add 60 ml of hydrochloric acid, 15 g of boric acid, and carry out the cupferron precipitation and ignition as described in the previous paragraph. Weigh as Ta_2O_5. $Ta_2O_5 \times 0{\cdot}819 = Ta$. Very rarely, the sample may contain bismuth, and this element will require elimination by hydrogen sulphide as described above for tin, before the cupferron precipitation of tantalum.

The resin can be regenerated by passing through two column volumes of the 5 : 4 : 11 mixture.

If desired, the initial decomposition with hydrochloric–hydrofluoric acid can be replaced by a fusion. In a quartz crucible, fuse the material with about 10 times its weight of sodium bisulphate for 30 minutes. Dissolve the cooled melt in the 5 : 4 : 11 mixture and transfer to the resin column. Alternatively, fuse the sample in a nickel or zirconium crucible with sodium peroxide. Leach the cooled melt in a polyethylene beaker with water, neutralize with 1 : 1 sulphuric acid to a *p*H of 5–7, dilute with water to 110 ml, add 50 ml hydrochloric acid, 40 ml of hydrofluoric acid, and heat on a steam bath until a clear solution results. If a black residue of nickel oxide remains, add a few drops of hydrogen peroxide, and heat until the solution is clear. Cool and transfer to the resin column.

(2) *Steels and High-Temperature Alloys.* Decompose the sample in a polyethylene beaker with 10 hydrochloric acid : 1 nitric acid mixture. Add

1–5 ml hydrofluoric acid to prevent hydrolysis of Nb, Ta, W, etc, and evaporate to dryness on the steam bath. Remove nitric acid by adding 5 ml hydrochloric acid and 3 ml hydrofluoric acid, and repeating the evaporation. Dissolve the salts in the 5 : 4 : 11 mixture and continue the determination as outlined above for niobium and tantalum in ores and concentrates.

Some alloys are more readily decomposed in hydrofluoric–nitric mixtures; the nitric acid can be removed by repeated evaporations with hydrofluoric acid, and the residue dissolved in the 5 : 4 : 11 mixture.

B. Other Gravimetric Methods

Niobium and tantalum can be determined by other gravimetric procedures, using precipitation with tannin [14, 16, 20, 27, 29], separation in a cellulose column of ammonium fluoro-salts with methyl ethyl ketone [20, 25, 27], a combination of tannin and cellulose column techniques [20, 25, 27], or double hydrolysis with sulphurous acid [1]. Most of these are subject to interference from titanium or tungsten, but in their absence can be used in place of A.

2. Volumetric Procedure

When niobium and tantalum are separated from other elements by hydrolysis with sulphurous acid and weighed as the combined oxides, they can be dissolved, passed through a Jones reductor, and the niobium titrated with potassium permanganate. Tantalum is not reduced by zinc in acid solution and is not titrated with potassium permanganate. The method is used in steel analysis [1, 16, 20]. Unfortunately, the combined oxides from the hydrolytic separation are contaminated with tungsten, titanium, and molybdenum. Tungsten must be removed by an ammonium hydroxide separation; titanium and molybdenum are determined colorimetrically and their weights as oxides deducted from that of the combined $Nb_2O_5 + Ta_2O_5$.

3. Colorimetric Procedures

A. Niobium with Hydroquinone

Isolate niobium by ion exchange and a cupferron precipitation as described earlier under Gravimetric Procedure A. If niobium is very low, use 10 mg of zirconium as zirconium sulphate and only 10 ml of 6% cupferron solution to collect the niobium. Fuse the oxide in a quartz crucible with potassium bisulphate and dissolve the cooled melt in 4% ammonium oxalate solution.

Make to a convenient volume and withdraw an aliquot containing about 200 μg of niobium and 1 g of potassium bisulphate to a beaker. Add nitric, hydrochloric, 10 ml of 1 : 1 sulphuric, acids, and evaporate to strong fumes of the latter to destroy all organic matter. To the cool solution, add 1 drop of 10% stannous chloride in hydrochloric acid, and 25 ml of 5% hydroquinone in 1 : 1 sulphuric acid. Transfer to a 50 ml flask with hydroquinone solution and use the latter to dilute to volume. Allow to stand for ten minutes and measure at 490 nm [20].

B. TANTALUM WITH PYROGALLOL

Isolate tantalum by ion exchange and cupferron precipitation, as already described under Gravimetric Procedure A. Fuse the oxide in a quartz crucible with potassium bisulphate, and dissolve the cooled melt in 15 ml of hot ammonium citrate–ammonium oxalate solution. The latter is made by dissolving 3 g of ammonium citrate and 3 g ammonium oxalate in 100 ml of 0·45M sulphuric acid.

Transfer the solution, or an aliquot containing about 300 μg of tantalum, to a 25 ml volumetric flask. If an aliquot is used, add sufficient potassium bisulphate to provide about 0·5 g. Add 3 g pyrogallol, swirl to dissolve, and dilute to the mark with the ammonium citrate–ammonium oxalate solution. Allow to stand 10 minutes, and measure the absorbance at 398–420 nm [20].

C. OTHER COLORIMETRIC PROCEDURES

(1) *Niobium.* Other colorimetric methods are useful for niobium in many materials. Niobium can be determined in high-tantalum samples by an extraction of its thiocyanate complex in hydrochloric acid with diethyl ether [5, 9, 15, 20, 22].

When hydrogen peroxide is added to a strong sulphuric acid solution of niobium, a yellow peroxyniobate complex is formed, which can be determined photometrically [2, 16, 20, 31]. Small quantities of tantalum can be tolerated, but of course other elements give a colour with hydrogen peroxide.

Bromopyrogallol red gives a blue colour with niobium; the latter can be isolated by ion exchange [30], or many elements, including tantalum, may be masked by reagent additions [4].

PAR, 4-(2-pyridylazo)-resorcinol, has been used for niobium in steels [26] and other metals [32]; tantalum interferes.

(2) *Tantalum.* Various colorimetric procedures for tantalum are listed [20], but do not appear to have gained acceptance for general work.

4. Optical Spectrography

Niobium and tantalum are rarely determined by emission spectroscopy, but a few examples have been listed [20]. The spectral sensitivity of tantalum is poor.

5. X-Ray Spectrography

Niobium can be determined in many materials by X-ray fluorescence, but this technique is not suitable for tantalum [20, 21].

6. Atomic-Absorption Spectrophotometry

Using a fuel-rich oxy-acetylene flame, and lines 4058·94, 4079·73 and 4100·92, the sensitivity of niobium is about 250 μg/ml [12]. The sensitivity of tantalum has been reported to be 6 μg/ml [28].

REFERENCES

1. A.S.T.M., *Chemical Analysis of Metals. Sampling and Analysis of Metal Bearing Ores*, Philadelphia, Pa., American Society for Testing and Materials, 1969.
2. BAKES, J. M., GREGORY, G. R. E. C., and JEFFERY, P. G., *Anal. Chim. Acta* **27,** 540–4 (1962).
3. BANDI, W. R., BUYOK, E. G., LEWIS, L. L., and MELNICK, L. M., *Anal. Chem.* **33**, 1275–8 (1961).
4. BELCHER, R., RAMAKRISHNA, T. V., and WEST, T. S., *Talanta* **12**, 681–90 (1965).
5. BERGSTRESSER, K. S., *Anal. Chem.* **31**, 1812–14 (1959).
6. BRITISH IRON AND STEEL RESEARCH ASSOCIATION, *J. Iron Steel Institute* **187**, 341–3 (1957).
7. BURSTALL, F. H., and WILLIAMS, A. F., *Analyst* **77**, 983–99 (1952).
8. CABELL, M. J., and MILNER, I., *Anal. Chim. Acta* **13**, 258–67 (1955).
9. CANADA, D. C., *Anal. Chem.* **39**, 381–3 (1967).
10. COCKBILL, M. H., *Analyst* **87,** 611–29 (1962).
11. ELWELL, W. T., and WOOD, D. F., *Anal. Chim Acta* **26**, 1–31 (1962).
12. FASSEL, V. A., and MOSSOTTI, V. G., *Anal. Chem.* **35**, 252–3 (1963).
13. FERRARO, T. A., *Talanta* **15,** 923–30 (1968).
14. FURMAN, N. H., ed., *Scott's Standard Methods of Chemical Analysis*, 6th ed., Vol. 1, Princeton, N.J., Van Nostrand, 1962.
15. HASTINGS, J., and MCCLARITY, T. A., *Anal. Chem.* **26,** 683–5 (1954).
16. HILLEBRAND, W. F., LUNDELL, G. E. F., BRIGHT, H. A., and HOFFMAN, J. I., *Applied Inorganic Analysis*, 2nd ed., New York, Wiley, 1953.
17. HUNT, E. C., and WELLS, R. A., *Analyst* **79**, 345–50 (1954).
18. HUNT, E. C., and WELLS, R. A., *Analyst* **79**, 351–9 (1954).
19. KALLMANN, S., OBERTHIN, H., and LIU, R., *Anal. Chem.* **34**, 609–13 (1962).
20. KOLTHOFF, I. M., and ELVING, P. J., *Treatise on Analytical Chemistry*, Part II, Vol. 6, 177–406, New York, Interscience Publishers, 1964.
21. LUKE, C. L., *Anal. Chem.* **35,** 56–8 (1963).
22. LUKE, C. L., *Anal. Chim. Acta* **34**, 165–8 (1966).
23. MCKAVENEY, J. P., *Anal. Chem.* **35**, 2139–44 (1963).
24. MERCER, R. A., and WELLS, R. A., *Analyst* **79**, 339–45 (1954).
25. MOSHIER, R. W., *Analytical Chemistry of Niobium and Tantalum*, London, Pergamon, 1964.
26. PAKALNS, P., *Anal. Chim. Acta* **41,** 283–92 (1968).
27. SCHOELLER, W. R., and POWELL, A. R., *The Analysis of Minerals and Ores of the Rarer Elements*, 3rd ed., London, Charles Griffin, 1955.
28. SLAVIN, W., *Atomic Absorption Newsletter* **5**, 42–5 (1966).
29. WEBB, H. W., ASHWORTH, V., and HILLS, J. M., *Analyst* **88,** 142–4 (1963).
30. WILLIAMS, A. I., *Analyst* **92**, 43–6 (1967).
31. WOOD, D. F., and ADAMS, M. R., *Anal. Chim. Acta* **31,** 153–8 (1964).
32. WOOD, D. F., and JONES, J. T., *Analyst* **93**, 131–41 (1968).

26 Nitrogen

Most analytical chemists, unless they are in laboratories concerned with food and agriculture, are not often required to determine nitrogen. The element is so common, however, that any laboratory may occasionally be requested to determine nitrogen in soils, fertilizers, waters, nitrogenous compounds, and even in metals and alloys.

Nitrogen in gases is usually found by difference, after removal of all other constituents by absorption or combustion; details may be consulted in standard references on gas analysis [6].

In certain laboratories, such as those in the explosives field, where a large number of determinations of nitrogen in nitrates, nitrites, or mixed acids are required, a nitrometer is often used. This instrument measures the nitric oxide evolved when sulphuric acid in the presence of mercury acts on a nitrate:

$$2KNO_3 + 4H_2SO_4 + 3Hg \rightarrow K_2SO_4 + 3HgSO_4 + 4H_2O + 2NO$$

Details of the operation of a nitrometer are given in standard references on gas analysis. The titration of nitric acid with ferrous sulphate or ferrous ammonium sulphate has replaced to some extent the nitrometer method of determining nitric acid or nitrogen in mixed and refuse acids and in nitric esters [9].

Apart from the specialized procedures for nitrogen in metals, there are three types of nitrogen determinations which the ordinary mining laboratory may encounter.

1. Total Nitrogen

The standard method for determining nitrogen is to oxidize organic substances with sulphuric acid in the presence of catalysts and convert the nitrogen to ammonia, which is held by the acid as ammonium sulphate [1, 2, 3, 11, 13, 17, 21]. The ammonia is found by making the solution alkaline with sodium hydroxide, heating to drive off the ammonia, and absorbing it in standard acid. Total nitrogen includes all the nitrogen present in the sample; if nitrates are present they must be reduced with salicylic acid, zinc dust, etc. prior to the above oxidation process.

Place 0·5–5 g, depending on the nitrogen content of the material to be analysed, in a Kjeldahl digestion flask. Add 30 ml sulphuric acid containing 1 g salicylic acid. Shake until thoroughly mixed and allow to stand, shaking frequently, for at least 30 minutes, or until complete solution results. Add 5 g sodium thiosulphate and heat the solution for 5 minutes, cool, add 0·7 g mercuric oxide, 10 g potassium sulphate or anhydrous sodium sulphate, and heat very gently until foaming ceases.

Digest for a time after the mixture is colourless or nearly so, or until oxidation is complete, which usually requires about 2 hours. Cool, dilute with about 150 ml water and add a few pieces of granulated zinc, pumice stone, or Carborundum to prevent bumping. Add 25 ml of 4% potassium or sodium sulphide solution and mix to precipitate mercury.

Add sufficient sodium hydroxide, usually 60 ml of 45% solution, to make the solution strongly alkaline; pour it down the side of the flask so that it does not immediately mix with the acid. Connect the flask to the condenser by means of a Kjeldahl connecting bulb, taking care that the lower tip of the condenser extends below the surface of standard 0·1 N sulphuric acid in a 300 ml Erlenmeyer receiving flask. An adapter to give fine bubbles is frequently attached to the end of the condenser.

Mix the contents of the Kjeldahl flask by shaking, and distil until all ammonia has passed over into the measured quantity of standard acid. The first 100 ml of the distillate usually contains all the ammonia. From 10 to 50 ml of standard acid, diluted to 50–75 ml with distilled water, are generally used in the receiving flask. Place 2–3 drops of methyl red indicator in the standard acid at the outset. If the sample is high in nitrogen, and the colour changes from red to a pale yellow before the end of the distillation, more standard acid can be added quickly from a pipette without loss of the determination. Ammonia is held in an aqueous solution fairly well until the saturation point is reached.

Titrate the excess acid with 0·1N sodium hydroxide. From the quantity of acid neutralized by ammonia the nitrogen content of the sample is calculated. One ml of 0·1N H_2SO_4 = 0·0017 g NH_3 or 0·0014 g N. A blank, of course, should be determined on the reagents. If large numbers of nitrogen analyses are required, it is convenient to adjust the standard acids and alkalies to N/14 so that 1 ml of acid = 0·001 g N, or 0·10% on a 1 g sample. The reactions in this procedure may be expressed:

$$KNO_3 + 4H_2 \rightarrow NH_3 + KOH + 2H_2O$$
$$N_2 + 3H_2 \rightarrow 2NH_3$$
$$2NH_3 + H_2SO_4 \rightarrow (NH_4)_2SO_4$$
$$(NH_4)_2SO_4 + 2NaOH \rightarrow 2NH_3 + 2H_2O + Na_2SO_4$$

2. Organic and Ammoniacal Nitrogen

Organic and ammoniacal nitrogen does not include nitrates, and the sample is simply oxidized by digesting with sulphuric acid and carrying out the regular Kjeldahl–Gunning–Arnold method given below [2]. This nitrogen determination is of great importance, because protein is calculated from it.

For cereal foods such as wheat, flour, bread, macaroni, etc, organic and ammoniacal nitrogen $\times$ 5·7 = protein. In calculating protein of milk and its products, the factor used is 5·38. For nearly all other organic materials like hays and fodders, fruits, nuts, vegetables, meat, beer and wine, tea, etc, organic and ammoniacal nitrogen $\times$ 6·25 = protein.

Place 0·5–5 g, according to the nitrogen content of the sample to be analysed, in a Kjeldahl flask. Add 15–18 g potassium sulphate or anhydrous sodium sulphate, 1 g copper sulphate plus 0·2 g selenium, or 0·7 g mercuric oxide plus 0·2 g selenium, and 25 ml of sulphuric acid. Heat the mixture gently until frothing ceases, then boil briskly and continue the digestion for a time after the sample is colourless or nearly so, or until oxidation is complete. This usually requires about 2 hours.

Cool, carefully add 150 ml water, a few pieces of pumice stone or Carborundum, and if mercuric oxide has been used, add 25 ml of 4% potassium or sodium sulphide solution. Make strongly alkaline with 60 ml of 45% sodium hydroxide, and complete the determination as outlined in the previous section.

3. Nitrate Nitrogen

The determination of nitrate nitrogen is of importance in agricultural and sanitary chemistry in soils, waters, wastewaters, fertilisers, plants, etc, and is usually carried out colorimetrically with phenoldisulphonic acid [1, 2, 4, 10, 14, 15, 17, 19, 24].

A. Soils

Place 100 g of —20 mesh air-dried soil and 500 ml water in a suitable container and agitate for 5 minutes. Add 1 g calcium oxide or 2 g of precipitated calcium carbonate, agitate thoroughly, and allow to stand for 10–20 minutes to obtain a clear filtrate. If the latter contains less than 6 ppm of chloride, proceed under (a) below; if it contains more than 6 ppm chloride use procedure (b). In both cases use 25 ml or a quantity that will contain not more than 0·1 mg nitrogen in the form of nitrate.

B. Waters

(a) *For water of low chloride content*, < 6 ppm. To 100 ml or a quantity that contains less than 0·05 mg nitrogen as NO_3, add sufficient silver sulphate solution to precipitate all but about 0·5 mg chloride. The silver sulphate solution is made by dissolving 4·397 g Ag_2SO_4 in 1 litre of water; this gives a solution of 1 ml = 1 mg of chloride.

Heat to boiling and allow to settle, or add a little alumina cream or carbon black; filter and wash well with small quantities of hot water. Evaporate the filtrate to dryness in a porcelain dish on a steam bath. When cool, treat with 2 ml of phenoldisulphonic acid, rubbing with a glass rod to ensure intimate contact. Dilute with water and slowly add ammonium hydroxide until the maximum yellow colour is developed. Filter if necessary, transfer to a colorimeter, and compare with standards in the usual manner. Alternatively, measure the colour intensity in a spectrophotometer at 410 nm.

(b) *For water of high chloride content*, > 6 ppm. To 100 ml, or a quantity that contains less than 0·1 mg of nitrogen as NO_3, in a 300 ml casserole add 2 ml of the sodium hydroxide solution and concentrate by boiling to about one-third of the original volume. Transfer to a 100-ml test tube and dilute to a volume of about 75 ml. Prepare a blank by placing about 75 ml of ammonia-free water and 2 ml of sodium hydroxide solution in a 100-ml test tube. Place a strip of aluminium foil in each tube. Close the ends of the test tubes with other test tubes containing about 50 ml of ammonia-free water. These latter tubes serve as traps to prevent the escape of ammonia and at the same time permit the free evolution of hydrogen.

Allow the sample and blank to stand at room temperature for 12 hours or until reduction is complete. Add 2 ml Nessler's reagent to the traps. If high in ammonia, discard the determination. Disregard the traps if they contain only 0·01–0·02 mg of nitrogen as NH_3. Transfer the sample and blank to distillation flasks, using 250 ml of ammonia-free water for each; distil into Nessler tubes containing 2 ml Nessler's reagent until ammonia ceases to be evolved, and compare with standards as in the determination of free ammonia. Alternatively, measure the colour intensity in a spectrophotometer at 400–425 nm. Subtract the quantity of nitrogen found in the blank from that found in the sample. Calculate to ppm nitrate nitrogen.

REAGENTS

Aluminium foil. Pure strips 10 cm long, weighing about 0·5 g.

Sodium hydroxide solution. Dissolve 250 g of sodium hydroxide in 1250 ml of water. Add 2 strips of aluminium foil and allow to stand 12 hours; concentrate the solution to 1 litre by boiling.

Nessler's reagent. Dissolve 143 g sodium hydroxide in 750 ml water and filter through asbestos. Add 50 g red mercuric iodide to the filtrate and dilute with water to 1 litre. Mix thoroughly, allow to settle, and use the supernatant liquid; store in the dark.

Standard nitrate solution. Dissolve 0·607 g of sodium nitrate in 1 litre of nitrate-free water. Evaporate 50 ml of this to dryness in a porcelain dish. Cool, add 2 ml phenoldisulphonic acid, rubbing to ensure intimate contact, and dilute to 500 ml. This solution, which is stable, has a value of 1 ml = 0·01 mg of nitrogen, or 0·0443 mg NO_3. Prepare standards for comparison in a colorimeter by adding ammonium hydroxide to a measured volume of this solution.

4. Nitrogen in Metals and Alloys

A. DISTILLATION–VOLUMETRIC

The nitrogen content of steels, which is usually around 0·1%, can be determined by digesting the sample in a sulphuric–phosphoric acid mixture, with the aid of hydrochloric or perchloric acid if necessary, followed by neutralization with sodium hydroxide and distillation of ammonia as in the regular Kjeldahl determination [3].

For titanium and its alloys, where the nitrogen content ranges from 0·005 to 0·2%, the sample is dissolved in a mixture of fluoboric and hydrochloric acids, thus converting the nitrides to ammonium salts. The ammonia is distilled from an alkaline solution into boric acid and titrated with standard sulphuric acid [3].

Copper and its alloys are dissolved in a mixture of sulphuric acid and hydrogen peroxide. Free ammonia is evolved by treating the sample solution with an excess of sodium hydroxide, and steam-distilling into a dilute solution of boric acid. Standard hydrochloric acid, and a modified methyl red indicator, are used in the titration [12].

Nitrogen in nitrides can be determined by fusing with sodium hydroxide, sweeping the ammonia by means of argon into boric acid, and titrating with standard hydrochloric acid [20].

B. Distillation—Photometric

In zirconium and its alloys, nitrogen may be from 2 to 120 ppm. It is determined by solution of the sample in hydrochloric–hydrofluoric acids, addition of sodium hydroxide, and separation of ammonia by steam distillation [3]. To the distillate is added Nessler's reagent, and the yellow ammonium complex is measured photometrically at 430 nm. The same procedure is used for determining 20–1000 ppm of nitrogen in niobium and its alloys [3].

When the nitrogen content of copper or its alloys is less than 0·02%, the conventional distillation of ammonia is followed by photometric determination, based on the reaction of ammonia with sodium phenoxide in the presence of sodium hypochlorite to form a blue complex [12].

It has been shown that various refractory metals and alloys can be satisfactorily decomposed for the nitrogen determination by a mixture of hydrofluoric acid, phosphoric acid, and potassium dichromate [16]. Coarse samples are dissolved in 1–3 hours in a Teflon bottle at a moderate boiling rate; steam distillation is followed by final colorimetric measurement using the sodium phenoxide–sodium hypochlorite method [16].

The low nitrogen content of aluminium and its alloys can be determined by dissolving 10 g in sodium hydroxide solution, distilling off the ammonia, and measuring it colorimetrically with Nessler's reagent [5].

C. Vacuum Fusion

Small quantities of nitrogen, as well as hydrogen and oxygen, in metals and alloys can be determined by fusion of the sample in a vacuum in the presence of carbon. All oxides present are reduced to carbon monoxide, and nitrogen and hydrogen are liberated. The evolved gases are collected and analysed by low-pressure techniques. Hydrogen is oxidized to water and absorbed in anhydrous magnesium perchlorate. Carbon monoxide is converted to carbon dioxide and frozen out in liquid nitrogen. The residual gas is nitrogen. This method has been adopted as standard, for instance, for the 0·0001–0·01% nitrogen which is present in electronic nickel [3, 18].

5. Nitrogen Analyser

Equipment is available for determining nitrogen in a wide variety of materials, based on automation of the Dumas method [7]. The latter consists of decomposing the sample by heat, in the absence of air, and in intimate contact with copper oxide powder. In practice, an atmosphere of carbon dioxide is maintained in the system, the carbon dioxide is absorbed in potassium hydroxide, and the volume of nitrogen evolved is measured directly.

6. Oxides of Nitrogen in Air

The determination of oxides of nitrogen in air is occasionally required in the mining industry. The subject is fully discussed in several references [8, 22].

7. Micro Amounts of Nitrogen

A spectrophotometric determination of nitrogen has been described, based on the reaction of ammonia with hypochlorite and phenol, producing a blue indophenol which is measured at 625 nm [23].

REFERENCES

1. AMERICAN PUBLIC HEALTH ASSOCIATION, *Standard Methods for the Examination of Water and Wastewater*, New York, American Public Health Association, 1965.
2. A.O.A.C., *Official Methods of Analysis of the Association of Official Analytical Chemists*, Washington, D.C., Association of Official Analytical Chemists, 1965.
3. A.S.T.M., *Chemical Analysis of Metals. Sampling and Analysis of Metal Bearing Ores*, Philadelphia, Pa., American Society for Testing and Materials, 1969.
4. BEATTY, R. L., BERGER, L. B., and SCHRENK, H. H., *U.S. Bur. Mines Rept. Invest.*, 3687, 1943.
5. BRITISH ALUMINIUM CO., *Analysis of Aluminium and its Alloys*, London, British Aluminium Co., 1961.
6 BURRELL CORP., *Burrell Manual for Gas Analysts*, Pittsburgh, Pa., Burrell Corp., 1951.
7. COLEMAN INSTRUMENTS, INC., MAYWOOD, Ill., *Bulletin*, 258–A, 1961.
8. DILLON, V. S., *Assay Practice on the Witwatersrand*, Johannesburg, Transvaal and Orange Free State Chamber of Mines, 1955.
9. EASTERBROOK, W. C., *J. Appl. Chem.* **9**, 410–17 (1959).

10. Eastoe, J. E., and Pollard, A. G., *J. Science Food and Agriculture* **1**, 266–9 (1950).
11. Ekpete, D. M., and Cornfield, A. H., *Analyst* **89**, 670–3 (1964).
12. Elwell, W. T., and Scholes, I. R., *Analysis of Copper and its Alloys*, Oxford, Pergamon, 1967.
13. Furman, N. H., ed., *Scott's Standard Methods of Chemical Analysis*, 6th ed., Vol. 1, Princeton, N.J., Van Nostrand, 1962.
14. Hora, F. B., and Webber, P. J., *Analyst* **85**, 567–9 (1960).
15. Johnson, C. M., and Ulrich, A., *Anal. Chem.* **22,** 1526–9 (1950).
16. Kallmann, S., Hobart, E. W., Oberthin, H. K., and Brienza, W. C., *Anal. Chem.* **40**, 332–5 (1968).
17. Kolthoff, I. M., and Elving, P. J., *Treatise on Analytical Chemistry*, Part II, Vol. 5, 217–316, New York, Interscience Publishers, 1961.
18. Lewis, C. L., Ott, W. L., and Sine, N. M., *The Analysis of Nickel*, Oxford, Pergamon, 1966.
19. Middleton, K. R., *J. App. Chem.* **8**, 505–9 (1958).
20. Passer, R. G., Hart, A., and Julietti, R. J., *Analyst* **87**, 501–3 (1962).
21. Potrafke, K. A., Kroll, M., and Blom, L., *Anal. Chim. Acta* **31**, 128–38 (1964).
22. Rabson, S. R., Quilliam, J. H., and Goldblatt, E., *J. South African Institute Mining Met.* **61**, 152–82 (1960).
23. Rommers, P. J., and Visser, J., *Analyst* **94**, 653–8 (1969).
24. Ungar, J., *J. Appl. Chem.* **6**, 245–8 (1956).

27 Oxygen

Oxygen is seldom determined in mining laboratories, except in gases, a few metals and alloys, and as dissolved oxygen in waters, effluents, or cyanide solutions. In gas analysis, oxygen is usually measured in an Orsat apparatus by absorption in alkaline pyrogallol or chromous chloride, after the prior removal of carbon dioxide [12, 14, 17].

1. Oxygen in Metals and Alloys

A. Vacuum Fusion

When metals are fused in a vacuum in the presence of excess carbon, all oxides and dissolved oxygen are reduced to carbon monoxide, while nitrogen and hydrogen are liberated. The evolved gases are collected and analysed by low pressure techniques. Carbon monoxide is converted to carbon dioxide and frozen out in liquid nitrogen.

This vacuum fusion procedure is used for 0·001–0·1 % oxygen in electronic nickel [5, 22], for 0·0005–0·01% in molybdenum metal which is melted in an iron-tin bath, and for approximately 0·03–0·5% in titanium and its alloys, where a platinum or tin flux is employed [5]. Vacuum fusion methods for other metals have been described [13, 16, 19]. One variation for tungsten and other metals uses vacuum fusion in a molten platinum bath in a graphite crucible, followed by separation and measurement of oxygen by gas chromatography [38].

A simplified vacuum fusion procedure for copper and its alloys has been outlined; it is recommended in the presence of a stable oxide, a metal which strongly absorbs oxygen, or a large amount of an element which is volatile in hydrogen [15]. The sample is melted in a graphite combustion boat in an evacuated system of known volume, and oxygen present in the sample is reduced to carbon monoxide. The oxygen content is calculated from the increase in pressure due to carbon monoxide. Sulphur does not interfere, because any sulphur dioxide formed is reduced by the excess of graphite to sulphur and carbon monoxide.

B. Conductometric Oxygen Analyser

For 0–0·035% oxygen in a 1-g sample of many ferrous and non-ferrous metals and alloys, a conductometric method can be employed [20]. A purified inert gas is passed over an inductively heated graphite crucible containing the sample. Oxides are reduced to carbon monoxide and are carried through an oxidizing agent. The carbon dioxide thus formed is then measured in a conductivity cell where the change in resistance is read. This change is correlated with percentage oxygen on a graph; the latter is prepared by using a standard oxygen sample.

C. Oxygen and Sulphur in Refined Copper

Oxygen and sulphur in refined copper can be determined simultaneously by passing purified hydrogen over copper drillings heated in a combustion tube, and catching the resulting hydrogen sulphide and water in ammoniacal cadmium chloride solution. The sulphur is determined by titration of the sulphide with the iodine–thiosulphate procedure; oxygen is obtained by difference from the loss in weight of the copper drillings after expulsion of oxygen + sulphur [36].

This procedure is not applicable in the presence of significant amounts of elements which are volatile in hydrogen, such as arsenic, bismuth, phosphorus, selenium, tellurium, and zinc. The content of these elements in nearly all refined copper, however, is so low in relation to the oxygen, that the method can be applied to most samples.

Purify commercial cylinder hydrogen by passing it through the following train :

1. Mercury check valve
2. Empty trap bottle
3. 500 ml of 10% potassium hydroxide + 15 g potassium permanganate
4. Empty trap bottle
5. 400 ml sulphuric acid
6. Empty trap bottle
7. 500 ml of 80% potassium hydroxide + 25 g pyrogallic acid
8. Trap bottle with a few sticks of potassium hydroxide
9. Drying bottle half full of stick potassium hydroxide
10. Calcium chloride drying bottle
11. Drying bottle containing glass wool sprinkled with phosphorus pentoxide
12. Distribution bottle. Two outlet tubes to furnaces.

Screen the copper drillings through a $\frac{3}{32}$ in (2·4 mm) sieve and use only the coarse. Weigh out 50 g, place in a combustion boat, and introduce into the silica tube of the furnace. Pass the hydrogen, with resulting hydrogen sulphide and water vapour, from the combustion tube into sulphur absorption bottles consisting of two Erlenmeyer flasks containing ammoniacal cadmium chloride. Insert an empty bottle as a trap between the absorption flasks and the combustion tube.

Pass hydrogen through the system for 5 minutes, turn on the current and bring the furnace slowly up to a temperature of 850°C, with hydrogen passing through at the rate of about 3 bubbles a second. Maintain this temperature for four hours, turn off the current, and allow the furnace to cool. When the combustion tube has cooled nearly to room temperature, turn off the hydrogen gas, transfer the sample to a desiccator and weigh. The difference in weight is oxygen + sulphur.

Before commencing the stream of hydrogen through the system, place 25 ml ammoniacal cadmium chloride solution in the sulphur absorption flask. All the hydrogen sulphide is caught in the first flask; the second is merely a safeguard. When the heating is over, detach the flask, wash down, and add 25 ml hydrochloric acid. Cool, add 10 ml of 0·01N iodine from a burette, and titrate the excess with 0·01N sodium thiosulphate, using starch indicator. The quantity of sulphur determined is subtracted from the loss in weight of the copper sample to obtain oxygen.

The stoppers on the end of the quartz combustion tube are protected by asbestos and a spiral of water-cooled copper tubing. The quantity of sample may vary from 20 to 200 g, but a weight of 50 g in a quartz bulb is a convenient size; in this case the whole bulb is weighed.

The quantity of oxygen in different types of copper may vary appreciably. Vacuum-melted copper may contain less than 2 ppm oxygen, the electrolytic tough pitch grade has usually 0·02–0·04% but sometimes exceeds 0·1%, and copper powders often have more than 1% oxygen.

Reagents

Ammoniacal cadmium chloride. Dissolve 12 g cadmium chloride in 50 ml of water. Add 70 ml ammonium hydroxide and dilute to 1 litre with water; transfer to a large stock bottle. To 5 ml ammonium hydroxide add 995 ml water, and add this solution to the original in the stock bottle.

Sodium thiosulphate solution, 0·01N. Dissolve 2·482 g $Na_2S_2O_3.5H_2O$ in water and dilute to 1 litre.

Iodine solution, 0·01N. Dissolve together in the minimum quantity of water 5–7 g potassium iodide and 1·28 g iodine; dilute to 1 litre. One ml of 0·01N iodine = 0·0001603 g sulphur. Standardize against arsenious acid.

Arsenious acid, 0·01N. Dissolve 0·4946 g As_2O_3 in 20 ml of 20% potassium hydroxide solution. Neutralize the excess alkali with dilute sulphuric acid, add 500 ml of water containing 25 g sodium bicarbonate, and dilute to 1 litre.

The reactions in this determination may be represented :

$$H_2 + Cu_2O \rightarrow H_2O + 2Cu$$

$$H_2 + Cu_2S \rightarrow H_2S + 2Cu$$

$$H_2S + CdCl_2 + 2NH_4OH \rightarrow CdS + 2NH_4Cl + 2H_2O$$

$$CdS + 2HCl \rightarrow H_2S + CdCl_2$$

$$H_2S + I_2 \rightarrow 2HI + S$$

$$2H_3AsO_3 + 4NaHCO_3 + 2I_2 \rightarrow As_2O_5 + 4NaI + 5H_2O + 4CO_2$$

In the absence of sulphur, the procedure described in detail above can be used for other metals, if their oxides are reduced by hydrogen under these conditions, by absorbing the evolved water vapour in magnesium perchlorate or phosphorus pentoxide. For example, when applied to a steel it will give a measure of the oxygen which is combined with iron but not the oxides of Mn, Al, or Si which are not reduced by hydrogen.

A modification of this procedure has been described, which permits the presence of about 1% of arsenic, selenium, and tellurium; large amounts of phosphorus or zinc cannot be tolerated [15]. The sample is heated in a closed system in contact with an excess of hydrogen, the water formed is condensed, then vaporized into an evacuated system of known volume, and finally determined by differential-pressure measurement.

D. Other Methods for Oxygen

In aluminium, oxygen is determined by volatilizing the metal as aluminium chloride by heating at 400–450°C in a stream of dry hydrogen chloride. The residue of alumina and other non-volatile impurities is collected, and its aluminium content, equivalent to the oxygen present in the sample, is determined gravimetrically or colorimetrically [8].

A method has been described for oxygen in the oxide coating of powdered copper [21]. Hot dilute hydrochloric acid dissolves cupric and cuprous oxides, in an atmosphere of carbon dioxide. In the presence of metallic copper, cupric oxide goes to the cuprous state. An excess of ferric ammonium sulphate is added, and the ferrous iron formed is titrated with standard potassium permanganate. A similar procedure is outlined for other metallic oxides; the oxide material is dissolved in hydrochloric acid containing ferrous iron, and the residual ferrous chloride is titrated with standard ceric sulphate [37].

Oxygen in some common rock-forming minerals has been determined by extracting with bromine pentafluoride, and measuring in a constant-volume manometer [35].

The use of solid electrolyte oxygen concentration cells for the determination of oxygen in molten steel has been described [23].

2. Dissolved Oxygen

The importance of dissolved oxygen in the purification of sewage or trade wastes, in cyanidation of gold, in boiler waters, and corrosion problems in general, places this determination among those encountered from time to time in nearly every mining and metallurgical laboratory. Most procedures for dissolved oxygen are based on the Winkler method, which utilizes the ability of manganese to function as an oxygen carrier or temporary fixative. In alkaline solution, manganous ion reacts with dissolved oxygen to form a higher oxide of manganese. If the solution is acidified in the presence of a readily-oxidizable substance such as iodide, the manganese is reduced to the divalent state and iodine is liberated in an amount equivalent to the oxygen originally present. The free iodine is titrated with a standard solution of sodium thiosulphate, using visual, potentiometric, or amperometric detection of the endpoint. The determination of iodine or of the iodine-starch complex may also be made by various colorimetric means. Other substances, such as *o*-tolidine or 3 : 3′-dimethylnaphthidine may be used in place of iodine; the final colorimetric measurement is then carried out on the oxidized form of the organic substance.

A. For Streams, Reservoirs, Effluents, and Sewage

Dissolved oxygen in most waters, apart from boiler waters, can be determined by the following procedure [2]. Collect the sample in a narrow-mouth, glass-stoppered bottle of about 300-ml capacity, taking all precautions to avoid entrainment or solution of air. For example, in sampling from a line under pressure, lead the water from rubber tubing to the bottom of the bottle and allow it to overflow for five minutes before carefully withdrawing the tubing from the bottle, leaving the latter full to the top, and inserting the stopper so that no air is entrapped. Equipment for sampling streams or reservoirs is described in authoritative references [2, 17]. If the water temperature exceeds 21°C, the sample should be passed through a cooling coil before collecting in the 300-ml glass-stoppered bottle [24].

(1) If the sample contains not more than 1 mg/litre of ferrous iron, add 2 ml manganous sulphate followed by 2 ml alkali–iodide–azide reagent. All solution additions to the sample are conveniently made by means of 10 ml burettes to which tubes, drawn out to fine jets, long enough to reach the bottom of the sample bottle, are attached. The greatest attention must be paid to exclude all air bubbles. Stopper with care and mix by inverting the bottle several times; allow to stand for five minutes. Add 2 ml sulphuric acid to the bottom of the bottle, replace the stopper carefully, and mix by inversion several times. Allow to stand 5 minutes, pour the sample into a porcelain casserole and titrate with 0·025N sodium thiosulphate, adding 2 ml starch indicator when the colour becomes a faint yellow and continuing the titration to the first disappearance of the blue colour. Disregard a slow return of colour which sometimes occurs, due to traces of nitrite or iron. Correct the volume of sample for the loss by displacement with the reagents. 1 ml of 0·025N sodium thiosulphate = 0·0002 g oxygen. If conversion to volume is desired, mg of dissolved oxygen per litre × 0·698 = ml of dissolved oxygen per litre at 0°C and 760 mm pressure. Dissolved oxygen is often expressed as ppm by weight; ml O_2 per litre × 1·43 = ppm O_2 by weight. Sometimes, results are expressed as μg, or γ, per litre; one μg O_2 per litre = 0·001 ppm O_2, by weight.

(2) If the sample contains more than 1 mg per litre of ferrous iron, it is necessary to oxidize the latter by adding potassium permanganate, then destroy the excess permanganate, before proceeding with the fixation of dissolved oxygen by means of manganous sulphate. Because these samples frequently contain ferric iron, the effect of even such high concentrations as

occur in mine waters can be overcome by adding potassium fluoride initially.

To the sample in the 300-ml bottle, collected as outlined previously, add below the surface exactly 0·7 ml sulphuric acid, followed by 1 ml potassium permanganate solution and 1 ml potassium fluoride solution. Stopper carefully and mix by inversion. Add sufficient potassium permanganate to obtain a violet tinge which persists for at least 5 minutes. If the colour is destroyed in a shorter time, add more permanganate, or preferably a stronger solution, but avoid a large excess. Remove the permanganate colour by adding 0·5–1 ml potassium oxalate solution, mix well, and allow to stand in the dark for 5–10 minutes. Add only an amount of oxalate which completely decolorizes the solution without having an excess of more than 0·5 ml; otherwise low results are obtained.

Add 2 ml manganous sulphate solution, followed by 3 ml alkali–iodide–azide reagent. Stopper, mix, and allow to stand for 5 minutes. Add 2 ml sulphuric acid, stopper, mix thoroughly, and immediately titrate with 0·025N sodium thiosulphate as described in the preceding section. Correct the volume of sample for the loss incurred through displacement with reagents.

Reagents

Manganous sulphate solution. Dissolve 480 g $MnSO_4.4H_2O$ in water and dilute to 1 litre.

Alkali–iodide–azide reagent. Dissolve 500 g sodium hydroxide and 150 g potassium iodide in water and dilute to 1 litre. To this solution add 10 g sodium azide dissolved in 40 ml of water.

Starch solution. Prepare an emulsion of 5 g soluble starch with a small quantity of water, pour carefully into a litre of boiling water, boil, and cool. It may be preserved, if desired, by the addition of a few drops of toluene or of 1·25 g salicylic acid per litre.

Sodium thiosulphate, 0·025N. Dissolve 6·205g $Na_2S_2O_3.5H_2O$ in water and dilute to 1 litre. Standardize against 0·025N potassium bi-iodate by weighing 0·8124 g of the latter and diluting to 1 litre; dissolve 2 g potassium iodide in 100 ml water, add 10 ml 1 : 9 sulphuric acid, 20·00 ml of the standard bi-iodate solution, dilute to 200 ml, and titrate with standard thiosulphate, adding starch towards the end of the titration.

Potassium fluoride solution. Dissolve 40 g $KF.2H_2O$ in water and make up to 100 ml.

Potassium permanganate solution. Dissolve 6·3 g potassium permanganate in water and dilute to 1 litre.

Potassium oxalate solution. Dissolve 2 g $K_2C_2O_4.H_2O$ in 100 ml water; 1 ml of this solution is sufficient for the reduction of about 1·1 ml of the above permanganate solution.

B. For Boiler Waters

The low oxygen contents of boiler waters in modern power plants have necessitated modifications of the Winkler procedure to improve its accuracy and sensitivity. The subject has been reviewed in two papers, one covering the earlier work to 1949 [40] and the other discussing subsequent developments up to 1960 [39].

The procedure for boiler waters may take one of several forms, depending upon the dissolved oxygen content and the precision demanded.

(1) *Titration of iodine with sodium thiosulphate using an amperometric end-point.* In an amperometric titration, potential is applied between two smooth platinum electrodes immersed in the water sample. At the end point of the reaction there is an abrupt change in the flow of current, due to the removal of all iodine. The current stops; hence the term "dead stop" end-point. Since it is easier to detect with a microammeter the commencement of a flow of current, in practice a known excess of sodium thiosulphate is added to the water sample and back-titration is carried out with a standard iodine solution. The amperometric finish detects the appearance of the current due to iodine, and can have sufficient sensitivity to easily give a precision for dissolved oxygen of $\pm 0{\cdot}001$ ppm. The presence of at least 2 ppm of ferrous, ferric, or cupric ions does not interfere.

For these low oxygen concentrations, a special type of sampling vessel has been developed [26, 27]. This procedure, together with the sampling vessel, is recommended for work with feedwater to high pressure boilers, where the tolerated dissolved oxygen is below the low value of 0·01 ppm [21, 22, 23, 24, 25, 29, 30, 31, 32, 33].

(2) *Titration of iodine with sodium thiosulphate using the reversed-reagents method.* This procedure, originally proposed by Schwartz and Gurney [34], is one of the most reliable for the determination of medium concentrations of dissolved oxygen in boiler water. Two water samples of equal volume are used. One, A, is processed in the usual way; in the other, B, called a "blank", the order of addition of reagents is reversed, so that in it the dissolved oxygen is not fixed by the manganous salt. The interfering substances in the water and in reagents react in both the sample and the blank. $A-B$ represents the oxygen in the sample plus that in the reagents [6].

If a considerable quantity of interfering reductants are present in the water, oxidizing additions are made in equal quantity to ensure that iodine is present before the final analytical determination.

The dissolved oxygen in reagents can be determined by the technique of "double reagents". Four sampling vessels of equal size are used. One pair, A, is processed by the reversed-reagents method using the ordinary quantities of reagents; the other pair, B, has double these amounts. The difference between A and B measures the dissolved oxygen in the ordinary increments of reagents, and is used to correct the result obtained for A.

Another procedure to correct for the dissolved oxygen in reagents is the "triple-sample" technique. One 500-ml, A, and two 250-ml water samples, B and C are obtained. A and B are processed normally, using the same reagent quantities. No reagents are added to C and this sample is mixed with B before the final titration. The dissolved oxygen in reagents and the interfering substances present in the water both react in the two final samples, and the titration difference $A - (B + C)$ is the dissolved oxygen content of a 250-ml sample.

3. Colorimetric Determination of Iodine

For routine laboratory and works practice, a useful procedure is to remove the iodine which is liberated in the Winkler test by several successive extractions with a small quantity of carbon tetrachloride, and determine it in a spectrophotometer or with standard discs in a Lovibond Nessleriser [3, 4, 9, 10]. Other colorimetric determinations of iodine have been described [39].

4. Colorimetric Determination of the Oxidized Forms of an Organic Reagent

A. Absorption of dissolved oxygen by manganous salt in alkaline solution and acidification as in the classic Winkler test

(1) *o-tolidine.* In this procedure, free chlorine, proportional to the oxygen in the sample, which is liberated on acidification of manganous chloride, reacts with *o*-tolidine to give a yellow oxidized *o*-tolidine; the latter is measured by standard discs in a Nessleriser, or in a spectrophotometer [18].

(2) 3 : 3′-*dimethylnaphthidine* (*DMN*). This compound forms a suitable coloured solution for determining dissolved oxygen absorptiometrically when it is added, with sodium acetate, to the regular Winkler reagents of manganous chloride, alkaline potassium iodide, and sulphuric acid [7].

B. Direct determination of colour based on reaction of indigo carmine with oxygen

When reduced indigo carmine reacts with oxygen a colour change is observed through orange, red, purple, blue, and finally to a blue-green in the completely oxidized form. A procedure has been developed for low quantities of dissolved oxygen by using a 1 : 1 solution of glycerol in water as the solvent for indigo carmine, adding 1% potassium hydroxide for *p*H control, and 1% glucose as the reducing agent [11]. The colour can be measured spectrophotometrically at 555 nm; for plant tests, liquid standards [11] or permanent colour discs [1] can be used.

C. For Cyanide Solutions

Dissolved oxygen plays an important role in the leaching of gold ores with cyanide solutions. Its determination in the latter may be carried out in several ways.

(1) The modification of the Winkler procedure used in the presence of ferrous iron, wherein potassium permanganate solution is added, and the excess later destroyed with potassium oxalate, as outlined in **2**A(2), may be

employed. The strength of the potassium permanganate must be higher than that used normally, in order that 1 ml of permanganate will destroy the cyanide and give a pink colour to the solution which persists for at least five minutes.

(2) For works control, colorimetric comparison with a series of standards from 0·5 to 7 mg per litre can be made by carefully adding to a 300 ml sample, 0·12 g of pyrogallic acid, followed by 1 ml of 2N sodium hydroxide, and mixing by inversion. The usual precautions must be taken to avoid the introduction of an air bubble to the sample bottle. The brown colour fades gradually, and permanent colour standards must be made up with a dye which is matched against standard oxygen concentrations. A solution containing a known content of dissolved oxygen can be made by pouring water into a Winchester bottle until it is approximately half filled, shaking vigorously for five minutes, or passing a stream of air through it, until the water is saturated with air, taking the temperature and barometric pressure, and reading the solubility of oxygen in water from standard charts [14].

(3) For work of higher accuracy, the Weinig–Bowen titration procedure with standard sodium hydrosulphite in the presence of indigotin-disulphonate as indicator, may be used [14].

REFERENCES

1. Alcock, G. P., and Coates, K. B., *Chemistry and Industry*, **1958**, May 10 554–5.
2. American Public Health Association, *Standard Methods for the Examination of Water and Wastewater*, New York, American Public Health Association, 1965.
3. Arnott, J., and McPheat, J., *Engineering* **176**, 103 (1953).
4. Arnott, J., McPheat, J., and Ling, F. B., *Engineering* **169,** 553 (1950).
5. A.S.T.M., *Chemical Analysis of Metals. Sampling and Analysis of Metal Bearing Ores*, Philadelphia, Pa., American Society for Testing and Materials, 1969.
6. *A.S.T.M. Standards*, Part 7, 1155–62 (1952).
7. Banks, J., *Analyst* **79**, 170–3 (1954).
8. British Aluminium Co., *Analysis of Aluminium and its Alloys*, London, British Aluminium Co., 1961.
9. British Standards Institution, *Tests for Water Used In Steam Generation*, B.S. 1427, London, British Standards Institution, 1949.
10. British Standards Institution, *Methods for Testing Water Used in Industry*, B.S. 2690, London, British Standards Institution, 1956.
11. Buchoff, L. S., Ingber, N. M., and Brady, J. H., *Anal. Chem.* **27**, 1401–4 (1955).
12. Burrell Corporation, *Manual for Gas Analysts*, Pittsburg, Pa., Burrell Corp., 1951.
13. Covington, L. C., and Bennett, S. J., *Anal. Chem.* **32**, 1334–7 (1960).
14. Dillon, V. S., *Assay Practice on the Witwatersrand*, Johannesburg, Transvaal and Orange Free State Chamber of Mines, 1955.

15. Elwell, W. T., and Scholes, I. R., *Analysis of Copper and its Alloys*, Oxford, Pergamon, 1967.
16. Elwell, W. T., and Wood, D. F., *Analysis of the New Metals Titanium, Zirconium, Hafnium, Niobium, Tantalum, Tungsten, and their Alloys*, Oxford, Pergamon, 1966.
17. Furman, N. H., ed., *Scott's Standard Methods of Chemical Analysis*, 6th ed., Vol. 1, Princeton, N. J., Van Nostrand, 1962.
18. Haslam, J., and Moses, G., *J. Soc. Chem. Ind.* **57**, 344–7 (1938).
19. Horton, W. S., and Brady, J., *Anal. Chem.* **25**, 1891–8 (1954).
20. Laboratory Equipment Corp., *Leco Conductometric Oxygen Analyzer*, St. Joseph, Mich., 1960.
21. Julietti, R. J., and Hart, A., *Analyst* **87**, 452–4 (1962).
22. Lewis, C. L., Ott, W. L., and Sine, N. M., *The Analysis of Nickel*, Oxford, Pergamon, 1966.
23. Pargeter, J. K., and Fourschou, D. K., *J. Metals* **21**, No. 3, 46–8 (1969).
24. Pinkney, E. T., and Young, R. S., *Combustion* **18**, 35–7 (1947).
25. Potter, E. C., *J. Appl. Chem.* **7**, 285–97 (1957).
26. Potter, E. C., *J. Appl. Chem.* **7**, 297–308 (1957).
27. Potter, E. C., and Everitt, G. E., *J. Appl. Chem.* **9**, 642–5 (1959).
28. Potter, E. C., and Everitt, G. E., *J. Appl. Chem.* **9**, 645–50 (1959).
29. Potter, E. C., and Everitt, G. E., *J. Appl. Chem.* **10**, 48–56 (1960).
30. Potter, E. C., and White, J. F., *J. Appl. Chem.* **7**, 309–17 (1957).
31. Potter, E. C., and White, J. F., *J. Appl. Chem.* **7**, 317–28 (1957).
32. Potter, E. C., and White, J. F., *J. Appl. Chem.* **7**, 459–67 (1957).
33. Potter, E. C., and Whitehead, G., *J. Appl. Chem.* **7**, 629–39 (1957).
34. Schwartz, M. C., and Gurney, W. B., *Am. Soc. Testing Materials Proc.* **34**, Part 2, 798–820 (1934).
35. Sharma, T., and Clayton, R. N., *Anal. Chem.* **36**, 2001–2 (1964).
36. Smith, G. A., *Trans. Institution Mining Met.* **72**, Part 7, 469–96 (1962–63).
37. Van Oosterhout, G. W., and Visser, J., *Anal. Chim. Acta* **33**, 330–32 (1965).
38. Wood, D. F., and Wolfenden, G., *Anal. Chim. Acta* **38**, 385–402 (1967).
39. Young, R. S., *British Power Engineering* **1**, No. 6, 46–50 (1960).
40. Young, R. S., Pinkney, E. T., and Dick, R., *Power Engineering* **54**, No. 7, 62–5, 110, 112, 114–16 (1950).

28 Phosphorus

The determination of phosphorus may be required in analyses of rocks, minerals, ores, concentrates, slags, metals and alloys, soils, fertilizers, waters, and many other materials examined in a mining and metallurgical laboratory. In nature, phosphorus occurs as the phosphate. Many phosphate minerals occur, but the most important sources of phosphorus are fluorapatite $Ca_5F(PO_4)_3$, hydroxyapatite $Ca_5OH(PO_4)_3$, and chlorapatite $Ca_5Cl(PO_4)_3$; these minerals are the principal constituents of phosphate rock.

For most work, an oxidizing attack with nitric acid should be used initially, to prevent possible loss of phosphorus as phosphine. Samples should not be fumed for extended periods at a high temperature with sulphuric acid. Most materials containing phosphorus yield readily to acid attack; a few will require a fusion of an insoluble residue with sodium carbonate or sodium peroxide.

ISOLATION OR SEPARATION OF PHOSPHORUS

Members of the hydrogen sulphide group can be separated from phosphorus by precipitation of the former with hydrogen sulphide in 5–10% hydrochloric or sulphuric acid. Treatment of the ammonium hydroxide and ammonium sulphide groups with sodium hydroxide will leave phosphorus in the filtrate with Be, Al, V, Mo, W, and Zn, while Fe, Ti, Ni, Co, Zr, Mn, Cr, Th, and U are precipitated; if sodium peroxide is used, Cr and U will be thrown into the filtrate. Phosphorus, with Al, Be, Ti, V, Zr, W, and U, is not deposited at a mercury cathode in dilute sulphuric acid and can thereby be separated from Fe, Ni, Co, Zn, Cr, Mo, etc. A cupferron separation in cold 10% sulphuric acid will leave phosphorus in the filtrate with Al, Be, Mn, Ni, Co, and Cr, whereas Fe, Ti, Zr, V, Nb, Ta, and W are precipitated. Co, Ni, and Mn are separated from phosphorus and nearly all other elements by a zinc oxide separation. Small amounts of phosphorus may often be conveniently collected from large quantities of Cu, Ni, or Co in a ferric hydroxide precipitate.

Aluminium can be separated from phosphorus by precipitation as the oxinate from ammoniacal solution. This separation is also satisfactory in acetic acid–acetate solution, provided that the precipitation is made by first adding the oxine to a N hydrochloric acid solution and then raising the *p*H to about 5 with ammonia and ammonium acetate [19].

1. Separation of Ammonium Phosphomolybdate Prior to Final Gravimetric or Volumetric Determination

When a substantial quantity of phosphorus is present, it is almost invariably separated in the presence of a number of other elements by precipitation with ammonium molybdate in nitric acid [2, 3, 4, 10, 13, 14].

Dissolve 0·25–5 g in nitric acid, with the addition of hydrochloric, hydrofluoric, or perchloric acids if necessary. Evaporate to dryness, moistening several times with nitric acid. Dissolve in hot 5–10% nitric acid, filter, and wash thoroughly with hot water. It may be preferable for some samples to fuse with sodium carbonate in a platinum crucible the residue which is initially insoluble in nitric acid alone, rather than add mixed acids to effect decomposition.

To the clear filtrate of 100–200 ml in a flask containing 5–10% nitric acid, 5–15% ammonium nitrate, and about 0·05 g of phosphorus, which is at 40–50°C, add a 15–25 fold excess of ammonium molybdate reagent, or 14 g for 0·05 g of phosphorus, stopper, and shake vigorously for 10 minutes. If the amount of phosporus is small, or substances that retard the precipitation are present, such as V, Ti, Zr, As, Si, H_2SO_4, HCl, or organic substances, use a larger excess of the molybdate reagent. In routine analyses allow the precipitate to settle 30 minutes; for work of high accuracy allow the sample to stand overnight.

In the presence of vanadium, before the addition of the molybdate reagent cool the solution to 15°C, add sufficient ferrous sulphate to reduce the vanadium, followed by a few drops of sulphurous acid. If the solution does not contain iron, at this point add about 1 g of iron as ferric nitrate, free from phosphorus, to prevent subsequent reduction of the molybdenum. Add a 20–30 fold excess of cool molybdate reagent and proceed as directed in the previous paragraph.

The separation of ammonium phosphomolybdate may be represented:

$$H_3PO_4 + 12(NH_4)_2MoO_4 + 21HNO_3 \rightarrow (NH_4)_3PO_4.12MoO_3 + 21NH_4NO_3 + 12H_2O$$

From this point, the determination may be completed gravimetrically or volumetrically.

A. Gravimetric Determination

The yellow ammonium phosphomolybdate is dissolved in ammonium hydroxide, phosphorus is precipitated as magnesium ammonium phosphate, ignited and weighed as $Mg_2P_2O_7$ [2, 4, 10, 13, 14].

Filter the ammonium phosphomolybdate precipitate on Whatman No. 42 paper and wash thoroughly with 5% ammonium nitrate solution. Dissolve the precipitate in dilute ammonium hydroxide and wash the paper successively with hot water, dilute hydrochloric acid, and again with dilute ammonium hydroxide. To the ammoniacal solution add 0·5 g citric acid and filter if not clear.

Make the solution slightly acid with hydrochloric acid, cool, add 10 ml of magnesia mixture and then 1 ml additional for every 0·0005 g of phosphorus that is expected. Add dilute ammonium hydroxide slowly with constant stirring until the solution is alkaline and the white magnesium ammonium phosphate appears. Add a further 10 ml of dilute ammonium hydroxide per 100 ml of solution, and set aside for 4 hours or preferably overnight.

Filter on Whatman No. 42 paper and wash the precipitate several times with dilute ammonium hydroxide. Wash the precipitate into the original beaker with dilute hydrochloric acid; wash the filter thoroughly with dilute hydrochloric acid and finally with hot water. Dilute the solution to 100 ml and add 2 ml magnesia mixture. Add dilute ammonium hydroxide slowly with constant stirring as before, and finally add 10 ml excess. Allow the solution to stand several hours or preferably overnight.

Filter on Whatman No. 42 paper, and wash with 5% ammonium hydroxide until free from chlorides. Transfer paper and precipitate to a weighed platinum crucible, dry, char, burn the carbon at the lowest possible temperature with free access of air, and ignite to constant weight at 1000°C. Weigh as $Mg_2P_2O_7$; $Mg_2P_2O_7 \times 0{\cdot}2783 = P$.

The reactions may be represented:

$$H_3PO_4 + MgCl_2 + 3NH_4OH \rightarrow MgNH_4PO_4 + 2NH_4Cl + 3H_2O$$

$$2MgNH_4PO_4 \rightarrow 2NH_3 + H_2O + Mg_2P_2O_7$$

B. Volumetric Determination

For routine purposes, it is common practice to dissolve the ammonium phosphomolybdate in a measured excess of standard sodium hydroxide and back-titrate with standard nitric acid [2, 4, 10, 13, 14].

Allow the yellow ammonium phosphomolybdate to settle for 15 minutes after the 10-minute shaking period, following the procedure outlined previously. Filter through Whatman No. 42 paper and wash thoroughly with successive small portions of 1% potassium nitrate until no further acid is present in the washings. Return the paper and precipitate to the flask, and add sufficient 0·1N sodium hydroxide to dissolve the precipitate and to leave about 2 ml excess. Dilute with 25 ml water, and stir or shake until the precipitate has completely dissolved. Dilute to 100 ml with water, add 3–4 drops of phenolphthalein, and titrate to a colourless endpoint with 0·1N nitric acid. It is convenient to have the latter exactly equivalent to the 0·1N sodium hydroxide; the endpoint may be checked by adding 1 drop of the standard 0·1 N sodium hydroxide, which should restore the pink colour.

Reagents

Ammonium molybdate solution. Add 65 g of 85% molybdic acid to 142 ml of water, and to this, add with stirring 143 ml of ammonia. When the molybdic acid has dissolved, cool the solution and pour slowly with stirring into 715 ml of nitric acid of s.g. 1·20. The latter is made by mixing 476 ml of concentrated nitric acid with 524 ml of water. Add 1–2 drops of a 10% solution of ammonium phosphate and allow the solution to stand for 12 hours. Decant the clear liquid or filter through an asbestos pad as required for use.

Magnesia mixture. Dissolve 100 g $MgCl_2.6H_2O$ and 200 g NH_4Cl in 500 ml of water, add a slight excess of ammonium hydroxide, let the solution stand overnight, filter, make just acid with hydrochloric, and dilute to 1 litre.

Standard 0·1N NaOH. Dissolve 4 g of sodium hydroxide in 1 litre of CO_2-free water, add 1 ml of a standard $Ba(OH)_2$ solution, mix thoroughly and allow to stand for 24 hours. Protect the solution from carbon dioxide with a trap of Ascarite or solid potassium hydroxide. Titrate against potassium acid phthalate; 1 ml of 0·1N NaOH = 0·0204 g potassium acid phthalate. Check the sodium hydroxide solution against a standard steel of known phosphorus content. One ml of 0·1N NaOH = 0·000135 g P.

Standard 0·1N HNO_3. Mix 6·8 ml of nitric acid with 1000 ml of CO_2-free water. Titrate against standard sodium hydroxide, using phenolphthalein, and make it equivalent to the sodium hydroxide by adding distilled water.

For some routine determinations, it is convenient to use solutions of sodium hydroxide and nitric acid of such a strength that 1 ml of NaOH = 0·01% P on a 2-g sample. These can be made by dissolving 6·6 g sodium hydroxide in 1 litre and adjusting to an exact titre; an equivalent nitric acid solution contains approximately 10 ml nitric acid per litre.

2. Colorimetric Determination

A. Molybdovanadophosphoric Acid Method

When an excess of ammonium molybdate is added to an acidified mixture of a vanadate and an orthophosphate, a yellow colour is produced which can be measured at 460–470 nm. Because interferences are relatively few, this procedure can be used for a wide variety of materials in the mineral industries [1, 2, 4, 7, 8, 10, 14, 17]. In this and other colorimetric procedures for small amounts of phosphorus, it is important to remember that most detergents contain phosphorus; all traces of these must be removed from glassware by hot hydrochloric acid and thorough rinsing.

Decompose the sample by conventional means in acids, fusing any insoluble residue with sodium carbonate, and dehydrate silica by evaporation with hydrochloric, sulphuric, or perchloric acids. Filter off silica and wash thoroughly; add 10 ml 1 : 1 sulphuric acid if not already present, and evaporate the filtrate to fumes of the latter. Alternatively, silica can be removed by treatment with hydrofluoric and sulphuric acids in a platinum dish.

Adjust the acidity to 10% by volume of sulphuric acid, cool to 10°C, transfer the sample to a separatory funnel, add sufficient cold 6% cupferron solution to precipitate all the iron and other insoluble cupferrates, and shake vigorously. Add successive small portions of cold chloroform until all the cupferrates of Fe, Ti, V, Zr, Mo, Nb, Ta, and W have been extracted into the lower chloroform layer. Phosphorus, with Al, Be, Ni, Co, Cr, Mn, U(VI), and

Zn, remains in the upper aqueous layer. Transfer the latter to a beaker, place on the edge of the hot plate until traces of chloroform have been driven off, add 10 ml nitric acid and evaporate to fumes of sulphuric acid. If all organic matter has not been destroyed, add more nitric acid and repeat the treatment, adding a few drops of perchloric acid, if necessary, to assist the action. Do not fume excessively or anhydrous aluminium sulphate will be formed, which is difficult to get into solution. Also, prolonged fuming with sulphuric acid at elevated temperatures can lead to losses of phosphorus. Filter and wash with hot water. To the filtrate add 5 g ammonium chloride and precipitate Al, P, etc. with ammonium hydroxide. Dissolve the first precipitate in hydrochloric acid and re-precipitate with ammonia. Filter and wash well with hot 2% ammonium chloride.

If aluminium is desired, ignite the precipitate at 1200°C to constant weight, and weigh as Al_2O_3 plus P_2O_5. Fuse the ignited precipitate with sodium carbonate for 20 minutes, dissolve the solidified melt in water and make slightly acid with nitric acid. When alumina is calcined at a high temperature, it is often refractory towards sodium carbonate and one fusion will still leave considerable residue. Sodium or potassium pyrosulphate is a much more satisfactory flux for such material.

When aluminium is not required, omit the ignition step and dissolve the aluminium hydroxide containing the phosphorus directly in the minimum quantity of nitric acid. If the sample contains more than 2 mg of phosphorus, make it up to a known volume in a graduated flask and transfer an aliquot containing less than 2 mg to a 50 ml volumetric flask for the final colorimetric determination.

To about 25 ml of the clear, nearly neutral or faintly acid solution contained in a 50 ml volumetric flask, add 5 ml of 1 : 2 nitric acid and 5 ml of the ammonium vanadate solution. Add 5 ml of the ammonium molybdate solution, dilute to the mark, and mix. After 5 minutes, measure the absorbance at approximately 470 nm in a spectrophotometer. Calculate the phosphorus by reference to a calibration curve prepared from suitable aliquots of a standard phosphate solution.

The optimum acidity is about 0·5N in nitric acid; hydrochloric, sulphuric, or perchloric acid can be used in place of nitric acid. The following ions do not interfere: Al, Be, Zn, Na, K, NH_4, Sr, Ba, Hg(II), Sn(II), Mn(II), uranyl, zirconyl, acetate, arsenite, benzoate, bromide, carbonate, chlorate, citrate, cyanide, formate, iodate, lactate, molybdate, nitrate, nitrite, oxalate, perchlorate, periodate, pyrophosphate, salicylate, selenate, sulphate, sulphite, tartrate, and tetraborate. The following quantities, in p.p.m., can be tolerated: Cu and Ni 1000, Bi 400, Fe(II), Co, sulphide, thiosulphate, arsenate, and thiocyanate 100, F 50, Cr(III) 10. Cerium (IV), Sn(IV), iodide, dichromate, and permanganate should be absent. If chloride is present, silver should be absent; in the absence of chloride, silver even in the substantial quantities found in some brazing alloys is without effect.

REAGENTS

Ammonium vanadate solution. Dissolve 2·5 g of NH_4VO_3 in 500 ml of boiling water. Cool, add 20 ml nitric acid, dilute to 1 litre, and store in a polyethylene bottle.

Ammonium molybdate solution. Dissolve 50 g of $(NH_4)_6Mo_7O_{24}.4H_2O$ in 1 litre of warm water, and store in a polyethylene bottle.

Standard phosphate solution. Dissolve 0·4395 g of potassium dihydrogen phosphate in water, dilute to 1 litre, and store in a polyethylene bottle. Each ml of this solution contains 0·1 mg phosphorus.

The procedure described above is intended for complex materials, or for silicate analyses where aluminium and phosphorus are conveniently isolated and determined together. For many samples the procedure can be considerably shortened. In aluminium, for example, after removal of silica following a perchloric acid dehydration, the colorimetric measurement of the molybdovanadophosphoric acid is made directly [8]. For copper and its alloys, the optical density of the molybdovanadophosphoric acid can be measured, without separations, against a compensating solution containing an equal amount of copper. There is no interference from 1% silicon and 0·1% of As, Sb, Sn, Pb, Ni and Fe [4, 9].

In another variation of the molybdovanadophosphoric acid method for phosphorus, interference from other elements that form coloured ions in the aqueous solution is eliminated by extracting the molybdovanadophosphoric acid into isoamyl alcohol. The procedure for copper and its alloys has been outlined [9]. If more than 0·1% arsenic is present, eliminate it with antimony, selenium, and tin by hydrobromic–perchloric acid evaporation. If chromium is present, reduce the dichromate with sulphurous acid, and boil to remove the excess of this reagent. After the addition of the ammonium vanadate and ammonium molybdate solutions, transfer the sample to a separatory funnel, and dilute to 80 ml. Add 20·0 ml of isoamyl alcohol, shake for several minutes, allow to settle, and discard the lower aqueous layer. Measure the optical density of the organic layer at 435 nm.

An automated procedure for the rapid determination of phosphorus in copper, utilizing the molybdovanadophosphoric acid method, has been described [5].

B. Molybdenum Blue Method

Phosphorus reacts with ammonium molybdate to form a phosphomolybdenum complex; the latter can be reduced with substances such as hydrazine sulphate to form molybdenum blue, the intensity of which can be measured in a spectrophotometer at 700 or 830 nm [2, 4, 7, 14, 17].

Decompose the sample as outlined in the previous section and obtain a neutral solution containing no more than about 0·06 mg phosphorus per 25 ml; transfer this to a 50 ml volumetric flask. Add 5 ml of the molybdate solution and 2 ml of the hydrazine sulphate solution. Dilute to the mark, mix, immerse the flask in a boiling water bath for 10 minutes, cool, shake, and again adjust to volume if necessary. Measure the absorbance at 830 nm and determine phosphorus by comparison with a graph prepared with appropriate aliquots of a standard phosphate solution.

The following ions do not interfere: Al, NH_4, Cd, Cr(III), Cu, Co, Ca, Fe(II), Mg, Mn(II), Ni, Zn, chloride, bromide, acetate, citrate, silicate, fluoride, vanadate, and borate. Fe(III) should not exceed 200 ppm and tungstate 10 ppm; Pb, Bi, Ba, and Sb interfere by giving a turbidity with sulphuric acid, and Sn(II), nitrate, and arsenate should be absent. Arsenic, antimony, selenium, and tin can be volatilized with a mixture of hydrobromic acid and either sulphuric or perchloric acid.

REAGENTS

Molybdate solution. Dissolve 25 g of $Na_2MoO_4.2H_2O$ in 10N sulphuric acid and dilute to 1 litre with 10N sulphuric acid.

Hydrazine sulphate solution. Dissolve 1·5 g of hydrazine sulphate in water and dilute to 1 litre.

Standard phosphate solution. Dissolve 0·2197 g of potassium dihydrogen phosphate in water and dilute to 1 litre. Each ml is equivalent to 0·05 mg phosphorus.

The basic procedure can be readily modified to suit certain materials. For instance, it is used for irons and steels, where no element usually present will interfere with the exception of arsenic and tungsten. Decomposition with aqua regia and evaporation to fumes with perchloric acid is followed by addition of 15 ml of 10% sodium sulphite solution and boiling prior to the addition of ammonium molybdate and hydrazine sulphate. Higher phosphorus contents, 0·05–0·3%, are measured at 700 nm.

The range and specificity of the molybdenum blue method can be improved by extracting phosphorus as the yellow molybdophosphoric acid into n-butyl or isobutyl alcohol, or butyl acetate, and reducing it to the heteropoly blue complex with stannous chloride or ascorbic acid [6, 15, 16, 18]. Typical of such procedures is the one outlined below, by which as little as 1 ppm of phosphorus in nickel can be determined [15].

To 0·5 g of nickel add a little nitric acid, 6 ml of perchloric acid, and evaporate to dense fumes of the latter. Cool, add 1 ml of 10% hydrobromic acid, and evaporate to dense fumes of perchloric acid to eliminate arsenic. Cool, add 5 ml of 10% sodium molybdate solution, and 30 ml of water. Transfer to a small separatory funnel and extract with 20 ml butyl alcohol for 1 minute. Allow the layers to separate, and discard the aqueous phase. Shake for 1 minute with successive 20, 20, and 10 ml portions of water, discarding the washings. Add 15 ml of chlorostannous acid made by dissolving 2·38 g of $SnCl_2.2H_2O$ in 170 ml of concentrated hydrochloric acid and diluting to 1 litre with water. Shake for 15 seconds, and discard the aqueous layer. Transfer the organic phase to a 25 ml volumetric flask, dilute to volume with butyl alcohol, and read the absorbance at 630 nm.

SPECIAL PROCEDURES

1. An indirect method for the determination of the low concentrations

of sodium hexametaphosphate used in threshold water treatment has been described [20]. It depends on the fact that hexametaphosphate does not respond to the ordinary colorimetric procedures for orthophosphate, but it can be converted into orthophosphate by maintaining just below the boiling point 100 ml of water with 12 ml of 4N sulphuric acid for 4 hours. Metaphosphate is determined as the difference between the orthophosphate found in the water as it is sampled and that found after conversion of the metaphosphate to orthophosphate.

In some treated water, tannins and their oxidation products interfere in the colorimetric determination of phosphate. Tannin can be removed, with practically no absorption of phosphate, by adding to 50 ml of water, in turn, two 0·7 g portions of a decolorizing charcoal, shaking, filtering, and discarding the first 15 ml of the filtrate [11].

2. A gravimetric procedure for phosphorus in the presence of arsenic has been described [12]. After an extraction of molybdophosphoric acid with n-butyl acetate, phosphorus is transferred to an aqueous layer by shaking with dilute ammonium hydroxide, added to an ammonium molybdate solution, digested for an hour at 50°C, filtered through a sintered glass crucible, dried at 280°C, and weighed as the ammonium-12-molybdophosphate, $(NH_4)_3$-$PMo_{12}O_{40}$.

REFERENCES

1 ABBOTT, D. C., EMSDEN, G. E., and HARRIS, J. R., *Analyst* **88**, 814–16 (1963).
2. A.O.A.C., *Official Methods of Analysis of the Association of Official Analytical Chemists*, Washington, D.C., Association of Official Analytical Chemists, 1965.
3. ARCHER, D. W., HESLOP, R. B., and KIRBY, R., *Anal. Chim. Acta* **30**, 450–9 (1964).
4 A.S.T.M., *Chemical Analysis of Metals. Sampling and Analysis of Metal Bearing Ores*, Philadelphia, Pa., American Society for Testing and Materials, 1969.
5. BARABAS, S., and LEA, S. G., *Anal. Chem.* **37**, 1132–5 (1965).
6. BAUER, G. A., *Anal. Chem.* **37**, 154–5 (1965).
7. BOLTZ, D. F., *Colorimetric Determination of Nonmetals*, New York, Interscience Publishers, 1958.
8. BRITISH ALUMINIUM CO., *Analysis of Aluminium and its Alloys*, London, British Aluminium Co., 1961.
9. ELWELL, W. T., and SCHOLES, I. R., *Analysis of Copper and its Alloys*, Oxford, Pergamon, 1967.
10. FURMAN, N. H., ed., *Scott's Standard Methods of Chemical Analysis*, 6th ed., Vol. 1, Princeton, N.J., Van Nostrand, 1962.
11. GOLDMAN, L., and LOVE, R. N., *U.S. Bur. Mines Rept. Invest.* 3983 (1946).
12. HESLOP, R. B., and PEARSON, E. F., *Anal. Chim. Acta* **39**, 209–21 (1967).
13. HILLEBRAND, W. F., LUNDELL, G. E. F., BRIGHT, H. A., and HOFFMAN, J. I., *Applied Inorganic Analysis*, 2nd ed., New York, Wiley, 1953.

14. KOLTHOFF, I. M., and ELVING, P. J., *Treatise on Analytical Chemistry*, Part II, Vol. 5, 317–402, New York, Interscience Publishers, 1961.
15. LEWIS, C. L., OTT, W. L., and SINE, N. M., *The Analysis of Nickel*, Oxford, Pergamon, 1966.
16. PAKALNS, P., *Anal. Chim. Acta* **40,** 1–12 (1968).
17. SNELL, F. D., and SNELL, C. T., *Colorimetric Methods of Analysis*, Vol. II and IIA, New York, Van Nostrand, 1949, 1959.
18. THEAKSTON, H. M., and BANDI, W. R., *Anal. Chem.* **38**, 1764–7 (1966).
19. WILSON, A. D., *Analyst* **88**, 18–25 (1963).
20. YOUNG, R. S., and GOLLEDGE, A., *Industrial Chemist* **XXVI**, 13–14 (1950).

29 Platinum Metals

The platinum metals may be found in ores, concentrates, mattes, speiss, anode copper, lead bullion, refinery slimes, jewellery, catalysts, and a variety of miscellaneous industrial products. The most important sources of the platinum metals are copper-nickel sulphide ores, where minerals such as sperrylite $PtAs_2$, cooperite PtS, braggite Pt_2PdNiS_4, laurite RuS_2, stibiopalladinite Pd_3Sb, michenerite and froodite PdBi, and others may occur. Platinum is also found in alluvial deposits in the form of an alloy with smaller quantities of the other platinum metals. Osmiridium or iridosmine, a native alloy containing as principal constituents iridium and osmium, occurs in some gold ores and in alluvial deposits containing platinum.

Platinum and palladium are encountered much more frequently than the other members of the platinum group. Rhodium, iridium, ruthenium, and osmium are rarely determined outside the precious metals industry, except in certain electrolytic slimes and enriched mattes and residues of base metal refineries.

ISOLATION OR SEPARATION OF PLATINUM METALS

The platinum metals belong to the hydrogen sulphide group; all can be precipitated by prolonged treatment with hydrogen sulphide in hot acid solution. With the exception of iridium, these metals will precipitate in hot 0·5N hydrochloric acid on passing in hydrogen sulphide for about 30 minutes; iridium requires about 3N hydrochloric acid and a 3-hour gassing period. If the quantity of iridium is small, an addition of 0·5 g of $AlCl_3.6H_2O$ per 100 ml will aid the coagulation of colloidal iridium sulphide.

Separations of the platinum metals from other elements of the hydrogen sulphide group can be made in several ways. Silver and Hg(I) may be removed as chlorides. Mercury may be volatilized by igniting or fusing with sodium carbonate, or even by repeated evaporation in a shallow, uncovered beaker with hydrochloric acid. Selenium and tellurium, with gold, can be precipitated with sulphur dioxide in hydrochloric acid solution. Gold can also be precipitated by oxalic acid at *p*H 4–6·5. Arsenic, antimony, and germanium can be separated by distillation in hydrochloric acid. Arsenic, antimony, tin, and selenium can be eliminated by several evaporations with sulphuric and hydrobromic acids. Molybdenum and gold can be removed with an ether extraction; gold can also be separated from platinum metals by precipitation of the former with sodium nitrite or hydroquinone. The other elements of the acid sulphide group, Cu, Cd, In, Sn, Pb, and Bi can be separated by hydrolysis with sodium nitrite. The platinum group metals, except palladium, form soluble complex nitrites which are not precipitated when the solution is carefully made neutral or slightly alkaline with sodium hydroxide and sodium bicarbonate, whereas the base metals are precipitated as hydroxides or carbonates. A re-precipitation of the base metals frees them from traces of the platinum metals, except palladium. The latter is easily removed, subsequently, by precipitation with dimethylglyoxime in dilute acid solution. It should be emphasized that the excellent determination of palladium with dimethylglyoxime must be carried out on a solution which does not contain gold; the reagent precipitates gold as metal.

In ores and other products of extractive metallurgy, the platinum metals are usually isolated from a number of other elements by collection in lead, tin, or a copper-nickel-iron alloy by means of a fire assay [5, 30]. In the classical fire assay, gold and silver in the final bead from the cupellation of a lead button are accompanied by platinum, palladium, rhodium and iridium; most of the ruthenium and osmium are lost in cupellation.

Platinum, palladium, rhodium, iridium, and ruthenium in the form of chloro complexes can be separated from large amounts of copper, iron, and nickel by adsorption of the base metals on a cation exchanger, such as Dowex 50–X8 [5, 10, 16, 30, 33, 37, 38, 46, 47, 52].

1. Determination in Ores, etc. by Fire Assay

Weigh 1–2 assay tons of —100 mesh material; when the ore is low grade a number of fusions may be made, and the lead buttons afterwards combined [9, 12, 35, 45]. If the sulphur content is over 2%, roast the sample before commencing the fusion. Mix with a suitable flux, depending on the type of ore, to obtain a lead button of 50–60 g. Add sufficient pure silver foil to give a final silver content greater than 10 times the combined precious metals.

Fuse in an assay crucible until a quiet melt is obtained, and pour into a cast iron mould. Hammer the lead button free from slag, and scorify until base metals have been removed and the button has been reduced to the proper size for cupellation. Cupel at a temperature of 1000°C.

Place the bead in a 50 ml beaker, add 5 ml sulphuric acid and take to fumes of the latter. The residue from this parting should contain all the gold and members of the platinum group except palladium; practically all of the latter, with silver, dissolves. Wash the residue thoroughly by decantation, saving the washings. Transfer the residue finally, by washing, to a small glazed porcelain cup, dry, ignite, and weigh. This weight represents Au, Pt, Rh, Ir, small quantities of Os and Ru, but only traces of palladium.

Cool the decanted washings from the above separation, dilute to about 100 ml and adjust the acidity to approximately 3%. Add sufficient 0·1% solution of dimethylglyoxime to precipitate all the palladium. It requires about 8 mg of precipitant per mg of palladium. Allow to stand at least an hour, filter off the yellow palladium dimethylglyoxime in a micro Gooch crucible, wash with 1% hydrochloric acid and finally with hot water, and dry at 110°C to constant weight. The weight of palladium dimethylglyoxime × 0·3161 = Pd.

Dissolve the residue of Au + Pt, etc. in aqua regia, and evaporate twice to dryness with hydrochloric acid to eliminate nitric acid. Dissolve in 1–2 ml hydrochloric acid and water, precipitate the gold with sodium nitrite or sulphur dioxide as described in the chapter on Go ld, filter, and wash. Place filter paper and contents in lead foil and add silver to the extent of 3–4 times the weight of gold. Dry, char the paper, and scorify at a temperature of 1000°C. Cupel, clean the bead well, flatten, and anneal. Part in a porcelain cup first with 1 : 6 nitric acid, then with 1 : 1 nitric acid. Wash three times by decantation with hot water, dry, anneal, and weigh as metallic gold.

If it is necessary to determine the individual platinum metals, combine the filtrates, after the precipitation of gold by sodium nitrite, obtained from a number of separate assay fusions. Add 50 ml hydrochloric acid, boil until red fumes are driven off, add small portions of formic acid with continued boiling until evolution of red fumes has ceased. Cool, cement out the platinum metals with several additions of magnesium. Filter, wash with 1% sulphuric acid, transfer to a porcelain crucible, dry, and ignite at 800°C. Cool, digest in 1 : 9 sulphuric acid, filter, wash, transfer to an annealing cup, dry, and ignite. Place the annealing cup in a large nickel crucible, pack the space between cup and crucible walls with charcoal, heat in muffle and weigh as "total platinum metals". Determine the individual metals by chromatographic separation on a cellulose column followed by spectrophotometric measurement [10, 16, 30, 33, 35, 38].

Alternatively, the platinum metals can be determined by spectrographic analysis of the fire assay bead [1, 5, 23, 31, 32].

For concentrates and mattes which may contain large quantities of copper, nickel, or other base metals it is convenient to dissolve a large sample by treatment with boiling hydrochloric acid containing 10–20% ammonium chloride. The acid extract will be free from platinum metals and may be discarded, leaving a small precipitate to be treated by a crucible fusion. A concentrate which contains pyrite should be heated at about 600°C in a closed graphite crucible and cooled out of contact with air. This will convert the pyrite to ferrous sulphide; the latter is soluble in hydrochloric acid.

Separation of platinum metals from a large quantity of base metals can also be carried out by the use of thiourea. The latter, in hot concentrated sulphuric acid, will precipitate platinum metals from most elements which are not precipitated by hydrogen sulphide in acid solution. Add excess thiourea to a cold, concentrated sulphuric acid solution of the platinum metals; heat at 240°C for 30 minutes, cool, dilute, allow to stand hot for 30 minutes, filter, and wash. Repeat the thiourea treatment on the filtrate. A considerable part of the silver, and a part of the gold, are held in solution by the thiourea; the remainder is precipitated with the platinum metals. It is possible, however, to separate small quantities of platinum metals completely from silver by cooling and diluting, without subsequent heating, prior to filtering. Silver is not precipitated, except in traces, in a cool, dilute solution. Filter the precipitate of platinum metals, dissolve in sulphuric and nitric acids, evaporate to fumes of sulphuric acid, and repeat the thiourea separation to remove traces of silver. A similar double precipitation with thiourea will separate nearly all copper from platinum metals when the former occurs in moderate quantities.

The treatment of osmiridium concentrates is described in authoritative reference books [5, 12, 41].

The platinum metals may also be collected in an iron-copper-nickel alloy [5, 10, 29, 36, 37, 38, 47, 49, 50], or in tin [15, 16, 17, 19].

2. Determination in High Grade Materials by Wet Methods

A. Decomposition

There are several ways to decompose a high grade sample of platinum metals [7, 20, 21, 22, 24, 35]. Platinum and palladium dissolve in aqua regia; rhodium, iridium, and ruthenium will dissolve if they are alloyed with about ten times their weight of platinum or palladium. The sulphide and arsenide of platinum must be roasted before they will dissolve in aqua regia.

A good method of bringing a mixture of platinum metal residues into solution is to treat with hydrofluoric acid in a nickel crucible to remove silica, and fuse the silica-free residue with sodium peroxide [35]. The solubility of nickel in hydrofluoric acid is low unless evaporation is prolonged at elevated temperatures.

Another procedure for bringing into solution a complex mixture of platinum metal residues is a combination of aqua regia treatment and barium peroxide fusion. Digest the sample in hot aqua regia for several hours, dilute, filter, wash, dry and ignite the precipitate, and evaporate the filtrate to dryness. Mix the precipitate with barium peroxide in a porcelain crucible and sinter at 800°C for an hour. Cool, dissolve in 1 : 1 hydrochloric acid, and evaporate to dryness. Dissolve in a little dilute hydrochloric acid, precipitate the barium with sulphuric acid, filter, and combine this filtrate with the original filtrate from the aqua regia treatment.

B. Isolation of the Platinum Group

The platinum metals may be separated from base metals, silver, and gold in a variety of ways, depending on the composition of the sample [7, 21, 24, 30]. Prolonged treatment with hydrogen sulphide in acid solution, with the precautions mentioned at the beginning of this chapter, will separate the platinum group from Fe, Ni, Co, etc. [24, 27, 39, 41]. Further isolation of the platinum metals from other members of the acid sulphide group can be carried out as described earlier.

Another isolation procedure consists in distilling off osmium and ruthenium, treating the raffinate with sodium nitrite in faintly alkaline solution to precipitate gold and some of the base metals, and finally passing the filtrate through a resin column such as Zeo-Karb 225 to remove the remaining base metals. In the effluent are Pt, Pd, Rh, and Ir [35].

A further method of separation is to remove gold with sodium nitrite from hydrochloric acid solution by adding sodium carbonate until the solution is alkaline to cresol red, boiling, and filtering. To the filtrate add hydrochloric acid, boil until red fumes are driven off, and add small portions of formic acid until red fumes are no longer evolved from the boiling solution. Cement out the platinum metals with magnesium, as described earlier.

C. Final Determination of Platinum Metals

Depending on the number and quantity of platinum metals present, the final determination of the individual members may take several forms.

Osmium and ruthenium can be determined spectrophotometrically in their distillates by means of thiourea at 480 and 675 nm respectively [35]. If more than 10 mg of ruthenium is present, the distillate can be evaporated to dryness and ruthenium ignited to metal under hydrogen. The remaining metals can be separated chromatographically in a cellulose column and determined as follows :

(a) Palladium can be further isolated by dimethylglyoxime and ignited under hydrogen to metal, or determined photometrically by measurement of the chloride at 470 nm. It can also be determined with alpha-furildioxime [5, 34]. Palladium can also be precipitated with potassium iodide, reduced by placing in a fireclay annealing cup which is placed inside a nickel crucible packed with charcoal, and weighed as metal.

(b) Platinum can be reduced with magnesium from a diluted perchlorate solution and weighed as metal, or determined by photometric measurement in hydrochloric acid solution with stannous chloride at 403 nm. Alternatively, platinum may be precipitated from dilute sulphuric acid solution by formic acid, ignited and weighed as metal.

(c) Iridium may be precipitated with sodium bromate, reduced under hydrogen, and weighed as metal; for less than 10 mg a photometric finish can be used, measuring the absorption at 490 nm of an iridium chloride solution oxidized with chlorine water [35]. Alternatively, iridium can be precipitated from sulphuric acid solution by thiourea, reduced, and weighed as metal.

(d) Rhodium can be precipitated with thiourea in sulphuric acid solution, reduced, and weighed as metal. Small quantities can be determined by the absorption at 515 nm of a dilute hydrochloric acid solution which has been oxidized with a little chlorine water [35].

3. Colorimetric Determination of Platinum and Palladium in Assay Beads

Platinum and palladium are frequently the only members of the platinum group which occur in significant quantities in the gold–silver bead obtained from the fire assay of ores, concentrates, mattes and other products of the mining and metallurgical industry. The determination of these small quantities may be carried out by colorimetric, spectrographic, polarographic, atomic absorption, or other techniques outlined in this chapter or cited in references [1, 2, 3, 4, 5, 6, 8, 11, 13, 14, 18, 26, 34, 37, 43].

The colorimetric methods described below are simple, specific, and reasonably sensitive; they can be used for most samples where a quantitative measure of traces of platinum and palladium in ore materials is required.

Collect the platinum and palladium with the gold and silver by fire assay into a silver bead, containing enough silver to allow solution of platinum and palladium by subsequent parting in 1 : 1 nitric acid. The usual ratio is about 15 parts of silver to 1 part of total noble metals. Dissolve the silver, platinum, and palladium in 1 : 1 nitric acid; the residue is gold, rhodium, and iridium. Most of the ruthenium and osmium have been lost in the previous cupelling step. Any residual ruthenium remains with the precipitate; traces of osmium dissolved in nitric acid are disregarded. Dilute, filter, and wash with water.

Platinum

Dilute the filtrate to about 100 ml, heat to boiling, and precipitate silver with a slight excess of hydrochloric acid. Allow the silver chloride to coagulate, filter, and wash with hot water. Retain the filtrate. Dissolve the silver chloride in a little hot 1 : 1 ammonium hydroxide, dilute to about 100 ml, acidify with nitric acid, and add 1 drop of hydrochloric acid. Boil, filter, and wash as before. Combine the two filtrates from the silver chloride precipitations and evaporate to dryness. Add to the beaker containing the salts 25 ml each of hydrochloric and nitric acids; evaporate to dryness.

If palladium is present in more than traces, remove it in 0·25 M solution of hydrochloric or nitric acid by an excess of an aqueous sodium dimethylglyoximate solution, using about 8 mg of precipitant for every mg of palladium. Stir, place on the steam bath for 15 minutes, and allow to stand 45 minutes to coagulate the orange–yellow precipitate. Filter through a small porcelain filtering crucible, wash with water, dry at 110°C, and weigh. Palladium dimethylglyoximate $\times$ 0·3161 $=$ palladium. If a re-precipitation is necessary, indicated by a bronze or green colour of the precipitate, dissolve the contents of the crucible in 2 ml sulphuric acid and a few drops of nitric acid. Evaporate to fumes of sulphuric acid, cool, add more nitric acid, and again evaporate to fumes of sulphuric acid. Adjust the acidity to about 0·25M in hydrochloric or nitric acid, and reprecipitate the palladium as described above.

To the filtrate after the removal of palladium, add 10 ml of 1 : 1 sulphuric acid and fume nearly to dryness. Add 3 drops of hydrochloric acid, 1 drop of nitric acid, heat to boiling, dilute with a little water, and filter if necessary, washing with water. If necessary, evaporate the solution to below 25 ml, cool, and transfer to a 50-ml volumetric flask. Add 5 ml of hydrochloric acid, and 10 ml of a stannous chloride solution made by dissolving 23 g $SnCl_2.2H_2O$ in 100 ml of 3·5M hydrochloric acid. Dilute, cool to 20°C, and dilute to the mark with water. Mix, and read the absorbance at 403 nm against a blank. The colour is very stable.

For the standard graph, use 0·25–10 ml of a dilute platinum solution containing 0·1 mg of platinum per ml. This can be made by dilution of a chloroplatinic acid solution containing 1 mg of platinum per ml. The latter can be prepared by quickly weighing 0·25 g of commercial chloroplatinic acid containing 40% platinum, dissolving in 50 ml of water and 10 ml of hydrochloric acid, and diluting to 100 ml in a volumetric flask. The exact platinum value of the stronger solution can be found by determining platinum gravimetrically with formic acid [24].

PALLADIUM

Dilute the filtrate from the parting of the silver bead, containing silver, palladium, and platinum to a convenient volume. Withdraw an aliquot containing less than 75μg of palladium; for some samples it will be necessary to take the whole solution.

Add 5 ml of 1 : 1 sulphuric acid, a few drops of nitric acid, and evaporate to fumes of the former. Cool, dilute, again cool, transfer to a small separatory funnel, and dilute to about 50 ml. Add 2 ml of 1% dimethylglyoxime solution in 0·5M sodium hydroxide, shake, and allow to stand for 10 minutes. Extract three times with 5-ml portions of chloroform, shaking for 1 minute each time. Transfer the chloroform layers to a 25-ml volumetric flask, and dilute to the mark with chloroform. Mix, and measure the absorbance at 375 nm against a blank. The chloroform extract is stable for many hours.

The standard curve can be constructed by using aliquots of a palladium solution made in the following manner. Dissolve 0·1042 g of a commercial palladium chloride solution, containing 60% palladium, in 0·1M hydrochloric acid, and dilute to 250 ml with 0·1M hydrochloric acid. Find the exact palladium value of this solution by the dimethylglyoxime gravimetric procedure. Dilute 5 ml of this strong solution to 250 ml with 0·1M hydrochloric acid; 1 ml of this solution will contain about 5 μg of palladium.

4. Other Chemical Procedures

Many other methods for individual metals, which are suitable for certain materials, have been described [5, 28, 30, 42, 44, 51].

5. Optical Spectrography

Small quantities of the platinum metals in many materials can be determined by emission spectrography [1, 5, 23, 30, 31]. The traces occurring in ores must first be concentrated by fire assay [5, 23, 31]. Procedures have been outlined for the determination of the platinum metals in fire assay beads [5, 23] and in pure metals of the platinum group [5, 32].

6. X-Ray Spectrography

The platinum metals can be determined by X-ray fluorescence [1, 5, 30].

This technique covers a higher concentration range than does optical spectrography, which is usually below 1%.

7. Atomic-Absorption Spectrophotometry

Platinum, palladium, and rhodium have been determined by atomic absorption [14, 40, 48]. For platinum, line 2659·4 has a sensitivity of 5 μg/ml; other lines of lower sensitivity are 2598·0, 3064·7, 2646·9, and 2174·7. The most sensitive line of palladium is 2447·9, with a sensitivity of 0·3 μg/ml; others are 2476·4, 3404·6, 2763·1, 3242·7, 3634·7, and 3610·0. With a sensitivity of 0·3 μg/ml the most sensitive line of rhodium is 3434·9; others are 3692·4, 3502·5, 3396·9, 3658·0, and 3700·9. No interferences have been reported for palladium; conflicting statements have appeared for platinum and rhodium, but with the air-acetylene flame it seems unlikely that interferences are serious.

REFERENCES

1. Beamish, F. E., *Talanta* **2,** 244–65 (1959).
2. Beamish, F. E., *Talanta* **12,** 743–72 (1965).
3. Beamish, F. E., *Talanta* **13,** 773–801 (1966).
4. Beamish, F. E., *Talanta* **13,** 1053–68 (1966).
5. Beamish, F. E., *The Analytical Chemistry of the Noble Metals*, Oxford, Pergamon, 1966.
6. Beamish, F. E., *Anal. Chim. Acta* **44,** 253–86 (1969).
7. Beamish, F. E., and McBryde, W. A. E., *Anal. Chem.* **25,** 1613–17 (1953).
8. Bell, C. F., and Rose, D. R., *Talanta* **12,** 696–700 (1965).
9. Bugbee, E. E., *A Textbook of Fire Assaying*, New York, Wiley, 1940.
10. Coburn, H. G., Beamish, F. E., and Lewis, C. L., *Anal. Chem.* **28,** 1297–300 (1956).
11. Dagnall, R. M., El-Ghamry, M. T., and West, T. S., *Talanta* **15,** 1353–7 (1968).
12. Dillon, V. S., *Assay Practice on the Witwatersrand*, Johannesburg, Transvaal and Orange Free State Chamber of Mines, 1955.
13. El-Ghamry, M. T., and Frei, R. W., *Talanta* **16,** 235–43 (1969).
14. Elwell, W. T., and Gidley, J. A. F., *Atomic-Absorption Spectrophotometry*, 2nd ed., Oxford, Pergamon, 1966.
15. Faye, G. H., *Anal. Chem.* **37,** 259–61 (1965).
16. Faye, G. H., and Inman, W. R., *Anal. Chem.* **33,** 278–83 (1961).
17. Faye, G. H., and Inman, W. R., *Anal. Chem.* **34,** 972–4 (1962).
18. Faye, G. H., and Inman, W. R., *Anal. Chem.* **35,** 985–8 (1963).
19. Faye, G. H., Inman, W. R., and Moloughney, P. E., *Anal. Chem.* **36,** 366–8 (1964).
20. Gabbe, D. R., and Hume, D. N., *Anal. Chim. Acta* **30,** 308–9 (1964).
21. Gilchrist, R., *Anal. Chem.* **25,** 1617–21 (1953).
22. Grimaldi, F. S., and Schnepfe, M. M., *U.S. Geol. Survey Prof. Paper* No. 600B, B99–103 (1968).
23. Haffty, J., and Riley, L. B., *Talanta* **15,** 111–17 (1968).

24. HILLEBRAND, W. F., LUNDELL, G. E. F., BRIGHT, H. A., and HOFFMAN, J. I., *Applied Inorganic Analysis*, 2nd ed., New York, Wiley, 1953.
25. HOFFMAN, I., and BEAMISH, F. E., *Anal. Chem.* **28,** 1188–93 (1956).
26. HOLLAND, W. J., and BOZIC, J., *Anal. Chem.* **40,** 433–4 (1968).
27. JACKSON, D. S., and BEAMISH, F. E., *Anal. Chem.* **22,** 813–17 (1950).
28. JAMES, G. S., *South African Industrial Chemist* **15,** No. 4, 62–8 (1961).
29. KAVANAGH, J. M., and BEAMISH, F. E., *Anal. Chem.* **32,** 490–91 (1960).
30. KOLTHOFF, I. M., and ELVING, P. J., *Treatise on Analytical Chemistry*, Part II, Vol. 8, 379–522, New York, Interscience Publishers, 1963.
31. LEWIS, C. L., *Can. Mining Met. Bull.* **50,** No. 539, 163–7 (1957).
32. LINCOLN, A. J., and KOHLER, J. C., *Anal. Chem.* **34**, 1247–51 (1962).
33. MACNEVIN, W. M., and CRUMMETT, W. B., *Anal. Chem.* **25,** 1628–30 (1953).
34. MENIS, O., and RAINS, T. C., *Anal. Chem.* **27**, 1932–4 (1955).
35. PAYNE, S. T., *Analyst* **85,** 698–714 (1960).
36. PLUMMER, M. E. V., and BEAMISH, F. E., *Anal. Chem.* **31,** 1141–3 (1959).
37. PLUMMER, M. E. V., LEWIS, C. L., and BEAMISH, F. E., *Anal. Chem.* **31,** 254–8 (1959).
38. SANT, B. R., and BEAMISH, F. E., *Anal. Chem.* **33,** 304–5 (1961).
39. SANT, S. B., CHOW, A., and BEAMISH, F. E., *Anal. Chem.* **33**, 1257–60 (1961).
40. SCHNEPFE, M. M., and GRIMALDI, F. S., *Talanta* **16,** 591–5 (1969).
41. SCHOELLER, W. R., and POWELL, A. R., *The Analysis of Minerals and Ores of the Rarer Elements*, 3rd ed., London, Charles Griffin, 1955.
42. SEGAR, G. A., *Analyst* **87,** 230–2 (1962).
43. SEN GUPTA, J. G., *Anal. Chem.* **39,** 18–22 (1967).
44. SEN GUPTA, J. G., and BEAMISH, F. E., *Anal. Chem.* **34,** 1761–4 (1962).
45. SMITH, E. A., *The Sampling and Assay of the Precious Metals*, London, Charles Griffin, 1947.
46. TERTIPIS, G. G., and BEAMISH, F. E., *Anal. Chem.* **32,** 486–9 (1960).
47. TERTIPIS, G. G., and BEAMISH, F. E., *Anal. Chem.* **34,** 108–10 (1962).
48. VAN LOON, J. C., *Z. anal. Chemie* **246,** 122–4 (1969).
49. VAN LOON, J. C., and BEAMISH, F. E., *Anal. Chem.* **36,** 872–5 (1964).
50. VAN LOON, J. C., and BEAMISH, F. E., *Anal. Chem.* **37,** 113–6 (1965).
51. YOUNG, R. S., *Analyst* **76,** 49–52 (1951).
52. ZACHARIASEN, H., and BEAMISH, F. E., *Anal. Chem.* **34,** 964–6 (1962).

30 Potassium and Sodium

Potassium and sodium are found in rocks, minerals, ores, slags, flue dusts, soils, fertilizers, brines, building materials, glass, refractories, and a variety of important tonnage chemicals such as salt, caustic soda, soda ash, etc. Their determination is a common occurrence in many laboratories serving the mineral industries.

ISOLATION OR SEPARATION OF POTASSIUM AND SODIUM

Potassium and sodium, with the other members of the alkali metal group — lithium, rubidium, and caesium — can be separated from members of preceding groups by precipitation of the latter with acids, with hydrogen sulphide in acid and ammoniacal solution, with ammonium carbonate, and with ammonium phosphate. They are consequently found in the final filtrates of a complete analysis. Lithium has already been discussed in a brief chapter; rubidium and caesium are extremely rare and methods for their determination may be found elsewhere [7, 10, 12].

If a chemical procedure is used, potassium and sodium are usually weighed together as chlorides, potassium is separated as the perchlorate or chloroplatinate, and sodium obtained by difference. When potassium and sodium are determined on a routine basis, flame photometry or atomic absorption are commonly employed. Careful blanks must be carried along with analyses for small quantities of these elements, because traces of sodium and potassium are common in reagents, glassware, detergents, and waters.

In many industrial chemicals the sodium or potassium content is obtained indirectly; examples are an acid titration for alkali hydroxide or carbonate, sulphur determination for sulphate or sulphide, and a chlorine determination for chlorides of the alkalies.

1. Gravimetric Method

A standard method for potassium and sodium in ores and similar samples is their conversion to chlorides by the J. Lawrence Smith decomposition, removal of calcium and sulphate, and weighing as combined chlorides. Potassium is then separated by the insolubility of its chloroplatinate or perchlorate in ethyl alcohol, and sodium is obtained by difference [7, 10]. If the sample cannot be fused directly in a platinum crucible, it is treated with acids, filtered, washed, and the residue decomposed with ammonium chloride and calcium carbonate as described. The original filtrate from the acid treatment is treated with hydrogen sulphide, ammonium sulphide, ammonium carbonate, and ammonium phosphate as required, to remove all members of other groups from the alkalies; the final filtrate is combined with the aqueous extract of the residue after fusion of the latter with ammonium chloride and calcium carbonate.

Triturate 0·5–1·0 g of the finely ground sample, or the residue from acid treatment, with an equal quantity of pure ammonium chloride in an agate mortar over glazed paper, add 3–6 g pure calcium carbonate, and mix intimately. Use the special, low-alkali grades of ammonium chloride and calcium carbonate which are made for these fusions. Transfer the mixture to a J. Lawrence Smith platinum crucible, rinse the mortar and pestle with 1–2 g of calcium carbonate and add it to the contents of the crucible. Line the bottom of the crucible initially with a little calcium carbonate.

Place the covered crucible in a slightly inclined position with the bottom projecting through a hole in an asbestos board, and the top protected from the heat of the flame or element by the board. Heat the crucible gradually for about 15 minutes until no more ammonia is evolved, but avoid heating sufficiently to cause the evolution of ammonium chloride. Raise the temperature gradually until the lower three-quarters of the crucible is brought to red heat, and maintain this temperature for 1 hour. The following reactions occur:

$$CaCO_3 + 2NH_4Cl \rightarrow CaCl_2 + 2NH_3 + H_2O + CO_2$$

$$2KAlSi_3O_8 + CaCl_2 + 6CaCO_3 \rightarrow 6CaSiO_3 + Ca(AlO_2)_2 + 2KCl + 6CO_2$$

Allow the crucible to cool, and remove the sintered cake by digesting in warm water in a porcelain or platinum dish. Heat the covered dish with 50–75 ml of water for 30 minutes, reduce the large particles to a fine powder by rubbing with a pestle in the dish, and decant the clear solution through a filter. Wash the residue four times by decantation, transfer to a filter, and wash with hot water until free of chlorides. To make sure that decomposition of the mineral has been complete, treat the residue with hydrochloric acid; no trace of undecomposed mineral should remain.

Treat the aqueous extract obtained in the above operation with ammonium hydroxide and ammonium carbonate to remove the calcium. Boil, filter, wash, re-dissolve in hydrochloric acid, and repeat the precipitation. If sulphur is present, remove it with 10% barium chloride solution before the second precipitation of calcium. Remove the excess of barium by ammonium carbonate with the second precipitation of calcium.

Evaporate the combined filtrates to dryness, preferably in a large platinum dish, and carefully drive off the ammonium salts by gentle ignition. Dissolve the residue in a little water and precipitate the last traces of calcium by the addition of a few drops of ammonium hydroxide and ammonium oxalate. Allow to stand several hours and filter off any calcium oxalate, washing with 0·1% ammonium oxalate.

Evaporate the filtrate to dryness in a weighed platinum dish and gently ignite to remove ammonium salts. Moisten the cooled mass with hydrochloric acid to transform any carbonate to chloride, evaporate to dryness and ignite gently. Conduct this evaporation and ignition carefully over a steam bath, in an oven at 110–130°C, and finally over a hot plate. Remove the last traces of ammonium salts by waving a flame back and forth beneath the platinum dish. Do not heat the latter to more than faint redness in any one spot; the salt must not be heated higher than the first sign of incipient fusion.

Cool the dish in a desiccator and weigh the combined chlorides, NaCl + KCl. Dissolve in water; if an insoluble residue remains, filter, weigh, and deduct from the weight of the alkali chlorides.

A. Potassium as Chloroplatinate

In an atmosphere which is free of ammonia fumes, treat the aqueous solution of the combined alkali chlorides in a small glass or porcelain vessel with slightly more than enough chloroplatinic acid to convert all the chlorides present into the corresponding chloroplatinates [2, 7, 10]. The reaction may be expressed:

$$H_2PtCl_6.H_2O + 2KCl \rightarrow K_2PtCl_6 + 2HCl + H_2O$$

The chloroplatinic acid solution is made by dissolving a weighed quantity of platinum foil in aqua regia diluted with an equal volume of water. Evaporate the solution nearly to dryness, add a large excess of hydrochloric acid, and again evaporate almost to dryness. Repeat the evaporation with hydrochloric acid 3–4 times to get rid of any nitrogen compounds; dilute with 1% hydrochloric acid so that the resulting solution will contain 10% platinum.

Evaporate the solution of alkali chloroplatinates on the steam bath to a syrupy consistency or until solidification occurs on cooling. Flood the cooled residue with a small quantity of 80% ethyl alcohol, grind thoroughly with the flattened end of a stirring rod, and allow to stand 30 minutes. Pour the liquid, which contains the sodium chloroplatinate, through a previously weighed filtering crucible. Rub the residue again with the glass pestle, adding more alcohol, and wash by decantation with small portions of alcohol until the wash liquid becomes colourless. Transfer the precipitate to the crucible and wash several times with alcohol. Dry in an oven for 1 hour at 110°C, cool in a desiccator, and weigh as K_2PtCl_6. Calculate the KCl equivalent to the K_2PtCl_6 and subtract this from the original residue of KCl + NaCl to give the NaCl.

$$K_2PtCl_6 \times 0{\cdot}3067 = KCl$$

$$K_2PtCl_6 \times 0{\cdot}1608 = K$$

B. Potassium as Perchlorate

In an atmosphere free of ammonia fumes, add to the neutral or slightly acidified solution of combined alkali chlorides twice as much perchloric acid as is required to convert the bases into perchlorates [2, 7, 10]. Evaporate on the steam bath, with stirring, to a syrupy consistency; add a little hot water, and continue stirring until all the hydrochloric acid is expelled and heavy fumes of perchloric acid are given off. Cool, stir the crystallized mass with 20 ml of 97% ethyl alcohol to which 0·2% perchloric acid has been added; do not break up the crystals of potassium perchlorate so fine that they pass through the filter. Allow the mixture to settle and decant the alcohol through a filtering crucible. Repeat the washing by decantation once, and then warm to remove the alcohol.

Dissolve the residue in hot water, add 1 ml perchloric acid, and evaporate again until fumes of the latter are evolved. Using 97% ethyl alcohol containing 0·2% perchloric acid, wash the residue once by decantation and then several times on the filter. Remove the adhering wash liquid by washing with 97% ethyl alcohol, dry in an oven at 110°C, and finally heat in a muffle for 15 minutes at 350°C. Cool in a desiccator and weigh as $KClO_4$. Subtract the equivalent weight of potassium chloride from the combined KCl + NaCl residue to obtain the sodium.

$$KClO_4 \times 0{\cdot}5381 = KCl$$

$$KClO_4 \times 0{\cdot}2822 = K$$

In both the perchlorate and chloroplatinate methods, lithium accompanies sodium, whereas rubidium and caesium are precipitated with the potassium.

When a small amount of sodium occurs in the presence of a large quantity of potassium, or *vice-versa*, it may be advantageous to concentrate one before a final separation is attempted. Evaporate the aqueous solution as far as possible, without causing deposition, set the vessel in ice water, and saturate the solution with hydrogen chloride gas. Wash the delivery tube and the inside of the flask into the solution, and stir to break up lumps. Filter, using gentle suction; transfer the precipitate to the filter with ice-cold water saturated with hydrogen chloride gas, and wash several times with small portions of this liquid. When a small quantity of potassium is present in a large amount of sodium, the potassium will be found in the filtrate along with some sodium and salts of other metals originally present. When a small quantity of sodium is present in a large amount of potassium, the sodium will be found in the filtrate along with some potassium.

C. Sodium

(1) *By difference*

Usually sodium and potassium are weighed together as chlorides, and potassium determined by the chloroplatinate or perchlorate methods. The sodium chloride is determined by difference; NaCl × 0·3934 = Na.

(2) *Directly on the filtrate from the potassium determination*

(a) Evaporate the final filtrate from the chloroplatinate method for potassium on a water bath to remove most of the alcohol. Dilute to 100 ml, add 5 ml formic acid, and boil until a brown colour indicating initial reduction of platinum is obtained. Add a slight excess of ammonium hydroxide and continue boiling for a few minutes. Allow the sample to stand in a warm place until the platinum has settled; filter hot and wash free of chloride. Evaporate the filtrate to dryness on the water bath and ignite gently to remove ammonium salts and organic matter. Dissolve in water, filter into a weighed platinum dish, and determine the sodium as NaCl in the same manner as described earlier for the mixed chlorides.

(b) Evaporate the alcohol from the final filtrate of the perchlorate method for potassium on a hot plate, and cool. Add from a burette with constant stirring 1·5 ml of n-butyl alcohol containing 20% of hydrogen chloride gas, and then 7 ml more to form a 6% hydrochloric acid solution. Heat to boiling to coagulate the sodium chloride, cool, filter on a weighed filtering crucible and wash 8–10 times with small portions of n-butyl alcohol containing 7% hydrogen chloride gas. Dry the crucible at 250°C for an hour and ignite at dull red heat. Cool and weigh as impure sodium chloride. Wash the precipitate through the crucible with hot water, dry, ignite, cool, and weigh. The loss in weight represents pure sodium chloride.

(3) *As sodium zinc uranyl acetate*

This method is applicable to small quantities of sodium; moderate amounts of K, NH_4, Mg, Ca, and Ba do not interfere, but silicates, phosphates, oxalates, arsenates, and molybdates should be absent [2, 7, 10, 20]. The solution should preferably be a chloride, but sulphate may be present in the absence of Ca, Sr, and Ba.

Separate any interfering elements with ammonium sulphide and ammonium carbonate, and take an aliquot if necessary so that the final sample will not contain more than 0·02 g sodium and 0·05 g potassium. Evaporate the neutral solution to a low volume; add at least 10 ml zinc uranyl acetate reagent for each ml of the solution. Stir, and allow to stand at 20°C for 45 minutes, with occasional stirring. Filter through a weighed Gooch crucible, and wash 5–10 times with 2 ml portions of the zinc uranyl acetate reagent. Wash 5 times with 2 ml portions of 95% ethyl alcohol saturated with sodium zinc uranyl acetate at 20°C. Finally wash thoroughly with ether to remove the alcohol, and draw dry air through the filter to remove the ether. Dry the precipitate at 40°C for 15 minutes, place in a desiccator for a few minutes, and weigh as sodium zinc uranyl acetate. $NaZn(UO_2)_3.(C_2H_3O_2)_9 \times 0{\cdot}01495 =$ Na.

REAGENTS

Zinc uranyl acetate

Solution A. Weigh 77 g $UO_2(C_2H_3O_2)_2.2H_2O$ into a 600-ml beaker, add 13·3 ml acetic acid and 410 ml of water. Heat to 70°C and stir until solution is complete.

Solution B. Weigh 231 g $Zn(C_2H_3O_2)_2.3H_2O$ into a 600-ml beaker. Add 6·6 ml acetic acid, 262 ml water, warm to 70°C, and stir to complete solution.

Mix solutions *A* and *B* at about 70°C and allow to stand 24 hours. Filter into a clean dry bottle. The reagent is stable but should be filtered before use if not perfectly clear.

Uranyl zinc acetate is available from some chemical supply houses. It requires approximately 6·6 g of this reagent to react with 0·1 g sodium. The reaction may be expressed:

$$3UO_2(C_2H_3O_2)_2.2H_2O + 2Zn(C_2H_3O_2)_2.3H_2O + 2NaCl \rightarrow$$
$$NaZn(UO_2)_3(C_2H_3O_2)_9.6H_2O + ZnCl_2 + NaC_2H_3O_2.3H_2O + 3H_2O$$

2. Flame Photometric Method

Potassium and sodium can be determined rapidly in routine work by flame photometry, measuring the light intensities of atomized solutions in a spectrophotometer or similar instrument, in comparison with a set of standards [1, 2, 5, 6, 9, 11, 12, 13, 14, 16, 17, 18, 19, 22]. Nearly all potassium and sodium determinations are now carried out by flame photometry or atomic absorption.

To 0·5–2 g sample in a platinum dish, add 5 ml 1 : 1 sulphuric acid and 15 ml hydrofluoric acid; evaporate to dryness, ignite the dish at dull redness for 30 minutes, and cool. Transfer the residue by washing to a beaker, dilute to about 200 ml with water, and digest on the hot plate for 30 minutes. Add a little hydrogen peroxide, boil, add ammonium hydroxide to precipitate Fe, Al, Ti, etc. together with the residue of calcium sulphate and ignited oxides. Boil, filter, wash with hot 2% ammonium chloride which has been made neutral to methyl red with ammonium hydroxide. Acidify the filtrate with hydrochloric acid, cool, make up to a definite volume, and aspirate in the flame photometer. Determine sodium at 588 nm and potassium at 765 nm against a series of standards from the following solutions:

Dissolve 2·5421 g sodium chloride in water and make up to 1 litre; 1 ml = 1 mg Na.

Dissolve 1·9066 g potassium chloride in water and make up to 1 litre; 1 ml = 1 mg K.

Potassium gives a straight-line relation between concentration and transmittancy up to about 2 mg potassium per 100 ml. Sodium, above small quantities, does not give a straight-line relation, and it is necessary to use a different set of standards for sodium concentrations which extend over any appreciable range.

For some samples, such as coal ash, it has been found that calcium interference with the sodium determination in flame photometry can be eliminated by having aluminium in an amount such that the aluminium-calcium ratio is at least 3. In this case, there is no need to remove elements after the filtration of the residue from the hydrofluoric–sulphuric acid decomposition [11].

3. Optical Spectrography

Low concentrations of potassium and sodium in many materials can be determined by emission spectroscopy [2, 3, 4, 12, 13, 15]. Detailed procedures have been published for agricultural products [2, 15], 5–500 ppm of sodium in aluminium [3], and 25–200 ppm in nickel [13].

4. X-Ray Spectrography

Potassium can be determined in some samples by X-ray fluorescence; the technique is not suitable for sodium [12].

5. Atomic – Absorption Spectrophotometry

Atomic absorption may be used as an alternative technique to flame photometry for potassium and sodium. The same procedures already described under **2** above, for decomposition, separation, and preparation of the sample solution are used for atomic absorption.

The most sensitive potassium line is 7664·9, with a sensitivity of 0·03 μg/ml; line 4044·1 is sometimes used. The sodium doublet, 5890·0 and 5895·9, with a sensitivity of 0·03 μg/ml, is generally used; the doublet 3302·3 and 3302·9 may also be employed [18].

Potassium can be determined by atomic absorption in the presence of a large excess of sodium; in the determination of sodium by this technique, a quantity of potassium equal to that of sodium has no effect but larger amounts enhance the absorbance reading.

6. Sodium – Ion Electrode

Low levels of sodium in water can be determined by the use of a glass electrode which is responsive to sodium ions [8, 21].

REFERENCES

1. Adams, P. B., *Anal. Chem.* **33,** 1602–5 (1961).
2. A.O.A.C., *Official Methods of Analysis of the Association of Official Analytical Chemists*, Washington, D.C., Association of Official Analytical Chemists, 1965.
3. A.S.T.M., *Chemical Analysis of Metals. Sampling and Analysis of Metal Bearing Ores*, Philadelphia, Pa., American Society for Testing and Materials, 1969.
4. A.S.T.M., *Methods for Emission Spectrochemical Analysis*, 5th ed., Philadelphia, Pa., American Society for Testing and Materials, 1968.
5. Burriel-Marti, F., and Ramirez-Munoz, J., *Flame Photometry*, Amsterdam, Elsevier, 1957.
6. Dean, J. A., *Flame Photometry*, New York, McGraw-Hill, 1960.
7. Furman, N. H., ed., *Scott's Standard Methods of Chemical Analysis*, 6th ed., Vol. 1, Princeton, N.J., Van Nostrand, 1962.

8. HAWTHORN, D., and RAY, N. J., *Analyst* **93,** 158–65 (1968).
9. HERRMANN, R., and ALKEMADE, C. T. J., *Chemical Analysis by Flame Photometry*, New York, Wiley, 1963.
10. HILLEBRAND, W. F., LUNDELL, G. E. F., BRIGHT, H. A., and HOFFMAN, J. I., *Applied Inorganic Analysis*, 2nd ed., New York, Wiley, 1953.
11. JACKSON, P. J., and SMITH, A. C., *J. Applied Chem.* **6,** 547–59 (1956).
12. KOLTHOFF, I. M., and ELVING, P. J., *Treatise on Analytical Chemistry*, Part II, Vol. 1, 301–460, New York, Interscience Publishers, 1961.
13. LEWIS, C. L., OTT, W. L., and SINE, N. M., *The Analysis of Nickel*, Oxford, Pergamon, 1966.
14. MAVRODINEANU, R., and BOITEUX, H., *Flame Spectroscopy*, New York, Wiley, 1965.
15. MITCHELL, R. L., *Tech. Comm.* No. 44A, Commonwealth Bureau of Soils, Harpenden, England, 1964.
16. OSBORN, G. H., and JOHNS, H., *Analyst* **76**, 410–15 (1951).
17. PUNGOR, E., *Flame Photometry Theory*, London, Van Nostrand, 1967.
18. RAMIREZ-MUNOZ, J., *Atomic-Absorption Spectroscopy and Analysis by Atomic-Absorption Flame Photometry*, New York, Elsevier, 1968.
19. ROY, N., *Anal. Chem.* **28**, 34–9 (1956).
20. SITTIG, M., *Sodium: Its Manufacture, Properties and Uses, A.C.S. Monograph* No. 133, New York, Reinhold, 1956.
21. WEBBER, H. M., and WILSON, A. L., *Analyst* **94**, 209–20 (1969).
22. WEBBER, H. M., and WILSON, A. L., *Analyst* **94**, 569–74 (1969).

31 Selenium and Tellurium

Selenium and tellurium are found in some minerals and ores, concentrates, mattes, blister copper, flue dusts, refinery slimes, glasses, a few steels and non-ferrous alloys, photoelectric cells, rectifiers, tellurium lead, and various other products. Most copper ores contain small amounts of selenium and lesser quantities of tellurium; in fact the commercial source of these elements is the anode slimes of copper refineries. Selenium is also found in pyrite, and in rare minerals with copper, lead, mercury, bismuth, and silver. Tellurium occurs in a few minerals with gold, silver, lead, bismuth, iron, and copper. Because these elements frequently occur together, especially in the extractive industries, they will be discussed together in this chapter.

ISOLATION OR SEPARATION OF SELENIUM AND TELLURIUM

Although selenium and tellurium are members of the hydrogen sulphide group, precipitation by hydrogen sulphide in acid solution is not entirely quantitative unless the elements are in their lower state of oxidation and the gassing is prolonged. The sulphides of selenium and tellurium, with As, Sb, Sn, and Mo, can be separated from those of Cu, Cd, Bi, Pb, and Hg by solution of the former in cold ammonium sulphide; in the presence of copper the separation is not complete. Selenium and tellurium, like arsenic, can be separated from antimony and tin by precipitating the former with hydrogen sulphide in 2 parts hydrochloric acid to 1 part of water. Selenium and tellurium, with As, Sb, Bi, Sn, Pb, and P, can be separated from Cu and Mo by collection of the former in a precipitate of ferric hydroxide.

The outstanding separation for these elements is precipitation by sulphur dioxide in hydrochloric acid. In strong acid, only selenium is precipitated; in dilute acid tellurium comes down, and in an intermediate range of acidity both can be precipitated free from nearly all interfering elements. Gold is also precipitated by sulphur dioxide, but not the platinum metals.

1. Gravimetric Method

Selenium and to a lesser extent tellurium are volatilized from strong hydrochloric acid solutions heated above 100°C. Selenium is eliminated, unlike tellurium, by a potassium bisulphate fusion or by repeated evaporation with sulphuric and hydrobromic acids, or even sulphuric and nitric acids.

The following directions are designed for a complex material containing 1–15% selenium or tellurium [1, 15, 16, 25, 30]. If the product is lower or higher in these elements it may be necessary to vary the size of the sample or take a suitable aliquot. When decomposing mattes or sulphide concentrates, care must be taken that the addition of nitric acid to large quantities of chlorate in the presence of sulphide does not give a violent reaction. For anode or blister copper, selenium and tellurium may be collected from a 50–100 g sample in a ferric hydroxide precipitate which is formed by adding an iron salt to the copper nitrate solution, making ammoniacal, and filtering off the ferric hydroxide together with Se, Te, As, Sb, Bi, Sn, Pb, and P. The quantity of iron should be 20 times that of the combined Se + Te etc.

Weigh 0·5–2 g into a 400-ml beaker, add 2–4 g potassium chlorate, 10 ml water, 20–30 ml nitric acid, and allow to stand 10 minutes. Place on a steam bath and evaporate to dryness. Add 15 ml 1 : 1 hydrochloric acid and again evaporate on the steam bath to dryness. Dissolve in 15 ml hydrochloric acid, 30 ml water, and warm on the steam bath until all soluble salts are in solution. Filter off the insoluble residue of SiO_2, AgCl, etc. on Whatman No. 40 paper and wash thoroughly with hot water.

Heat the clear filtrate nearly to boiling, remove from the hot plate, and add hydrochloric acid equal to one-third the volume of solution present. For reduction with sulphur dioxide, selenium and tellurium must be in the quadrivalent state and nitric acid must be absent. The sexavalent compounds are easily reduced to the quadrivalent state by heating with dilute HCl, < 6 N, at a temperature below 100°C until chlorine has been expelled. Nitric acid can be eliminated by careful evaporation with dilute hydrochloric acid at a temperature below 100°C.

Pass in sulphur dioxide gas from a commercial cylinder at a temperature of 15–20°C; red selenium is precipitated first and finally tellurium appears as a black precipitate. Continue the stream of sulphur dioxide until the precipitate coagulates, wash down the gas tube, and allow the beaker to stand at room temperature for at least an hour. Filter through Whatman No. 40 paper and wash several times with cold water. Avoid a prolonged passage of sulphur dioxide into solutions of selenium and tellurium; it is often preferable to add a saturated solution of sulphur dioxide and allow the sample to stand overnight. Selenium and tellurium are precipitated together, free from nearly all interfering elements except gold, in a solution containing 30–40% by volume of concentrated hydrochloric acid. Selenium alone is not quantitatively precipitated unless the concentration is over 28% hydrochloric acid; tellurium is not precipitated quantitatively if the hydrochloric acid concentration is over 42%.

Place the paper and precipitate in the original beaker, add 15 ml nitric acid, and break up the paper by means of a pair of glass rods. Heat on the steam bath for 15 minutes or until the precipitate of selenium and tellurium is entirely dissolved. Dilute to 30 ml with water, filter, and wash thoroughly with numerous small portions of hot water. Evaporate the filtrate to dryness on the steam bath. This treatment separates selenium and tellurium from gold, which would also be precipitated by sulphur dioxide.

Add 100 ml hydrochloric acid, wash down the sides of the beaker with not more than 5–7 ml hot water, warm on the steam bath for several minutes, cool to 15–20°C, and saturate with sulphur dioxide gas. Allow the red precipitate of selenium to settle for several hours, and filter through a weighed Gooch crucible having a thick mat of asbestos. Obtain the tare weight of the crucible by washing the latter with concentrated hydrochloric acid, and heating in the muffle at 725°C before placing in the drying oven.

Wash the precipitate several times with hydrochloric acid, scrubbing the beaker thoroughly, and then wash with cold water until free of chlorides. Retain the filtrate and washings for tellurium. Finally wash several times with alcohol, discarding the alcohol washings. Dry at 110°C for 1–2 hours and weigh as selenium. After weighing, this figure may be checked by igniting the crucible at 725°C to expel the selenium, and re-weighing.

Dilute the filtrate and washings to 350 ml if they are not already up to that volume, so that the concentration of hydrochloric acid is about 25–30% by volume. Heat just to boiling, remove from the hotplate, pass in sulphur dioxide until the solution is saturated, add 10 ml of a 15% solution of hydrazine hydrochloride and allow to stand for several hours on a steam bath. Filter on a weighed Gooch crucible, wash with 1 : 3 hydrochloric acid, then with hot water, and finally with alcohol. Wash tellurium rapidly, because it oxidizes more readily when moist than does selenium. Dry at 110°C for 1–2 hours and weigh as tellurium.

The above procedure is applicable to selenium and tellurium in nearly all materials. For glass, it is necessary to decompose with hydrofluoric and sulphuric acids in a platinum dish. Lead slimes can be fused at a low heat with sodium carbonate and sodium nitrate in a nickel crucible, acidified with hydrochloric and sulphuric acids, filtered to remove silver and lead, and the precipitate re-treated if necessary to recover selenium and tellurium. For copper anode slimes rich in silver, it is advisable to set free occluded selenium and tellurium from silver chloride by solution of the latter in ammonia and re-precipitation [30, 31].

For tellurium in lead, dissolve 100 g in 200 ml water and 100 ml nitric acid, add 70 ml 1 : 1 sulphuric acid, cool, filter, and wash thoroughly with hot water. Retain the precipitate to recover occluded tellurium; evaporate the filtrate to fumes, dilute, filter, and wash, retaining the filtrate. To the combined precipitates add hydrochloric acid, heat, filter, adjust the strength of hydrochloric acid in the filtrate to about 20%, and precipitate tellurium with sulphur dioxide and hydrazine hydrochloride as described. Evaporate the combined filtrates to dryness, add hydrochloric acid, filter if necessary, adjust the concentration of hydrochloric acid to about 20%, and precipitate tellurium as previously outlined.

Small quantities of selenium and tellurium may often be conveniently isolated in a ferric hydroxide precipitate, before final precipitation with sulphur dioxide. If sufficient iron is not present initially in the sample so that it is about 20 times the weight of the combined Se + Te, add a little ferric nitrate to provide this quantity of iron. Make ammoniacal, boil, filter, wash well with hot water, dissolve the precipitate in hydrochloric acid and re-precipitate. For low concentrations of selenium and tellurium, allow the ammoniacal solution to stand for at least an hour, with occasional stirring, before filtering. By this means, small quantities of these elements can be isolated from large amounts of copper, silver, molybdenum, nickel, and cobalt.

2. Volumetric Methods

As a general rule, volumetric methods for selenium and tellurium do not possess advantages over the gravimetric procedure. The elements must be isolated, and separated from one another, in most volumetric methods. The latter have application, however, for certain materials.

A. Iodometric Determination of Selenium

Selenious acid reacts with iodine in acid solution according to the equation:

$$H_2SeO_3 + 4KI + 4HNO_3 \rightarrow 4KNO_3 + Se + 3H_2O + 2I_2$$

The iodine is titrated with sodium thiosulphate, using starch indicator, in the usual manner.

Decompose the sample by appropriate means, taking all precautions to avoid loss of selenium. With stainless steels, for example, dissolve 5 g in 30 ml nitric acid and 30 ml hydrochloric acid, add 35 ml perchloric acid, and heat until a deep red colour appears, which indicates oxidation of chromium [2]. Cool, dilute with 50 ml water, warm, add 200 ml hydrochloric acid, 100 ml 6% sulphurous acid and collect the precipitated selenium on a Gooch as described previously. Dissolve in nitric acid, cool, add 10 ml of 30% potassium iodide, and allow to stand 2–3 minutes. Add 5 ml of 1% starch solution and titrate with 0·01N sodium thiosulphate. One ml of 0·01N sodium thiosulphate = 0·0001974 g Se. The sodium thiosulphate may be standardized against a selenious acid solution containing 0·490 g H_2SeO_3 in 500 ml of water; one ml of this solution contains 0·0006 g Se. Alternatively, the sodium thiosulphate may be standardized by any other means, and the selenium equivalent calculated on the theoretical basis of 1 ml of N $Na_2S_2O_3$ = 0·01974 g Se.

The iodometric determination of selenium in ores and flue dusts in a solution of 7 ml sulphuric acid and 5 g ammonium chloride per 100 ml has been described; it is preceded by distillation of selenium with hydrobromic acid, precipitation of the element and separation from tellurium with sulphur dioxide in strong hydrochloric acid, solution in nitric acid, and destruction of nitric acid with urea in presence of sulphuric acid [26].

B. Permanganate Determination of Selenium

Dilute a solution of selenious acid, or tellurous acid, in 25 ml of 40%

sulphuric acid, to 150 ml and add 12 g of sodium phosphate to prevent the precipitation of manganese dioxide. Add a measured excess of 0·1N potassium permanganate, allow to stand about 30 minutes, and titrate the excess permanganate with 0·1N ferrous ammonium sulphate solution. One ml of 0·1N $KMnO_4$ = 0·003948 g Se, or 0·006381 g Te. The procedure can be used for refined selenium, sodium selenite, sodium selenate, and iron selenide [4]. With *o*-phenanthroline ferrous sulphate as indicator, the colour changes at the end from pale blue to bright red.

C. Dichromate Determination of Tellurium

In an acid solution, tellurous ions are gradually oxidized to the sexavalent state by an excess of standard potassium dichromate. The excess of the latter is determined with standard ferrous ammonium sulphate, potentiometrically or visually. Selenium does not interfere with this titration. One ml of 0·1N $K_2Cr_2O_7$ = 0·006381 g Te. This method is employed directly, for example, with copper-tellurium alloys in nitric–sulphuric acid solution [2], but for most materials tellurium must be isolated.

D. Other Titrimetric Methods

A differential potentiometric determination, using stannous chloride, has been described for selenium and tellurium in refined selenium products and for tellurium in refined tellurium materials [3, 5].

3. Colorimetric Methods

A. Selenium

Selenium can be determined colorimetrically in a variety of samples by measuring the absorbance of a complex formed in dilute acid solution with 3,3′-diaminobenzidine [6, 8, 9, 11, 12, 14, 17, 18, 19, 20, 24, 27, 28]. In a few cases it is possible to determine selenium directly by masking such elements as iron and copper; for most samples it is necessary to isolate selenium by distillation, ion exchange, collection using arsenic or ferric hydroxide as co-precipitant, extraction of the complex with toluene, or by other techniques.

The procedure described below, for traces of selenium in nickel [19], is typical of those employing 3,3′-diaminobenzidine. Dissolve a 5-g sample in 25 ml of a 5 : 1 mixture of perchloric–nitric acid. Evaporate to dense fumes of perchloric acid to remove all the nitric acid. Cool, add 50 ml water, and warm to dissolve salts. Add 50 ml hydrochloric acid, and 2 ml of a sodium arsenite solution made by dissolving 0·25 g of arsenious oxide and 10 pellets of sodium hydroxide in 10 ml of water and diluting to 200 ml. Add 15 ml of 50% hypophosphorous acid, heat to boiling, and boil for 5 minutes.

Cool to 70–80°C and filter carefully through Whatman No. 42 paper: the precipitate has a strong tendency to creep. Wash 4–5 times with 6N hydrochloric acid at room temperature. Return the paper and precipitate to the original beaker; if necessary, wipe the funnel with a piece of filter paper and add to the beaker. Add 20 ml of nitric acid, 10 ml of perchloric acid, and evaporate to about 2 ml to destroy all organic matter. Cool, transfer to a 200-ml beaker, and dilute to 25 ml with water. Add 2 ml of 10% formic acid and 10 ml of 2% EDTA solution.

Adjust the *p*H to 2–3 with 1 : 1 ammonium hydroxide, using a *p*H meter. Add 2 ml of a 0·5% aqueous solution of 3,3′-diaminobenzidine hydrochloride, and allow to stand at room temperature for 30–50 minutes. Use rubber gloves when handling 3,3′-diaminobenzidine; such reagents have carcinogenic properties. Adjust the *p*H to 6–7 with 1 : 1 ammonium hydroxide, and transfer to a 125-ml separatory funnel, keeping the final volume to about 60 ml. Add 20·0 ml toluene from a pipette, shake the funnel for 1 minute, and allow the phases to separate. Drain off and discard the lower aqueous layer and a small amount of the organic layer. Pour the organic layer into a 50-ml Erlenmeyer flask containing about 1 g of anhydrous sodium sulphate. Stopper and swirl to dry the organic phase. Measure the absorbance at 420 nm against a blank.

Similar procedures are used for determining selenium in copper [12, 14, 18, 20], lead [20], biological materials [11], sea water, silicates, and marine organisms [8], metal salts [24], soils and sediments [27], and many other materials [15, 17, 18, 28].

B. Tellurium

Tellurium in minute amounts can be determined in many materials by the procedure described below, which is used for nickel [19, 20]. The tellurium is collected with arsenic after reduction with hypophosphorous acid, complexed with sodium diethyldithiocarbamate in the presence of potassium cyanide, extracted with chloroform, and measured at 415 nm.

Dissolve a 5-g sample in nitric and perchloric acids, as described for selenium under 3A above. Continue with the same steps of collection with arsenic in hydrochloric and hypophosphorous acids, filtration, and destruction of filter paper by evaporation to a low volume with nitric and perchloric acids, which were detailed for selenium.

Add 2 ml sulphuric acid, and evaporate to a volume of about 1 ml. Cool, add 50 ml of water, and again cool. Adjust the *p*H to 7 with 1 : 1 ammonium hydroxide. Add 5 ml of 5% potassium cyanide and continue the addition of 1 : 1 ammonium hydroxide to *p*H 8·5. Pour the solution into a separatory funnel of low actinic glass, and add 2 ml of a 1% aqueous solution of sodium diethyldithiocarbamate. Add 20·0 ml of chloroform, and shake vigorously for 30 seconds. Allow the layers to separate, and drain most of the chloroform phase into a 50 ml conical flask, of low actinic glass, containing approximately 1 g of anhydrous sodium sulphate. Swirl gently and measure the absorbance immediately at 415 nm against a blank.

C. Other Colorimetric Procedures

Selenium has been determined, with few interferences, by means of 2-mercaptobenzoic acid [10]. Tellurium has been determined in copper with bismuthiol II [23], and with thiourea [12]; it has also been measured in metal salts with sym-diphenylthiourea [24].

4. Optical Spectrography

Selenium is rarely determined by emission spectrography [18, 29], but tellurium in some materials, for instance 0·01–0·1% in pig lead, can be readily measured by this technique [2, 18].

5. X-Ray Spectrography

Selenium and tellurium are not particularly suitable for X-ray fluorescence, and are not normally determined by this means [7, 18].

6. Atomic-Absorption Spectrophotometry

Both elements can be determined by atomic absorption, with a maximum sensitivity of about 2 μg/ml. The most sensitive selenium line is 1960·3, with others at 2039·9 and 2062·8; the only tellurium line is 2142·7 [13, 21, 22].

Selenium and Tellurium Development Committee

Chemists who have a special problem in the analysis of selenium or tellurium may enlist the assistance of The Selenium and Tellurium Development Committee, Battelle Memorial Institute, 505 King Avenue, Columbus, Ohio. This organization, sponsored by the leading North American producers of base metals containing small quantities of selenium and tellurium, was established to disseminate knowledge of, and promote sales for, these by-products.

REFERENCES

1. A.O.A.C., *Official Methods of Analysis of the Association of Official Analytical Chemists*, Washington, D.C., Association of Official Analytical Chemists, 1965.
2. A.S.T.M., *Chemical Analysis of Metals. Sampling and Analysis of Metal Bearing Ores*, Philadelphia, Pa., American Society for Testing and Materials, 1969.
3. Barabas, S., and Bennett, P. W., *Anal. Chem.* **35**, 135–8 (1963).
4. Barabas, S., and Cooper, W. C., *Anal. Chem.* **28**, 129–30 (1956).
5. Bennett, P. W., and Barabas, S., *Anal. Chem.* **35**, 139–41 (1963).
6. Broad, W. C., and Barnard, A. J., *Chemist-Analyst* **50**, 124–6 (1961).

7. BURKE, K. E., YANAK, M. M., and ALBRIGHT, C. H., *Anal. Chem.* **39**, 14–18 (1967).
8. CHAU, Y. K., and RILEY, J. P., *Anal. Chim. Acta* **33**, 36–49 (1965).
9. CHENG, K. L., *Anal. Chem.* **28**, 1378–42 (1956).
10. CRESSER, M. S., and WEST, T. S., *Analyst* **93**, 595–600 (1968).
11. CUMMINS, L. M., MARTIN, J. L., MAAG, G. W., and MAAG, D. D., *Anal. Chem.* **36**, 382–4 (1964).
12. DOZINEL, C. M., *Modern Methods of Analysis of Copper and its Alloys*, 2nd ed., London, Elsevier, 1963.
13. ELWELL, W. T., and GIDLEY, J. A. F., *Atomic-Absorption Spectrophotometry*, 2nd ed., Oxford, Pergamon, 1966.
14. ELWELL, W. T., and SCHOLES, I. R., *Analysis of Copper and its Alloys*, Oxford, Pergamon, 1967.
15. FURMAN, N. H., ed., *Scott's Standard Methods of Chemical Analysis*, 6th ed., Vol. 1, Princeton, N. J., Van Nostrand, 1962.
16. HILLEBRAND, W. F., LUNDELL, G. E. F., BRIGHT, H. A., and HOFFMAN, J. I. *Applied Inorganic Analysis*, 2nd ed., New York, Wiley, 1953.
17. HOSTE, J., and GILLIS, J., *Anal. Chim. Acta* **12**, 158–61 (1955).
18. KOLTHOFF, I. M., and ELVING, P. J., *Treatise on Analytical Chemistry*, Part II, Vol. 7, 137–205, New York, Interscience Publishers, 1961.
19. LEWIS, C. L., OTT, W. L., and SINE, N. M., *The Analysis of Nickel*, Oxford, Pergamon, 1966.
20. LUKE, C. L., *Anal. Chem.* **31**, 572–4 (1959).
21. MARIEC, M. V., KINSON, K., and BELCHER, C. B., *Anal. Chim. Acta* **41**, 447–51 (1968).
22. NAKAGAWA, H. M., and THOMPSON, C. E., *U.S. Geol. Survey Prof. Paper*, No. 600B, B123–5 (1968).
23. POLLOCK, E. N., *Anal. Chim. Acta* **40**, 285–90 (1968).
24. RUSSELL, B. G., LUBBE, W. V., WILSON, A., JONES, E., TAYLOR, D. J., and STEELE, T. W., *Talanta* **14**, 957–66 (1967).
25. SCHOELLER, W. R., and POWELL, A. R., *The Analysis of Minerals and Ores of the Rarer Elements*, 3rd ed., London, Charles Griffin, 1955.
26. SILL, C. W., and PETERSON, H. E., *U.S. Bur. Mines Rept. Invest.* 5047 (1954).
27. STANTON, R. E., and MCDONALD, A. J., *Analyst* **90**, 497–9 (1965).
28. VEALE, C. R., *Analyst* **85**, 133–9 (1960).
29. WARING, C. L., WORTHING, H. W., and HAZEL, K. V., *Anal. Chem.* **30**, 1504–6 (1958).
30. YOUNG, R. S., *Anal. Chim. Acta* **4**, 366–85 (1950).
31. YOUNG, R. S., and GOLLEDGE, A., *Ind. Eng. Chem.* **39**, 1299–1300 (1947).

32 Silicon

Silicon is an abundant and widespread element, occurring always in the combined form in nature, and its determination is required in rocks, minerals, ores, tailings, residues, concentrates, slags, flue dusts, fluxes, refractories, metals and alloys, and many other materials in mining and metallurgical laboratories. In irons and steels, and refined metals, silicon is reported as the element; in nearly all other materials of extractive metallurgy it is expressed as silica, SiO_2.

ISOLATION OR SEPARATION OF SILICON

Appreciable quantities are normally separated as an insoluble precipitate of silica which is obtained on dehydration with acids, and determined by the loss in weight after treatment with hydrofluoric acid. Tungsten, niobium, and tantalum accompany silicon in the precipitate after dehydration with acids; only silicon, however, is volatilized by heating with hydrofluoric acid. In the usual silica determination, the only elements which interfere are boron and fluorine. Procedures to eliminate these interferences are given at the end of the gravimetric procedure.

1. Gravimetric Determination

A. True Silica

Weigh out 0·5–5 g, depending on the silicon content, moisten with water, add 10–20 ml nitric acid, 10–20 ml hydrochloric acid, 10 ml 1 : 1 sulphuric acid, and evaporate to strong fumes of the latter [1, 9, 10, 12]. For sulphide samples, add a few drops of bromine or KBr–Br mixture prior to the initial addition of acid. If an insoluble residue remains after acid decomposition, filter, wash, ignite, fuse with sodium carbonate in a platinum crucible, and dissolve the cooled melt in dilute hydrochloric acid. If the sample does not contain sulphides or metals that would injure platinum, decompose directly by a carbonate fusion and dissolve in dilute hydrochloric acid. Silicon carbide is fused with sodium carbonate in a platinum crucible; ferrosilicon and elemental silicon are fused with sodium peroxide in an iron or zirconium crucible.

Dehydrate the silica, to transfer it into a form amenable to filtration and washing, by evaporating to dryness twice with hydrochloric acid, moistening with the latter each time. Evaporation must be continued until the residue is dry; if this is carried out on a steam bath, break up crusts with a glass rod.

Heat the residue in an oven at 110°C for an hour, after initial evaporation on the hot plate or steam bath, to ensure complete dehydration.

Irons and steels are usually dehydrated with sulphuric or perchloric acids in preference to hydrochloric; strong fuming with sulphuric or perchloric acid for 10–15 minutes will effect complete dehydration. Perchloric acid has several advantages in dehydrating silica: it does not give a sparingly soluble ferric sulphate as does sulphuric acid on prolonged dehydration, it decomposes resistant carbides better than sulphuric acid, and the perchlorates of Pb, Ca, Ba, and Sr are soluble, unlike the corresponding sulphates [14].

Following dehydration with hydrochloric, sulphuric, or perchloric acid, cool the sample, add 5 ml hydrochloric acid unless there is sulphuric or perchloric acid present, 50 ml water and boil. Filter through Whatman No. 40 paper, with pulp, wash thoroughly with hot 2% hydrochloric acid and finally with hot water. Scrub the beaker, platinum dish, or casserole thoroughly with a rubber policeman to remove all adhering traces of silica. Retain the filtrate and again evaporate this to dryness twice, moistening twice with hydrochloric acid, and finally heating in an oven at 110°C for an hour. If sulphuric or perchloric acid had been used initially to dehydrate silica, add 5 ml of these respective acids and evaporate the filtrate to strong fumes for 10–15 minutes.

Cool, dissolve the soluble salts in dilute hydrochloric acid if this has been used for dehydration; add water to the samples if sulphuric or perchloric acids were employed for dehydration. Boil, filter, wash, combine the two precipitates in a platinum crucible, dry, ignite slowly, and finally at a temperature of 1200°C. Cool, weigh, add 2 drops of 1 : 1 sulphuric acid, 10 ml hydrofluoric acid, and evaporate on asbestos on the hot plate to dryness. Silica is volatilized in accordance with the equation:

$$SiO_2 + 4HF \rightarrow 2H_2O + SiF_4$$

Ignite slowly in a muffle to 1200°C, cool, and re-weigh. The difference in weights represents SiO_2; $SiO_2 \times 0{\cdot}4675 = Si$.

If dehydration is effected by evaporating to dryness with hydrochloric acid, this is best done on a water bath; some samples have a strong tendency to spatter if evaporated on a hot plate. An efficient water bath can be provided by placing a casserole or large platinum dish containing the sample on a 400-ml beaker which is half filled with water and boiled on the hot plate. Several Hengar granules, or a few pieces of granular lead or Carborundum, in the beaker will prevent bumping. When the sample has evaporated to dryness on the water bath, transfer to an oven at 110°C for an hour to complete the dehydration.

To evaporate a silica sample containing a high iron content rapidly to dryness without spattering, clamp a quartz radiant heater directly above the casserole containing the sample; evaporation by top heating minimizes spattering.

The presence of boron in the silica precipitate will introduce an error in the determination, because it will volatilize with the hydrofluoric acid treatment.

Boron may be volatilized as $B(OCH_3)_3$ with methyl alcohol in hydrochloric acid before the expulsion of silica with hydrofluoric acid. A more convenient procedure is to dehydrate silicic acid in the presence of glycerol and precipitate silica with gelatin [18].

In the presence of fluorine, the silica determination must be modified as follows. Treat 0·5 g sample with 15 ml of 20% perchloric acid saturated with boric acid at 50°C. Heat until fumes of perchloric acid have been evolved for 5 minutes, cool, add a few ml of water, and repeat the fuming. Dilute with 75 ml water, boil, filter, wash with dilute perchloric acid and finally with hot water, and ignite in a platinum crucible. Weigh, volatilize silica with hydrofluoric acid, and re-weigh as usual. The method is based on the fact that when a silicate containing a fluoride is decomposed in the presence of boric acid, the fluorine is expelled as boron trifluoride and no loss of silica occurs [5, 9, 10, 12].

For ferrotungsten or tungsten metal, fuse with sodium peroxide in an iron or zirconium crucible. To the dissolved melt add sulphuric, phosphoric, and perchloric acids; evaporate to fumes of perchloric. Cool, add tartaric acid, filter off the silica as usual, and ignite at 1100–1150°C. Weigh, treat with hydrofluoric acid, etc. ignite and re-weigh [1].

B. "Insoluble" or Crude Silica

In rapid routine analyses of many products of mineral technology, the residue obtained by dehydrating an acid solution of the sample is weighed as crude silica or "insoluble". The latter is usually reported directly, but it may be converted by means of a correction factor to true silica. For slags or other products of fairly constant composition, the method is often satisfactory for works control because the loss of silica through a single evaporation is balanced by the gain from weighing small quantities of substances other than silica in the final precipitate.

Fuse 0·5–1 g sample with sodium carbonate in a small platinum dish or crucible, pour the molten material onto a depression in a Monel metal plate [27], and place the solidified button together with the dish or crucible in a 800-ml beaker. Add a little hydrochloric acid to dissolve the fusion, rinse and remove the platinum dish, and evaporate the contents of the beaker rapidly on the hot plate. Remove, moisten with hydrochloric acid, place on the hot plate and evaporate to dryness; repeat this operation, remove, add 5–10 ml hydrochloric acid, 50–75 ml water and boil. Filter, wash, transfer paper and precipitate to a fireclay crucible, ignite in the muffle at about 800–900°C for 15 minutes, transfer to desiccator plate [28], cool, brush contents onto the watch glass of the balance pan, and weigh.

Various modifications of this rapid silica determination are employed. For example, in iron ores and iron sinter, fusion may be carried out in platinum with a mixture of 10 parts sodium carbonate to 7 of zinc oxide; the resulting button comes away easily from the crucible and dissolves readily.

Silica can be determined rapidly for control purposes, on many products such as metals and alloys, without dehydration by employing gelatin to flocculate silicic acid [4, 11, 18, 20].

2. Colorimetric Determination

There are two common colorimetric procedures for the determination of small quantities of silicon.

A. Molybdisilicic Acid Method

A slightly acid solution of silicic or fluosilicic acid treated with ammonium molybdate forms yellow molybdisilicic acid, which is measured at approximately 400 nm [1, 9, 12]. Phosphorus interferes unless citric acid is added to destroy molybdiphosphoric acid.

A general procedure for a wide variety of materials is to bring the sample into solution by acid treatment or fusion, or both, and dilute to a volume of 250 ml. Withdraw a 10-ml aliquot containing not more than 0·5 mg of silica and pass it through a column of Amberlite IR-120 hydrogen-form cation exchange resin in a 50 ml burette. Catch the effluent in a 100-ml flask and wash the column with six 10-ml portions of water. Iron and other coloured cations are removed by the resin.

Add a drop of phenolphthalein to the flask, then 1 : 1 ammonium hydroxide until pink, followed by 1 : 1 hydrochloric acid until colourless and finally 2 ml excess acid. Add 2 ml of 20% ammonium molybdate solution, dilute to about 90 ml, allow to stand for 10 minutes, and add 4 ml of 10% citric acid solution. Dilute to 100 ml, mix, and read without delay at 400 nm.

The standard curve can be prepared by fusing 0·0759 g of pure silicic acid containing 79·1% silica with 2 g sodium carbonate in a platinum crucible for 15 minutes; dissolve the melt in about 150 ml of water containing 10 ml hydrochloric acid, boil out carbon dioxide, cool, and dilute to 1 litre. Store in a polyethylene bottle. One ml of this solution contains 0·06 mg SiO_2. Prepare a blank at the same time, using 2 g of sodium carbonate.

Modifications of this basic procedure, without removing iron or other elements, are described for silica in aluminium, copper, magnesium, and alloys based on these metals [1].

B. Molybdenum Blue Method

In dilute acid solution, ammonium molybdate reacts with silica to form molybdisilicic acid; if to the latter is added a reducing agent such as sodium bisulphite, ascorbic acid, or stannous chloride a blue complex is formed, termed "molybdenum blue", which can be measured photometrically [1, 2, 3, 8, 9, 12, 13, 21].

For 0·001–0·3% silicon in electronic nickel, dissolve the sample in nitric acid, adjust the acidity, reduce the molybdisilicic acid with stannous chloride and measure the molybdenum blue colour at 765 nm [1]. For 0·005–0·1% silicon in titanium and its alloys, dissolve the sample in hydrofluoric acid, precipitate the titanium by hydrolysis, develop the molybdenum blue colour in the presence of the precipitate, filter, and measure the intensity at 700 nm [1].

From 10 to 200 ppm silicon in zirconium and its alloys can be determined by dissolving in hydrofluoric acid and potassium chlorate; add boric acid, molybdic acid, adjust *p*H to 1·2–1·3, add tartaric acid and a reducing solution of sodium bisulphite + sodium hydroxide + 1-amino-2-naphthol-4-sulphonic acid. Dilute to volume and keep at 23 °C for 20 minutes; measure the blue complex at approximately 815 nm [1].

A similar procedure is used for traces of silicon in copper and its alloys [8]. After solution in aqua regia and addition of ammonium molybdate and citric acid, the molybdisilicic acid is reduced by sodium sulphite, 1-amino-2-naphthol-4-sulphonic acid, and sodium metabisulphite. Measurement is made at 820 mμ; for very low silicon contents the copper is removed initially by electrodeposition.

3. Optical Spectrography

Silicon in some materials can be determined advantageously by emission spectrography. For example, methods have been described for 0·1–0·8% in plain carbon and low alloy steels, 0·02–14% in aluminium and its alloys, and 20–500 ppm in zirconium and its alloys [1]. Silicon is also determined routinely in copper and its alloys [6], and in nickel in the range of 6–4000 ppm [13].

4. X-Ray Spectrography

Silicon is occasionally determined by X-ray spectrography [12, 15].

5. Atomic-Absorption Spectrophotometry

Using the nitrous oxide–acetylene flame, silicon can be determined with a sensitivity of 3 μg/ml, using the line 2516·1; the line 2582·5 may also be employed [7]. Information on interferences is not complete at the time of going to press. The determination of silicon in iron and steel, aluminium alloys, and cement by this technique has been described [19].

SPECIAL PROCEDURES

1. Free Silica, or Quartz

Methods for the differentiation of free silica in rocks and dusts, or quartz in the presence of silicates, have been described [12, 22, 23]. They are based on the fact that hot pyrophosphoric acid dissolves silicate minerals, but has little solvent action on quartz. These techniques vary with the type of material, and results should be checked by microscopic and X-ray investigations.

2. Other Methods

A volumetric determination of silica in ferromanganese and silicomanganese analyses, based on isolation of potassium fluorosilicate and titration with sodium hydroxide, has been described [16].

The colorimetric determination of the traces of silicon in boiler waters has been discussed in a series of papers [17, 24, 25, 26].

REFERENCES

1. A.S.T.M., *Chemical Analysis of Metals. Sampling and Analysis of Metal Bearing Ores*, Philadelphia, Pa., American Society for Testing and Materials, 1969.
2. Bagshawe, B., and Truman, R. J., *Analyst* **79**, 17–23 (1954).
3. Baker, P. M., and Farrant, B. R., *Analyst* **93**, 732–6 (1968).
4. Cronkright, W. A., *Anal. Chem.* **35**, 766–7 (1963).
5. Czech, F. W., Hrycyshyn, T. P., and Fuchs, R. J., *Anal. Chem.* **36**, 2026–7 (1964).
6. Dozinel, C. M., *Modern Methods of Analysis of Copper and its Alloys*, 2nd ed., London, Elsevier, 1963.
7. Elwell, W. T., and Gidley, J. A. F., *Atomic-Absorption Spectrophotometry*, 2nd ed., Oxford, Pergamon, 1966.
8. Elwell, W. T., and Scholes, I. R., *Analysis of Copper and its Alloys*, Oxford, Pergamon, 1967.
9. Furman, N. H., ed., *Scott's Standard Methods of Chemical Analysis*, 6th ed., Vol. 1, Princeton, N.J., Van Nostrand, 1962.
10. Hillebrand, W. F., Lundell, G. E. F., Bright, H. A., and Hoffman, J. I., *Applied Inorganic Analysis*, 2nd ed., New York, Wiley, 1953.
11. Jenkins, M. H., and Webb, J. A. V., *Analyst* **75**, 481–5 (1950).
12. Kolthoff, I. M., and Elving, P. J., *Treatise on Analytical Chemistry*, Part II, Vol. 2, 107–206, New York, Interscience Publishers, 1962.
13. Lewis, C. L., Ott, W. L., and Sine, N. M., *The Analysis of Nickel*, Oxford, Pergamon, 1966.
14. Methods of Analysis Committee of the Metallurgy (General) Division of the British Iron and Steel Research Association, *Metallurgia* **38**, 346–52 (1948).
15. Meyer, J. W., *Anal. Chem.* **33**, 692–6 (1961).
16. Morris, A. G. C., *Analyst* **90**, 325–34 (1965).
17. Morrison, I. R., and Wilson, A. L., *Analyst* **94**, 54–61 (1969).
18. Pasztor, L. C., *Anal. Chem.* **33**, 1270–2 (1961).
19. Price, W. J., and Roos, J. T. H., *Analyst* **93**, 709–14 (1968).
20. Stross, W., *Metallurgia* **38**, 63–5 (1948).
21. Sturton, J. M., *Anal. Chim. Acta* **32**, 394–7 (1965).
22. Talvitie, N. A., *Anal. Chem.* **23**, 623–6 (1951).
23. Talvitie, N. A., *J. Am. Hyg. Assoc.* **25**, 169–78 (1964).
24. Webber, H. M., and Wilson, A. L., *Analyst* **89**, 632–41 (1964).
25. Webber, H. M., and Wilson, A. L., *Analyst* **94**, 110–20 (1969).
26. Wilson, A. L., *Analyst* **90**, 270–7 (1965).
27. Young, R. S., *Ind. Eng. Chem. Anal. Ed.* **16**, 590 (1944).
28. Young, R. S., *Chemist-Analyst* **35**, 66–8 (1946).

33 Silver

Silver is found in minerals, ores, concentrates, smelter and refinery products, jewellery, electroplated articles, brazing alloys, and many other materials. The principal silver minerals are sulphides, chlorides, and tellurides, usually associated with other sulphides. More than three-quarters of the world's silver production is derived from the mining of copper, lead, and zinc.

ISOLATION OR SEPARATION OF SILVER

Silver is usually determined gravimetrically by precipitating it as chloride in nitric acid solution. Few elements interfere, the only common ones being silicon, lead, mercury, and to a lesser degree bismuth and antimony. Silica and lead may be separated by dehydrating with sulphuric acid, and interference from mercury is obviated by a preliminary oxidation to put the latter into the bivalent form before addition of hydrochloric acid. Mercury can also be volatilized by several evaporations with sulphuric and hydrochloric acids, or by a sodium carbonate fusion. Bismuth and antimony, which hydrolyse in the dilute acid solution that is necessary for the complete precipitation of silver, may be removed by solution of the mixed chlorides in ammonium hydroxide and re-precipitation of silver with hydrochloric acid.

Silver chloride is virtually insoluble in dilute nitric acid containing a small excess of the chloride ion; it is appreciably soluble in strong nitric or hydrochloric acids, and in water. Palladium contaminates a precipitate of silver chloride, imparting to it a pink tint. It can be entirely removed by solution in ammonium hydroxide and re-precipitation with hydrochloric acid in nitric acid solution, repeated if necessary.

Silver is precipitated by hydrogen sulphide in acid and alkaline solution; the sulphide is not soluble in alkali sulphides or polysulphides.

Small quantities of silver in ores and many products of extractive metallurgy are usually determined by fire assay methods. It is important in fire assay work for silver to keep the cupellation temperature as low as possible to minimize losses.

Silver in traces can be collected and isolated as the dithizonate in strong sulphuric acid solution; it will be accompanied by Pd, Au, Hg, and Cu.

1. Gravimetric Determination

Weigh out a 0·5–5 g sample, depending on the silver content, to obtain a final weight not exceeding about 0·2 g silver [8, 9, 11]. Decompose with

10–20 ml nitric acid; if silica and lead are present, add 10 ml 1 : 1 sulphuric acid and evaporate to strong fumes of the latter. If part of the silver is initially present as chloride, treat the nitric acid-insoluble portion with ammonium hydroxide to dissolve silver chloride and liberate occluded substances, filter, and add to the main solution. If an insoluble residue remains after treatment with nitric acid, fuse it with sodium carbonate, cool, add water, filter off the insoluble silver carbonate from chlorides, etc, wash, and dissolve the silver carbonate in nitric acid.

Bring the solution of approximately 1% nitric acid, containing about 0·2 g of silver in 400 ml, to 70°C, and add sufficient 10% hydrochloric acid slowly and with constant stirring to precipitate all the silver. Avoid the addition of a large excess, and stir vigorously to coagulate the fine silver chloride particles which otherwise tend to pass through the filter. Work out of direct sunlight. In accordance with the reaction $AgNO_3 + HCl \rightarrow HNO_3 + AgCl$, 5 ml of 10% hydrochloric acid will precipitate approximately 0·5 g of silver.

Warm until the precipitate settles and test the supernatant liquid with several drops of 10% hydrochloric acid to make sure that precipitation is complete. Allow the covered solution to stand in a dark place for several hours or preferably overnight. Pour the supernatant liquid through a weighed Gooch crucible, wash the precipitate by decantation several times with 0·1% hydrochloric acid and transfer the precipitate to the crucible, washing several times with hot water. Dry the crucible and contents at 130°C in an oven for 1–2 hours to a constant weight. Cool, weigh as AgCl; $AgCl \times 0{\cdot}7527 = Ag$.

When analysing certain materials for silver by the wet method, it is sometimes desirable to express the result in ounces per ton as well as in percentage. This can be done by utilizing the relation: 100 oz per ton = 0·3429%. In some countries, gold and silver are reported in penny-weights; 20 dwt = 1 oz Troy.

It is not generally realized that for many materials a sample weight may be selected which will enable a satisfactory silver determination to be made by weighing as silver chloride on the standard analytical balance sensitive to 0·0001 g. The limit of accuracy in weighing is of the order of 0·3–0·4 oz silver/ton of ore when using 10-g samples on materials ranging from 10 to 1000 oz silver/ton. If a 100-g sample is taken, the maximum weighing inaccuracy is only 0·04 oz silver/ton on materials varying from 1 to 1000 oz silver/ton. It is thus possible for occasional silver determinations, even on fairly low-grade material, to be made gravimetrically without recourse to a fire assay procedure.

Silver may be determined in lead by precipitating as silver iodide in the presence of tartaric acid to hold up impurities which might otherwise precipitate with potassiun iodide [1]. Pig lead usually contains 0·001–0·05% silver. To a 100-g sample add 1 g tartaric acid, 400 ml 1 : 4 nitric acid, and heat to dissolve. When reaction ceases, add 10 ml nitric acid, boil, cool, filter through a Gooch crucible, and discard the precipitate. Dilute to 400 ml, heat to 50°C, add slowly with stirring 5 ml of 1% potassium iodide solution, and digest at 50°C for 15 minutes, avoiding direct sunlight. Filter through a weighed Gooch crucible, wash twice with a nitric–tartaric acid wash solution made by dissolving 20 g of tartaric acid in 200 ml of 1 : 9 nitric acid and diluting to 1 litre. Wash 6 times with hot water, and dry at 110°C to constant weight; AgI × 0·46 = Ag.

For silver in copper refinery slimes and similar heterogeneous materials, decompose a large sample, dilute to volume, and withdraw a suitable aliquot [16]. Decompose the sample with nitric acid and potassium chlorate, and evaporate to a low volume. If large amounts of arsenic, antimony, tin, or selenium are present, volatilize these by several evaporations with hydrobromic and perchloric acids. Cool, dilute, add 10 ml hydrochloric acid and allow to settle at least 4 hours. Filter through Whatman No. 42 paper and wash thoroughly with 2% nitric acid; discard the filtrate. Dissolve silver chloride off the paper with hot 1 : 1 ammonium hydroxide and wash well. Heat the filtrate, cool, and if any turbidity appears, filter before making acid. Add nitric acid to about 1%, and precipitate silver by the addition of hydrochloric acid; this gives a precipitate free of Au, Si, Pb, etc. Do not boil ammoniacal solutions of silver salts because explosive compounds of silver and nitrogen can be formed.

2. Volumetric Determination

This method is based on the greater affinity of silver ions than ferric for thiocyanate [8, 9, 11]. When all the silver has been precipitated as thiocyanate, the ferric indicator reacts with ammonium thiocyanate to give the characteristic red colour of ferric thiocyanate.

$$AgNO_3 + NH_4CNS \rightarrow NH_4NO_3 + AgCNS$$
$$Fe(NO_3)_3 + 3NH_4CNS \rightarrow 3NH_4NO_3 + Fe(CNS)_3$$

The titration is carried out in the cold in a dilute nitric acid solution, usually 1–10% by volume. Nitrous fumes, chlorides, sulphates, mercury, palladium, and salts that impart a strong colour to the solution must be absent.

Prepare about 200 ml of a solution containing approximately 0·3 g silver in 5% nitric acid, free of the interferences listed above. Add 4 ml ferric alum indicator solution, and then nitric acid drop by drop until the solution becomes practically colourless. Titrate with the standard ammonium thiocyanate solution until a faint pink tint persists after vigorous shaking.

$$1 \text{ ml of } 0{\cdot}1\ \text{N}\ NH_4CNS = 0{\cdot}016989 \text{ g } AgNO_3$$
$$= 0{\cdot}010788 \text{ g Ag}$$

0·1 N NH_4CNS. Dissolve 9 g ammonium thiocyanate in 1 litre of water, and standardize against 0·1 N silver nitrate solution which contains 16·989 g silver nitrate per litre. If a number of volumetric determinations are to be made, it is usual to adjust the ammonium thiocyanate solutions so that 1 ml = 0·005 g silver. In this case about 3·7 g ammonium thiocyanate per litre are required; adjust the solution against a standard silver solution containing 0·005 g/ml, which is made by dissolving 1·25 g silver foil in 10 ml nitric acid and diluting to 250 ml.

Ferric alum. Dissolve enough ferric ammonium sulphate or ferric sulphate to make a saturated solution, and then add enough boiled nitric acid to bleach the brown colour of the solution. Use the same quantity of 2 ml per 100 ml of solution in all titrations.

3. Colorimetric Determination

Small quantities of silver may be determined by the reddish colour they impart to a dilute nitric acid solution containing *p*-dimethylaminobenzylidene-rhodanine [10, 11, 12, 14]. Hg, Pd, Pt, and Au interfere; Hg can be overcome by hydrochloric acid, and Pd, Pt and Au by addition of potassium cyanide. Ten μg of silver can be determined in the presence of 100 μg Cu, Pb, Fe, Al, Ca, SO_4, CO_3, and CN.

Pipette 20 ml of the neutral silver solution, containing 1–20 μg of silver, into a 25-ml flask, add 0·5 ml nitric acid, dilute to about 24 ml and shake. Add 0·5 ml of the 0·02% *p*-dimethylaminobenzylidenerhodanine solution in ethyl alcohol, dilute to 25 ml, and mix. Allow to stand for 20 minutes and compare with the colour obtained by a series of standards at 495 nm.

A variation of this procedure, using EDTA to complex heavy metals, is described below for the determination of 440 ppm of silver in nickel [12]. Dissolve 0·25 g of nickel in a minimum of 1 : 1 nitric acid, and dilute to 100 ml in a beaker. Add 17 ml of 10% EDTA solution. Adjust the *p*H to 8–10 by addition of 1 : 1 ammonium hydroxide, and then add 3 ml of 0·02% *p*-dimethylaminobenzylidenerhodanine in ethanol. Adjust the *p*H to 6–7 with dilute acetic acid. Allow the precipitation to proceed for 2 hours in the dark.

Filter the solution through a fine-porosity sintered glass filter crucible, and wash with a little water. Place the crucible in the beaker in which the precipitation was carried out; add 40 ml of ethanol partly to the crucible and partly to the beaker. Cover, heat gently, and reflux for 15 minutes. Wash the outside of the crucible with a stream of alcohol, and filter the contents of the beaker through the crucible. Wash the beaker and crucible carefully with cold alcohol until the filtrate is completely colourless. The crucible now contains the silver rhodanate.

Clean and re-assemble the suction filtration flask and crucible holder, placing a test tube of about 40 ml capacity immediately beneath the crucible holder. Pour 5 ml of 0·5% potassium cyanide in 0·001N sodium hydroxide solution into the original beaker and thence into the crucible. Draw the solution through the filter into the test tube. Rinse the beaker with 5 ml of water and draw through the crucible. Repeat this operation with 4 ml of potassium cyanide solution and 4 ml of water. Transfer the solution in the test tube to a 25 ml volumetric flask, washing with a minimum of water, and make up to volume. Measure the absorbance at 460 nm against a blank.

4. Fire Assay

Small quantities of silver in ores and many other materials are usually determined by fire assay. The procedure is similar to that outlined previously for gold and the platinum metals: fusion and collection of the silver in a lead button, removal of the lead by scorification and cupellation, weighing the silver bead, parting the silver and gold, and re-weighing [1, 2, 3, 7, 15].

For many products, a combination of chemical solution and fire assay is employed. The determination of silver in blister or anode copper is typical. To a 1–4 assay ton sample in a 1000-ml beaker add 20 ml of 2·5% mercuric sulphate solution and allow to stand about 30 minutes. Add 150 ml sulphuric acid, evaporate to strong fumes of this acid, cool, dilute with 300 ml water, and boil. Add 20 ml of 5% sodium chloride solution, stir thoroughly, and allow to stand hot for about 30 minutes. Filter through double Whatman No. 40 papers, wash well with hot water, and wipe out the beaker with a filter paper. Sprinkle 10 g of test lead on the residue, wrap it in 10 g of sheet lead, and place in a 3-in (76·2-mm) scorifier. Dry overnight and char at 350°C in a muffle. Add 40 g of test lead and scorify to a lead button of approximately 25 g. Pour, cool, clean the button, place in a 600 ml beaker, and dissolve in 1 : 4 nitric acid. Filter, wash with hot water, and finally with dilute nitric acid, retaining the filtrate. Wrap the residue in 10 g of sheet lead, scorify as before, and dissolve the lead button in 1 : 4 nitric acid. Filter, catching the filtrate in the beaker containing the previous filtrate; discard the residue.

Precipitate silver in the combined filtrate with 5–10 ml of 5% sodium chloride, stir, and allow to stand overnight. Heat and stir the solution to coagulate the silver chloride, filter through double Whatman No. 40 papers, wash with warm water and finally with 0·1% nitric acid. Wipe the beaker, wrap the residue in sheet lead, and place in a 3-in (76·2-mm) scorifier. Dry overnight, scorify to a lead button of about 15 g, pour, and clean. Cupel at the lowest possible temperature, for instance starting at 870°C and decreasing to 775°C, clean and weigh the silver bead.

A radiochemical investigation of the fire assay method for silver has indicated that a crucible fusion at 1030°C for 45 minutes usually gives a loss of about 0·7% in the slag and crucible walls; losses are higher for charges rich in sulphides of copper or nickel, or charges producing viscous slags. Scorification at 980°C for 30–40 minutes gave a loss of about 0·2%, whereas cupellations conducted at 895–1010°C led to a loss of approximately 2% of the silver present [7].

5. Optical Spectrography

Small quantities of silver in many materials can be determined by emission spectrography, usually with the lines 3280·68 and 3382·89. In refined copper, silver is normally less than 20 ppm, and is almost invariably determined spectrographically. Directions have been published for the spectrographic determination of 0·4–70 ppm silver in nickel [12], and for 5–200 ppm in pig lead [1].

6. X-Ray Spectrography

Silver can be determined in some materials by X-ray fluorescence [4, 11].

7. Atomic-Absorption Spectrophotometry

Small amounts of silver can be determined in many samples rapidly and accurately by atomic absorption [5, 6, 13]. The sensitivity is 0·1 μg/ml for the most sensitive line 3280·7; another line used is 3382·9. The air–coal gas flame is adequate; interference has been reported in the presence of aluminium.

8. Other Procedures

Silver can be determined in copper alloys by amperometric titration of silver with polarized silver/silver chloride electrodes [6].

In pig lead, silver can be separated in nitric acid solution by electrolysis, together with copper and bismuth, and finally determined turbidimetrically by hydrochloric acid addition to the nitric acid solution of the electrodeposit [1]. The same procedure has been used for copper alloys [4].

REFERENCES

1. A.S.T.M., *Chemical Analysis of Metals. Sampling and Analysis of Metal Bearing Ores*, Philadelphia, Pa., American Society for Testing and Materials, 1969.
2. Beamish F. E., *The Analytical Chemistry of the Noble Metals*, Oxford, Pergamon, 1966.
3. Dillon, V. S., *Assay Practice on the Witwatersrand*, Johannesburg, Transvaal and Orange Free State Chamber of Mines, 1955.
4. Dozinel, C. M., *Modern Methods of Analysis of Copper and its Alloys*, 2nd ed., London, Elsevier, 1963.
5. Elwell, W. T., and Gidley, J. A. F., *Atomic-Absorption Spectrophotometry*, 2nd ed., Oxford, Pergamon, 1966.
6. Elwell, W. T., and Scholes, I. R., *Analysis of Copper and its Alloys*, Oxford, Pergamon, 1967.
7. Faye, G. H., and Inman, W. R., *Anal. Chem.* **31**, 1072–6 (1959).
8. Furman, N. H., ed., *Scott's Standard Methods of Chemical Analysis*, 6th ed., Vol. 1, Princeton, N.J., Van Nostrand, 1962.
9. Hillebrand, W. F., Lundell, G. E. F., Bright, H. A., and Hoffman, J. I., *Applied Inorganic Analysis*, 2nd ed., New York, Wiley, 1953.
10. Hopkin and Williams Ltd., *Organic Reagents for Metals*, Chadwell Heath, Essex, Hopkin and Williams, 1955.
11. Kolthoff, I. M., and Elving, P. J., *Treatise on Analytical Chemistry*, Part II, Vol. 4, 1–69, New York, Interscience Publishers, 1966.
12. Lewis, C. L., Ott, W. L., and Sine, N. M., *The Analysis of Nickel*, Oxford, Pergamon, 1966.
13. Rawling, B. S., Amos, M. D., and Greaves, M. C., *Trans. Inst. Mining Met.* **71**, Part 1, 15–26 (1961–62).
14. Sandell, E. B., *Colorimetric Determination of Traces of Metals*, 3rd ed., New York, Interscience Publishers, 1959.
15. Smith, E. A., *The Sampling and Assay of the Precious Metals*, London, Charles Griffin, 1947.
16. Young, R. S., *Anal. Chim. Acta* **4**, 366–85 (1950).

Sodium

See Potassium and Sodium

34 Sulphur

Sulphur is one of the commonest elements determined in most laboratories of the mining and metallurgical industries. It occurs as sulphides, sulphates, and many other sulphur-containing compounds; its analysis is required in minerals, ores, concentrates, tailings, mattes, slags, calcines, flue dusts, residues, metals and alloys, gases, soils, waters, foodstuffs, building products, and many other materials.

Sulphur is usually determined in three ways: gravimetrically by precipitation as barium sulphate in dilute hydrochloric acid solution; by evolution of hydrogen sulphide on treatment with hydrochloric acid; or by combustion in oxygen, absorption of sulphur dioxide in starch-iodide solution, and titration with standard potassium iodate. The combustion method, following the absorption of sulphur dioxide in a suitable liquid, may be completed photometrically in place of the titrimetric iodate finish. Few elements interfere in the general procedures; steps are outlined to overcome the effect of Pb, Ba, Sb, Ca, and Sr in the gravimetric method.

1. Gravimetric Method

In this procedure, oxidation of all sulphide sulphur by addition of a solution of bromine or chlorate must be carried out before decomposition is effected with nitric and hydrochloric acids [1, 2, 10, 16, 18, 19]. Silica is then dehydrated with hydrochloric acid, filtered off, and a solution of barium chloride added to the filtrate to precipitate sulphur as barium sulphate. In most industrial analyses, no elements are removed other than silica by dehydration and tungsten by cinchonine; for work of higher accuracy, iron is removed or reduced to the ferrous state.

Weigh out 0·5–10 g, depending on the sulphur content, into a 400-ml beaker and add 10–30 ml potassium bromide–bromine mixture. The latter is made by dissolving 160 g potassium bromide in a small quantity of water in a 2-litre beaker, adding 100 ml of bromine under a hood with a good draught, stirring vigorously, and making up with water to a 1-litre mark on the beaker.

Allow the sample to stand for about 10 minutes in a fume cupboard, add 10–25 ml nitric acid and allow to stand for 15 minutes at room temperature. Place on asbestos at the edge of the hot plate and evaporate slowly to dryness. Moisten with hydrochloric acid, evaporate to dryness, and repeat to dehydrate silica completely. Add 5 ml hydrochloric acid, 50–75 ml hot water, boil, filter through Whatman No. 40 paper with pulp and wash. To obtain a perfectly clear filtrate from highly siliceous ores, tailings, or slags it may be necessary to use Whatman No. 42 with paper pulp, or double Whatman No. 40 papers with pulp.

Dilute the filtrate to about 300 ml, boil, and cautiously add sufficient 10% barium chloride solution to precipitate all the sulphur; 10 ml of 10% $BaCl_2.2H_2O$ solution will precipitate 0·13 g sulphur. Allow to stand in a warm place for two hours or preferably overnight. Filter through Whatman No. 40 paper, using pulp, wash several times by decantation with hot 1% hydrochloric acid, and finally wash thoroughly with hot water. Transfer paper and precipitate to a weighed platinum crucible, dry, char, and finally ignite for 30 minutes or to constant weight at 800–900°C, cool, and weigh. $BaSO_4 \times 0{\cdot}1373 = S$.

Sometimes sulphur is expressed as a combined form; $BaSO_4 \times 0{\cdot}343 = SO_3$, and $BaSO_4 \times 0{\cdot}4119 = SO_4$.

In many industrial laboratories, for routine work the paper and precipitate of barium sulphate is placed in a fireclay crucible, ignited, cooled, the contents brushed on to the watch glass of the balance pan, and weighed. For this and other routine gravimetric determinations, it is convenient to have a copper or transite tray with holes to accommodate fireclay or other crucibles, and provided with a vertical handle for lifting in and out of a large desiccator. The latter is kept in the balance room, and only the tray carried from muffle or oven to the desiccator.

If the sample is insoluble in nitric acid, add hydrochloric acid after the initial decomposition with nitric in the presence of bromine has proceeded for 10–15 minutes. This will bring into solution the sulphur from nearly all materials which have been ground to pass 100 mesh; the residue usually will not contain sulphur. If the sample contains appreciable tungsten, dissolve in hydrofluoric and nitric acids, add perchloric acid, fume, add cinchonine and hydrochloric acid, and filter off the tungstic acid with the silica before precipitating sulphur.

If there is reason to believe that the residue insoluble in nitric acid, or in nitric and hydrochloric acid, contains sulphur, it should be filtered, washed, dried, ignited, fused with a mixture of 6 parts sodium carbonate to 1 part of potassium nitrate, and added to the main part of the sample. With the exception of ores of antimony, barium, calcium, lead, and strontium, which are discussed below, the conventional acid treatment outlined previously will bring sulphur completely into solution in all mineral samples. Only a trace will be found in the discarded residue of silica, etc.

For the great majority of materials encountered in extractive metallurgy, the solubility of calcium sulphate in dilute hydrochloric acid is such that, with the usual sample weights and volumes employed, no sulphur will be lost with the removal of silica by the conventional acid treatment described above. For example, an ore containing 56% calcium sulphate gave identical sulphur results whether simple acid decomposition or fusion with sodium carbonate plus potassium nitrate were used.

The gravimetric determination of sulphur in antimony ores, concentrates, or metal is complicated by the hydrolysis of antimony in the dilute hydrochloric acid solution which is required for precipitation of barium sulphate. Remove antimony by heating for an hour in 20% hydrochloric acid with 5 g of iron ingot filings; repeat the treatment if necessary. Filter, wash, and proceed with addition of barium chloride.

If the sample contains Pb, Ba, or large quantities of Ca and Sr, insoluble sulphates will be removed with silica in the conventional gravimetric sulphur method. To overcome this, digest the sample, after initial decomposition and dehydration of silica if necessary, in a strong solution of sodium carbonate for 30 minutes at boiling temperature. Filter, wash thoroughly with hot dilute sodium carbonate and then with hot water, discard the precipitate of water-insoluble carbonates of Pb etc, and determine sulphur on the filtrate in the usual manner by precipitation with barium chloride in dilute hydrochloric acid. This procedure gives excellent results on lead sulphide ores and concentrates.

Some materials which are insoluble in acid solution, such as the sulphates of Pb, Ba, Ca, and Sr, can be fused in a platinum crucible directly with a mixture of 6 parts sodium carbonate to 1 part potassium nitrate, extracted with water, filtered, washed with dilute sodium carbonate and then with hot water. Sulphur is in the filtrate, free of Pb, Ba, etc, and may be determined by boiling, acidifying with hydrochloric acid, and proceeding with the precipitation of sulphur by barium chloride. The solubility of the carbonates of Pb, Ca, and Sr is much less than that of the corresponding sulphates; although the solubility of barium carbonate exceeds that of barium sulphate it is still sufficiently low to ensure the accuracy of this transposition technique for all routine work.

If the sample contains considerable ferric iron, the precipitate of barium sulphate will be slightly contaminated with iron. For most routine analyses this can be neglected; for work of the highest accuracy remove the iron by two ammonia precipitations, using an excess of ammonium hydroxide and ammonium chloride to minimize the occlusion of sulphur in the ferric hydroxide precipitate. The sulphate ion is adsorbed by ferric hydroxide to only a slight extent, and a single precipitation will suffice unless the iron precipitate is large; phosphate separation of iron likewise gives low adsorption of sulphur, but a zinc oxide precipitate may adsorb sulphur strongly [24].

If iron is reduced to the ferrous state it is less likely to contaminate a precipitate of barium sulphate. For many samples, this practice is followed instead of an iron separation. Boil the dilute hydrochloric acid solution for 5 minutes with a little powdered aluminium or zinc, filter, wash, and proceed with the determination of sulphur in the colourless filtrate.

From a long experience with sulphur determinations on a wide variety of materials in extractive metallurgy, the writer believes that, for many products, no significant difference in sulphur will be found whether iron is reduced, removed by double precipitation with ammonia, or allowed to remain in the solution. For example, careful routine work on a sample of pyrite containing

39·8% iron gave the following results for sulphur, in percentages:
Iron not removed or reduced 43·61, 43·53
Iron removed by double precipitation with ammonia 43·48, 43·60
Iron reduced by boiling with aluminium 43·55, 43·59

2. Evolution of Sulphur as Hydrogen Sulphide

For some metallurgical products such as certain irons and steels, mattes, titanium, and antimony, where all the sulphur is present as sulphide, a rapid and satisfactory procedure involves treatment with hydrochloric acid, absorption of the evolved hydrogen sulphide in cadmium chloride solution, addition of an excess of potassium iodate, acidification, and titration of the excess iodine with standard sodium thiosulphate solution [2, 8, 16, 19, 20, 21].

Weigh 0·5–5 g into a 500-ml Erlenmeyer flask which is fitted with a dropping funnel and a reflux condenser; the top of the latter is connected to a litre flask which serves as the absorption vessel. Add 100 ml hydrochloric acid rapidly through the dropping funnel, heat the Erlenmeyer flask and boil until the sample is decomposed. Solution of pure nickel may be accelerated by addition of a few drops of platinic chloride solution [21]. Remove the absorption flask, to which had been originally added 50–100 ml ammoniacal cadmium chloride solution diluted with water to about 400 ml, and add 10–25 ml of standard potassium iodate–potassium iodide solution. Acidify the solution with about 50 ml hydrochloric acid, stopper, and shake the flask vigorously. Add starch solution and titrate the excess iodine with standard sodium thiosulphate.

In the presence of selenium, a part of which is evolved as H_2Se, insert between the reflux condenser and the absorption flask a bottle containing 200 ml of an acid zinc-potassium chloride solution; this retains selenium but allows hydrogen sulphide to pass through.

Reagents

Potassium iodate–potassium iodide solution. Dissolve 5·599 g KIO_3 and 24 g KI in water and dilute to 1 litre; 1 ml = 0·0025 g S. Standardize this solution against arsenious acid; 25 ml of 0·1N arsenious acid is equivalent to 16·20 ml of the potassium iodate–potassium iodide solution. A 0·1N arsenious acid solution is made by dissolving 4·95 g As_2O_3 in 15 g sodium carbonate dissolved in a minimum quantity of water, and diluting to 1 litre.

$$As_2O_3 + KIO_3 + 2HCl \rightarrow As_2O_5 + KCl + H_2O + ICl$$

Sodium thiosulphate solution. Dissolve 39·10 g $Na_2S_2O_3.5H_2O$ in 1 litre; 1 ml = 1 ml of the above potassium iodate–potassium iodide solution.

Ammoniacal cadmium chloride solution. Dissolve 22·5 g $CdCl_2$ in 1 litre of 1 : 1 ammonium hydroxide; 50 ml of this solution will absorb the hydrogen sulphide from a sample containing 0·17 g sulphur.

Acid zinc-potassium chloride solution. Mix 10 g zinc oxide with water and dissolve in a slight excess of hydrochloric acid. Add 50 g potassium chloride and dilute to 1 litre. Add a solution of potassium hydroxide until a yellow colour is obtained with thymol blue, then add dilute hydrochloric acid until the colour is red.

The reactions for this evolution procedure are:

$$2HCl + FeS \rightarrow H_2S + FeCl_2$$
$$H_2S + CdCl_2 \rightarrow CdS + 2HCl$$
$$H_2S + I_2 \rightarrow 2HI + S$$
$$KIO_3 + 5KI + 6HCl \rightarrow 6KCl + 3H_2O + 3I_2$$
$$I_2 + 2Na_2S_2O_3 \rightarrow 2NaI + Na_2S_4O_6$$

Sulphur can also be determined in irons and steels by a hydriodic acid distillation–iodometric titration method [22].

3. Direct Combustion-Iodate Method

In this rapid procedure, the sample is heated to a high temperature in a resistance or induction furnace in a stream of purified oxygen [2, 7, 11, 15, 16, 19, 20, 25, 28]. The evolved sulphur dioxide is passed into starch–iodide solution and titrated with standard potassium iodate solution. An accelerator is added to the sample before it is inserted into the furnace. This is usually metallic iron, tin, or copper, or all three, but vanadium(V) oxide which has been heated in a stream of air at 670°C has been recently shown to be a better accelerator for nickel-, iron-, and copper-base alloys [6].

In practice, the gas enters a titration vessel containing hydrochloric acid, potassium iodide, starch, and enough potassium iodate to release some free iodine and turn the starch blue. The sulphur dioxide bleaches the blue colour, and potassium iodate is added manually or automatically to maintain the blue colour. When all the sulphur dioxide has been released from the sample, the blue colour no longer fades, and the amount of potassium iodate used is a measure of the sulphur in the sample.

$$SO_2 + 2H_2O + I_2 \rightarrow H_2SO_4 + 2HI$$
$$KIO_3 + 5KI + 5HCl \rightarrow 6KCl + 3H_2O + 3I_2$$

Theoretically, a solution containing 0·2225 g potassium iodate in 1 litre has a value of 1 ml = 0·0001 g sulphur. Because some materials do not yield all their sulphur as sulphur dioxide by this procedure, it is usually necessary to standardize the titrating solution against a standard sample. With stainless steels, for example, it is recommended to use the figure of 1 ml KIO_3 = 0·0001 g S for a potassium iodate solution containing 0·2069 g potassium iodate per litre [2].

4. Combustion-Photometric Method

In this procedure, the sample is heated in a stream of oxygen as in **3** above, and the sulphur dioxide is absorbed in the liquid. The amount of sulphur dioxide is measured photometrically in place of the titrimetric determination with standard potassium iodate solution. The usual absorbent for sulphur dioxide is 0·2 M sodium tetrachloromercurate; after the addition of 0·04% pararosaniline and 0·2% formaldehyde, and allowing the solution to stand for 30 minutes, the absorbance is read at 560 nm. This procedure has been found suitable for 0–500 ppm sulphur in blister and refined copper [4], nickel and its alloys [7], rocks [28], and air [26].

SPECIAL PROCEDURES

1. Forms of Sulphur in Ores etc.

Many materials contain sulphides and sulphates, both of which may include water-soluble forms like Na_2S or $CoSO_4$, and insoluble compounds like CuS or $CaSO_4$. When soluble sulphides and sulphates occur together, the former may be expelled as hydrogen sulphide by acidifying with hydrochloric acid and boiling. Total sulphur is determined by oxidizing the solution initially with bromine to convert sulphides to sulphates; total sulphur minus sulphate sulphur = sulphide sulphur.

If an insoluble sulphate like $CaSO_4$ must be differentiated from an insoluble sulphide, there are several procedures available. A dilute ammonium chloride solution will dissolve calcium sulphate in ores, but has virtually no effect on sulphides of Cu, Fe, Co, Zn, and Pb [34]. Alternatively, boil a 1–3 g sample of ore for 15 minutes with 50 ml of 10% sodium carbonate, filter, and wash. The filtrate contains the sulphur derived from the calcium sulphate and other insoluble sulphates, and the precipitate contains the insoluble base metal sulphides [33]. Sulphur may then be determined on either fraction, or on both. Another procedure is to evolve sulphur as hydrogen sulphide from a sulphide by the addition of hydrochloric acid, in the presence of either soluble or insoluble sulphate, and to absorb and determine it as described earlier. Unlike pyrrhotite, pyrite cannot be determined in this manner, so the method is not applicable to ores or concentrates containing the latter mineral. Differentiation of pyrite from pyrrhotite has been discussed in the chapter on Iron.

A differentiation of the forms of sulphur occurring in oil shale has been described [29]. Sulphate sulphur is extracted with 10% perchloric acid, pyrite sulphur is decomposed by a treatment with lithium aluminium hydride, and organic sulphur is determined on the residue by an Eschka fusion.

2. Sulphur in Coal

Sulphur in coal is usually determined on the residue in the bomb after a measurement of Btu (kJ) in a calorimeter. It may also be determined by mixing 3 parts of Eschka's mixture to 1 part of coal, placing in a crucible in a cold muffle which is raised to 800°C and kept at that temperature for 1½ hours. Cool, digest with water, filter, and determine sulphur on the filtrate. Eschka's mixture consists of 2 parts magnesium oxide to 1 part anhydrous sodium carbonate.

3. Sulphur in Flue Gases and Air

The determination of sulphur dioxide and trioxide in flue gases and air has been described by many workers [5, 9, 12, 13, 17, 19, 26, 27, 31, 32].

4. Analysis of Sulphur

The analysis of commercial sulphur is discussed in the data books and manuals of the leading sulphur producers [14, 30].

5. Other Methods

Trace amounts of sulphur in iron and steel have been determined by combustion in oxygen at 1500°C and a final coulometric measurement [3].

Small quantities of sulphur in tin have been determined by combustion in oxygen, absorption of the sulphur dioxide in an extremely dilute sulphuric acid solution, and measurement of the change in electrical conductivity [23].

REFERENCES

1. A.O.A.C., *Official Methods of Analysis of the Association of Official Analytical Chemists*, Washington, D.C., Association of Official Analytical Chemists, 1965.
2. A.S.T.M., *Chemical Analysis of Metals. Sampling and Analysis of Metal Bearing Ores*, Philadelphia, Pa., American Society for Testing and Materials, 1969.
3. Bandi, W. R., Buyok, E. G., and Straub, W. A., *Anal. Chem.* **38**, 1485–9 (1966).
4. Barabas, S., and Kaminski, J., *Anal. Chem.* **35**, 1702–5 (1963).
5. Berk, A. A., and Burdick, L. R., *U.S. Bur. Mines Rept. Invest.* 4618, 1950.
6. Burke, K. E., *Anal. Chem.* **39**, 1727–31 (1967).
7. Burke, K. E., and Davis, C. M., *Anal. Chem.* **34**, 1747–51 (1962).
8. Codell, M., Norwitz, G., and Clemency, C., *Anal. Chem.* **29**, 1496–9 (1957).
9. Corbett, P. F., *J. Soc. Chem. Ind.* **67**, 227–30 (1948).
10. Dillon, V. S., *Assay Practice on the Witwatersrand*, Johannesburg, Transvaal and Orange Free State Chamber of Mines, 1955.
11. Elwell, W. T., and Scholes, I. R., *Analysis of Copper and its Alloys*, Oxford, Pergamon, 1967.
12. Fletcher, A. W., *Chem. and Industry*, **1956**, April 28, pp. 303–4.
13. Flint, D., *J. Soc. Chem. Ind.* **67**, 2–5 (1948).
14. Freeport Sulphur Co., *The Sulphur Data Book*, New York, McGraw-Hill, 1954.
15. Fulton, J. W., and Fryxell, R. E., *Anal. Chem.* **31**, 401–5 (1959).
16. Furman, N. H., ed., *Scott's Standard Methods of Chemical Analysis*, 6th ed., Vol. 1, Princeton, N.J., Van Nostrand, 1962.
17. Gillham, E. W. F., *J. Soc. Chem. Ind.* **65**, 370–2 (1946).
18. Hillebrand, W. F., Lundell, G. E. F., Bright, H. A., and Hoffman, J. I., *Applied Inorganic Analysis*, 2nd ed., New York, Wiley, 1953.
19. Kolthoff, I. M., and Elving, P. J., *Treatise on Analytical Chemistry*, Part II, Vol. 7, 1–135, New York, Interscience Publishers, 1961.

20. Lewis, C. L., Ott, W. L., and Sine, N. M., *The Analysis of Nickel*, Oxford, Pergamon, 1966.
21. Luke, C. L., *Anal. Chem.* **29**, 1227–8 (1957).
22. Luke, C. L., *Anal. Chem.* **31**, 1393–4 (1959).
23. Pell, E., Malissa, H., Murphy, N. A., and Chamberlain, B. R., *Anal. Chim. Acta* **43**, 423–8 (1968).
24. Pinkney, E. T., Dick, R., and Young, R. S., *J. Am. Chem. Soc.* **68**, 1126–8 (1946).
25. Rice-Jones, W. G., *Anal. Chem.* **25**, 1383–5 (1953).
26. Scaringelli, F. P., Saltzman, B. E., and Frey, S. A., *Anal. Chem.* **39**, 1709–19 (1967).
27. Seidman, E. B., *Anal. Chem.* **30**, 1680–2 (1958).
28. Sen Gupta, J. G., *Anal. Chem.* **35**, 1971–3 (1963).
29. Smith, J. W., Young, N. B., and Lawlor, D. L., *Anal. Chem.* **36**, 618–22 (1964).
30. Texas Gulf Sulphur Co., *Sulphur Manual*, New York, Texas Gulf Sulphur Co., 1961.
31. Urone, P., Evans, J. B., and Noyes, C. M., *Anal. Chem.* **37**, 1104–7 (1965).
32. Van Straten, H. A. C., *Anal. Chim. Acta* **14**, 325–8 (1956).
33. Williamson, J. E., *Can. Mining Met. Bull.* **L**, 138–200 (1947).
34. Young, R. S., and Hall, A. J., *J. Soc. Chem. Ind.* **66**, 375 (1947).

Tantalum

See Niobium and Tantalum

Tellurium

See Selenium and Tellurium

35 Thorium

Thorium is important in radioactivity and atomic energy developments; it may also be encountered in a few mining laboratories. The most important source of thorium is monazite sand, (Ce, Y, La, Th) PO_4, but a number of other thorium-bearing minerals occur in nature.

ISOLATION OR SEPARATION OF THORIUM

Thorium may be separated from members of the acid sulphide group by precipitation of the latter with hydrogen sulphide, and from the alkaline earths, magnesium, and the alkalies by precipitation with ammonium hydroxide in the presence of ammonium chloride. Separations of thorium from other members of the ammonium hydroxide and ammonium sulphide groups can be made in several ways. Treatment with sodium hydroxide will precipitate Th, with Fe, Ti, Ni, Co, Mn, Cr, Zr, and U from Al, Be, V, Zn, W, and P; sodium peroxide will throw Cr and U into the filtrate.

Thorium is not precipitated by hydrogen sulphide in ammoniacal tartrate solution and can thereby be separated from Fe, Ni, Co, and Zn; iron may also be separated by ether, or a mercury cathode in dilute sulphuric acid. Chromium may be eliminated as chromyl chloride by boiling hydrochloric and perchloric acids, or it may be deposited on a mercury cathode. The following elements are quantitatively removed from thorium by a mercury cathode separation in dilute sulphuric acid: Fe, Co, Ni, Cu, Zn, Ga, Ge, Rh, Pd, Ag, Cd, In, Sn, Ir, Pt, Au, Hg, Tl, Cr, Mo, Re, Po, Bi, As, Se, Te, Pb, and Os; thorium is accompanied by U, Nb, Ta, W, Al, Be, Ti, Ce, V, and Zr. A zinc oxide separation will precipitate thorium, together with most other elements from Co, Ni, and Mn.

Thorium can be separated from uranium by oxidizing the solution with potassium permanganate, and carrying out a cupferron separation in cold 10% sulphuric acid which precipitates thorium along with Ti, V, Zr, Fe, Nb, Ta, and W; Cr, Mn, Ni, Co, Zn, Al, and Be accompany uranium. It should be emphasized that thorium is a catalyst for the explosive decomposition of cupferron, and the temperature of the solution must not exceed 10–15°C.

Separations of thorium are discussed at length in a monograph on the analytical chemistry of this element [14].

1. Gravimetric Method

The following procedure is applicable to all types of thorium-containing samples. It combines the principal separations of thorium, which are based on the insolubility in acid solution of its oxalate and of its iodate [2, 3, 6, 7, 8, 9, 11, 13, 14, 16].

Transfer 0·5–5 g sample to a silica evaporating dish, add 8 ml sulphuric acid, cover, heat on a sand bath at 200°C for two hours, stirring frequently with a glass rod. Take precautions to avoid overheating, boiling, or spattering. Cool, add 50 ml ice-cold water with constant stirring, decant the turbid liquid into a 400 ml beaker, and treat the residue with 3 ml sulphuric acid for an additional three hours. Cool, add water as before, wash all the material into the 400 ml beaker, and stir thoroughly to dissolve all soluble salts. Warm to about 40°C, add 0·5 ml of 1% gelatin solution, filter on Whatman No. 42 paper, and wash thoroughly. Transfer the insoluble residue of silica and undecomposed minerals to a platinum dish, dry, calcine, cool, mix with about three times its weight of potassium acid fluoride, fuse, and stir with a platinum wire. Cool, add 2 ml sulphuric acid, fume strongly; fuse any insoluble residue with sodium carbonate, leach with water, and acidify with sulphuric acid. Filter off any precipitate and add the sulphuric acid extracts to the main solution.

To a 600-ml beaker that contains the clear solution of about 100 ml from the above acid treatments and fusions, add, with stirring, 50 ml nitric acid. If ceric salts are present, reduce them to the cerous state by adding hydrogen peroxide to the nitric acid solution. Now add slowly to the acid solution, with stirring, an iodate solution containing 16 g potassium iodate dissolved in 50 ml nitric acid and 30 ml water. Cover, cool in ice water to room temperature, and filter through Whatman 541 paper to obtain a clear filtrate. Wash the last traces of thorium iodate on to the paper with 30–50 ml of a wash solution containing 8 g potassium iodate and 50 ml nitric acid per litre. Transfer the precipitate back to the original beaker with the wash solution, stir to break up lumps, and re-filter through the same paper. Wash the precipitate back into the original beaker with water, heat nearly to boiling, and add slowly 30–90 ml nitric acid until the solution is clear. Prepare a solution of potassium iodate in boiling water such that 8 g potassium iodate is present for every 60 ml of nitric acid that is used to dissolve the precipitate. Add this potassium iodate solution, with brisk stirring, to the hot thorium iodate solution. Cool in ice water to room temperature, filter, wash off the paper and re-filter as was done with the first iodate precipitation. Finally wash the precipitate back into the original beaker with water, washing both sides of the paper thoroughly; retain the paper.

Warm the suspension of thorium iodate, which should have a volume of about 100 ml, to 40°C and add 25 ml hydrochloric acid. Pass in sulphur dioxide until a clear, faintly yellow solution which smells of sulphur dioxide is obtained. Filter through the same Whatman 541 paper that was used previously. Remove all iodate from the paper by passing the filtrate through the paper a second time. Wash thoroughly with water and retain the filter paper. Make the filtrate alkaline with ammonium hydroxide, and allow the white hydroxide to stand for an hour. Filter through the same paper used previously; test the filtrate with ammonium hydroxide to be sure that precipitation of thorium was complete. Wash the precipitate with 50–100 ml of 5% ammonium chloride which has been made ammoniacal with a few drops of ammonium hydroxide. Dissolve the thorium hydroxide off the paper by pouring through it a solution of 25 ml hydrochloric acid in 100 ml of water. Refilter the filtrate through the same paper, and wash the latter well with hot water.

Heat the filtrate, which should have a volume of about 400 ml, to boiling. Add, with constant stirring, a boiling solution of 25 g oxalic acid in 200 ml of water. Stir the thorium oxalate precipitate, and allow to settle overnight. Filter through a Whatman No. 42 paper and wash thoroughly with hot water. Remove all traces of precipitate from the walls of the beaker; if necessary, wipe carefully with a small filter paper. Place paper and precipitate in a weighed platinum crucible, dry, char, ignite for 2 hours with free access of air at 900–1000°C. Weigh as ThO_2; $ThO_2 \times 0{\cdot}8788 = Th$.

The reactions in this analysis are:

$$Th(NO_3)_4 + 4KIO_3 \rightarrow 4KNO_3 + Th(IO_3)_4$$

$$Th(IO_3)_4 + 4NH_4OH \rightarrow 4NH_4IO_3 + Th(OH)_4$$

$$Th(OH)_4 + 4HCl \rightarrow ThCl_4 + 4H_2O$$

$$ThCl_4 + 2H_2C_2O_4 \rightarrow 4HCl + Th(C_2O_4)_2$$

$$Th(C_2O_4)_2 + O_2 \rightarrow ThO_2 + 4CO_2$$

An alternative method for the separation of thorium is a chromatographic one; thorium nitrate can be eluted from a cellulose column by ether containing 12·5% by volume of nitric acid, whereas rare earths and most elements associated with thorium move very slowly, if at all, down the column. Uranium, zirconium, and scandium are eluted with thorium; uranium can be removed initially by extracting with 3% nitric acid in ether which does not remove the thorium, and the addition of tartrate or phosphate ions renders zirconium and scandium non-extractable in the ether–12·5% nitric acid solution used to elute thorium. For some materials, an alumina column, followed by a cellulose column to remove small quantities of ferric and aluminium nitrates which pass through the alumina, should be employed. Details have been described [11, 12, 16, 20].

From 0·2 to 25% thorium in magnesium alloys is determined by solution in hydrochloric acid, reduction of cerium with hydroxylamine hydrochloride, precipitation of thorium and zirconium with benzoic acid in faintly ammoniacal solution, solution of the benzoate in hydrochloric acid, precipitation of thorium with oxalic acid, and ignition to ThO_2 [2]. The procedure is applicable in the presence of Al, Cu, Fe, Pb, Mn, Ni, Si, Sn, Zn, Zr, and rare earths.

2. Colorimetric Method

Small quantities of thorium may be determined colorimetrically at *p*H 0·4–1·2 by thorin or thoron, 2-(2-hydroxy-3,6-disulpho-1-naphthylazo) benzene arsonic acid [4, 7, 10, 11, 13, 14, 15]. The final thorium concentration should not exceed 30 ppm, and measurement is made at 545–555 nm. A number of elements interfere, including Fe, Ce, Zr, U, Ca, Ti, Ta, Nb, Sn, and F; if not removed, their effect can be often overcome by additions or adjustments. Thorium may also be determined colorimetrically by the arsenazo(III) reaction in a perchlorate medium at 660 nm [1].

3. Volumetric Determination

A wide range of thorium contents can be titrated with EDTA at *p*H 2·3–3·4, using Alizarin Red S as internal indicator [5, 20]. The following ions interfere and must be removed: Ti, Zr, Ce, V, Fe, Ni, Cu, Sn, Pb, and Bi.

4. Optical Spectrography

Thorium can be determined by emission spectrography, but the technique is not widely used [11, 14].

5. X-Ray Spectrography

Thorium has been determined in uranium ore concentrates by X-ray fluorescence in the range 0·01–1% [19], but the use of X-ray spectrography in thorium analysis is not widespread [11, 14].

6. Other Procedures

Traces of thorium may be determined in a wide variety of materials by a fluorometric method using morin in an alkaline solution of diethylenetriamine-pentaacetic acid (DTPA). Thorium is precipitated by barium sulphate, and the latter is dissolved in alkaline DTPA [17, 18].

REFERENCES

1. Abbey, S., *Anal. Chim. Acta* **30**, 176–87 (1964).
2. A.S.T.M., *Chemical Analysis of Metals. Sampling and Analysis of Metal Bearing Ores*, Philadelphia, Pa., American Society for Testing and Materials, 1969.
3. Dillon, V. S., *Assay Practice on the Witwatersrand*, Johannesburg, Transvaal and Orange Free State Chamber of Mines, 1955.
4. Fletcher, M. H., Grimaldi, F. S., and Jenkins, L. B., *Anal. Chem.* **29**, 963–7 (1957).
5. Fritz, J. J., and Ford, J. J., *Anal. Chem.* **25**, 1640–2 (1953).
6. Furman, N. H., ed., *Scott's Standard Methods of Chemical Analysis*, 6th ed., Vol. 1, Princeton, N.J., Van Nostrand, 1962.
7. Grimaldi, F. S., et al, *Collected Papers on Methods of Analysis for Uranium and Thorium*, U.S. Geol. Survey Bull. 1006, 1954.

8. GRIMALDI, F. S., JENKINS, L. B., and FLETCHER, M. H., *Anal. Chem.* **29**, 848–51 (1957).
9. HILLEBRAND, W. F., LUNDELL, G. E. F., BRIGHT, H. A., and HOFFMAN, J. I., *Applied Inorganic Analysis*, 2nd ed., New York, Wiley, 1953.
10. JOHNSTON, M. B., BARNARD, A. J., and BROAD, W. C., *Thorin: An Interesting Chromogenic Agent and Chelatochrome Indicator*, Phillipsburg, N.J., J. T. Baker Chemical Co., 1961.
11. KOLTHOFF, I. M., and ELVING, P. J., *Treatise on Analytical Chemistry*, Part II, Vol. 5, 139–216, New York, Interscience Publishers, 1961.
12. NIETZEL, O. A., WESSLING, B. W., and DE SESA, M. A., *Anal. Chem.* **30**, 1182–5 (1958).
13. RODDEN, C. J., *Analytical Chemistry of the Manhattan Project*, New York, McGraw-Hill, 1950.
14. RYABCHIKOV, D. I., and GOL'BRAIKH, E. K., *The Analytical Chemistry of Thorium*, Oxford, Pergamon, 1963.
15. SANDELL, E. B., *Colorimetric Determination of Traces of Metals*, 3rd ed., New York, Interscience Publishers, 1959.
16. SCHOELLER, W. R., and POWELL, A. R., *The Analysis of Minerals and Ores of the Rarer Elements*, 3rd ed., London, Charles Griffin, 1955.
17. SILL, C. W., and WILLIS, C. P., *Anal. Chem.* **34**, 954–64 (1962).
18. SILL, C. W., and WILLIS, C. P., *Anal. Chem.* **36**, 622–30 (1964).
19. STOECKER, W. C., and MCBRIDE, C. H., *Anal. Chem.* **33**, 1709–13 (1961).
20. STRELOW, F. W. E., *Anal. Chem.* **33**, 1648–50 (1961).

36 Tin

Tin is found in a few minerals and ores, but, apart from actual tin recovery operations, its determination is not a frequent occurrence in mining laboratories. The principal tin mineral is cassiterite, SnO_2. Complex sulphides are another source of tin, but are of importance only in Bolivia. An accurate chemical determination of tin in economic deposits is more difficult than that of other base metals, because the content of tin is very low. For example, a rich tin ore in Malaysia, the leading producer, contains less than 0·05% tin. In the pure and alloyed state, tin is an important metal; consequently its analysis is a common one in many laboratories serving the fabricating metallurgical industries.

ISOLATION OR SEPARATION OF TIN

In tin determinations it must be kept in mind that stannic chloride is appreciably volatile; hydrochloric acid or aqua regia solutions should not be evaporated to dryness. No volatilization occurs when dilute hydrochloric acid solutions are boiled in covered beakers, or when hydrochloric–sulphuric acid solutions are evaporated to fumes of sulphur trioxide. Separation of silica from tin should be done by dehydrating the former with sulphuric or perchloric acid, or eliminating it with sulphuric and hydrofluoric acids.

Tin, of course, is a member of the hydrogen sulphide group; it can be separated from the ammonium hydroxide, ammonium sulphide, and ammonium carbonate groups by precipitation with hydrogen sulphide in 5–10% hydrochloric or sulphuric acids. Separations of tin from other elements of the hydrogen sulphide group may take a variety of forms. Lead is removed by evaporation with sulphuric acid; silver and mercurous mercury can be precipitated with hydrochloric acid. Tin sulphide, like those of As, Sb, and Mo, is soluble in alkaline sulphide, whereas the sulphides of Cu, Bi, Cd, Pb, and Hg(II) are insoluble. Tin, with As, Sb, Bi, Se and Te, is collected and removed from a solution containing copper and molybdenum by a ferric hydroxide precipitate in the presence of ammonia and ammonium salts. Tin can be separated from arsenic by distillation of the latter as the trichloride, or by precipitation of arsenic in strong hydrochloric acid with hydrogen sulphide; antimony and copper can be removed by precipitation on metallic iron. Selenium and tellurium can be removed by sulphur dioxide in strong and weak hydrochloric acid solutions, respectively.

For some purposes it is convenient to separate tin by distilling it with a mixture of 4 hydrobromic acid: 1 bromine, or of sulphuric and hydrobromic acids; arsenic and antimony are likewise distilled. Tungsten sometimes accompanies tin in a sample; small amounts can be separated by precipitating tin with hydrogen sulphide in dilute hydrochloric acid solution containing tartaric acid. If considerable tungsten is present, the separation should be done by repeated precipitation with ammonium hydroxide in the presence of ammonium salts, followed by solution of the final precipitate in hydrochloric and tartaric acids, and precipitation of the tin by hydrogen sulphide.

Tin may be separated from the following elements by the cupferron–chloroform extraction in cold 10% sulphuric acid: Al, Be, P, B, Mn, Ni, Co, Ca, Mg, Pb, Zn, Cr, Cd, Se, Te, In, Ge, As, Pt, and Au. Tin is deposited on a mercury cathode in dilute sulphuric acid, together with Fe, Co, Ni, Cu, Zn, Ag, Cd, Pt, Au, Cr, Mo, Pd, and several other elements; the technique is accordingly of little value in a determination of this metal. Tin(II), but not tin(IV), is extracted from mineral acid solutions by a chloroform solution of diethylammonium diethyldithiocarbamate; a first extraction separates tin(IV) from Cu, Bi, Hg, As(III), and Sb(III), and a second extraction separates tin(II) from a large number of metals which remain after the first extraction. Tin(IV) can be extracted as the iodide from aqueous solutions by benzene, toluene, carbon tetrachloride and other solvents.

1. Volumetric Method

For a wide range of materials and of tin contents, the best method is reduction of tin to the bivalent state by boiling in hydrochloric acid with lead, nickel, iron, or aluminium, followed by oxidation to the quadrivalent condition by a standard solution of iodine [3, 4, 7, 10, 13, 14, 16, 28]. Usually this iodimetric method can be applied without the complete isolation of tin. It is necessary to ensure that the tin has been entirely reduced and is not subsequently re-oxidized by air; for the latter it is essential to keep an atmosphere of carbon dioxide over the sample during titration.

A. Decomposition of Metals and Alloys

As a general rule, decompose 0·5–5 g sample with 5–20 ml hydrochloric acid and 5–20 ml nitric acid, add 10 ml 1 : 1 sulphuric acid and evaporate to strong fumes of the latter. A few samples, such as ferrotungsten, should be decomposed in a platinum dish with hydrofluoric, nitric, and sulphuric acids, and evaporated to fumes of the latter. Titanium metal can be dissolved in 1 : 1 sulphuric acid and fluoboric acid [10, 27].

Filter off silica and lead, wash thoroughly, and precipitate tin in the filtrate with hydrogen sulphide after adjusting to about 5–10% sulphuric or hydrochloric acid; this will separate it from Fe, Al, Zn, Ni, Cr, Be, U, Zr, Ti,

Mn, Th, Co, V, etc. If substantial quantities of copper are present, they may be removed in several ways:

(a) Electrolysis in a nitric–sulphuric acid solution, for brasses, etc.

(b) Cementation, together with antimony, by boiling the solution of the sample for 30 minutes with 5 g of iron in 50 ml hydrochloric acid, and repeating if necessary. This is used for bearing metal and similar materials.

(c) Collection and precipitation of the tin in ammonia and ammonium salts with ferric hydroxide, re-solution of the precipitate in acid, and re-precipitation to pass all copper into the filtrate. This is employed for bronzes, some brasses, and in the final separation after hydrogen sulphide treatment for steels, ferrotungsten, etc.

(d) Treatment of the mixed sulphide precipitate with potassium hydroxide and potassium sulphide to put tin into the filtrate and leave Cu, Bi, etc. as a precipitate. This is one of the separations used in some complex materials such as ferrotungsten, etc.

The sample should be finally obtained in a solution containing not more than 0·2 g tin, about 10–15 ml sulphuric acid, and 100 ml hydrochloric acid in 300 ml solution. Nitrates, tungsten, molybdenum, and vanadium should be absent. Proceed with the reduction of the tin and its iodimetric titration as outlined below in C.

For some metals, of course, the decomposition procedure can be varied. Aluminium is brought into solution with 1 : 2 hydrochloric acid, and magnesium with hydrochloric acid; for lead, the tin is collected with manganese dioxide in nitric acid, arsenic removed by distillation, antimony titrated in the residual solution, and tin finally determined on the latter after reduction with lead [3].

B. Decomposition of Ores, Concentrates, etc.

Weigh 0·5 g of medium or high grade tin concentrate, or 1 g low grade, into a 50 ml high-form iron or nickel crucible. Add 4 g sodium carbonate, mix thoroughly, and cover with 8 g sodium peroxide. Heat the crucible over a small burner or element to expel any water in the flux until the charge starts to melt. Carefully continue the fusion by revolving the crucible around the edge of the flame until the mixture melts down quietly. Heat at full temperature of the burner or element for a further two minutes.

Let the melt solidify and cool, place the crucible in a 400 ml beaker, add 60 ml water, and allow the melt to disintegrate. Remove the crucible from the beaker, rinse thoroughly, add 50 ml hydrochloric acid to the beaker, heat, clean the crucible with 50 ml hydrochloric acid and add this to the main solution. Transfer the latter to a 500-ml Erlenmeyer flask, and dilute to about 250 ml.

C. Titration of Tin

To the flask containing the solution of tin from A or B, introduce a nickel strip or foil weighing about 5 g, and boil the solution gently for 30 minutes until all ferric chloride has been reduced. Introduce a second nickel strip and close the flask with a rubber stopper containing a glass tube which extends on the outside nearly to the bottom of the flask. Boil the solution gently for another 20 minutes; seal the outside end of the glass tube with a hot solution of sodium bicarbonate in a 250-ml beaker. Remove the flask from the hot plate, cool to 15°C, and titrate with 0·125 N iodine solution, using starch as indicator.

For work of high accuracy, exclude air during titration by inserting the tip of the burette through a small hole in the stopper of the flask, which is kept closed until the addition of the standard iodine solution. If the sample contains small quantities of molybdenum or vanadium, or appreciable quantities of tungsten, lead is preferable to nickel as a reducing agent.

The iodine solution may be made by transferring 80 g iodine and 160 g potassium iodide to a large beaker and covering with water. When all has dissolved, filter through asbestos into a dark bottle, and dilute with water to 5 litres. Allow to stand for two weeks, and standardize against a tin concentrate or tin alloy which has been carried through the steps outlined above. The strength of this iodine solution is such that all concentrates can be run on 0·5–1 g samples without using more than 50 ml iodine solution. One ml of 0·125N iodine = 0·007419 g Sn.

If desired, a solution of potassium iodate–iodine may be used instead of iodine. The strength of the iodate solution may be adjusted to the tin content as follows: for concentrates, 1 ml = 0·0075 g Sn, 4·51 g KIO_3, 100 g KI, and 4 g KOH per litre; for ores, 1 ml = 0·005 g Sn, 3·005 g KIO_3, 100 g KI, and 2 g KOH per litre; for tailings, 1 ml = 0·0025 g Sn, 1·503 g KIO_3, 100 g KI, and 2 g KOH per litre [4].

For some metals and alloys, weaker solutions are employed. A solution of potassium iodate, approximately 0·02N, and having a titre of about 1 ml = 0·001 g tin, can be made by dissolving 0·7134 g KIO_3, 10 g KI, and 1 g NaOH in 1 litre of water. An approximately 0·02N iodine solution with a titre of 1 ml = 0·001 g tin, is made by dissolving 2·54 g iodine and 40 g potassium iodide in 25 ml water, and diluting to 1 litre. In all cases, it is best to standardize against a sample of known tin content which has been carried through the procedure described.

Reduction of tin can be effected by boiling in hydrochloric acid solution with sheets, pellets, or powder of aluminium, iron, or lead, in place of nickel. For the determination of tin in aluminium, the latter metal is naturally employed as a reducer; for tin in lead, this metal is used as a reducer, and for bearing metal, iron is the preferred reducing agent. Lead is used with tungsten and ferrotungsten, and most steels; iron or nickel are usually employed for titanium and its alloys; magnesium and most magnesium-base alloys do not require any reducing agent.

For some ores and concentrates, both aluminium and nickel can be used to reduce the tin and precipitate copper, etc. Add 4 g aluminium turnings to the sample in hydrochloric acid solution, warm cautiously for 20 minutes, add 100 g nickel pellets and another 60 ml hydrochloric acid, place on the edge of the hot plate until all aluminium and tin have dissolved, and boil gently for an hour [4].

Cassiterite can also be decomposed by heating in a stream of hydrogen at 800–850°C for 20–30 minutes; most associated oxides are not attacked. In a highly glazed porcelain crucible prepare a "cup" of calcium oxide by placing a smaller crucible within the larger one and packing calcium oxide around the space between the two [4]. Into another porcelain crucible weigh out 0·5 g calcium oxide and 0·5 g tin concentrate, mix, and transfer to the prepared crucible with its "cup" of calcium oxide. Cover with a thin layer of calcium oxide and reduce in hydrogen for 20–30 minutes at 800–850°C. Cool for 10 minutes with hydrogen flowing, transfer to a 600-ml beaker, add hydrochloric acid to dissolve, and reduce as usual. For refractory samples of tin associated with silica or tantalum, replace half the lime in the crucible fusion with sodium carbonate.

Tin in blister copper may also be determined by first reducing the sample in hydrogen before proceeding with isolation and titration of tin.

Some ore samples may be fused satisfactorily with 15 g sodium hydroxide to 1 g ore in a nickel crucible; heat carefully for the first five minutes and then at full heat for a further 10 minutes. Reduce with nickel and titrate as usual.

It is useful to remember that tin can often be determined, with arsenic and antimony, on the same sample. Distillation at a low temperature from hydrochloric acid will isolate arsenic, and treatment of the filtrate with hydrogen sulphide will precipitate antimony and tin; after determination of antimony in the reduced solution with potassium bromate or permanganate, tin can be reduced and titrated iodimetrically as usual.

2. Gravimetric Method

For some metals and alloys, such as silver solders, it is common practice in control analyses to determine tin by simply dissolving the sample in 1 : 4 nitric acid, filtering the hot solution rapidly on double ashless papers, washing with hot water, and igniting the precipitate of metastannic acid in a porcelain crucible under good oxidizing conditions. $SnO_2 \times 0{\cdot}7877 = Sn$.

3. Colorimetric Methods

A. Dithiol

For small quantities of tin, a colorimetric procedure can be employed, based on the red colour imparted to solutions of stannous salts in acid solution by toluene-3,4-dithiol (4-methyl-1,2-dimercaptobenzene), commonly known as dithiol.

Decompose the sample as outlined in previous sections, with acids and fusion of any tin-bearing residue with sodium peroxide; transfer the solution to a distilling flask with 25 ml of 1 : 1 sulphuric acid, 5 g potassium dihydrogen phosphate or 3 ml phosphoric acid, 20 ml of 6 N hydrochloric acid, and 1 g of hydrazine sulphate [2, 8, 15, 24, 25]. Pass a slow stream of carbon dioxide through the solution and boil; when the temperature reaches 160°C, add 20 ml 6 N hydrochloric acid at the rate of about 1 drop every 4 seconds while maintaining the temperature at 155–165°C. Discard the distillate and washings, which contain any arsenic and antimony in the sample.

Place a fresh beaker of cold water under the outlet of the condenser; add a mixture of 15 ml of 6 N hydrochloric acid and 7 ml hydrobromic acid dropwise while maintaining the temperature at 145–160°C and passing in carbon dioxide. After 20 minutes, discontinue the distillation, add to the distillate 0·5 ml of 1 : 2 sulphuric acid, 5 ml nitric acid, and evaporate to fumes of the former.

Add 4 ml of water, transfer to a 10-ml volumetric flask, and make up to about 7 ml with water. Add 1 drop of thioglycollic acid, mix, add 0·05 ml of 1 : 19 Santomerse S, swirl, add 0·5 ml of dithiol solution, and mix. Dithiol solution is made by dissolving 0·1 g in 50 ml of 1% sodium hydroxide, with about 0·4 ml of thioglycollic acid as stabilizer; it will keep for a week in a refrigerator. Heat at 50°C for 5 minutes, add 0·5 ml of 1 : 3 Santomerse S and mix. Cool, dilute to the mark with water, mix, and measure the absorbance at 530 nm without delay. Obtain the standard curve by taking 0–60 μg of tin from dilution of a standard tin solution. The latter is made by dissolving 0·100 g of tin in 10 ml hydrochloric acid, adding 0·5 ml of thioglycollic acid, and diluting to 100 ml with water; 1 ml = 1 mg Sn. From this solution take 10 ml and dilute to 1 litre with 1 : 19 hydrochloric acid; 1 ml = 10 μg Sn.

Selenium is partially volatilized with the tin in this procedure, and will give a white turbidity in the final photometric stage if present in quantities greater than 10 μg. It should be removed by sulphur dioxide and hydroxylamine hydrochloride in the distillate before proceeding with the photometric determination.

The disadvantages inherent in the use of dithiol itself may be overcome by employing the zinc-dithiol complex, which is quite stable in the air but generates dithiol immediately when sprinkled upon the surface of an acid solution [8].

B. PHENYLFLUORONE

Stannic tin reacts with phenylfluorone, 2,3,7-trihydroxy-9-phenyl-6-fluorone, to give a red colloidal solution. Germanium, zirconium, iron, and many other elements interfere. Depending on the kind and amount of interfering elements, collect the traces of tin by co-precipitation with ferric hydroxide in ammoniacal solution, with hydrous manganese dioxide in nitric acid solution, or with hydrogen sulphide in dilute hydrochloric or sulphuric acid to which copper has been added as collector. Further separation of tin can often be made by means of cupferron–chloroform in cold 10% sulphuric acid solution. When tin has been isolated from major amounts of interfering elements by one or more of the above separations, there are several alternatives to the phenylfluorone method. For copper-base and lead-base alloys, extract with diethylammonium diethyldithiocarbamate in chloroform to remove Cu, Bi, Hg, As(III), and Sb(III) from tin(IV), reduce the latter to the stannous state, and again extract with diethylammonium diethyldithiocarbamate in chloroform to separate tin from all remaining metals except molybdenum. Destroy organic matter by treatment with sulphuric and perchloric acids, and to the final solution containing 1 ml sulphuric acid in a volume of 10 ml, add 1 ml of 3% hydrogen peroxide, 10 ml of an acetate buffer, 1ml of 1% gum arabic, and 10 ml of phenylfluorone. The latter is made by dissolving 0·05 g in 50 ml methanol and 1 ml hydrochloric acid, and diluting to 500 ml. Allow to stand for 5 minutes, add 16 ml of 1 : 9 hydrochloric acid, and transfer to a 50 ml volumetric flask. Dilute to volume with the 1 : 9 acid, and measure the absorbance at 510 nm. The *p*H adjustment and some other variables are critical; for full details of the determination, consult the references [18, 19, 25].

For 0·001–0·1% tin in low alloy steel, precipitate it as sulphide in 5% hydrochloric acid containing 1 g tartaric acid and 1 mg copper as chloride. Filter, wash, destroy organic matter by sulphuric, nitric, and perchloric acids, and obtain the tin in 10% sulphuric acid. Make a double carbamate-chloroform extraction, destroy organic matter, and dissolve tin sulphate in dilute hydrochloric acid. Add ammonium oxalate, hydrogen peroxide, sodium chloride, and phenylfluorone solution. Allow to stand, extract with methyl isobutyl ketone, and measure the tin-phenylfluorone compound at 530 nm [20].

Tin in a number of other metals and alloys can be determined by another modification [21]. Dissolve in acid, add sodium iodide, and sulphurous acid until all the iodine is reduced. Extract with benzene, discard the lower layer, and wash the benzene layer with acid. Evaporate the benzene and the iodine, and take to fumes with sulphuric and perchloric acids. Dissolve in hydrochloric acid and determine tin by the methyl isobutyl ketone–phenylfluorone method, omitting the addition of ammonium oxalate and hydrogen peroxide.

Traces of tin in silicate rocks have been determined by getting the sample into sulphuric acid solution and extracting with 8-hydroxyquinoline in chloroform. Discard the chloroform layer, and evaporate the aqueous layer to strong fumes with nitric and sulphuric acids. Dissolve in water, add phenylfluorone solution, a buffer, gelatin, and measure the absorbance at 510 nm [23]. Titanium interferes, but not Al, Fe, Cu, or Ni.

C. Catechol Violet

For 0–30 μg of tin in organic matter, the latter can be subjected to wet oxidation and the tin finally obtained in 9 N sulphuric acid. Add potassium iodide, and extract tin(IV) iodide into toluene. Return tin to an aqueous solution by shaking with sodium hydroxide. Acidify, remove free iodine with ascorbic acid, and add catechol violet at *p*H 3 maintained by an acetate buffer. Measure the absorbance at 552 nm [1, 22].

4. Optical Spectrography

In a number of materials, tin is routinely determined by emission spectrography. Methods have been published for the determination of 0·001–7·5% tin in aluminium and its alloys [3], 1 ppm to about 10% in copper and its alloys [9], 1–100 ppm in nickel [17], 0·001–0·1% in lead [3], 0·001–0·5% in zinc-base alloys [3], and 20–2000 ppm in zirconium and its alloys [3].

5. X-Ray Spectrography

Tin can be determined in many samples by X-ray fluorescence [9, 16].

6. Atomic-Absorption Spectrophotometry

This technique offers another useful method for tin. The most sensitive line is 2863·3, with a sensitivity of 5 μg/ml; other lines are 2246·0, 2840·0, 2706·5, 3034·1, 3009·1, and 2546·5 [11]. In ores and concentrates, tin has been determined in the nitrous oxide–acetylene flame by heating the sample with ammonium iodide and dissolving the resulting tin iodide in dilute hydrochloric acid [5].

7. Other Methods

Tin in copper [12] and aluminium [6] can be determined by polarography.

A turbidimetric procedure, in which tetravalent tin reacts with phenylarsonic acid, is sometimes used for copper [12].

Tin in stream sediment surveys can be determined by heating with ammonium iodide, leaching in boiling hydrochloric acid, buffering, adding gallein, and comparing with standards [26].

Tin Research Centre

The Tin Research Institute, Fraser Road, Greenford, Middlesex, England, an international producers' organization established to assist users of tin, is always helpful to chemists who may have analytical problems with this metal.

REFERENCES

1. Analytical Methods Committee, *Analyst* **92**, 320–3 (1967).
2. Analytical Methods Committee, *Analyst* **93**, 414–16 (1968).
3. A.S.T.M., *Chemical Analysis of Metals. Sampling and Analysis of Metal Bearing Ores*, Philadelphia, Pa., American Society for Testing and Materials, 1969.
4. Blain, I. R., and Van der Spuy, R. C. M., Rooiberg Minerals Development Co., Rooiberg, Tvl., South Africa, personal communication, 1961.

5. BOWMAN, J. A., *Anal. Chim. Acta* **42**, 285–91 (1968).
6. BRITISH ALUMINIUM CO., *Analysis of Aluminium and its Alloys*, London, British Aluminium Co., 1961.
7. BRITISH STANDARDS INSTITUTION 3338, *Methods for the Sampling and Analysis of Tin and Tin Alloys*, London, British Standards Institution, 1961.
8. CLARK, R. E. D., *Analyst* **82**, 182–5 (1957).
9. DOZINEL, C. M., *Modern Methods of Analysis of Copper and its Alloys*, 2nd ed., London, Elsevier, 1963.
10. DUPRAW, W. A., *Anal. Chem.* **26**, 1642–5 (1954).
11. ELWELL, W. T., and GIDLEY, J. A. F., *Atomic-Absorption Spectrophotometry*, 2nd ed., Oxford, Pergamon, 1966.
12. ELWELL, W. T., and SCHOLES, I. R., *Analysis of Copper and its Alloys*, Oxford, Pergamon, 1967.
13. FURMAN, N. H., ed., *Scott's Standard Methods of Chemical Analysis*, 6th ed., Vol. 1, Princeton, N.J., Van Nostrand, 1962.
14. HILLEBRAND, W. F., LUNDELL, G. E. F., BRIGHT, H. A., and HOFFMAN, J. I., *Applied Inorganic Analysis*, 2nd ed., New York, Wiley, 1953.
15. HOPKIN and WILLIAMS LTD., *Organic Reagents for Metals*, Chadwell Heath, Essex, Hopkin and Williams, 1955.
16. KOLTHOFF, I. M., and ELVING, P. J., *Treatise on Analytical Chemistry*, Part II, Vol. 3, 327–71, New York, Interscience Publishers, 1961.
17. LEWIS, C. L., OTT, W. L., and SINE, N. M., *The Analysis of Nickel*, Oxford, Pergamon, 1966.
18. LUKE, C. L., *Anal. Chem.* **28**, 1276–9 (1956).
19. LUKE, C. L., *Anal. Chem.* **31**, 1803–4 (1959).
20. LUKE, C. L., *Anal. Chim. Acta* **37**, 97–101 (1967).
21. LUKE, C. L., *Anal. Chim. Acta* **39**, 404–6 (1967).
22. NEWMAN, E. J., and JONES, P. D., *Analyst* **91**, 406–10 (1966).
23. POLLOCK, E. N., and ZOPATTI, L. P., *Anal. Chem.* **37**, 290–1 (1965).
24. RAVEN, T. W., *Analyst* **87**, 827–8 (1962).
25. SANDELL, E. B., *Colorimetric Determination of Traces of Metals*, 3rd ed., New York, Interscience Publishers, 1959.
26. STANTON, R. E., and MCDONALD, A. J., *Trans. Institution Mining Met.* **71**, Part 1, 27–9 (1961–62).
27. SULLIVAN, T. A., LEWIS, R. W., CARPENTER, L., and BOYLE, B. J., *U.S. Bur. Mines Rept. Invest.* 5639, 1960.
28. WRIGHT, P. A., *Extractive Metallurgy of Tin*, London, Elsevier, 1966.

37 Titanium

Titanium occurs in rocks, minerals, ores, concentrates, tailings, soils, slags, ferrous and non-ferrous alloys, cemented carbides, paints, pigments, papers, floor coverings, rubbers, and in the pure metallic state; it is an important element in many mining and metallurgical laboratories. The principal titanium minerals are rutile TiO_2, ilmenite $FeTiO_3$, perovskite $CaTiO_3$, and titanite $CaTiSiO_5$.

ISOLATION OR SEPARATION OF TITANIUM

Titanium belongs to the ammonium hydroxide group, and is separated from members of the acid sulphide group by precipitation of the latter with hydrogen sulphide in 5–10% hydrochloric or sulphuric acid. Together with Fe, Al, Be, Cr, Zr, Th, and U, titanium is isolated with ammonium hydroxide and ammonium chloride from elements of succeeding groups such as Mn, Co, Ni, Zn, Tl, Ca, Sr, Ba, Mg, and the alkalies. Titanium, with Fe, Zr, and Th, is precipitated with sodium hydroxide + sodium peroxide and can be thereby separated from Al, P, V, Cr, U, and Zn. Together with Fe, V, Zr, Mo, Nb, Ta, and W, titanium is precipitated by cupferron in cold 10% sulphuric acid, and may be isolated from Al, Be, Ni, Co, P, Mn, Cr, U(VI), and Zn. Titanium, like Al, Zr, P, V, Be, and U, is not deposited at the mercury cathode in dilute sulphuric acid, and can be thus separated from Fe, Sn, Cd, Bi, Cr, Ni, Co, Mo, Cu, and Zn [4, 18, 21, 25]. When ammonium sulphide is added to an ammoniacal tartrate solution of titanium, the latter remains in solution, together with Al, V, Cr, Nb, Ta, and W, whereas Fe, Ni, Co, and Zn are precipitated. It is sometimes convenient to separate large quantities of iron from titanium by an ether separation. Titanium, with iron and most other metals, can be separated from Co, Ni, and Mn by zinc oxide. Chromium can be separated from titanium by volatilization of the former with boiling perchloric and hydrochloric acids.

1. Colorimetric Methods

A. With Hydrogen Peroxide

Moderate amounts of titanium can be determined in most samples encountered in extractive metallurgy, with a minimum number of separations, by utilizing the yellow to orange colour of titanium sulphate in the presence of hydrogen peroxide [1, 2, 3, 5, 6, 9, 10, 11, 12, 13, 15, 16, 19, 21]. Interferences are Fe, Ni, Co, Cu, V, Mo, Cr, and Nb because their solutions are coloured or they form coloured compounds with hydrogen peroxide; fluorine, large amounts of phosphate or alkali salts bleach the colour. The procedure outlined below is designed for complex materials; it can be shortened appropriately for most samples.

Decompose 0·5–5 g, containing about 0·005–0·01 g titanium, in 10–20 ml hydrochloric acid, 10–20 ml nitric acid, and 10 ml 1 : 1 sulphuric acid. Evaporate to strong fumes of sulphuric acid, cool, dilute, filter, wash, dry and ignite the precipitate, fuse with sodium carbonate or potassium bisulphate and add to the main filtrate. Alternatively, decompose the sample in a platinum dish with hydrofluoric, nitric, and sulphuric acids; evaporate to fumes of sulphur trioxide, cool, wash down, add a little more sulphuric acid if necessary and again evaporate to fumes to get rid of all fluorine. Metallic titanium is soluble in hot concentrated hydrochloric or sulphuric acids, but is usually dissolved in a sulphuric + hydrofluoric acid mixture; it is not soluble in nitric acid. Other decomposition procedures for titanium-bearing materials have been reviewed [8].

If members of the acid sulphide group are present, precipitate them with hydrogen sulphide in 5–10% sulphuric acid, filter and wash well. To the filtrate add 1 g tartaric acid, and then about 2 ml excess ammonium hydroxide per 100 ml. Add 1 g ammonium bisulphite, pass in hydrogen sulphide for five minutes, filter off the precipitated Fe, Ni, Co, and Zn, and wash thoroughly. If the precipitate is very heavy, dissolve in acid and re-precipitate to recover any occluded titanium. Boil the filtrate to the complete expulsion of hydrogen sulphide.

Adjust the volume of the solution to 150 ml and add 30 ml of 1 : 1 sulphuric acid. Cool to 10°C, and precipitate titanium with a cold freshly prepared 6% solution of cupferron until an excess of the latter is indicated by the formation of a white precipitate which rapidly disappears. Filter the precipitated titanium, together with Zr, V, Nb, Ta, and W; wash thoroughly with cold 5% sulphuric acid, transfer to a platinum dish or large crucible, ignite, cool, fuse with potassium bisulphate, or potassium pyrosulphate, and extract the solidified melt with dilute sulphuric acid.

If vanadium and large amounts of phosphorus are present in the sample, they will still be with the titanium at this stage, and will interfere in the subsequent colorimetric method. In this case, it is better to separate iron and titanium with sodium hydroxide at an earlier stage because the presence of ferric iron is necessary for the complete precipitation of titanium and its separation from V, P, etc. After removing elements of the acid sulphide group, precipitate Fe, Ti, Zr, etc. with sodium hydroxide, filter through Whatman No. 50 paper, re-precipitate if necessary, dissolve in acids, and proceed with the cupferron separation.

Transfer the final solution, which has been freed of coloured ions, V, P, and other interferences, and which contains 5–10% by volume sulphuric acid, to a 50 ml volumetric flask. Add 5 ml of 3% hydrogen peroxide, make to volume, mix, and compare in a photometer with standards at 449 nm.

For standardization, dissolve 0·5 g titanium dioxide in sulphuric and hydrofluoric acids in a platinum dish, evaporate to fumes of sulphur trioxide, cool, wash down and again evaporate to fumes. Dilute to 1 litre; 1 ml = 0·0005 g TiO_2. Withdraw 2·5, 5, and 10 ml of solution, dilute to 50 ml in a flask containing 5 ml sulphuric acid and 5 ml hydrogen peroxide. This will give values of absorbance for 1·25, 2·50, and 5 mg TiO_2 per 50 ml.

In rock and ore analysis, titanium is usually expressed as TiO_2. For most other materials, titanium is expressed as metal. In this case, for the standardization, weigh 0·834 g titanium dioxide and proceed as directed above to obtain a solution having a value of 1 ml = 0·0005 g titanium.

B. With Tiron

The colorimetric determination of titanium with Tiron, 1,2-dihydroxy-benzene-3,5-disulphonate, is more sensitive than the method using hydrogen peroxide [2, 19, 23]. Iron is rendered harmless by thioglycollic acid, and the following are without effect: SiO_2, Al_2O_3, CaO, MgO, Na_2O, K_2O, P_2O_5, MnO, S, Cl, F, BaO, SrO, and ZrO_2. The tolerance limits for 0·5% titanium dioxide on the following interferences are, in %, Cr_2O_3 0·84, V_2O_3 0·13, CuO 0·06, WO_3 0·19, MoO_3 0·01 and U_3O_8 0·60.

Decompose a 0·5 g sample, which should not contain more than 4 mg titanium dioxide, with hydrofluoric and sulphuric acids in a platinum dish. Evaporate to fumes of sulphur trioxide, and, if necessary to remove organic matter, cool, add nitric acid and refume with sulphuric acid. Filter through Whatman No. 42 paper into a 100-ml volumetric flask, wash thoroughly, make up to the mark, and mix. Withdraw a 5-ml aliquot to a 50-ml volumetric flask, and add the following: 25 ml of a buffer solution made from a mixture of 1000 ml of 1M sodium acetate solution and 390 ml of glacial acetic acid, 5 ml of a 5% aqueous solution of Tiron, and 2 ml of 20% thioglycollic acid solution. Make up to the mark and mix.

Allow to stand for one hour and measure the absorbance in a photometer at 380 nm. Prepare the calibration graph as follows: Heat 0·5 g titanium dioxide with 10 g ammonium sulphate and 25 ml sulphuric acid in a Pyrex "copper" flask on a hot plate, and finally over a flame until the oxide is completely dissolved. Dilute to 1 litre with 1N sulphuric acid. From a burette, measure 20 ml of this solution and dilute to 1 litre with 0·1N sulphuric acid. The last solution will contain 10 μg of titanium dioxide per ml. Add 1–15 ml of this solution to 25 ml of buffer, 5 ml of Tiron, and 2 ml of thio-glycollic acid solution in 50 ml volumetric flasks to establish the standard graph.

C. Other Colorimetric Methods

Titanium has been determined in ores [17], and in metals [27], with diantipyrlmethane, sometimes called diantipyrinylmethane.

2. Volumetric Determination

A. Jones Reductor

The volumetric determination of titanium depends on its reduction from the quadrivalent state by amalgamated zinc, lead, cadmium, or aluminium, and subsequent oxidation by ferric ammonium sulphate; potassium thiocyanate is used as the indicator [3, 5, 6, 7, 10, 11, 12, 15, 16, 19, 22, 24, 26]. Interfering elements are members of the acid sulphide group like As, Sb, Sn, Mo, and Cu, and Cr, V, U, W, Nb, and Ta.

Decompose a suitable weight of sample containing not more than 0·2 g titanium by one of the following methods:

(a) Treat with hydrofluoric, nitric, and sulphuric acids in a platinum dish, and evaporate to fumes of sulphur trioxide, twice if necessary, to eliminate all hydrofluoric acid.

(b) Treat with hydrochloric, nitric, and sulphuric acids in a beaker and follow by fusing any residue with sodium carbonate or potassium bisulphate.

(c) To the sample in a conical flask add 10 g sodium or ammonium sulphate and 10 ml sulphuric acid. Digest on the hot plate and heat over a burner until solution is complete.

(d) Add 0·5 ml sulphuric acid and about 20 g of sodium or potassium bisulphate to the sample in a conical flask, and heat over a burner to complete decomposition.

Dilute with water, boil, and, if members of the acid sulphide group are present, adjust the acidity to 5–10% and precipitate them with hydrogen sulphide; filter, wash, remove all hydrogen sulphide from the filtrate by vigorous boiling, oxidize with hydrogen peroxide or bromine, and boil to remove the excess of oxidant.

If Cr, U, V, or W are present, separate titanium together with Fe, Zr, and Th by precipitation with sodium hydroxide + sodium peroxide. Filter and wash. Dissolve the precipitate in hydrochloric, nitric, and sulphuric acid, and evaporate to fumes of the latter; alternatively, ignite the precipitate and fuse with sodium bisulphate. In the absence of Cr, U, V, and W, which interfere in the volumetric determination, isolate titanium with other members of this group by ammonia and ammonium chloride, and wash thoroughly. Dissolve the precipitate in about 100 ml of 1 : 1 hydrochloric acid and warm until solution is complete.

If niobium and tantalum are present, add tartaric acid to make its concentration in the solution 25%. Only titanium will then be reduced in the subsequent stage with zinc amalgam.

Add 70 g of amalgamated zinc shot and keep hot for 10 minutes. Flush out a Jones reductor with carbon dioxide and continue the flow throughout the determination. Drain the liquid in the reductor to the level of the zinc, pour the reduced titanium solution into the reductor and rinse with 1 : 9 hydrochloric acid. Allow the solution to pass down the reductor over a period of about 4 minutes. Rinse the original container and the reductor with 4–5 successive 50-ml portions of 1 : 9 hydrochloric acid, taking care that the level of solution drops to the top of the zinc before adding the next rinse.

Add 25 ml of 40% potassium thiocyanate through a tap funnel to the solution which has passed through the reductor, agitate the solution magnetically, and titrate with standard ferric ammonium sulphate until a faint reddish brown colour persists for at least one minute.

Reagents

Amalgamated zinc shot. Transfer 1 kg of zinc shot into a 2-litre beaker. Wash twice with water, and cover with water to a depth of about $\frac{1}{2}$ inch. Digest 24 g of mercuric chloride in 220 ml of hydrochloric acid and add to the zinc; stir with a wooden paddle for 2 minutes. Decant the amalgamating solution, then wash, by decantation with water, about ten times.

Amalgamated zinc for Jones reductor. Wash 300 g of 20-mesh zinc several times with water. Cover to a depth of $\frac{1}{2}$ inch with water, and add a solution of 14 g of mercuric chloride in 120 ml of hydrochloric acid in six portions; shake for 15 seconds after each addition, and finally shake for $1\frac{1}{2}$ minutes. Decant this amalgamating solution and wash, by decantation, about ten times. Pour the amalgamated zinc into a Jones reductor filled with water and allow to settle for several hours. Before using the reductor, activate it by washing with three 300-ml portions of 1 : 9 hydrochloric acid, and finally add water until the zinc is covered.

Ferric ammonium sulphate. Dissolve 30 g of $FeNH_4(SO_4)_2.12H_2O$ in about 300 ml water, add 20 ml sulphuric acid, and cool. Add 0·1 N potassium permanganate dropwise until a slight excess is present, indicated by a faint pink colour. Transfer to a litre flask and dilute with water to the mark. This solution has an approximate value of 1 ml = 0·005 g titanium dioxide; standardize against 0·2 g of pure titanium dioxide.

Alternatively, especially for titanium in metals and alloys, a 0·1 N solution of ferric ammonium sulphate may be used. Dissolve 48·4 g of $FeNH_4$-$(SO_4)_2.12H_2O$ in about 300 ml of water, add 30 ml sulphuric acid, cool, and continue as outlined in the previous paragraph. This solution has an approximate titre of 1 ml = 0·00479 g titanium; it can be standardized against 0·2 g of pure titanium.

B. Other Methods of Reduction

Alternative procedures to reduce titanium are equally satisfactory. The Nakozono reductor is a modified spherical separatory funnel provided with two inlets and stopcocks at the top of the vessel for maintaining a flow of carbon dioxide; the chloride solution is shaken for five minutes with zinc amalgam, the latter drawn off, and the solution titrated with ferric ammonium sulphate, using potassium thiocyanate as indicator, in the usual manner.

In place of zinc, aluminium may be used as a reducing agent [19, 22]. To the solution of titanium in a flask containing about 100 ml of 1 : 1 hydrochloric acid, add 1–3 g aluminium metal and heat until all titanium is reduced. Exclude air by inserting one end of a delivery tube in the flask, while the other end dips beneath a solution of sodium bicarbonate in a beaker. Cool, and titrate with a standard ferric salt, using potassium thiocyanate as indicator.

Reduction of titanium has also been accomplished by cadmium, followed by potentiometric titration with ferric ammonium sulphate [7].

For routine purposes on high grade titaniferrous materials, reduction is sometimes effected by boiling the solution containing about 100 ml 1 : 1 hydrochloric acid with 10 g test lead for 15 minutes. Remove from the hot plate, decant the solution immediately into a beaker, wash the test lead rapidly 5–6 times by decantation, add 5 ml saturated potassium thiocyanate, and quickly titrate with standard ferric ammonium sulphate to a faint pink colour. Because air is not excluded during reduction and titration, a slight error is inevitable; this error, however, may be well below the precision demanded.

3. Gravimetric Methods

In comparatively few materials is titanium determined gravimetrically. A procedure for titanium in ferrotitanium is outlined briefly as an example [2]. Dissolve the sample in a platinum dish with hydrofluoric, sulphuric, and nitric acids; evaporate to dryness, dissolve in 30 ml of 1 : 4 sulphuric acid, and transfer to a beaker. Add 20 ml of 50% tartaric acid solution, and pass in hydrogen sulphide for 20 minutes; filter and wash thoroughly with 1% sulphuric–tartaric acid solution. To the filtrate, or to the clear solution if filtration was unnecessary, add ammonium hydroxide to slight excess, and pass in hydrogen sulphide for 10 minutes. Filter, using paper pulp, and wash thoroughly with 1% ammonium sulphide–ammonium tartrate solution.

Boil the filtrate for 10 minutes, add 25 ml of 1 : 1 sulphuric acid and reduce the volume by boiling to about 150 ml. Cool to 15°C, add a slight excess of cold 6% solution of cupferron, filter, and wash thoroughly with 0·1% cupferron in 1 : 9 sulphuric acid. Ignite the paper and precipitate in a weighed platinum dish, initially at a low temperature and finally at 1050–1100°C for 20 minutes. Cool, and weigh as $TiO_2 + ZrO_2 + V_2O_5$. Fuse the oxides with 5 g of potassium pyrosulphate, dissolve the cold melt in 70 ml of 1 : 9 sulphuric acid, add 5 ml hydrogen peroxide and 25 ml of 16% di-ammonium phosphate solution. Allow to stand overnight in a warm spot, filter, wash with 2% hydrochloric acid and finally with 5% ammonium nitrate. Place in a platinum crucible, dry, char, ignite at 1050–1100°C to constant weight, cool, and weigh as ZrP_2O_7; $ZrP_2O_7 \times 0{\cdot}4647 = ZrO_2$.

Correct for vanadium by dissolving a separate sample in a platinum dish with hydrofluoric, sulphuric, and nitric acids, and evaporating to strong fumes of sulphur trioxide. Cool, transfer to a 600-ml beaker containing 25 ml of 4% boric acid, dilute to 200 ml, and add 40 ml nitric acid. Boil for an hour or longer until the final volume is 100–125 ml, dilute to 300 ml with cold water, cool to 10°C, and titrate potentiometrically with 0·02N ferrous ammonium sulphate. Calculate any vanadium found to V_2O_5 by the relation $V \times 1{\cdot}7851 = V_2O_5$.

With the exception of Nb, Ta, and W, no other element will interfere in the above procedure for titanium. Niobium and tantalum may be separated from titanium by chromatography or ion exchange as described in the chapter on Niobium and Tantalum. If tungsten is present it may be separated from titanium in the following manner [24]. Fuse the mixed oxides in a platinum crucible with sodium carbonate for 20 minutes. Transfer the crucible and contents to a nickel dish containing 20% sodium hydroxide solution, and digest on the steam bath for several hours with occasional stirring and addition of water to replace evaporation losses. Withdraw and rinse the crucible, cool the liquid, filter, and wash the residue with half-saturated sodium chloride solution. To the filtrate add phenolphthalein, dilute hydrochloric acid to discharge the red colour, and heat on the steam bath; add additional hydrochloric acid from time to time to discharge the red colour. When the small quantity of hydrous oxide has flocculated, filter, wash as before, and combine this with the first precipitate. Pulp the filters containing the two precipitates with hot dilute hydrochloric acid, make the liquid ammoniacal, filter, wash, and ignite to titanium dioxide.

4. Optical Spectrography

Titanium can be determined in many samples by emission spectrography. Methods have been published for 0·001–0·5% in aluminium and its alloys [2], 0·05–0·30% in plain carbon and low alloy steel [2], 20–200 ppm in zirconium and its alloys [2], and 1–1000 ppm in nickel [20].

5. X-Ray Spectrography

For a few materials this technique can be used for the determination of titanium [19].

6. Atomic-Absorption Spectrophotometry

A paper has been published on the determination by atomic absorption of 0·1–1·2% titanium in steels, magnet alloys, and cast iron using a nitrous oxide–acetylene flame and hydrofluoric acid solution [14].

REFERENCES

1. Allport, N. L., and Brocksopp, J. E., *Colorimetric Analysis*, London, Chapman and Hall, 1963.
2. A.S.T.M., *Chemical Analysis of Metals. Sampling and Analysis of Metal Bearing Ores*, Philadelphia, Pa., American Society for Testing and Materials, 1969.
3. Barksdale, J., *Titanium: Its Occurrence, Chemistry and Technology*, 2nd ed., New York, Ronald Press, 1964.
4. British Iron and Steel Research Assoc., *J. Iron Steel Institute* **176**, Part 1, 29–36 (1954).
5. Codell, M., *Analytical Chemistry of Titanium Metals and Compounds*, New York, Interscience Publishers, 1959.

6. CODELL, M., NORWITZ, G., and MIKULA, J. J., *Anal. Chem.* **27**, 1379–83 (1955).
7. DENTON, C. L., and WHITEHEAD, J., *Analyst* **91**, 224–36 (1966).
8. DOLEZAL, J., POVONDRA, P., and SULCEK, Z., *Decomposition Techniques in Inorganic Analysis*, London, Iliffe, 1968.
9. ELWELL, W. T., and SCHOLES, I. R., *Analysis of Copper and its Alloys*, Oxford, Pergamon, 1967.
10. ELWELL, W. T., and WOOD, D. F., *The Analysis of Titanium, Zirconium and Their Alloys*, London, Wiley, 1961.
11. ELWELL, W. T., and WOOD, D. F., *Analysis of the New Metals Titanium, Zirconium, Hafnium, Niobium, Tantalum, Tungsten, and their Alloys*, Oxford, Pergamon, 1966.
12. FURMAN, N. H., ed., *Scott's Standard Methods of Chemical Analysis*, 6th ed., Vol. 1, Princeton, N.J., Van Nostrand, 1962.
13. HAYWOOD, F. W., and WOOD, A. A. R., *Metallurgical Analysis by Means of the Spekker Photoelectric Absorptiometer*, 2nd ed., London, Hilger and Watts, 1956.
14. HEADRIDGE, J. B., and HUBBARD, D. P., *Anal. Chim. Acta* **37**, 151–5 (1967).
15. HILLEBRAND, W. F., LUNDELL, G. E. F., BRIGHT, H. A., and HOFFMAN, J. I., *Applied Inorganic Analysis*, 2nd ed., New York, Wiley, 1953.
16. IMPERIAL CHEMICAL INDUSTRIES LTD., *The Analysis of Titanium and its Alloys*, 3rd ed., London, I.C.I., 1959.
17. JEFFERY, P. G., and GREGORY, G. R. E. C., *Analyst* **90**, 177–9 (1965).
18. JOHNSON, H. O., WEAVER, J. R., and LYKKEN, L., *Anal. Chem.* **19**, 481–3 (1947).
19. KOLTHOFF, I. M., and ELVING, P. J., *Treatise on Analytical Chemistry*, Part II, Vol. 5, 1–60, New York, Interscience Publishers, 1961.
20. LEWIS, C. L., OTT, W. L., and SINE, N. M., *The Analysis of Nickel*, Oxford, Pergamon, 1966.
21. PAPUCCI, R. A., *Anal. Chem.* **27**, 1175–6 (1955).
22. RAHM, J. A., *Anal. Chem.* **24**, 1832–3 (1952).
23. RIGG, T., and WAGENBAUER, H. A., *Anal. Chem.* **33**, 1347–9 (1961).
24. SCHOELLER, W. R., and POWELL, A. R., *The Analysis of Minerals and Ores of the Rarer Elements*, 3rd ed., London, Charles Griffin, 1955.
25. STEINMETZ, M. H., *Ind. Eng. Chem. Anal. Ed.* **14**, 109 (1942).
26. THOMPSON, J. M., *Anal. Chem.* **24**, 1632–4 (1952).
27. WOOD, D. F., and JONES, J. T., *Anal. Chim. Acta* **47**, 215–24 (1969).

38 Tungsten

Tungsten is found in many alloy steels, non-ferrous alloys, and cemented carbides; its determination is occasionally or frequently required in most laboratories serving the fabricating metallurgical industries. Apart from producers of tungsten, and of some ores of niobium, tantalum, and tin, few laboratories in the field of extractive metallurgy will encounter the element in more than traces. The important tungsten minerals are scheelite $CaWO_4$, and wolframite $(Fe,Mn)WO_4$; others occurring to a lesser extent are tungstenite WS_2, ferberite $FeWO_4$, hubnerite $MnWO_4$, and cuproscheelite $(CaCu)WO_4$.

ISOLATION OR SEPARATION OF TUNGSTEN

Tungsten accompanies silicon, niobium, and tantalum in the general scheme of analysis, tungstic acid being largely insoluble on digestion with sulphuric, hydrochloric, perchloric, or nitric acids. Part of the tungsten passes into the filtrate, however, and must be recovered; a precipitate of tungstic acid is usually contaminated with one or more of the following: Si, Nb, Ta, Sn, Sb, P, Mo, V, Ag, and Fe.

Tungsten is left in the filtrate, with Ti, V, Nb, Ta, Al, and Cr when a solution is treated with ammonium sulphide and ammonium tartrate; Fe, Co, Zn, and a large part of the Ni and Mn are precipitated. Sodium hydroxide will precipitate most elements except W, Be, Al, V, Mo, Zn, Sn, Pb, As, Sb, Se, and Te; when sodium peroxide is used, Cr and U are thrown into the filtrate. Electrolysis at a mercury cathode in dilute sulphuric acid will deposit Fe, Co, Ni, Cu, Zn, Ag, Cd, Sn, Cr, and Mo, leaving in solution W, Nb, Ta, V, Ti, Zr, U, Al, Be, and P. Cupferron in cold 10% sulphuric or hydrochloric acid will precipitate W, Ti, V, Zr, Mo, Nb, Ta, Fe, Sn, Sb, and Bi from Al, Be, Cr, Mn, Co, Ni, U(VI), and Zn. Tungsten can be separated from iron by ether extraction, and from Ni, Co, and Mn by a zinc oxide separation. When separating tungsten from Cu, Mo, and other members of the acid sulphide group by hydrogen sulphide in dilute acid solution, tartrate should be present to prevent co-precipitation of tungsten. The presence of phosphoric acid prevents the precipitation of tungsten in acid solution; for example, 0·2 g of tungsten will not precipitate in 30 ml of a solution containing 15% v/v H_3PO_4 and 15% v/v H_2SO_4.

1. Gravimetric Methods

A. Ores, Concentrates, etc.

This acid digestion–cinchonine procedure will determine tungsten in any ore; for many samples it can be materially shortened [7, 8, 15, 16, 19]. Tungsten ores tend to be refractory to conventional acid or fusion treatments; their dissolution is greatly accelerated if the material is ground to −200 mesh. Transfer 1 g of the finely ground sample to a 400-ml beaker, add 100 ml hydrochloric acid and heat for an hour at 60°C. Increase the heat and boil to a volume of about 50 ml, stirring occasionally, and add 40 ml hydrochloric and 15 ml nitric acid. Boil until the volume is reduced to about 50 ml. Add 5 ml nitric acid, stir to break up all lumps, and boil to a volume of 10–15 ml. Add 150 ml of hot water, stir thoroughly, and allow to digest warm for 15 minutes. Add 5 ml of cinchonine solution, made by dissolving 125 g of cinchonine in 1 litre of 1 : 1 hydrochloric acid, and digest the sample at 60–70°C for at least 30 minutes.

Let the precipitate of tungstic acid settle, decant the supernatant liquid through Whatman No. 40 paper with pulp, and wash the precipitate four times by decantation with hot cinchonine wash solution made by diluting 25 ml of the cinchonine solution and 30 ml hydrochloric acid to 1 litre. Transfer the precipitate to the paper and wash the beaker, paper, and precipitate thoroughly. Add cinchonine solution to the filtrate and washings (No. 1), mix thoroughly, and set aside to be sure that recovery of tungsten is complete. Transfer the washed precipitate to the original beaker by 25 ml of water from a jet. Add about 6 ml ammonium hydroxide or enough to give a slight excess, cover, and warm gently for several minutes. Wash the sides of the beaker with warm 10% ammonium hydroxide containing 10 g ammonium chloride per litre. Stir the solution thoroughly, filter through the same paper that was used before, collect the filtrate in a 400-ml beaker, and wash the original beaker, the filter and the residue with warm dilute ammonium hydroxide. Reserve the residue (No. 2).

Evaporate the filtrate until most of the ammonia has been expelled. Add 20 ml hydrochloric and 10 ml nitric acid, and boil to a volume of 10–15 ml. Dilute to 150 ml with hot water, add 10 ml cinchonine solution, stir thoroughly, digest at 80–90°C for at least 30 minutes, and allow to cool. Add paper pulp, transfer the precipitate completely to a Whatman No. 40 paper, and wash with the hot cinchonine wash solution. This is the main precipitate; it is to be ignited together with any additional tungsten obtained. Thoroughly mix the filtrate and washings (No. 3), and set aside until precipitation is complete.

If the material was completely decomposed, any tungsten held in the reserved residue (No. 2) is usually combined with iron or aluminium; it can be dissolved by digesting the residue with warm 10% hydrochloric acid, filtering, washing in turn with hot 0·5% solution of ammonium chloride and the ammonium hydroxide wash solution. The tungsten (Recovery A) is then precipitated by further treatments with acid and cinchonine, as done with the boiled ammonium hydroxide extract, with which it should not be combined.

The residue (No. 4) still left after the treatments with dilute hydrochloric acid and ammonium hydroxide is usually free from tungsten, but may contain SiO_2, Nb, Ta, and Sn. Ignite the paper and residue (No. 4) at a low temperature in a porcelain crucible, transfer the ash to a platinum crucible, and volatilize silicon by treatment with hydrofluoric and sulphuric acids. Fuse the residue with a little sodium carbonate, cool, extract with water, filter, and acidify the filtrate with hydrochloric acid. Boil to expel carbon dioxide, test for tungsten by adding 5 ml cinchonine reagent and digesting warm for several hours and at room temperature overnight. If a precipitate appears in this solution, or in the two reserved filtrates and washings (No. 1 and 3), filter all through the same paper, and wash the combined precipitates with cinchonine wash solution. Extract with ammonium hydroxide, precipitate, and wash as in the preceding recovery (Recovery B).

Transfer the papers containing the main precipitate and the recoveries A and B to a large weighed platinum crucible, and heat at a low temperature until all carbon is destroyed. Cool, moisten the precipitate with a little hydrofluoric acid, evaporate to expel any silica, and then ignite at 750°C. Cool, weigh as impure tungstic oxide and repeat the ignition to constant weight. Because tungstic oxide commences to volatilize slowly at temperatures above 800°C, ignition must be not higher than 750°C.

Examine the tungstic oxide for possible contaminants by fusing with sodium carbonate, extracting the melt with water, filtering, washing with warm 1% sodium carbonate and then with hot water. Reserve the filtrate; ignite the paper and residue, and repeat the operation. Combine the filtrates and reserve this. Ignite the paper and residue, cool, weigh, and correct the weight of tungstic oxide for the oxides found. If silver is present, digest the first insoluble residue with ammonium hydroxide, filter, wash, and then proceed with the ignition and re-fusion with sodium carbonate. Treat the ammoniacal filtrate with ammonium sulphide, recover any precipitate and ignite it with that obtained in the combined water extracts.

Acidify the water extracts, boil to expel carbon dioxide and then add ammonium hydroxide in excess. Usually no precipitate will remain; if one does, filter, wash, ignite, and deduct its weight from that of the tungstic oxide. Add 3–5 g tartaric acid to the clear ammoniacal solution, saturate with hydrogen sulphide, add sulphuric acid until 1% by volume in excess, and digest for one hour at 40–60°C. Filter, wash with hydrogen sulphide water containing 0·5% each of tartaric and sulphuric acids. Reserve the filtrate if tests for phosphorus and vanadium are necessary. Since the sulphides may still contain tungsten, dissolve the precipitate in hot dilute nitric acid containing a little bromine, boil to expel the latter, add 1–2 g tartaric acid and make the solution ammoniacal. Again treat with hydrogen sulphide and proceed as before. Ignite carefully at a temperature not over 600°C, cool, weigh, and correct the tungstic oxide for the oxides found. $WO_3 \times 0{\cdot}793 = W$.

B. Steels, Ferrotungsten, Tungsten Metal, etc.

Transfer 1–5 g of steel to a 400-ml beaker, add 50 ml hydrochloric acid and warm until decomposition is complete [2, 5, 7, 9, 14, 16, 19]. Cautiously add 10 ml of nitric acid, and digest at 100°C with occasional stirring until the tungstic acid is bright yellow and free of black particles. Dilute to 150 ml with water.

Decompose 1 g of ferrotungsten or tungsten metal in a platinum dish with hydrofluoric, nitric, and sulphuric acids; evaporate to dense fumes of sulphur trioxide. Cool, transfer to a 600 ml beaker, rinse the dish several times with 1 : 1 hydrochloric acid and then with 1 : 1 ammonium hydroxide. Dilute to 150 ml with water.

To the sample solution, add 5–10 ml of 12·5% cinchonine solution in 1 : 1 hydrochloric acid, digest for at least 30 minutes and preferably overnight. Molybdenum retards the precipitation of tungsten with cinchonine; for high-molybdenum steels allow the solution to stand about 36 hours to obtain complete precipitation of tungsten. Filter, using pulp, and wash thoroughly with hot 0·3% cinchonine solution. Transfer to a platinum crucible, dry, char, ignite at about 600°C until all carbon has been destroyed; for steels add 1–2 drops of 1 : 1 sulphuric acid and 1–3 ml of hydrofluoric acid, evaporate to dryness, ignite at 750°C, and weigh as impure WO_3.

Add about 5 g of sodium carbonate, fuse, cool, dissolve in water, filter, wash the crucible and residue, and reserve the filtrate. Transfer the residue to the crucible, ignite, fuse with about 1 g of sodium carbonate, cool, and dissolve in water. Filter, wash, and combine the filtrate with the reserved filtrate. Again transfer the residue to the crucible and ignite. Cool, weigh, correct for the residue obtained from 6 g of sodium carbonate, and subtract the corrected weight from that of "impure WO_3".

If the combined filtrates show a yellow colour, evaporate to 100 ml and determine chromium colorimetrically. Calculate as Cr_2O_3 and deduct from the weight of "impure WO_3". Divide the solution into three equal parts. In one determine MoO_3 by the thiocyanate colorimetric method [2]; in the second, determine V_2O_5 colorimetrically [2]. In the third aliquot, separate Sn, Nb, Ta, etc. by acidifying, adding ammonium hydroxide in moderate excess, and boiling until the odour of ammonia is barely perceptible. Filter, wash thoroughly with hot 2% ammonium chloride, ignite, and weigh. Add the weights of the oxides found in the three aliquots, multiply by 3, and subtract from the weight of "impure WO_3".

Small quantities of tungsten in steels and similar materials can be collected and precipitated in sulphuric acid solution by adding a small amount of molybdenum, and precipitating the latter with the tungsten by α-benzoinoxime; tungsten is finally determined in the mixed oxides with cinchonine.

Cemented carbides are usually decomposed in a platinum dish with hydrofluoric and nitric acids, or by fusion with sodium peroxide in a nickel crucible. Separate the tungsten with cinchonine, weigh as impure WO_3, fuse with potassium bisulphate, and determine constituents like Ti, Nb, Ta with cupferron in the presence of tartrate [6, 18]. Determine molybdenum colorimetrically with thiocyanate and subtract its weight, also, from the impure tungstic oxide.

2. Colorimetric Methods

A. Thiocyanate Method

When thiocyanate and stannous chloride in sulphuric–hydrochloric acid are added to a tungstate solution, a yellow colour derived from a tungsten(V) complex slowly forms in the acid solution on standing [1, 2, 7, 13, 14, 15, 17]. The method is subject to interference from As, Mo, V, appreciable quantities of iron, and elements such as Ni, Co, and Cr which give coloured ions. The procedure described below has been used for many years by the writer on a wide variety of ores, tailings, and concentrates.

Nearly all interferences except arsenic, vanadium, and molybdenum are removed in the precipitate after a sodium peroxide–sodium carbonate fusion. Molybdenum thiocyanate fades very rapidly in strong acid; the absorbance of 5 mg of molybdenum is equivalent to about 0·05 mg of tungsten. If a large amount of molybdenum is present, it can be separated from tungsten by adding 2–5 g tartaric acid to the alkaline solution, acidifying, and passing in hydrogen sulphide until all molybdenum has been precipitated. Vanadium is a serious interference, but is fortunately rare in tungsten ores. If vanadium is present in the sample, it can be separated by the usual cupferron–chloroform extraction if the tungsten is first complexed with sodium fluoride. When arsenic is present in more than traces, volatilize it by evaporating the sample several times to dryness with hydrobromic and sulphuric acids.

Fuse 0·5–2 g, containing not more than 0·15 g tungsten, in an iron or zirconium crucible with 3 g sodium peroxide and 2 g sodium carbonate. Cool, place in a 250-ml beaker, cover, add 50 ml water, and when disintegration is complete rinse and remove the crucible.

Add 10 ml ethanol, boil for 3–4 minutes, cool, dilute with water, filter into a 100- or 250-ml volumetric flask, wash with 0·5% sodium hydroxide and make up to the mark. The volume will depend on the tungsten content. For about 0·1% tungsten, use a 100-ml flask and take a 25-ml aliquot, whereas if the tungsten content is around 1%, dilute to 250 ml and withdraw a 25-ml aliquot.

Transfer the aliquot containing 0·2–1·5 mg of tungsten to a 100-ml flask. Dilute if neccessary to 25 ml, cautiously add 10 ml sulphuric acid and 20 ml hydrochloric acid. Add 10 ml of 2M stannous chloride solution, and digest on the steam bath for 5 minutes. Remove, and stopper tightly with a rubber or glass stopper, preferably the former. Cool to 10°C or less in an ice-cold water bath, remove the stopper quickly and add 10 ml of 2M potassium thiocyanate solution. Dilute to the mark, mix well, and return to the cold water bath for about 3 minutes. Remove, and after 10 minutes measure the absorbance at 400 nm.

The reduced tungsten is easily oxidized by air, but after addition of potassium thiocyanate the colour is stable for at least an hour.

To prepare 2M stannous chloride, which is stable for about a month, dissolve 45·2 g $SnCl_2.2H_2O$ in 50 ml of warm hydrochloric acid, cool, and make to 100 ml with hydrochloric acid. The 2M potassium thiocyanate, which is also stable for about a month, is made by dissolving 19·4 g KCNS in 50 ml of warm water, cooling, and diluting to 100 ml.

Obtain the calibration graph by taking 2–15 ml aliquots of a standard tungsten solution, made as follows. Weigh 1·7940 g $Na_2WO_4.2H_2O$, add one pellet of sodium hydroxide, dissolve in water, and dilute to 1 litre. Transfer 10 ml of this solution to a 100-ml flask, and make to the mark; 1 ml of this final solution = 0·1 mg tungsten.

Various modifications of the procedure outlined above have been published [1, 2, 13, 14, 17]. In one, used by the U.S. Bureau of Mines [15], after the colour has developed an aliquot is pipetted into a small separatory funnel. A small quantity of 10% tributylphosphate in carbon tetrachloride is added, the funnel is shaken, and the layers allowed to separate. A portion of the organic layer is measured at 410 nm.

B. Dithiol Method

In the presence of stannous chloride in warm acid solution, toluene-3,4-dithiol reacts with tungsten to give a green to blue–green colour; the tungsten–dithiol complex can be extracted with isoamyl acetate and measured photometrically [2, 4, 8, 10, 17]. Molybdenum and copper interfere, and must be removed with hydrogen sulphide in dilute acid solution, in the presence of tartrate to prevent co-precipitation of tungsten.

A procedure recommended for titanium alloys is outlined here as an illustration of a dithiol application [10]. Dissolve 0·25 g in 25 ml 1 : 9 sulphuric acid and 0·5 ml of fluoboric acid. Warm gently, oxidize with a saturated solution of potassium permanganate until a permanent brown precipitate is formed, then add sulphurous acid drop-wise until the precipitate has dissolved. Boil for 2 minutes to remove most of the excess sulphur dioxide, add 15 ml of 50% tartaric acid, dilute to 150 ml, and heat to 80°C. Pass a rapid stream of hydrogen sulphide through the warm solution for 30 minutes, and allow to cool to room temperature. Transfer the solution and precipitate to a 500-ml calibrated flask and dilute to the mark.

Filter through a paper pulp pad into a dry beaker, then transfer a 50-ml aliquot to a 100-ml conical flask, add 2 ml sulphuric acid, 1 ml of an iron sulphate solution containing 15 mg iron, and evaporate to fumes of sulphur trioxide. Destroy carbonaceous matter by adding about three 2-ml portions of nitric acid cautiously to the hot fumed residue. Cool, add 5 ml of 10% stannous chloride solution in hydrochloric acid, place on a boiling water-bath and agitate at frequent intervals for 4 minutes. Add 10 ml of 1% toluene-3 : 4-dithiol solution in isoamyl acetate, and continue heating on the water-bath with frequent agitation for 10 minutes. Cool to about 30°C and transfer the solution to a separatory funnel. Rinse with three 2-ml portions of isoamyl acetate. Shake and allow to separate; draw off and discard the lower acid layer. Wash the isoamyl acetate layer twice with 10-ml portions of 4 : 1 hydrochloric acid, each time discarding the lower acid layer. Transfer the amyl acetate layer, containing the tungsten, to a dry 25 ml calibrated flask. Dilute to the mark with isoamyl acetate and mix well. Measure the optical density at 630 nm, using 4 cm cells.

Prepare the calibration graph by transferring five 25-ml aliquots of a titanium sulphate solution to each of five 100-ml conical flasks. The titanium sulphate solution is prepared by dissolving 0·625 g of tungsten-free titanium in 125 ml of 1 : 4 sulphuric acid and 1 ml of fluoboric acid, oxidizing with a slight excess of nitric acid, boiling to remove nitrous fumes, cooling, and diluting to 250 ml in a calibrated flask. To the five 100-ml conical flasks, add 1·25, 2·50, 3·75 and 5·0 ml of a standard tungsten solution in which 1 ml = 0·01 mg tungsten. The latter solution can be made by dissolving 0·1794 g $Na_2WO_4.2H_2O$ in 1 litre. To the standards in the flasks add 1 ml of the iron solution containing 15 mg iron; this solution is made by dissolving 1·5 g iron in 25 ml of 1 : 4 sulphuric acid and diluting to 100 ml. Evaporate the standards to fumes of sulphuric acid, cool, add 5 ml of the 10% stannous chloride solution and proceed as outlined above for the determination.

A geochemical field method for determining tungsten in soils and stream sediments has been described [3]. Fuse with sodium carbonate, potassium nitrate, and sodium chloride, dissolve in water, add stannous chloride, toluene-3,4-dithiol, and extract the blue–green tungsten dithiol complex with white spirit. Compare with a series of standards.

3. Volumetric Method

A procedure for 0·5–5% tungsten in tungsten–nickel alloys has been described [12]. Nickel is removed in a cation exchange resin column, titanium is removed in a hydroxide separation, tungsten is reduced with granular lead and a lead reductor, caught in a ferric iron solution and the ferrous iron produced is titrated with potassium dichromate.

4. Optical Spectrography

Tungsten is occasionally determined by emission spectrography. Methods have been published for 50–200 ppm in zirconium and its alloys [2], and for 50–600 ppm in nickel [11].

REFERENCES

1. Andrew, T. R., and Gentry, C. H. R., *Metallurgia* **LX**, 175 (1959).
2. A.S.T.M., *Chemical Analysis of Metals. Sampling and Analysis of Metal Bearing Ores*, Philadelphia, Pa., American Society for Testing and Materials, 1969.
3. Bowden, F., *Analyst* **89**, 771–4 (1964).
4. Chan, K. M., and Riley, J. P., *Anal. Chim. Acta* **39**, 103–13 (1967).
5. Elwell, W. T., and Wood, D. F., *Analysis of the New Metals Titanium, Zirconium, Hafnium, Niobium, Tantalum, Tungsten, and Their Alloys*, Oxford, Pergamon, 1966.
6. Furey, J. J., and Cunningham, T. R., *Anal. Chem.* **20**, 563–70 (1948).
7. Furman, N. H., ed., *Scott's Standard Methods of Chemical Analysis*, 6th ed., Vol. 1, Princeton, N.J., Van Nostrand, 1962.
8. Greenberg, P., *Anal. Chem.* **29**, 896–8 (1957).

9. Hillebrand, W. F., Lundell, G. E. F., Bright, H. A., and Hoffman, J. I., *Applied Inorganic Analysis*, 2nd ed., New York, Wiley, 1953.
10. Imperial Chemical Industries, *The Analysis of Titanium and its Alloys*, 3rd ed., London, I.C.I., 1959.
11. Lewis, C. L., Ott, W. L., and Sine, N. M., *The Analysis of Nickel*, Oxford, Pergamon, 1966.
12. Luke, C. L., *Anal. Chem.* **33**, 1365–8 (1961).
13. Luke, C. L., *Anal. Chem.* **36**, 1327–9 (1964).
14. Norwitz, G., *Anal. Chem.* **33**, 1253–7 (1961).
15. Peterson, H. E., Anderson, W. L., and Howcroft, M. R., Washington, D.C., *U.S. Bur. Mines Rept. Invest.* 6148, 1963.
16. Schoeller, W. R., and Powell, A. R., *The Analysis of Minerals and Ores of the Rarer Metals*, 3rd ed., London, Charles Griffin, 1955.
17. Sandell, E. B., *Colorimetric Determination of Traces of Metals*, 3rd ed., New York, Interscience Publishers, 1959.
18. Touhey, W. O., and Redmond, J. C., *Anal. Chem.* **20**, 202–6 (1948).
19. Wilson, C. L., and Wilson, D. W., *Comprehensive Analytical Chemistry*, Vol. 1C, London, Elsevier, 1962.

39 Uranium

In recent years, the great developments in the extractive metallurgy of uranium have led to a number of routine procedures for this element in minerals, ores, concentrates, tailings, residues, and compounds. Its importance in nuclear energy and other fields ensures that uranium will be encountered with increasing frequency in many industrial laboratories.

A number of uranium-bearing minerals contribute to the commercial sources of this element. Uraninite is a complex mixture of oxides, with an approximate composition of UO_2. Pitchblende is a variety of uraninite, containing only traces of thorium and rare earths; it is the uranium mineral of greatest economic importance. Autunite, probably the commonest uranium mineral, is a hydrated calcium uranyl phosphate. Tobernite, a hydrated copper uranyl phosphate, carnotite, a hydrated potassium uranyl vanadate, coffinite, a uranium hydroxy silicate, and thucolite, a uranium–thorium mineral containing large quantities of carbon and hydrogen, are all important in one or more of the uranium-producing areas of the world.

ISOLATION OR SEPARATION OF URANIUM

Uranium, a member of the ammonium hydroxide group, is separated from elements of the acid sulphide group by precipitation of the latter with hydrogen sulphide in dilute acid, and from members of the ammonium sulphide and succeeding groups by treatment with ammonium hydroxide which is free from carbon dioxide. Electrolysis at a mercury cathode in dilute sulphuric acid leaves uranium with Al, Be, P, Zr, Ti, V, Nb, Ta, and W in the filtrate, while Fe, Co, Ni, Cu, Zn, Cr, Bi, and Sn are deposited. A cupferron separation in cold 10% sulphuric or hydrochloric acid leaves hexavalent uranium in solution together with Cr, Mn, Co, Ni, Al, Be, Zn, and P, whereas Fe, Ti, V, Zr, Nb, Ta, W, Sn, Sb, and Bi are precipitated. Treatment with sodium hydroxide of the elements of the ammonium hydroxide group will precipitate uranium with Fe, Ti, Zr, Cr, and Th from Al, Be, V, and W; sodium peroxide, in place of sodium hydroxide, or in addition to it, will throw chromium and uranium into the filtrate.

The presence of ammonium carbonate prevents precipitation of uranium by ammonium hydroxide or ammonium sulphide; in this way uranium can be separated from Fe, Al, Be, Mn, Ni, Co, Zn, Ti, Zr, Th, Nb, Ta, and Cr. Iron can be removed from uranium by an ether extraction; Co, Ni, and Mn pass into the filtrate of a zinc oxide separation, leaving uranium with most other elements in the precipitate. Chromium can be volatilized from uranium solutions by evaporation with boiling perchloric and hydrochloric acids. Uranium can be separated from almost all other elements by extracting the dried nitrate with ethyl ether. When a nitric acid solution of uranium is transferred to the top of a column of cellulose, and eluted with ether containing nitric acid, uranium moves down the column whereas most other elements move only slowly or are retained at the top of the column [1, 2, 3, 5, 6, 7, 8, 14, 20, 22, 23, 25, 26, 28, 29].

Depending on the other elements in the material, and on the content of uranium, a wide choice of procedures is available. Usually, samples containing less than about 2% uranium are determined colorimetrically or fluorimetrically; larger quantities are analysed by volumetric or gravimetric methods. All procedures usually require a separation of uranium from interfering elements by hydrogen sulphide, cupferron, and mercury cathode isolations, a cellulose column, or an anion exchange resin.

1. Volumetric Procedure

A. Separation with Hydrogen Sulphide and Cupferron

Weigh out 0·5–3·5 g sample, containing 35–350 mg of U_3O_8. If the sample contains a large amount of organic matter, roast at 450°C for 30–60 minutes before proceeding with acid treatment. For materials in which the uranium is soluble in nitric and sulphuric acids, transfer to a beaker, add 25 ml 1 : 1 nitric acid and boil gently for 30 minutes; add 15 ml 1 : 1 sulphuric acid and evaporate to strong fumes of the latter [1, 2, 10, 13, 14, 21, 22, 24]. Dilute the solution with 100 ml of water, boil, transfer solution and residue to a 250-ml Erlenmeyer flask and bubble hydrogen sulphide through the hot solution for 15 minutes. Stopper the flask and allow to stand overnight.

For materials which require complete decomposition to bring all the uranium into solution, transfer the sample to a platinum dish, add 10–20 ml nitric acid, 20 ml hydrofluoric acid, and evaporate to dryness. Add 10–20 ml hydrofluoric acid and again evaporate to dryness. Add 15 ml 1 : 1 sulphuric acid, evaporate to strong fumes of the latter, cool, wash down with water, and again evaporate to fumes. Cool, dissolve soluble salts in hot water, filter, wash, ignite the residue in a platinum crucible, and fuse for 10–15 minutes with sodium carbonate. Extract the solidified melt with 1 : 9 sulphuric acid and add to the main filtrate. Adjust acidity and pass in hydrogen sulphide as outlined in the previous paragraph.

The solution of uranium metal and its alloys has been described [14, 15]; most are decomposed with nitric and hydrofluoric acid, or with hydrochloric acid plus oxidants.

If the material contains over 5% arsenic, it is preferable to remove this, with antimony and tin, by several evaporations with hydrobromic acid from a concentrated sulphuric acid solution, before proceeding with the removal of the hydrogen sulphide group. A heavy sulphide precipitate should be dissolved in nitric and sulphuric acids, fumed to sulphur trioxide, cooled, diluted, and re-precipitated to set free occluded uranium salts.

Filter the precipitate of acid sulphide elements, wash with acidulated hydrogen sulphide water, and boil the filtrate to eliminate all hydrogen sulphide. Add 10 ml of 1 : 1 sulphuric acid, evaporate the filtrate to about 80 ml, and add a 3% solution of potassium permanganate until a stable pink colour is obtained. Cool the solution in ice water to below 5°C, transfer to a 250 ml separatory funnel, and dilute with ice-cold water to 100 ml.

Add ice-cold 6% cupferron solution slowly until precipitation of Fe, Ti, etc. is complete, indicated by the appearance of white flakes of the excess reagent; add 5 ml excess cupferron. Add 10–20 ml chloroform, shake vigorously, allow the two layers to separate and draw off the lower layer containing the chloroform. Test for completion of precipitation by adding a further 2–3 ml of cupferron. If precipitation is complete, continue the extraction with successive 10 ml portions of chloroform until the chloroform layer is colourless.

Discard the chloroform extracts and return the aqueous solution to the beaker. Add 5 ml of nitric acid and 3 ml perchloric acid, evaporate until fumes appear, cover the beaker and continue heating strongly to oxidize organic matter. Cool, wash down the sides of the beaker, and again evaporate to strong fumes; repeat this operation once more. Cool, add sulphuric acid to bring the volume to 10 ml, add 100 ml water, boil, add a 3% potassium permanganate solution until the pink colour persists for at least one minute, and cool.

Pour the solution into a Jones reductor and allow to percolate through the zinc column at the rate of about 30 ml per minute. Wash the beaker and column with five 20-ml portions of 5% sulphuric acid, followed by four of water. Bubble air through the cold solution for ten minutes. Add 5 ml of 10% ferric sulphate solution, 20 ml of sulphuric–phosphoric acid solution, and 6 drops of sodium diphenylamine sulphonate indicator. Titrate with 0·05N potassium dichromate to a distinct purple colour. Subtract the indicator and reductor blank and calculate the percentage of uranium.

REAGENTS

Ferric sulphate. Dissolve 100 g of ferric sulphate in 500 ml of water containing 20 ml of 1 : 1 sulphuric acid, and dilute to 1 litre.

Sulphuric–phosphoric acid. Dilute 750 ml of 2 : 1 sulphuric acid with 250 ml of phosphoric acid.

Sodium diphenylamine sulphonate. Dissolve 0·3 g of the salt in 100 ml of water.

Potassium dichromate, 0·05N. Dissolve 4·9033 g in water and dilute to 2 litres. One ml = 0·00702 g U_3O_8, but it can be checked, if desired, against pure U_3O_8 dissolved in 1 : 1 sulphuric acid.

The reaction may be represented:

$$3UO^{+2} + Cr_2O_7^{-2} + 8H^+ \rightarrow 3UO_2^{+2} + 2Cr^{+3} + 4H_2O$$

$$\frac{K_2Cr_2O_7}{6} = \frac{U}{2} = \frac{U_3O_8}{6}$$

and 1 ml of N $K_2Cr_2O_7$ = 0·1404 g U_3O_8, or 0·1192 g U.

B. Separation with Mercury Cathode and Cupferron

Decompose the sample as outlined in A above, transfer the sulphuric acid solution to a mercury cathode cell, adjust to an acidity approximating 2 ml sulphuric acid per 100 ml, and electrolyse at 6–8 amperes until a spot test shows the absence of Fe, Cu, Ni, Co, etc. A convenient spot test consists of adding one drop of 1% zinc acetate solution and one drop of ammonium mercuric thiocyanate solution to a drop of the electrolyte. The ammonium mercuric thiocyanate solution is made by dissolving 8 g of mercuric chloride and 9 g of ammonium thiocyanate in 100 ml of water.

Filter the electrolyte and washings, but if a pale green precipitate of uranium phosphate is formed from a high phosphate content, add sufficient sulphuric acid to redissolve this precipitate before filtering. Add 20 ml of 1 : 1 sulphuric acid, boil, oxidize with potassium permanganate and proceed with cupferron separation as described in A. The final titration may be done with dichromate, or, for greater sensitivity, with ceric sulphate. If the latter is used, follow the procedure as given in the previous section to the addition of the sulphuric–phosphoric acid, add 2 drops of *o*-phenanthroline indicator and titrate with 0·01 N ceric sulphate until the indicator changes from red to colourless. Subtract the blank and calculate the U_3O_8.

Reagents

o-phenanthroline ferrous sulphate indicator, 0·025 M. This is the concentration of indicator furnished by most chemical supply houses.

Standard ceric sulphate solution, 0·01 N. Dissolve 6·3 g of $Ce(SO_4)_2.2(NH_4)_2SO_4.2H_2O$ or 3·3 g of $Ce(SO_4)_2$ in 500 ml of 1 : 9 sulphuric acid, heat and stir until the salt dissolves. Cool, dilute to almost 1 litre, stir, allow to stand overnight, filter, and make to volume. One ml of 0·01 N $Ce(SO_4)_2$ = 0·001404 g U_3O_8, but the value should be determined by standardizing against pure U_3O_8.

The reaction may be represented:

$$2Ce^{+4} + U^{+4} + 2H_2O \rightarrow UO^{+2} + 2Ce^{+3} + 4H$$

$$Ce = \frac{U}{2} = \frac{U_3O_8}{6}$$

and 1 ml of N $Ce(SO_4)_2$ = 0·1404 g U_3O_8, or 0·1192 g U.

C. Hydrogen Sulphide–Ammonium Sulphide Separation

Decompose the sample and remove members of the hydrogen sulphide group as outlined under A. To the boiled and oxidized filtrate add ammonia nearly to neutrality, add ammonium carbonate until a precipitate forms and then 5 g excess. Dilute to about 300 ml and pass hydrogen sulphide through the solution for 15 minutes; stopper the flask and allow to stand at least two hours. Filter, wash with 1% ammonium hydroxide–ammonium carbonate solution saturated with hydrogen sulphide, dissolve in 1 : 1 sulphuric acid and re-precipitate. The filtrate contains the uranium, because the latter is not precipitated by ammonium sulphide in the presence of carbonate. Acidify the filtrate with hydrochloric acid, add a solution of 10% bromine in hydrochloric acid until a slight excess is present, and boil to expel the excess.

Adjust the volume to 300 ml, add 5 g ammonium phosphate and introduce ammonium hydroxide to the boiling solution until it is just alkaline to methyl red. Add 3 ml of acetic acid and keep the solution in ice water for about one hour. Filter the precipitate of uranium phosphate, using pulp, wash with 3% ammonium nitrate solution, dissolve the precipitate in 1 : 1 sulphuric acid, oxidize with potassium permanganate, pass through a reductor, and complete as outlined in A.

D. Alumina–Cellulose Column Separation

Decompose the sample in a platinum dish with hydrofluoric and nitric acids, evaporate several times to dryness with nitric acid to eliminate hydrofluoric, cool, dilute, filter, wash, fuse any insoluble residue with sodium peroxide in a nickel crucible, dissolve the melt in nitric acid, and add to the main filtrate. Evaporate the latter to dryness and re-dissolve the residue in 6 ml of 3 : 7 nitric acid. If sulphates are present, add 3 g of $Ca(NO_3)_2.4H_2O$ for every 1 gram equivalent of sulphuric acid, and for every 1 gram of phosphate calculated as Na_2HPO_4 add approximately 4 g of $Fe(NO_3)_3.9H_2O$.

Pour the solution on to a column of approximately 6 g of cellulose in a height of about 6 cm, on top of which is placed about 15 g of 60–150 mesh activated alumina. Open the stopcock and allow the solution to descend to the top of the alumina. Rinse the solution down the column with 9–12 portions of 20 ml of an ether solvent made by adding 60 ml of nitric acid to 940 ml of ether. Add 50 ml of water to the effluent and evaporate the ether on a water bath. Place on a hot plate and evaporate to about 20 ml, remove, cool, add 20 ml 1 : 1 sulphuric acid and 3 ml perchloric acid. Evaporate to fumes, cover, and heat strongly for 10–20 minutes to oxidize organic matter. Cool, dilute, boil, add potassium permanganate solution to a pink colour, cool, and adjust the volume to 100 ml. Reduce, aerate, and titrate with 0·05 N potassium dichromate as described under A.

2. Gravimetric Procedure

A. Alumina–Cellulose Column Separation

While this procedure is particularly suitable for uranium in high grade concentrates, it can be used for many other materials. If the sample is low-grade and complex, final separation on alumina–cellulose can be preceded by treatments with hydrogen sulphide, cupferron, mercury cathode, ammonium sulphide, or by a combination of these, as described in **1**, A, B, C, and D.

Decompose the sample and make the necessary separations as previously outlined; pass the nitric acid solution of the sample through the alumina–cellulose column at the rate of about 10 ml per minute, and wash as described in **1** D, until about 180–200 ml of solvent has been collected. To the latter add 50 ml of water, evaporate the ether on a water bath, then evaporate the solution on the hot plate to about 20 ml. Transfer to a tared platinum or silica dish, evaporate on a water bath, place in a cold muffle and gradually increase the temperature to 500°C. Finally ignite at 900°C for 15 minutes, cool, and weigh as U_3O_8. Carry out a blank determination following the procedure employed, and subtract this from the weight of ignited U_3O_8; $U_3O_8 \times 0{\cdot}848 = U$.

3. Colorimetric Methods

A. With Sodium Hydroxide and Hydrogen Peroxide, after Cellulose Separation

Decompose the sample as in **1** A, evaporate to dryness, and dissolve it in about 10 ml of 3 : 7 nitric acid. Introduce a volume of paper pulp approximately equal to 25 ml and stir until the liquid is completely absorbed on the pulp. Add 40 ml of ether solvent, stir the contents of the beaker thoroughly and decant the solution rapidly into a 12·5 cm medium-speed paper in a silicone-treated filter funnel. Retain the pulp in the beaker and receive the solvent in a 600-ml beaker. Press the pulp with a decanting rod [7] to squeeze out the surplus solvent, and pour into the filter funnel. Repeat this extraction three times. Add 40 ml of water to the solvent, boil off the ether on a water bath, and then evaporate on a hot plate to a volume of about 20 ml. Remove, cool, add 5 ml of sulphuric–perchloric acid and evaporate until fumes appear; cover and boil the acid mixture for 5–6 minutes. Cool, wash down the sides of the beaker, add 0·5 ml of perchloric acid and evaporate to fumes. Cover the beaker and heat strongly for another 5–6 minutes or until all organic matter has been destroyed. Remove the cover, evaporate the solution to about 0·5 ml, cool, dilute with 10 ml of water, and boil.

Add 0·5 ml of 30% hydrogen peroxide, neutralize with 15% sodium hydroxide solution and add 5 ml in excess. Heat to boiling, cool, transfer to a 50-ml volumetric flask, make to volume, and mix. Filter through Whatman No. 54 paper, returning the first 10–20 ml of the filtrate through the same filter. Add one drop of 30% hydrogen peroxide to the filtrate. Measure the optical density of the solution at 390 nm from a calibration graph prepared with a standard uranium solution.

REAGENTS

Ether solvent. Add 60 ml of nitric acid to 940 ml of ether.

Sulphuric–perchloric acid. Dilute 600 ml of 1 : 1 sulphuric acid with 400 ml of perchloric acid.

Standard uranium solution. Dissolve 1 g of pure U_3O_8 in a small amount of nitric acid and dilute to 1 litre with water; 1 ml = 1·0 mg U_3O_8.

This colorimetric procedure can also be used after an ethyl acetate extraction of uranium [8, 11]; aluminium nitrate is employed as a salting-out agent.

B. WITH DIBENZOYL-METHANE AFTER EXTRACTION WITH TRIBUTYL PHOSPHATE IN ISOOCTANE

Uranium forms a yellow uranyl-dibenzoyl-methane complex when it is extracted with tributyl phosphate in isooctane and added to an acetone–water solution of dibenzoyl-methane and pyridine [9].

Weigh 2–5 g into a platinum dish and decompose with nitric and hydrofluoric acids, with repeated additions if necessary, until silica is removed. Evaporate to dryness, dissolve in 5 ml nitric acid and a few ml of water, transfer to a 100-ml volumetric flask, dilute to the mark and mix. Withdraw an aliquot not exceeding 5 ml and containing 0·05–0·50 mg of uranium into a 60 ml separatory funnel. If the aliquot taken was less than 5 ml, adjust to this volume with water; add 1 drop of 0·04% *m*-cresol purple indicator and by dropwise addition of dilute ammonium hydroxide or nitric acid adjust the *p*H to the yellow form of this indicator.

Add 8 ml of aluminium nitrate salting solution, mix, and add by pipette 3 ml of the tributyl phosphate extractant to the separatory funnel. Stopper the funnel and shake for 30 seconds. Allow the phases to separate and discard the aqueous phase. With a dry pipette, transfer 2 ml of the organic phase into a dry 25 ml volumetric flask, excluding any droplets of aqueous phase adhering to the walls of the separatory funnel. Fill the volumetric flask to the mark with the chromogenic reagent and mix. After 1 hour measure the absorbance of the solution at a wavelength of 410 nm, using a reagent blank which has been prepared in the same manner as the sample. Thorium interferes in this method.

REAGENTS

Aluminium nitrate salting solution. Dissolve 900 g of aluminium nitrate monohydrate in enough water to make 1 litre of solution.

Tributyl phosphate extractant. Mix 1 volume of tributyl phosphate with 10 volumes of isooctane.

Chromogenic reagent. Combine 17·5 ml of a 1% w/v acetone solution of dibenzoyl–methane, 809 ml of acetone, 43·5 ml of pyridine, and enough water to make 1 litre of solution.

4. Fluorimetric Method

Uranium in sulphuric acid solution exhibits a greenish yellow fluorescence when irradiated with ultraviolet light. A similar fluorescence is observed if uranium compounds are fused in sodium fluoride or in mixtures of alkali carbonate and fluoride, and the solidified products are exposed to ultraviolet light. This property can be employed for the rapid quantitative determination of small amounts of uranium [1, 6, 7, 14, 20, 21, 27].

Decompose the sample in 1 : 1 nitric acid; if the uranium is not entirely brought into solution by this treatment, filter, fuse the residue in sodium carbonate, dissolve in nitric acid, and add to the main solution. To the latter, containing about 10 ml nitric acid, add 1 g $Ca(NO_3)_2.4H_2O$ and warm until dissolved. Add 20 ml water, filter into a 50-ml volumetric flask, and rinse the filter paper with 3 : 7 nitric acid. Make to the mark with the latter solution and mix. Transfer 3 ml to a cellulose column and pass it through with 60 ml of ether solvent. Add 20 ml of water to the effluent, boil off the ether on a water bath, transfer the aqueous solution to a 50 ml volumetric flask, make up to the mark and mix.

Pipette 1 ml of the solution into a gold dish and dry under an infra-red lamp. Add 2·5 g of flux to the dish and fuse over a gas burner for a total time of 2·5 minutes. After one minute roll the flux twice around the rim of the dish to free the melt from bubbles and to collect any uranium. Set the fusion aside in a level position for 1·5 minutes to cool, and place in a desiccator. After approximately 15 minutes, measure the fluorescence in a fluorimeter. Do two blank determinations, following the above procedure, and subtract the average reading from the unknowns. Calibrate the fluorimeter frequently by means of suitable aliquots of a pure uranium solution carried through all steps of the procedure.

Clean the dishes by immersing in hot 1 : 3 nitric acid and rinsing in water. If the sample contains arsenic or molybdenum, place alumina on top of the cellulose column, as described in **1** D.

REAGENTS

Ether solvent. Add 60 ml of nitric acid to 940 ml of ether.

Flux. Mix, in a perfectly clean ball mill, 9·4% sodium fluoride, 39·4% sodium carbonate, and 51·2% potassium carbonate.

5. Polarographic Procedure

Moderate amounts of uranium can be determined polarographically [1, 3, 6, 7, 14, 16, 17]. One method is to elute a nitric acid–ether solution of uranium through a cellulose column, and remove nitric acid and organic matter by fuming with sulphuric acid. Cool, add 2·85 ml of supporting electrolyte containing 4·5% w/v oxalic acid, 0·1% v/v hydrochloric acid, and 0·015% gelatin, pass nitrogen through the polarograph cell for 15 minutes, and determine the uranium polarogram.

6. Physical Methods

Because uranium is one of the few naturally occurring radioactive elements, its radiations may be detected by various types of meters and counters. Comparison of unknown samples with a standard, by this physical means of analysis, provides a rapid procedure for uranium which is of great importance. It does not, however, distinguish uranium from thorium. A good introductory discussion of the subject has been published [7]; further information is provided in the trade literature of firms which provide this equipment.

7. Other Methods

Among the methods described for uranium are ion exchange and X-ray fluorescence [29], ultra-violet spectrophotometry [19], reversed-phase partition chromatography [12], anion exchange and reduction of certain ions [8], spectrophotometric determination with thiocyanate in butyl Cellosolve [18], reduction of uranium by titanous sulphate before final oxidimetric titration [30], and spectrophotometric determination as the tetrapropylammonium–uranium (IV) trinitrate complex [14].

A procedure has been described for determining uranium metal in the presence of uranium oxides and carbide by the solubility of the metal in bromine–ethyl acetate solution [4].

REFERENCES

1. Anonymous, *Analytical Chemistry of Uranium*, London, Oldbourne Press, 1964.
2. Arden, T. V., *J. Appl. Chem.* **4**, 539–53 (1954).
3. Arden, T. V., Harwell, England, *Atomic Energy Research Establishment Report* 2862, 1959.
4. Ashbrook, A. W., *Analyst* **87**, 751–4 (1962).
5. Burstall, F. H., and Wells, R. A., *Analyst* **71**, 396–409 (1951).
6. Burstall, F. H., Wells, R. A., and Williams, F. A., *Handbook of Chemical Methods for the Determination of Uranium in Mineral and Ores*, London, H.M.S.O., 1950.
7. Dillon, V. S., *Assay Practice on the Witwatersrand*, Johannesburg, Transvaal and Orange Free State Chamber of Mines, 1955.
8. Fisher, S., and Kunin, R., *Anal. Chem.* **29**, 400–2 (1957).

9. Francois, C. A., *Anal. Chem.* **30**, 50-4 (1958).
10. Furman, N. H., ed., *Scott's Standard Methods of Chemical Analysis*, 6th ed., Vol. 1, Princeton, N.J., Van Nostrand, 1962.
11. Guest, R. J., and Zimmerman, J. B., *Anal. Chem.* **27**, 931–6 (1955).
12. Hamlin, A. G., Roberts, B. J., Loughlin, W., and Walker, S. G., *Anal Chem.* **33**, 1547–52 (1961).
13. Hillebrand, W. F., Lundell, G. E. F., Bright, H. A., and Hoffman, J. I., *Applied Inorganic Analysis*, 2nd ed., New York, Wiley, 1953.
14. Kolthoff, I. M., and Elving, P. J., *Treatise on Analytical Chemistry*, Part II, Vol. 9, 1–188, New York, Interscience Publishers, 1962.
15. Larsen, R. P., *Anal. Chem.* **31**, 545–9 (1959).
16. Legge, D. I., *Anal. Chem.* **26**, 1617–21 (1954).
17. Legge, D. I., Chapter XXIV in *Assay Practice on the Witwatersrand*, by V. S. Dillon, Johannesburg, Transvaal and Orange Free State Chamber of Mines, 1955.
18. Nietzel, O. A., and De Sesa, M. A., *Anal. Chem.* **29**, 756–9 (1957).
19. Paige, B. E., Elliott, M. C., and Rein, J. L., *Anal. Chem.* **29**, 1029–32 (1957).
20. Rabbitts, F. T., *Canadian Bur. Mines Memorandum Series* 109 (1950).
21. Rodden, C. J., *Analytical Chemistry of the Manhattan Project*, New York, McGraw-Hill, 1950.
22. Schoeller, W. R., and Powell, A. R., *The Analysis of Minerals and Ores of the Rarer Elements*, 3rd ed., London, Charles Griffin, 1955.
23. Seim, H. J., Morris, R. J., and Frew, D. W., *Anal. Chem.* **29**, 443–6 (1957).
24. Sill, C. W., and Peterson, H. E., *U.S. Bur. Mines Rept. Invest.* 4882 (1952).
25. Steele, T. W., Chapter XXII in *Assay Practice on the Witwatersrand*, by V. S. Dillon, Johannesburg, Transvaal and Orange Free State Chamber of Mines, 1955.
26. Steele, T. W., *J. South African Inst. Mining Metallurgy* **57**, 144–52 (1956).
27. Thatcher, L. L., and Barker, F. B., *Anal. Chem.* **29**, 1575–8 (1957).
28. U.K. National Chemical Laboratory, *The Determination of Uranium and Thorium*, London, H.M.S.O., 1963.
29. Van Niekerk, J. N., De Wet, J. F., and Wybenga, F. T., *Anal. Chem.* **33**, 213–15 (1961).
30. Wahlberg, J. S., Skinner, D. L., and Rader, L. F., *Anal. Chem.* **29**, 954–7 (1957).

40 Vanadium

Vanadium is widely distributed in rocks, minerals, and ores; it is an important alloying element in irons and steels, and is often found in other metals, alloys, slags, and various materials.

There are many vanadium minerals, and the element is frequently found in small quantities in ores of titanium and iron. It also occurs in the ashes of many coals, petroleums, and asphalts. The principal vanadium minerals are patronite, a sulphide approximating VS_4, vanadinite, a lead chlorovanadate $3Pb_3(VO_4)_2.PbCl_2$, descloizite, a basic lead zinc orthovanadate $(Pb, Zn)_2$-$OH.VO_4$, and carnotite, a hydrous potassium uranyl vanadate, approximating $K_2O.2UO_3.V_2O_5.3H_2O$.

ISOLATION OR SEPARATION OF VANADIUM

Vanadium is separated from members of the acid sulphide group by precipitation of the latter with hydrogen sulphide in 5–10% hydrochloric or sulphuric acid; if the precipitate is bulky it must be re-dissolved and re-precipitated to recover the co-precipitated vanadium. The co-precipitation of vanadium with the acid sulphide group can be prevented by the addition of tartaric acid. If alone, vanadium is not precipitated by ammonium hydroxide, but in the presence of appreciable quantities of Fe, Al, etc. precipitation is quantitative. Electrolysis at a mercury cathode in dilute sulphuric acid will leave vanadium in solution, with Ti, Zr, Th, Nb, Ta, W, U, Al, Be, and P; Fe, Cr, Mo, Co, Ni, Cu, Zn, Sn, Cd, Tl, and Bi are deposited. Cupferron in cold 10% sulphuric or hydrochloric acid solution will precipitate vanadium, with Ti, Zr, Nb, Ta, Mo, W, Fe, Sn, Sb, and Bi; the following remain in solution: Al, Be, Cr, Mn, Co, Ni, Zn, Cd, and U(VI). Treatment with ammonium sulphide and ammonium tartrate will leave vanadium in solution, with Ti, Cr, Nb, Ta, W, and Al, whereas Fe, Co, Zn, and Tl are precipitated completely.

Precipitation with sodium hydroxide will leave vanadium in the filtrate with Al, Be, Zn, and W, whereas Ti, Zr, Th, U, Cr, Fe, Co, and Ni are precipitated; treatment with sodium peroxide throws Cr and U into the filtrate with vanadium. Fusion with sodium carbonate and potassium nitrate, and extraction of the melt with water, will separate vanadium with Al, Be, Zn, W, Cr, U, P, and Mo from Fe, Ti, Zr, Th, Ni, and Co. Vanadium may be isolated from iron by an ether extraction, and from Co, Ni, and Mn by a zinc oxide separation. Chromium can be separated from vanadium by volatilizing the former with boiling hydrochloric and perchloric acids. Ion exchange may also be used to separate vanadium from many metals [3, 6, 11].

1. Volumetric Methods

A. Separation with Mercury Cathode. Titration with Potassium Permanganate

Transfer 0·5–5 g, depending on the vanadium content of the sample, to a 400-ml beaker, add 10–20 ml nitric acid, 10–20 ml hydrochloric acid, and 10 ml 1 : 1 sulphuric acid [1, 8, 9, 11, 19]. Evaporate to heavy fumes of sulphur trioxide, cool, add 50 ml hot water, boil, filter, and wash. If the residue might contain vanadium, fuse it in a platinum crucible with sodium carbonate to which a little potassium nitrate has been added, and combine this with the original filtrate. If tungstic acid separates, filter it off, wash with 0·05% sulphuric acid, and determine vanadium in the precipitate by colorimetric means.

Adjust the acidity to 5–10% sulphuric acid, add 2–3 g tartaric acid, and pass in hydrogen sulphide until all members of the acid sulphide group have been precipitated. Filter and wash thoroughly with acidulated hydrogen sulphide water. If the precipitate is bulky, dissolve in nitric acid, add sulphuric acid and evaporate to fumes, and re-precipitate with hydrogen sulphide to recover the co-precipitated vanadium. Boil off all hydrogen sulphide from the filtrate, and cool.

For many materials on which vanadium is determined, such as irons and steels, some of the steps in the above decomposition procedure may be omitted. For example, most carbon steels, cast iron, open-hearth iron, wrought iron and vanadium steels can be brought into solution by dissolving 2–10 g in 30–100 ml of 1 : 9 sulphuric acid [1]. Other methods of decomposition can be employed. Silicates and other materials can be decomposed in a platinum dish with hydrofluoric and nitric or sulphuric acids. Many vanadium minerals and ores are conveniently decomposed by a pyrosulphate fusion in a quartz crucible. Metallic vanadium can be brought into solution by nitric or sulphuric acids, or by fusion with sodium hydroxide.

In the hands of experienced chemists, vanadium may be determined in routine analyses of many materials with few, if any, additional separations. It is usually preferable, however, to remove Fe, Cr, Ni, etc. by a mercury cathode, preceded if the iron content is very high by an ether extraction or a bicarbonate separation, or to remove chromium and nickel by cupferron, or volatilize chromium as chromyl chloride [2]. The presence of more than about 1% chromium in a sample renders difficult the detection of the end-point of a permanganate titration for vanadium.

For steels, dissolve the sample as outlined above, dilute to 100 ml, heat to boiling, add 8% sodium bicarbonate solution until a permanent precipitate is formed, and then 4 ml in excess. Cover, boil, filter, wash with hot water, and disregard a cloudiness in the filtrate due to oxidation and hydrolysis of iron. Vanadium and chromium are precipitated quantitatively, along with 10–15% of the nickel present and about 0·2–0·3 g of iron which becomes oxidized during the filtration and carries with it some elements such as manganese. Place the paper and precipitate in the original flask, add 5 ml sulphuric acid, 20 ml nitric acid, and heat over a free flame to dense white fumes. Cool, and if necessary repeat the treatment, wash down the flask, and again evaporate to fumes.

If the sample is high in chromium, it is preferable to eliminate it as chromyl chloride at this stage by repeated evaporations with perchloric and hydrochloric acids, followed by evaporation with sulphuric acid. Large amounts of chromium are awkward to remove by a mercury cathode.

Cool, dilute, neutralize with ammonium hydroxide, add an excess of 0·5 ml sulphuric acid per 100 ml of solution, and electrolyse in a mercury cathode to remove Fe, Cr, Ni, etc.

To the filtrate and washings from the mercury cathode separation, add 2–3 ml of 1 : 1 sulphuric acid, heat to 70°C, and add 2·5% potassium permanganate solution to the appearance of a deep pink colour. Heat to boiling, pass a stream of sulphur dioxide for 2–5 minutes into the solution until the vanadium is reduced. Continue boiling, and pass a stream of carbon dioxide through the solution until all sulphur dioxide has been removed; this stage may be ascertained by passing the gas issuing from the flask into 5 ml of water containing a drop of 1 : 1 sulphuric acid and enough potassium permanganate to give a faint pink colour. Cool the sample to 60–80°C and titrate with 0·03 N or 0·05 N potassium permanganate. Correct the titration by a blank determination on a solution of similar volume and acidity. The reactions may be expressed:

$$V_2O_5 + H_2SO_3 \rightarrow V_2O_4 + H_2SO_4$$

$$5V_2O_4 + 2KMnO_4 + 3H_2SO_4 \rightarrow 5V_2O_5 + K_2SO_4 + 2MnSO_4 + 3H_2O$$

$$1 \text{ ml of } 0{\cdot}1 \text{ N } KMnO_4 = 0{\cdot}005094 \text{ g V.}$$

Standardize the 0·1 N potassium permanganate, containing approximately 3·2 g potassium permanganate per litre, against pure sodium oxalate; 1 litre of 0·1 N potassium permanganate = 6·700 g sodium oxalate. The potassium permanganate solution may also be standardized against AnalaR ammonium vanadate; a solution made by dissolving 1·1482 g pure ammonium vanadate in water made slightly ammoniacal with ammonium hydroxide, and diluting to 500 ml, has a value of 1 ml = 0·001 g vanadium.

B. Separation with Cupferron. Titration with Potassium Permanganate

For some materials, it may be more convenient to remove elements like chromium and nickel by cupferron instead of a mercury cathode [8, 9, 11, 19]. Large quantities of iron can be removed by a bicarbonate separation as described in **1,** A; if a substantial amount of chromium is present it can be volatilized by repeated evaporation with boiling perchloric and hydrochloric acids [2].

To the cold 10% sulphuric acid solution add sufficient cold 6% cupferron solution to precipitate the vanadium, together with Fe, Ti, Zr, etc. Filter and wash 10 times with 1% sulphuric acid solution, transfer the paper and contents to the original beaker, and add 20 ml nitric acid and 10 ml sulphuric acid. Evaporate to fumes, cool, add 10 ml nitric acid, and again evaporate to fumes. Cool, dilute to 300 ml, add 3 ml phosphoric acid, and then potassium permanganate solution to a pink colour. Add sufficient 0·03N ferrous ammonium sulphate to reduce all the vanadium; 1 ml will reduce 0·0015 g V. If enough has been added, a drop of the solution will immediately give a blue colour with a drop of fresh dilute potassium ferricyanide solution.

Stir the solution thoroughly, add 8 ml of freshly prepared 15% ammonium persulphate solution, and stir for a minute [1]. Titrate with 0·03 or 0·05N potassium permanganate solution to a definite pink tint that does not fade upon continued stirring for 1 minute. Subtract a blank obtained by boiling the titrated solution for 10 minutes to destroy the slight excess of potassium permanganate, cooling, and titrating with permanganate to the same tint as was observed before.

This procedure depends on the fact that in cool acid solution containing no silver nitrate, ammonium persulphate oxidizes ferrous sulphate but does not react rapidly with quadrivalent vanadium, permanganate, bivalent manganese, or trivalent chromium. It is therefore possible to reduce vanadium by adding an excess of ferrous sulphate, to destroy the excess with ammonium persulphate, and finally to titrate the reduced vanadium with standard potassium permanganate. The reactions may be symbolized:

$$V_2O_5 + 2FeSO_4 + H_2SO_4 \rightarrow V_2O_4 + Fe_2(SO_4)_3 + H_2O$$

$$2FeSO_4 + (NH_4)_2S_2O_8 \rightarrow Fe_2(SO_4)_3 + (NH_4)_2SO_4$$

$$5V_2O_4 + 2KMnO_4 + 3H_2SO_4 \rightarrow 5V_2O_5 + K_2SO_4 + 2MnSO_4 + 3H_2O$$

$$1 \text{ ml of } 0{\cdot}1\ \text{N}\ KMnO_4 = 0{\cdot}005094 \text{ g V.}$$

C. Titration with Ferrous Ammonium Sulphate

Another common and satisfactory volumetric procedure for vanadium depends on the oxidation of the latter with potassium permanganate, reduction of the excess permanganate with sodium nitrite, destruction of the excess nitrite with urea, and final titration of the vanadate with standard ferrous ammonium sulphate, using diphenylaminesulphonate indicator or a potentiometric endpoint [2, 11, 20, 21].

Decompose the sample as outlined in **1,** A; alternative means of getting the sample into solution may, of course, be employed. Ores may be decomposed with sodium peroxide in a zirconium crucible, and the melt dissolved in dilute sulphuric acid [21]. Titanium-base alloys can be dissolved with 1 : 1 sulphuric acid and fluoboric acid [20]. Ferrovanadium is dissolved in 1 : 1 sulphuric acid, nitric acid, and a few drops of hydrofluoric acid [1].

As outlined previously, remove the hydrogen sulphide group, separate vanadium from most of the iron with a bicarbonate separation, or remove Fe, Cr, Ni, etc. with a mercury cathode; alternatively remove Cr and Ni by a cupferron precipitation, or volatilize chromium by repeated evaporations with perchloric and hydrochloric acids.

To the solution containing sulphuric acid, or sulphuric and nitric acids, add 3 ml phosphoric acid for each gram of sample, and 5 ml of 0·1N potassium permanganate until the solution is pink, then about 10 drops excess and allow to stand for 2 minutes; if the colour fades, add more potassium permanganate Add, with stirring, 0·35% sodium nitrite solution until the pink colour is discharged, and then 5 ml excess. Add 2 g of urea, stir, and allow to stand for 5 minutes. Add 5 g $NaC_2H_3O_2.3H_2O$ for each ml of excess sulphuric acid used to dissolve the sample; if too much acetate is added and a precipitate of ferric phosphate is formed, dissolve it in a few drops of 1 : 1 sulphuric acid.

Add a prepared solution of oxidized diphenylaminesulphonic acid indicator, and titrate with 0·01N ferrous ammonium sulphate with vigorous stirring until the purple colour changes to bluish-green. 1 ml of 0·01N ferrous ammonium sulphate = 0·0005094 g V. The above endpoint can be determined potentiometrically if desired.

In the presence of tungsten, evaporate the sample with sulphuric acid after the usual separations outlined in **1**, A; cool, dilute, and make alkaline with sodium hydroxide. Boil and stir until tungsten has dissolved, add 10 ml hydrofluoric acid, acidify with 1 : 1 sulphuric acid and add 10 ml excess. Dilute to 300 ml, add 1 g of ammonium persulphate, boil, add 10 ml of 0·1N ferrous sulphate, 0·1N potassium permanganate and complete the determination as outlined above, omitting the addition of phosphoric acid.

REAGENTS

Ferrous ammonium sulphate, 0·01N. Dissolve 4 g $FeSO_4.(NH_4)_2SO_4.6H_2O$ in 500 ml of water containing 20 ml of 1 : 1 sulphuric acid, and dilute to 1 litre. Standardize by transferring 25 ml to a 600-ml beaker containing 10 ml of 1 : 1 sulphuric acid and 5 ml phosphoric acid in a volume of 300 ml, adding diphenylaminesulphonic acid indicator, and titrating with 0·01N potassium dichromate. The latter standard is made by weighing 0·4904 g $K_2Cr_2O_7$ and diluting to 1 litre; 1 ml = 0·0005094 g vanadium.

Diphenylaminesulphonic acid indicator. To 0·32 g barium diphenylaminesulphonate add 100 ml of 10% sodium sulphate solution, stir, and allow the barium sulphate to settle. Pour the supernatant liquid through a dry filter into a dark bottle. Transfer approximately 0·3 ml of the sodium diphenylaminesulphonatesolution, which is th e amount required for each determination, to a 50 ml beaker, add 5 ml of water, 3–4 drops sulphuric acid, 5–6 drops of 0·01N ferrous ammonium sulphate and then 3–4 drops of 0·1N potassium dichromate. Add carefully from a burette the 0·01N ferrous ammonium sulphate until the purple colour of the solution just turns to bluish-green. As the purple colour begins to disappear, add the ferrous ammonium sulphate very cautiously so that no excess of the latter is present in the oxidized indicator solution. Add this bluish-green solution to the solution to be titrated. A stock solution of the oxidized indicator, which will keep for several weeks, can be prepared.

D. Amperometric Titration

Vanadium in vanadium metal has been determined by an amperometric titration [13]. After dissolution in nitric and hydrofluoric acids, and oxidation with ammonium persulphate, a few drops of a potassium permanganate solution are added to give an orange-tinted pink colour which persists for two minutes. Sodium nitrite is added until the pink colour is nearly discharged, then a small excess of the nitrite is added, followed by 1 g of urea. A weighed small excess of ferrous amonium sulphate is added, the solution stirred magnetically, and titrated amperometrically with standard 0·02N potassium dichromate.

2. Colorimetric Methods

A. Phosphotungstovanadic Acid

When phosphoric acid and sodium tungstate are added to an acid solution of vanadium(V) a yellow phosphotungstovanadic acid is formed; this is the basis of a colorimetric method for vanadium [4, 8, 20, 21]. Interferences include potassium, which forms slightly soluble phosphotungstates, and uranium, which can be tolerated up to about ten times the vanadium. The procedure can be used for small amounts of vanadium in steels after a mercury cathode separation; more than 0·5 mg of Ti, Zr, rare earths, and other elements which are not separated by the mercury cathode and form acid-insoluble phosphates must also be removed.

Decompose 2–5 g of ore in a platinum dish with hydrofluoric and nitric acids, evaporate to dryness, and repeat the additions of acids until all silica is removed. Dissolve in 5 ml nitric acid and a little water, transfer to a 100-ml volumetric flask, mix, and withdraw an aliquot containing 0·1–1 mg of vanadium. Decompose iron or steel with hydrochloric, nitric, hydrofluoric, and sulphuric acids, and evaporate to fumes of the latter. Make a mercury cathode separation, dilute the electrolyte and washings to a definite volume, and withdraw an aliquot containing less than 1 mg vanadium.

Add 5 ml nitric acid, evaporate nearly to dryness, dilute to 15 ml, heat, cool, add 10 drops of saturated bromine water and add 20% sodium hydroxide until the yellow bromine colour disappears. Add 7ml of 1 : 4 sulphuric acid, dilute to 30 ml, add 3 ml of 3 : 2 phosphoric acid and 5 ml of sodium tungstate solution, mix, and boil for five minutes. Cool, transfer to a 100-ml volumetric flask, make to volume, mix, and measure the absorbance at 400 or 410 nm using a reagent blank as reference solution.

REAGENTS

Sodium tungstate solution. Dissolve 16·5 g of pure $Na_2WO_4.2H_2O$ and one pellet of sodium hydroxide in water, and dilute to 500 ml in a plastic bottle.

Standard vanadium solutions. Dissolve 1·785 g of pure V_2O_5 in a slight excess of sodium hydroxide solution. Acidify with 1 : 1 sulphuric acid, add 10 ml excess, cool, dilute to 1 litre and mix; each ml contains 1 mg of vanadium. Dilute 10 ml of this stock solution to 1 litre after adding 10 ml of 1 : 1 sulphuric acid; each ml of this solution contains 0·01 mg of vanadium.

B. HYDROGEN PEROXIDE

When hydrogen peroxide is added to a sulphuric acid solution of vanadium(V), a reddish–yellow colour is formed which can be measured photometrically. Titanium, molybdenum, and iron which give a yellow colour with hydrogen peroxide, and coloured ions like chromium, nickel, cobalt, copper, etc., must be absent. The colour hue varies slightly with the ratio of hydrogen peroxide to sulphuric acid; the acid concentration and hydrogen peroxide addition should be kept approximately the same for standards and unknowns. The procedure given below is applicable to all ores; for some materials it may be suitably modified.

Decompose the sample with nitric, hydrochloric, and sulphuric acids, and evaporate to strong fumes of the latter. If the insoluble residue might contain vanadium, filter it off, wash, ignite, fuse with sodium carbonate, and add to the main portion. Any other suitable means of decomposition, such as hydrofluoric and sulphuric acid in a platinum dish, may be employed. If chromium is present, eliminate it as chromyl chloride by one or two evaporations in boiling hydrochloric and perchloric acids.

Adjust the acid concentration to 5–10%, add 1–3 g of tartaric acid, and pass in a rapid stream of hydrogen sulphide until all copper, molybdenum, and other members of this group have been precipitated. Wash with acidulated hydrogen sulphide water, and discard the precipitate. If the content of acid group elements or of vanadium is high, dissolve the residue and re-precipitate to recover occluded vanadium.

Boil out all hydrogen sulphide from the filtrate, oxidize with a few drops of 30% hydrogen peroxide, boil, cool, neutralize with sodium hydroxide and add a slight excess. Boil, filter, and wash with 1% sodium hydroxide. Dissolve the precipitate in sulphuric acid, reprecipitate with sodium hydroxide, filter and wash as before; discard the precipitate.

Acidify the filtrate with sulphuric acid, reduce the volume if necessary by evaporation, and adjust the acidity to approximately 5 ml sulphuric acid per 100 ml. Add 5 ml of 3% hydrogen peroxide and make to 100 ml in a volumetric flask. Measure the absorbance at 450 nm.

C. OTHER COLORIMETRIC METHODS

Vanadium from 0·025 to 25 g/litre in uranium–vanadium liquors is determined by measuring the absorbance at 775 nm of the blue vanadyl ion in 5% sulphuric acid solution containing about 5% sulphurous acid. Nitric acid must be absent, but small amounts of hydrochloric or perchloric acid do not interfere [20].

A simple method for vanadium in uranium materials, utilizing benzohydroxamic acid has been described; common impurities are tolerated at reasonable levels [12].

Vanadium has been determined in waters, biological materials, and silicates by a photometric determination with diaminobenzidine, after co-precipitation with ferric hydroxide, and separation by ion exchange [3].

Vanadox, 2,2′-dicarboxydiphenylamine, has been proposed for the photometric determination of vanadium in sulphuric acid at 610 nm for a variety of materials [7].

Vanadium forms a complex in hydrochloric acid with n-benzoyl-*o*-tolylhydroxylamine which can be extracted in carbon tetrachloride and measured at 510 nm [10].

A spectrophotometric procedure for vanadium in steel has been described, using 8-hydroxyquinoline in chloroform and measuring at 550 nm, after isolation with α-benzoinoxime [14].

3. Optical Spectrography

Vanadium can be satisfactorily determined in many materials by emission spectrography [1, 11]. Detailed procedures are outlined for 0·001–0·05% vanadium in aluminium and its alloys [1], 0·05–0·35% in plain carbon and low alloy steels [1], and 20–200 ppm in zirconium and its alloys [1].

4. X-Ray Spectrography

This technique can be used to determine vanadium in a few materials [11, 15], but its use is not as general as optical spectrography.

5. Atomic-Absorption Spectrophotometry

A solution of vanadium cupferrate in isobutyl methyl ketone [5], the latter with oleic acid [17], or in n-butyl acetate [16], can be used with a nitrous oxide–acetylene flame in atomic absorption. It has been reported that the absorption is enhanced by the addition of diethylene glycol [18].

REFERENCES

1. A.S.T.M., *Chemical Analysis of Metals. Sampling and Analysis of Metal Bearing Ores*, Philadelphia, Pa., American Society for Testing and Materials, 1969.

2. British Iron and Steel Research Association, Methods of Analysis Committee, *J. Iron Steel Institute* **182**, 156–9 (1956).
3. Chan, K. M., and Riley, J. P., *Anal. Chim. Acta* **34**, 337–45 (1966).
4. Cooper, M. D., and Winter, P. K., *Anal. Chem.* **21**, 605–9 (1949).
5. Crump-Wiesner, H. J., and Purdy, W. C., *Talanta* **16**, 124–9 (1969).
6. Fritz, J. S., and Abbink, J. E., *Anal. Chem.* **34**, 1080–2 (1962).
7. Frumina, N. S., Mustafin, I. S., Nikurashina, M. L., and Vechera, M. K., *Talanta*, **16**, 138–43 (1969).
8. Furman, N. H., ed., *Scott's Standard Methods of Chemical Analysis*, 6th ed., Vol. 1, Princeton, N.J., Van Nostrand, 1962.
9. Hillebrand, W. F., Lundell, G. E. F., Bright, H. A., and Hoffman, J. I., *Applied Inorganic Analysis*, 2nd ed., New York, Wiley, 1953.
10. Jeffery, P. G., and Kerr, G. O., *Analyst* **92**, 763–5 (1967).
11. Kolthoff, I. M., and Elving, P. J., *Treatise on Analytical Chemistry*, Part II, Vol. 8, 177–272, New York, Interscience Publishers, 1963.
12. Kuehn, P. R., Howard, O. H., and Weber, C. W., *Anal. Chem.* **33**, 740–4 (1961).
13. Lannoye, R. A., *Anal. Chem.* **35**, 558–61 (1963).
14. Luke, C. L., *Anal. Chim. Acta* **37**, 267–9 (1967).
15. Mitchell, B. J., *Anal. Chem.* **30**, 1894–1900 (1958).
16. Pearton, D. C. G., Taylor, J. D., Faure, P. K., and Steele, T. W., *Anal. Chim. Acta* **44**, 353–9 (1969).
17. Sachdev, S. L., Robinson, J. W., and West, P. W., *Anal. Chim. Acta* **37**, 12–19 (1967).
18. Sachdev, S. L., Robinson, J. W., and West, P. W., *Anal. Chim. Acta* **37**, 156–63 (1967).
19. Schoeller, W. R., and Powell, A. R., *The Analysis of Minerals and Ores of the Rarer Elements*, 3rd ed., London, Charles Griffin, 1955.
20. Union Carbide Metals Co., Niagara Falls, N.Y., Personal Communication, 1960.
21. Union Carbide Nuclear Co., Tuxedo, N.Y., Personal Communication, 1960.

41 Zinc

Zinc is one of the most important metals, and its determination is frequently required in many mining and metallurgical laboratories. The principal zinc mineral is sphalerite, ZnS, but others serving as a commercial source of zinc are marmatite, (Fe, Zn)S, smithsonite, $ZnCO_3$, willemite, hydrated zinc silicate, franklinite, (Fe,Mn,Zn) oxide, and zincite, ZnO. Zinc is frequently associated with lead in ores, and sometimes with copper.

ISOLATION OR SEPARATION OF ZINC

As a member of the ammonium sulphide group, zinc can be separated from elements of the acid sulphide group by precipitation of the latter with hydrogen sulphide in 5–10% hydrochloric or sulphuric acids. Precipitation with ammonium hydroxide and ammonium chloride serves to separate Fe, Al, Be, Cr, Ti, Zr, U, Th, and other members of the ammonium hydroxide group from zinc; the latter is accompanied by Mn, Ni, and Co. Precipitation of zinc with ammonium sulphide in ammoniacal solution, or with hydrogen sulphide in faintly acid solution, will remove it from the alkaline earths and alkalies.

At a mercury cathode in dilute sulphuric acid, zinc together with Fe, Cr, Mo, Co, Ni, Cu, Sn, Bi, Tl, Cd, Ag, and Au is deposited, and can be separated from Al, Be, Ti, Zr, P, V, U, W, Nb, and Ta which remain in solution. Cupferron in cold 10% sulphuric acid leaves in solution zinc, together with Al, Be, Cr, Ni, Co, U(VI), Cd, and Mn, whereas Fe, Ti, V, Zr, Mo, Th, Nb, Ta, W, Sn, Sb, and Bi are precipitated. Treatment with sodium hydroxide leaves zinc in solution, together with Al, Be, P, V, Mo, Sn, As, Sb, Pb, and W, whereas Fe, Mn, Cr, Co, Ni, Ti, Zr, Th, U, Cu, Ag, Tl, and Cd are precipitated; if sodium peroxide is used, U and Cr are thrown into the filtrate with the zinc.

The addition of ammonium tartrate to an ammonium sulphide treatment yields a precipitate of zinc, together with Fe, Co, Tl, and most of the Mn and Ni, whereas Ti, V, Al, Cr, Nb, Ta, and W remain in solution. Zinc can be separated from iron and nickel in a column of Amberlite IRA-400 or Dowex-1; nickel is not retained even in 12 N hydrochloric acid, whereas iron can be eluted by 0·5 N and zinc by 0·005 N hydrochloric acid [20]. Iron can be removed from zinc by an ether extraction, and chromium volatilized by boiling hydrochloric and perchloric acids.

1. Volumetric Methods

A. With Potassium Ferrocyanide

This method is employed for routine determinations of zinc in a wide variety of materials [6, 13, 15, 16, 17, 19, 28, 34]. The titration requires the absence of Cu, Ni, Co, V, Mo, W, U, Mn, Cd, and Fe.

Decompose 0·5–5 g with nitric and hydrochloric acid, with the usual initial addition of a few drops of bromine for high sulphides; add 10 ml 1 : 1 sulphuric acid and evaporate to strong fumes of the latter. Cool, dilute with hot water, boil, filter, and wash thoroughly with hot water. A precipitate will usually consist of silica, lead sulphate, etc. and will be free from zinc; if the sample is refractory, however, and the residue might contain zinc, ignite the paper and precipitate in a platinum crucible at a low heat, cool, add hydrofluoric acid and a few drops of sulphuric acid, evaporate to dryness several times with more additions of hydrofluoric and sulphuric acids if necessary to eliminate all silica, add dilute sulphuric acid, heat, filter, wash, and add the filtrate to the main portion. Any other convenient method of decomposition, to bring all zinc into solution, may be used.

Adjust the acidity to 10% by volume of concentrated sulphuric or hydrochloric acid and pass in hydrogen sulphide until all Group 2 elements have been precipitated. Filter and wash thoroughly with acidulated hydrogen sulphide water; if the precipitate is bulky, dissolve it in 1 : 1 hydrochloric acid, adjust the acidity, and re-precipitate with hydrogen sulphide to set free the small quantities of zinc which are carried down with sulphides of copper, etc. Boil the filtrate to eliminate all hydrogen sulphide, oxidize with hydrogen peroxide, and again boil.

In the absence of nickel and cobalt, zinc can now be satisfactorily separated from elements which interfere in the ferrocyanide titration by an ammonia precipitation in the presence of ammonium chloride, preceded by oxidation with bromine water or ammonium persulphate to throw manganese into the precipitate of Fe, Al, Ti, V, etc. Add 10–20 ml of saturated bromine water or 5–20 ml of 10% ammonium persulphate, 5 g ammonium chloride, make ammoniacal, boil, filter, and wash with dilute ammonium hydroxide–chloride solution. Dissolve the precipitate in hydrochloric acid, oxidize and re-precipitate as before, boil, filter, and wash; zinc is in the filtrate free from elements which interfere in the titration.

In the presence of nickel and cobalt, or where small amounts of zinc must be isolated from large quantities of members of the ammonium hydroxide group, precipitation of zinc as sulphide in faintly acid solution, with or without the addition of formic acid, yields a satisfactory separation. To the sulphuric acid solution of zinc, that is free from members of the hydrogen sulphide group and has been boiled and oxidized to remove hydrogen sulphide, add sodium hydroxide until the solution is neutral. Dilute so that the volume of solution is at least 200 ml and contains not more than 0·1 g zinc per 100 ml. Adjust the acidity carefully with 5% sulphuric acid so that the solution is as near 0·01N as possible; this is equivalent to approximately 1 ml of 5%

sulphuric acid per 200 ml of solution. Cool to room temperature and pass a rapid stream of hydrogen sulphide through the solution for 30 minutes. Allow to settle for 10 minutes, filter, and wash with cold water. Transfer the precipitate to the original beaker with hydrochloric acid, neutralize with ammonium hydroxide, and add 3 ml excess hydrochloric acid.

Alternatively, if zinc is to be precipitated in the presence of formic acid, add ammonium hydroxide just short of neutrality to the boiled and oxidized filtrate from the hydrogen sulphide precipitation of Group 2 sulphides. Add 25 ml of a 20% citric acid solution and make neutral to methyl orange with ammonium hydroxide. Add 25 ml of formic mixture, dilute to 200 ml for each 0·2 g zinc, pass in hydrogen sulphide for 5 minutes to the warm solution and allow to stand under pressure for 20 minutes. Filter and wash with cold 2·5% formic acid saturated with hydrogen sulphide. Dissolve the zinc sulphide off the paper with hydrochloric acid, wash thoroughly, neutralize the excess acid with ammonium hydroxide and add 3 ml excess hydrochloric acid.

If the precipitate of zinc sulphide in both the above separations is bulky, especially if the cobalt is more than 20% of the zinc, it is necessary to dissolve and re-precipitate to get rid of occluded impurities.

Heat the solution containing about 3 ml excess hydrochloric acid to 80–85°C, and add 0·3 ml of ferrous sulphate solution. Pour about one quarter of the sample solution into a 250 ml beaker, and add the standard potassium ferrocyanide solution, with stirring, to the remaining portion in the original beaker until the endpoint has been passed, which is indicated by a change in colour to a lighter shade. Add all but a few ml of the reserve solution and titrate carefully until the endpoint is again passed. Now add the final few ml of reserve solution and the rinsings, and titrate to the appearance of a red-brown tinge when a drop of solution is added to two drops of uranium nitrate or acetate external indicator on a spot plate.

Standardize the potassium ferrocyanide solution against 0·200 g pure zinc dissolved in hydrochloric acid, and containing the same volume, concentration of ammonium salts, etc. as the individual analyses.

The change in colour in the solution at the endpoint is due to the reaction between the small amount of ferrous iron present and the ferricyanide in the standard. As long as an excess of zinc is present it removes the ferrocyanide as precipitated potassium zinc ferrocyanide, $K_2Zn_3[Fe(CN)_6]$, and the iron remains as the blue ferro-ferricyanide. When potassium ferrocyanide is in excess, the blue ferro-ferricyanide is decomposed into the colourless ferro-ferrocyanide. This causes the sudden change in the colour of the solution to a lighter shade, or from blue to pea green.

$$2K_4Fe(CN)_6 + 3ZnCl_2 \rightarrow K_2Zn_3[Fe(CN)_6]_2 + 6KCl$$

$$3FeSO_4 + 2K_3Fe(CN)_6 \rightarrow Fe_3[Fe(CN)_6]_2 + 3K_2SO_4 \text{ (blue)}$$

$$K_4Fe(CN)_6 + FeSO_4 \rightarrow K_2Fe[Fe(CN)_6] + K_2SO_4 \text{ (white)}$$

This colour change may be difficult for the inexperienced, but the use of uranium nitrate or acetate as an outside indicator enables the endpoint to be detected readily. In fact, many chemists omit the addition of ferrous sulphate to the sample and of potassium ferricyanide to the standard potassium ferrocyanide, and rely entirely on the external indicator.

$$K_4Fe(CN)_6 + 2UO_2.(NO_3)_2 \rightarrow \underset{\text{red-brown}}{(UO_2)_2Fe(CN)_6} + 4KNO_3$$

The uranyl ion UO_2^{++} unites with any excess ferrocyanide ion to form the insoluble red-brown uranyl ferrocyanide. The solubility product of the latter is such that the endpoint is not stoichiometric, and potassium ferrocyanide must be standardized under conditions parallel to those used in the titration of zinc.

Reagents

Ferrous sulphate solution. Dissolve 1·25 g of $FeSO_4.7H_2O$ in water, add 3 ml of hydrochloric acid, and dilute to 250 ml.

Potassium ferrocyanide solution. Dissolve 43 g $K_4Fe(CN)_6.3H_2O$ and 0·5 g potassium ferricyanide in 1 litre of water; 1 ml = 0·01 g zinc.

Formic mixture. Dissolve 250 g of ammonium sulphate in 400 ml of water, add 30 ml ammonium hydroxide, 200 ml formic acid, and dilute to 1 litre with water.

Uranyl nitrate indicator. Dissolve 5 g of $UO_2.(NO_3)_2.6H_2O$ in water and dilute to 100 ml. Alternatively, dissolve 2·5 g uranyl acetate in 95 ml of water, boil, add 5 ml acetic acid and cool.

B. With Sodium Ethylenediamine Tetraacetate

Zinc in many materials can be determined by converting it to the zinc thiocyanate complex, extracting with methyl isobutyl ketone, stripping with ammonia, treating with cyanide and then with formaldehyde, and finally titrating with standard EDTA solution [3, 6, 12, 17].

The procedure recommended for zinc in brasses and bronzes, outlined below, is illustrative of this versatile method; it should be noted that the extraction procedure separates zinc from cadmium. Take either the whole electrolyte remaining from the determination of copper and lead, or an aliquot portion containing not more than 0·1 g of zinc. Neutralize with 20% sodium hydroxide, add 10 ml of 1 : 1 hydrochloric acid and cool. Transfer to a 500-ml separatory funnel and dilute to about 250 ml. Add 30 ml of 50% ammonium thiocyanate solution and mix; if the sample is high in iron or aluminium, add sufficient 20% ammonium fluoride to discharge the red colour of ferric thiocyanate. Add 50 ml of methyl isobutyl ketone and shake for 1 minute. Draw off the lower aqueous layer into a second separatory funnel, retaining the organic layer. Add an additional 50 ml of methyl isobutyl ketone to the second funnel and shake for 1 minute. Draw off and discard the lower aqueous layer and add the organic layer to that retained in the first separatory funnel. To the combined solvent extracts add about 100 ml of thiocyanate wash solution, shake, allow to separate, and draw off and discard the aqueous layer. To the organic layer add 40 ml of the buffer solution plus 60 ml of water, and shake well to strip the zinc from the organic phase. Again allow the layers to separate, and draw off the lower ammoniacal layer into a 600-ml beaker. Discard the organic layer.

Dilute to about 300 ml, insert into the beaker a tetrafluoroethylene stirring bar, add about 5 drops of Eriochrome black T indicator, and transfer the beaker to a magnetic stirring apparatus. Add 10% potassium cyanide solution in small portions until the red colour has changed to pure blue. Cautiously release the zinc from the cyanide complex by adding formaldehyde dropwise until the colour has changed again to wine red. Titrate with EDTA solution to a pure blue endpoint, making further additions of formaldehyde to ensure that all zinc has been released and titrated; avoid excessive amounts of formaldehyde.

REAGENTS

*Buffer solution, p*H 10. Dissolve 54 g of ammonium chloride in water, add 350 ml of ammonium hydroxide, and dilute to 1 litre.

Sodium ethylenediamine tetraacetate. Dissolve disodium ethylenediamine tetraacetate, $Na_2H_2C_{10}H_{12}O_8N_2.2H_2O$, in water as follows: 3·7225 g per litre = 0·01M, and 18·6125 g per litre = 0·05M. These solutions should be standardized against appropriate portions of copper-base alloys, containing known quantities of zinc, which have been carried through all steps of the procedure.

Eriochrome black T indicator solution. Dissolve 0·4 g of the sodium salt of 1-(1-hydroxy-2-naphtholazo)-5-nitro-2-naphthol-4-sulphonic acid in a mixture of 20 ml of ethyl alcohol and 30 ml of triethanolamine.

Thiocyanate wash solution. Add 100 ml of the 50% ammonium thiocyanate solution to 700 ml of water, add 9 ml hydrochloric acid, 5 ml nitric acid, and make up to 1 litre.

2. Gravimetric Procedures

A. AS ZINC OXIDE

Although zinc is normally determined in the mineral industries for routine work volumetrically or by atomic absorption, there are occasions when it is advantageous, for medium or high concentrations, to employ the gravimetric zinc oxide method [6, 17]. Decompose the sample, make the appropriate separations, and finally isolate zinc as sulphide by hydrogen sulphide treatment in 0·01 N sulphuric acid as outlined in **1,** A. Transfer the paper and precipitate to a weighed porcelain crucible, dry, ignite cautiously, and finally heat at 950°C to constant weight. ZnO × 0·8034 = Zn.

B. AS ZINC MERCURIC THIOCYANATE

If a gravimetric procedure is favoured, for contents of zinc below about 0·1% the element is preferably determined as zinc mercuric thiocyanate instead of zinc oxide [6, 8, 17]. After decomposition, the usual separations, and isolation of zinc sulphide in 0·01N sulphuric acid solution with hydrogen sulphide, described in **1,** A, dissolve the zinc sulphide with hot 1 : 3 hydrochloric acid, wash, add 5 ml sulphuric acid and evaporate to fumes of the latter. Cool, add water, boil, add 25 ml of sodium hydrosulphide solution, filter, and wash. Boil the filtrate to remove hydrogen sulphide, dilute to 120 ml, and oxidize the iron with 1% potassium permanganate. Add 25 ml of ammonium mercuric thiocyanate solution, stirring vigorously, and allow to stand overnight. Collect the precipitate on a weighed Gooch crucible, and wash with 1% ammonium mercuric thiocyanate wash solution. Dry at 105°C for an hour and weigh. Make a blank determination, following the same procedure and using the same amounts of reagents; subtract the blank from the weight of zinc mercuric thiocyanate. Zinc mercuric thiocyanate × 0·1289 = zinc.

REAGENTS

Sodium hydrosulphide solution. Dissolve 0·2 g of sodium hydroxide in 1 litre of water and saturate with hydrogen sulphide.

Ammonium mercuric thiocyanate solution. Dissolve 32 g of ammonium thiocyanate and 27 g of mercuric chloride in 500 ml of water, and filter to remove any residue.

3. Colorimetric Methods

A. DITHIZONE

Zinc forms a purple-red dithizonate in neutral, alkaline, and faintly acid solutions with dithizone. The interference of small quantities of many other metals which react under these conditions may be suppressed by certain complexing reagents. Sodium thiosulphate prevents reaction with Cu, Ag, Au, Hg, Pb, Cd, and Bi at *p*H 4·4–5·5; potassium cyanide suppresses interference by Ni and Co. Procedures for the determination of small quantities of zinc by dithizone have been described for waters [1, 30], metals [6, 8, 18, 25, 33], biological samples [2, 5, 22, 26, 30, 32], and many other materials [17, 21, 23, 30, 31].

The dithizone colorimetric procedure is used, of course, for trace amounts of zinc, and the isolation of the latter from large amounts of other elements may take a variety of forms. For zinc in most organic substances, dry ashing at 450 ± 20°C is just as good as wet oxidation [29]. Arsenic, antimony, tin, and selenium may be volatilized by evaporating several times with a mixture of hydrobromic and sulphuric acids. Copper may be removed by electrolysis, and lead as sulphate. All members of the acid sulphide group may be separated by hydrogen sulphide in 5% hydrochloric or sulphuric acid, with a re-precipitation to set free occluded zinc.

Very small amounts of zinc can be precipitated with hydrogen sulphide in faintly acid solution, as described earlier, if a small quantity of copper or mercury is used as a collector. Zinc can be separated from Fe, Ni, and Co by passing a 2N hydrochloric acid solution through Dowex 1 resin; zinc is retained and can be eluted with 0·001N hydrochloric acid. Micrograms of zinc can be extracted from milligrams of nickel in 2M hydrochloric acid by tri-iso-octylamine-xylene, and returned to the aqueous phase by back-extracting with 0·5M nitric acid solution [18]. Separations by cupferron, mercury cathode, ammonium hydroxide, and sodium hydroxide, discussed at the beginning of this chapter, may also be employed.

Obtain the zinc finally in a volume of 10–25 ml of about 0·1N hydrochloric acid, containing more than 5 μg of zinc, and preferably less than 1 mg of other metals such as Cu, Ag, Pb, Fe, Mn, Sn, Al, Cd, Ni, Bi, In, Te, Tl, Au, Pt, and Pd. Aluminium does not react with dithizone, but if present in amounts greater than 100 mg it hinders the extraction of zinc from slightly acid solution. It is preferable to reduce the acid concentration of the sample to about 0·1N by evaporation rather than neutralization, to avoid possible introduction of zinc. The presence of traces of zinc is widespread in reagents, and additions of the latter should be kept to a minimum; the procedure for unknowns and standards should be the same.

Add 0·5M sodium acetate solution, from which heavy metals have been removed by shaking with successive small portions of 0·005% dithizone in carbon tetrachloride, to give a *p*H of 5–5·5. Add sufficient 50% sodium thiosulphate solution to form complexes with interfering metals. For each mg of metal the following amounts, in milligrams, of $Na_2S_2O_3.5H_2O$ are required: copper 550, mercury 700, bismuth 325, silver 55, lead 36. Shake the solution vigorously with 2–3 ml of 0·005% dithizone in carbon tetrachloride for a minute or two, draw off the extract into another separatory funnel, and wash the first with 0·5–1 ml of carbon tetrachloride. Repeat the extraction until the last portion of dithizone remains unchanged in colour after shaking for 2–3 minutes.

Wash the combined carbon tetrachloride extracts 2–3 times with 5-ml portions of sodium thiosulphate wash solution. The latter is made by mixing 225 ml of 0·5M sodium acetate, 10 ml of 50% sodium thiosulphate, 40 ml of nitric acid, and making to 500 ml with water; remove traces of zinc by shaking with dithizone. Wash once with water and 2–3 times with 5 ml portions of a freshly prepared 0·04% sodium sulphide solution. If the last portion of sulphide wash solution is not colourless, continue washing until it remains colourless.

Dilute the washed zinc dithizonate solution to a definite volume with carbon tetrachloride, and measure the optical density at 535 nm against a blank carried through the whole procedure. Prepare the calibration curve from aliquots containing 1–100 μg of zinc, prepared from 99·999% zinc metal.

In the presence of appreciable nickel or cobalt it is necessary to use potassium cyanide as an additional complex-forming agent. Bring the sample solution to approximate neutrality with ammonium hydroxide, and add enough 5% potassium cyanide solution to dissolve the precipitate first formed. Add dilute hydrochloric acid dropwise until the *p*H is 3–4, and then adjust the *p*H to 5–5·5 with sodium acetate. Extract the clear solution with dithizone as described above. Fourteen ml of 5% potassium cyanide is sufficient to tie up 100 mg of either nickel or cobalt.

A rapid method for zinc in mill tailing, utilizing dithizone, will give a result in 10–15 minutes with an accuracy within 10% of the true value. Decompose 0·5 g in nitric acid and evaporate rapidly to dryness. Dissolve soluble salts in a small quantity of hot hydrochloric acid, and dilute to 500 ml with water. Pipette a 1-ml sample into a test tube to which has been added 10 ml of buffer solution. The latter contains, per litre, 136 g sodium acetate, 57 ml acetic acid, 5 g ammonium citrate, and 50 g sodium thiosulphate. Ammonium citrate is added to complex iron in the sample, and sodium thiosulphate eliminates interference from lead. Add 10 ml of 0·01% dithizone in carbon tetrachloride, shake the tube, and compare the colour which is developed with a series of standards.

B. Other Colorimetric Methods

Zinc has been determined photometrically in nickel alloys with 1-(2-pyridylazo)-2-naphthol [4].

4. Polarographic Procedures

Zinc in some materials can be conveniently determined by polarography. Detailed directions have been published for the polarographic determination of 0·001–0·05% zinc in lead- and tin-base solders [6], 0·001–0·01% in white metal bearing alloys [6], 0·02–1% in aluminium alloys [8], and 0·001–0·05% in nickel [18].

5. Optical Spectrography

Zinc is determined on a routine basis in some materials by emission spectrography. Methods have been outlined for 0·001–10% zinc in aluminium and its alloys [6, 9], 0·001–0·02% in pig lead [6], 50–800 ppm in nickel [18], and varying concentrations in copper and other materials [17].

6. X-Ray Spectrography

This technique has been utilized for determining zinc in a few products [17, 27].

7. Atomic-Absorption Spectrophotometry

Atomic absorption is an excellent procedure for zinc from traces to at least 5%. It is sensitive and specific, the line 2138·6 having a sensitivity of 0·03 μg/ml; another line 3075·9 may be used [11, 12].

8. Other Methods

For zinc in water, it has been reported that DTPA, diethylenetriamine-pentaacetic acid, is very much better than EDTA in its chelation of zinc over calcium [14].

Differentiation of the Forms of Zinc

1. *Oxide Zinc in Ores.* The following method is applicable to all types of zinc ores [7]. Weigh 0·5–1 g of –200 mesh sample into a 300-ml stoppered Erlenmeyer flask fitted with a Bunsen valve. Add 50 ml of 2% by volume sulphuric acid which is saturated with sulphur dioxide. Stopper the flask and allow to stand at 30–40°C for one hour, swirling the flask for a few seconds every 10 minutes. Filter, wash thoroughly with hot water, add 5 ml of 1 : 1 sulphuric acid to the filtrate and boil off all sulphur dioxide. Determine zinc in the filtrate by any convenient standard procedure; this figure represents the oxidized, or non-sulphide, zinc. If desired, zinc may also be determined on the residue, which contains the zinc sulphide.

Zinc oxide can be determined in zinc powder by allowing 2 g of the latter to stand for 2 hours in 100 ml of 30% ammonium acetate solution [24]. The oxide zinc in some simple ores may also be differentiated by adding to 1 g of ore 25 ml of 25% ammonium chloride solution and 10 ml of saturated ammonium acetate, boiling for 10 minutes, filtering , and determining in the filtrate the oxide zinc [19].

2. *Metallic Zinc in Fume or Dust.* To a dry 500 ml graduated flask add 7 g of pure powdered ferric sulphate and 0·5 g of the zinc dust; agitate the flask until they are thoroughly mixed [10]. Add 25 ml water, stopper the flask, and shake for 15 minutes. Add 300 ml of 1 : 1 sulphuric acid and make the solution up to the mark with water. Withdraw 100 ml of the solution and titrate the ferrous sulphate, which has been formed, with 0·1N potassium permanganate. Ferric sulphate is reduced by metallic zinc to ferrous sulphate:

$$Fe_2(SO_4)_3 + Zn \rightarrow ZnSO_4 + 2FeSO_4$$

$$1 \text{ ml of } 0{\cdot}1\text{N } KMnO_4 = 0{\cdot}005585 \text{ g Fe}$$

$$= 0{\cdot}003269 \text{ g Zn}$$

If significant quantities of iron are present in the zinc dust or fume, dissolve 1 g of the latter in sulphuric acid, determine the number of ml of 0·1N potassium permanganate required, and deduct a proportionate amount from the previous titration.

Zinc Development Association

Chemists who have special problems in zinc analysis are invited to contact the Zinc Development Association, 34 Berkeley Square, London W1X 6AJ; the Association is sponsored by leading producers to assist all users of zinc.

REFERENCES

1. American Public Health Association, *Standard Methods for the Examination of Water and Wastewater*, New York, American Public Health Association, 1965.
2. Analytical Methods Committee, *Analyst* **92**, 324–5 (1962).
3. Andrew, T. R., and Nichols, P. N. R., *Analyst* **86**, 676–7 (1961).

4. ANDREW, T. R., and NICHOLS, P. N. R., *Analyst* **90**, 161–4 (1965).
5. A.O.A.C., *Official Methods of Analysis of the Association of Official Analytical Chemists*, Washington, D.C., Association of Official Analytical Chemists, 1965.
6. A.S.T.M., *Chemical Analysis of Metals. Sampling and Analysis of Metal Bearing Ores*, Philadelphia, Pa., American Society for Testing and Materials, 1969.
7. BARKER, C. W., and YOUNG, R. S., *J. Soc. Chem. Ind.* **67**, 61 (1948).
8. BRITISH ALUMINIUM CO., *Analysis of Aluminium and its Alloys*, London, British Aluminium Co., 1961.
9. BRITISH ALUMINIUM CO., *Spectrochemical Analysis of Aluminium and its Alloys*, London, British Aluminium Co., 1961.
10. DILLON, V. S., *Assay Practice on the Witwatersrand*, Johannesburg, Transvaal and Orange Free State Chamber of Mines, 1955.
11. DONALDSON, E. M., and ROLKO, V. H. E., Ottawa, *Canadian Dept. Energy, Mines, Resources, Tech. Bull.* 93, 1967.
12. ELWELL, W. T., and SCHOLES, I. R., *Analysis of Copper and its Alloys*, Oxford, Pergamon, 1967.
13. FURMAN, N. H., ed., *Scott's Standard Methods of Chemical Analysis*, 6th ed., Vol. 1, Princeton, N.J., Van Nostrand, 1962.
14. HICKEY, J. J., and OVERBECK, C. J., *Anal. Chem.* **38**, 932–4 (1966).
15. HILLEBRAND, W. F., LUNDELL, G. E. F., BRIGHT, H. A., and HOFFMAN, J. I., *Applied Inorganic Analysis*, 2nd ed., New York, Wiley, 1953.
16. IMPERIAL SMELTING CORP., Avonmouth, England, Personal Communication, 1960.
17. KOLTHOFF, I. M., and ELVING, P. J., *Treatise on Analytical Chemistry*, Part II, Vol. 3, 95–169, New York, Interscience Publishers, 1961.
18. LEWIS, C. L., OTT, W. L., and SINE, N. M., *The Analysis of Nickel*, Oxford, Pergamon, 1966.
19. LOW, A. H., WEINIG, A. J., and SCHODER, W. P., *Technical Methods of Ore Analysis*, New York, Wiley, 1939.
20. LOWEN, J., and CARNEY, A. L., *Anal. Chem.* **27**, 1965 (1955).
21. MARGERUM, D. W., and SANTACANA, F., *Anal. Chem.* **32**, 1157–61 (1960).
22. MILLS, E. V., and BROWN, B. L., *Analyst* **89**, 551–3 (1964).
23. NISHIMURA, M., and SANDELL, E. B., *Anal. Chim. Acta* **26**, 242–8 (1962).
24. OSBORN, G. H., *Analyst* **76**, 114–15 (1951).
25. OTT, W. L., MACMILLAN, H. R., and HATCH, W. R., *Anal. Chem.* **36**, 363–4 (1964).
26. PAGE, E. R., *Analyst* **90**, 435–6 (1965).
27. RAWLING, B. S., and GREAVES, M. C., *Australasian Inst. Mining Met. Proc.* No. 211, 135–55 (1964).
28. RICHARDSON, M. R., and BRYSON, A., *Analyst* **78**, 291–9 (1953).
29. ROACH, A. G., SANDERSON, P., and WILLIAMS, D. R., *Analyst* **93**, 42–9 (1968).
30. SANDELL, E. B., *Colorimetric Determination of Traces of Metals*, 3rd ed., New York, Interscience Publishers, 1959.
31. VALLEE, B. L., *Anal. Chem.* **26**, 914–7 (1954).
32. WESTOO, G., *Analyst* **88**, 287–91 (1963).
33. YOUNG, R. S., *Metallurgia* **36**, 347–8 (1947).
34. ZINC CORPORATION LTD., Broken Hill, Australia, Personal Communication, 1960.

42 Zirconium

Zirconium is found in ores, irons and steels, alloys, refractories, compounds, and in the metallic state; its determination may be encountered in any laboratory serving the mineral industries. Zirconium is widely distributed, and is frequently associated with titanium, niobium, tantalum, and rare earths. The most abundant mineral is the silicate, zircon, $ZrSiO_4$; the other commercial source of zirconium is baddeleyite, essentially ZrO_2.

All chemical methods for the determination of zirconium actually give the amount of zirconium plus hafnium. The latter can be determined only by physical methods such as emission spectroscopy or X-ray spectrography. Hafnium occurs in nearly all zirconium ores and compounds to the extent of 1–3%.

ISOLATION OR SEPARATION OF ZIRCONIUM

Zirconium is a member of the ammonium hydroxide group, and can be separated from elements of the acid sulphide group by precipitation of the latter with hydrogen sulphide in 5–10% by volume of hydrochloric or sulphuric acid. Precipitation with ammonium hydroxide and ammonium chloride separates zirconium from Ni, Co, Mn, Zn, alkaline earths, and alkalies. A cupferron separation in cold 10% sulphuric acid will give zirconium in the precipitate with Ti, V, Nb, Ta, W, Mo, Fe, Sn, Sb, and Bi, whereas Al, Be, Cr, Mn, Co, Ni, P, B, U(IV), Zn, Cd, Se, and Te are left in solution. Large quantities of iron are often conveniently removed from zirconium by an ether extraction. Electrolysis with a mercury cathode in dilute sulphuric acid will deposit Cr, Mo, Fe, Co, Ni, Cu, Zn, Bi, Cd, and Sn, whereas zirconium remains in solution with Ti, B, V, Nb, Ta, W, Th, U, Be, Al, and P. A zinc oxide separation precipitates zirconium, with nearly all other elements, from Co, Ni, and Mn which pass into the filtrate. Treatment with sodium hydroxide precipitates zirconium, together with Ti, Th, U, Cr, Mn, Fe, Co, Ni, Cu, and Cd, while Al, Be, V, Zn, W, As, Sb, Sn, Se and Te remain in solution; addition of sodium peroxide throws Cr and U into the filtrate. Precipitation with ammonium sulphide and ammonium tartrate leaves zirconium in solution, with Ti, Cr, V, Nb, Ta, Th, W, and U, whereas Fe, Zn, Co, Tl, and most of the Mn and Ni are precipitated.

Probably the best separation of zirconium is effected by the addition of mandelic acid to a dilute hydrochloric or sulphuric acid solution. Zirconium tetramandelate is alone precipitated in the presence of all other cations.

Silica and phosphate will co-precipitate, a few other interfering anions can be easily removed, and a double precipitation is required when large amounts of readily hydrolysable cations are present. Full details are described later in the gravimetric procedure.

Another good separation of zirconium is made by precipitation of its phosphate in 10% by volume sulphuric acid containing hydrogen peroxide. If present in considerable amount, Sn, Bi, and Th may be partially precipitated, but they can be entirely removed by a re-precipitation. Niobium and tantalum are partially precipitated and must be removed initially by a tannin separation [18], or by fusing the ignited phosphate precipitate with potassium hydroxide, dissolving in water, filtering, igniting the residue, fusing with potassium pyrosulphate, dissolving in acid, and re-precipitating zirconium as ZrP_2O_7.

Cation exchangers such as Dowex 50 or Amberlite IRA–120 adsorb zirconium ions in 1–2M hydrochloric acid; the element can later be eluted with 4M hydrochloric acid. An anion exchanger like Dowex 1 will retain zirconium in 9M hydrochloric acid, whereas thorium passes through [11]. An anion exchange separation for handling a complex sample containing Zr, Ti, Nb, Ta, W, Mo, and V has been described [2]. Pass a bisulphate–0·5M oxalate solution through a column of Dowex 1–X8. Elute with a solution which is 1·5M in hydrochloric acid, 0·5M in oxalic acid, and 0·007M in hydrogen peroxide. Zirconium is in this eluate, with Ti and V; treatment with sodium hydroxide will separate Zr and Ti from V, and a final phosphate separation will remove Zr from Ti. Other ion exchange techniques have been described [3, 12].

1. Gravimetric Methods

A. Mandelic Acid Method

When mandelic acid is added to a solution of zirconium in dilute hydrochloric or sulphuric acid, zirconium tetramandelate is precipitated, which can be filtered and ignited to ZrO_2 [11, 16, 19]. The method is specific for zirconium, no other cation being precipitated.

Fluorides, oxalates, tartrates, permanganates, and nitrates interfere, but these can be eliminated by fuming with perchloric or sulphuric acids. High concentrations of sulphate produce low results, but the presence of low concentrations only requires a longer period of digestion for complete precipitation. Phosphates and silicates are coprecipitated; silica is nearly always removed initially after acid dehydration or by volatilization, and phosphorus can be removed by fusing the original material with sodium carbonate or sodium peroxide and extracting with water. Phosphorus can also be separated from zirconium by cupferron–chloroform extraction in cold 10% sulphuric or hydrochloric acid. If large amounts of easily hydrolysable ions are present, such as Ti, Sb, Bi, or Sn, dissolve the first zirconium tetramandelate precipitate in dilute ammonium hydroxide, acidify, and re-precipitate as described later.

Dissolve the sample by any appropriate method. Ores can be fused with borax, or with sodium carbonate followed by potassium bisulphate, either directly or after a preliminary acid treatment [4]. Monazite sand can be fused with potassium acid fluoride. Zirconium ores require a longer period of fusion than most other materials. Zirconium metal and its alloys can be dissolved in a platinum dish with hydrofluoric acid, together with sulphuric or perchloric acid. Irons and steels containing small amounts of zirconium may be dissolved in hydrochloric, nitric, or sulphuric acids, and any residue filtered, ignited, treated with hydrofluoric and sulphuric acids, and added to the reserved filtrate; alternatively fuse the residue with borax or potassium pyrosulphate and add to the filtrate.

In all cases, evaporate to a low volume with sulphuric or perchloric acid to eliminate interfering anions and to dehydrate silica. Cool, dilute, boil, filter, wash, and discard the precipitate. Add 30 ml of hydrochloric acid and enough water to make the final volume 125–150 ml. The solution should now be 2–4M in hydrochloric acid, and if the original solution was acid the amount of added hydrochloric acid should be decreased accordingly.

Add 50 ml of 15% mandelic acid solution, stir, and digest for at least 30 minutes at 80–90°C. Stir, remove from the hot plate, and let stand for 30 minutes. If the sample contains less than 10 mg of zirconium, or if sulphate ions are present, digest for 1–2 hours and allow to stand overnight before filtration.

Filter through Whatman No. 42 paper, using pulp, and wash with 50–100 ml of a wash solution prepared by dissolving 2 g of mandelic acid in 100 ml of 10% hydrochloric acid. Transfer the paper and precipitate to a tared crucible, and ignite to constant weight at 800–1000°C. Cool, and weigh as ZrO_2.

$$ZrO_2 \times 0{\cdot}7403 = Zr$$

If a large quantity of Ti, Sb, Bi, or Sn is present in the sample, carry out a double precipitation as follows. Transfer the precipitate back into the original beaker with a minimum quantity of water. Wash the paper with 25 ml of 15% ammonium hydroxide solution, and allow this solution to run into the beaker containing the precipitate. Stir, add 3–4 ml ammonium hydroxide, stir again, and filter through Whatman No. 42 paper. Wash the residue and paper 4–5 times with 15% ammonium hydroxide, combining these washings with the filtrate. Neutralize the filtrate with hydrochloric acid, and add about 20 ml excess. Add 50 ml of 15% mandelic acid, and carry out the procedure as outlined previously.

B. Precipitation with Cupferron and Phosphate

Zirconium hydrolyses readily; do not allow the acidity of solutions to fall too low. Another point should always be recalled in the determination of zirconium — the insolubility of its phosphate in acid solution; if phosphorus is present it will cause trouble, and the sample should be fused with sodium carbonate, extracted with water, and the residue and water extract handled separately. Alternatively, phosphorus and zirconium can be separated by a cupferron–chloroform extraction.

Transfer 0·5–5 g to a 400 ml beaker, add 10–20 ml hydrochloric acid, 10–20 ml nitric acid, 10 ml 1 : 1 sulphuric acid, and evaporate to dense fumes of the latter [1, 5, 6, 7, 8, 10, 11, 18]. Cool, dilute with a small quantity of hot water, boil, and filter. If a residue remains, which might contain zirconium, fuse it with sodium carbonate and dissolve in dilute sulphuric acid; if a small residue still persists, fuse this finally with potassium bisulphate, dissolve in dilute acid and add to the main portion. Evaporate to strong fumes of sulphur trioxide and filter off silica, etc. Other methods of decomposition may, of course, be employed, such as initial treatment with hydrofluoric and sulphuric acids in a platinum dish to eliminate all silica, followed by addition of other acids or by fusion to decompose any residue. The decomposition techniques mentioned previously under A, such as fusion with borax or potassium acid fluoride, may also be used. A good flux for zircon and similar refractory minerals is 1 g boric acid + 3 g sodium fluoride for a 0·5 g sample; fusion for 15–20 minutes in a platinum crucible at 850°C is usually sufficient.

Adjust the acidity to 5–10% by volume sulphuric or hydrochloric acid, and pass in a rapid stream of hydrogen sulphide for 15 minutes to precipitate all metals of the acid sulphide group. If a bulky precipitate is obtained, dissolve in acids and re-precipitate with hydrogen sulphide to liberate traces of occluded zirconium. Filter, wash thoroughly with acidulated hydrogen sulphide water, boil the filtrate until all hydrogen sulphide has been removed, oxidize with hydrogen peroxide and again boil. Adjust the acidity of the solution to 10% sulphuric acid by volume, cool to 10°C, and precipitate zirconium together with Fe, Ti, V, Nb, Ta, and W by an excess of cold 6% cupferron. Filter and wash with cold 10% sulphuric acid; transfer the paper and residue to a platinum dish or large platinum crucible, dry, ignite, and fuse with potassium pyrosulphate. Dissolve the cooled melt in dilute sulphuric acid, and add 2–10 ml hydrogen peroxide to oxidize the titanium.

If the sample contains vanadium, remove it after the cupferron separation and subsequent fusion by adding sodium hydroxide, filtering and washing; vanadium is in the filtrate with most of the tungsten, whereas zirconium, with Fe and Ti, is in the precipitate. Dissolve the precipitate in acids or fuse with potassium pyrosulphate.

To the clear solution containing about 10% sulphuric acid and an excess of hydrogen peroxide, add 25 ml of a 20% solution of diammonium phosphate, stir vigorously, and allow to stand for at least several hours in a warm place. A large excess of phosphate is required in this determination. Filter on Whatman No. 42 paper, with pulp, and wash thoroughly with cold 5% ammonium nitrate. Dry, ignite at a low temperature, and gradually heat to about 1050°C. Cool, and weigh as ZrP_2O_7; $ZrP_2O_7 \times 0{\cdot}344 = Zr$.

If more than traces of niobium and tantalum are present, fuse the ignited phosphate precipitate with potassium hydroxide in a nickel crucible, dissolve in water, filter and wash. Ignite the residue, fuse with potassium pyrosulphate, dissolve in dilute sulphuric acid and re-precipitate in 10% sulphuric acid solution in the presence of hydrogen peroxide with diammonium phosphate. Filter, wash, ignite, and weigh as ZrP_2O_7 in the manner outlined above.

The precipitate of zirconium pyrophosphate tends to hydrolyse and lose P_2O_5 on washing. When much zirconium is present, or when the highest accuracy is required, do not weigh the zirconium pyrophosphate but fuse it with sodium carbonate, extract the melt with water, filter, wash with 1% sodium carbonate and then with water, and ignite the residue. Fuse with potassium pyrosulphate, dissolve in dilute sulphuric acid, boil, precipitate with ammonium hydroxide, filter, wash with hot 2% ammonium nitrate, ignite and weigh as ZrO_2. It may be difficult to remove all phosphate by one carbonate fusion and extraction, and to eliminate all alkali salts by one precipitation with ammonium hydroxide; for the most accurate analyses each of the above operations should be repeated [10]. $ZrO_2 \times 0{\cdot}7403 = Zr$.

C. Mercury Cathode or Ammonium Sulphide–Ammonium Tartrate Separation, Cupferron and Phosphate Precipitation

Electrolyse a 0·3 N sulphuric acid solution in a mercury cathode, after hydrogen sulphide removal of Group 2 elements, to deposit Fe, Cr, Zn, Co, and Ni. Make a cupferron separation of the solution in 10% sulphuric acid solution to precipitate zirconium with Ti, V, Nb, Ta, and W. In the absence of the latter five metals, ignite the precipitate to ZrO_2; in their presence remove vanadium with sodium hydroxide as described in **1**, B, and precipitate zirconium as phosphate.

Alternatively, iron may be removed from zirconium by precipitating the former with ammonium sulphide in the presence of ammonium tartrate. Reduce the iron in acid solution with hydrogen sulphide, add tartaric acid equal to five times the approximate weight of the sulphides present, make ammoniacal, and pass in hydrogen sulphide until the precipitate has coagulated. Filter, wash with dilute ammonium sulphide, acidify the filtrate to 10% by volume sulphuric acid, boil out all hydrogen sulphide, and proceed directly to the cupferron separation. It is not necessary to destroy the tartrate. If other elements are present in the cupferron precipitate, take the latter into solution with acids or fusion, and precipitate zirconium as phosphate as outlined previously.

Iron may, of course, also be removed initially with an ether extraction. Zirconium may also be separated from Al, Fe, Th, U, and rare earths by extraction with 2-phenoyltrifluorone-xylene [15].

3. Colorimetric Methods

A. Alizarin Red S

Various colorimetric methods have been proposed for small quantities of zirconium, but one employing Alizarin Red S, also known as alizarin sulphonic acid or sodium alizarin sulphonate, is probably of widest application [1, 8, 11,

17, 21, 22]. This reagent forms a coloured solution with zirconium in dilute hydrochloric acid solution. The sensitivity increases with decreasing acid concentration, but interferences caused by foreign ions are increased. Oxidizing agents, fluorides, and oxalates must be absent. Phosphate can be tolerated up to 0·1 mg, and sulphates, silicates, and thorium in amounts up to 10 mg. Aluminium and tin cause slight positive errors, but they can usually be disregarded if their quantity is less than 25 mg. Iron in the ferric state interferes, but if reduced with thioglycollic acid is rendered harmless; a large excess of thioglycollic acid must be avoided. If substantial quantities of niobium and tantalum are present, they can be separated from zirconium by extracting the former in methyl isobutyl ketone from a solution 10M in hydrofluoric acid and 6M in sulphuric acid.

Traces of zirconium can be collected in an ammoniacal solution with the aid of a little titanium salt, or in a cupferron–chloroform extraction by the addition of a little iron.

By means of the decomposition procedures outlined previously in **1,** A, obtain the sample in hydrochloric acid solution and make to a convenient volume in a volumetric flask. If necessary, remove the interferences listed in previous paragraphs. Transfer a 10-ml aliquot containing 10–500 μg of zirconium to a 100-ml volumetric flask. Add additional hydrochloric acid, if necessary, to bring the total to 4 ml of this acid. Add 4 ml of 4% thioglycollic acid, and after reduction of iron add 10 ml of a 0·15% solution of Alizarin Red S. Mix, place the flask in a boiling water bath for exactly three minutes. Cool, dilute with water to 100 ml, and measure the absorbance at 550 nm against a blank.

B. Other Colorimetric Methods

Zirconium may be determined by pyrocatechol violet [11, 20] and other photometric reagents [3, 11].

4. Volumetric Method

Zirconium can be titrated in acid solution, directly or indirectly, with EDTA [11, 19]. A number of metals interfere, and the procedure finds its greatest use in zirconium salts and compounds.

5. Optical Spectrography

Zirconium can be determined in a few materials by emission spectrography. Methods have been published for 0·001–1% zirconium in aluminium and its alloys [1], up to 5·5% in uranium metal [9], and 17–200 ppm in nickel [13].

6. X-Ray Spectrography

X-ray fluorescence offers another technique for the determination of zirconium [11, 14].

REFERENCES

1. A.S.T.M., *Chemical Analysis of Metals, Sampling and Analysis of Metal Bearing Ores*, Philadelphia, Pa., American Society for Testing and Materials, 1969.
2. BANDI, W. R., BUYOK, E. G., LEWIS, L. L., and MELNICK, L. M., *Anal. Chem.* **33**, 1275–8 (1961).
3. CULKIN, F., and RILEY, J. P., *Anal. Chim. Acta* **32**, 197–210 (1965).
4. DOLEZAL, J., POVONDRA, P., and SULCEK, Z., *Decomposition Techniques in Inorganic Analysis*, London, Iliffe, 1968.
5. ELWELL, W. T., and SCHOLES, I. R., *Analysis of Copper and its Alloys*, Oxford, Pergamon, 1967.
6. ELWELL, W. T., and WOOD, D. F., *The Analysis of Titanium, Zirconium and Their Alloys*, London, Wiley, 1961.
7. ELWELL, W. T., and WOOD, D. F., *Analysis of the New Metals Titanium, Zirconium, Hafnium, Niobium, Tantalum, Tungsten, and their Alloys*, Oxford, Pergamon, 1966.
8. FURMAN, N. H., ed., *Scott's Standard Methods of Chemical Analysis*, 6th ed., Vol. 1, Princeton, N.J., Van Nostrand, 1962.
9. GOLEB, J. A., *Anal. Chem.* **28**, 965–7 (1956).
10. HILLEBRAND, W. F., LUNDELL, G. E. F., BRIGHT, H. A., and HOFFMAN, J. I., *Applied Inorganic Analysis*, 2nd ed., New York, Wiley, 1953.
11. KOLTHOFF, I. M., and ELVING, P. J., *Treatise on Analytical Chemistry*, Part II, Vol. 5, 61–138, New York, Interscience Publishers, 1961.
12. KORKISCH, J., and ORLANDINI, K. A., *Talanta*, **16**, 45–9 (1969).
13. LEWIS, C. L., OTT, W. L., and SINE, N. M., *The Analysis of Nickel*, Oxford, Pergamon, 1966.
14. MITCHELL, B. J., *Anal. Chem.* **32**, 1652–6 (1960).
15. MOORE, F. L., *Anal. Chem.* **28**, 997–1001 (1956).
16. NORWITZ, G., *Anal. Chim. Acta* **35**, 491–8 (1966).
17. SANDELL, E. B., *Colorimetric Determination of Traces of Metals*, 3rd ed., New York, Interscience Publishers, 1959.
18. SCHOELLER, W. R., and POWELL, A. R., *The Analysis of Minerals and Ores of the Rarer Elements*, 3rd ed., London, Charles Griffin, 1955.
19. SU, Y.-S., *Anal. Chem.* **37**, 1067–8 (1965).
20. WOOD, D. F., and JONES, J. T., *Analyst* **90**, 125–33 (1965).
21. WOOD, D. F., and MCKENNA, R. H., *Analyst* **87**, 880–3 (1962).
22. ZITTEL, H. E., and FLORENCE, T. M., *Anal. Chem.* **39**, 355–6 (1967).

Miscellaneous analyses and data

1. Gas Analyses

Nearly all mining and metallurgical laboratories must occasionally undertake an analysis of gases. This specialized field of analytical chemistry is fully covered in books devoted entirely to this subject [1, 5, 8, 9, 12, 13, 15, 16]; it is also discussed at some length in several general analytical reference works [6, 7, 11, 17]. Another source of up-to-date information is the section on Air Pollution in the annual reviews of the journal *Analytical Chemistry*.

Instruments for the detection and measurement of various toxic gases in air have been described [10, 13, 14, 15]. Some of the most useful information on gas analysis is provided in trade literature [3, 14]. Various papers discuss the simultaneous determination of CO_2 and H_2S [2], apparatus and procedures in absorption and combustion [4], and the determination of CO_2 and SO_2 in the Orsat apparatus [18].

REFERENCES

1. Ambrose, D., and Ambrose, B. A., *Gas Chromatography*, London, Newnes, 1961.
2. Blohm, C. L., and Riesenfeld, F. C., *Ind. Eng. Chem. Anal. Ed.* **18**, 373–6 (1946).
3. Burrell Corporation, *Manual for Gas Analysts*, 7th ed., Pittsburgh, Pa., Burrell Corporation, 1951.
4. Brooks, F. R., Lykken, L., Milligan, W. B., Nebeker, H. R., and Zahn, V., *Anal. Chem.* **21**, 1105–16 (1949).
5. Coates, V. J., Noebels, H. J., and Fagerson, I. S., *Gas Chromatography*, New York, Academic Press, 1958.
6. Dillon, V. S., *Assay Practice on the Witwatersrand*, Johannesburg, Transvaal and Orange Free State Chamber of Mines, 1955.
7. Hillebrand, W. F., Lundell, G. E. F., Bright, H. A., and Hoffman, J. I., *Applied Inorganic Analysis*, 2nd ed., New York, Wiley, 1953.
8. Jacobs, M. B., *The Chemical Analysis of Air Pollutants*, New York, Wiley-Interscience, 1960.
9. Keulemans, A. I. M., *Gas Chromatography*, 2nd ed., New York, Reinhold, 1959.
10. Kinnear, A. M., *Chemistry and Industry*, **1959**, March 14, 361–3.

11. KOLTHOFF, I. M., and ELVING, P. J., *Treatise on Analytical Chemistry*, Part 1, Vol. 3, 1657–1723, 1961, Vol. 7, 4317–40, New York, Interscience Publishers, 1967.
12. LITTLEWOOD, A. B., *Gas Chromatography*, New York, Academic Press, 1962.
13. MAGILL, P. L., HOLDEN, F. R., and ACKLEY, C., *Air Pollution Handbook*, New York, McGraw-Hill, 1956.
14. MINE SAFETY APPLIANCES COMPANY, Pittsburgh, Pa., trade literature, 1961.
15. MULLEN, P. W., *Modern Gas Analysis*, New York, Interscience Publishers, 1955.
16. STERN, A. C., *Air Pollution*, Vol. 2, 2nd ed., New York, Academic Press, 1968.
17. TREADWELL, F. P., and HALL, W. T., *Analytical Chemistry*, Vol. II, New York, Wiley, 1942.
18. YOUNG, R. S., BENFIELD, D. A., and STRACHAN, K. G., *Analyst* **78**, 320 (1953).

2. Water Analyses

Occasional or routine analyses of domestic or industrial waters, and of trade effluents or waste-waters, are required in nearly every laboratory serving the mineral industries. In addition to general reference books on the examination of water [1, 2, 3, 8, 9, 12, 13, 16, 18], numerous publications cover specialized segments of the field. These include boiler waters [3, 8, 12, 13, 16], irrigation waters [2], mine waters [5], oil well waters [4, 7], wastes and sewage [1, 9], and geochemical prospecting by determining traces of metals in streams [10, 17]. Other papers have discussed bromine in water [6], free acid in the presence of hydrolysable cations [11], and concentration methods for traces of metal [14, 15].

REFERENCES

1. AMERICAN PUBLIC HEALTH ASSOC., *Standard Methods for the Examination of Water and Wastewater*, New York, American Public Health Assoc., 1965.
2. A.O.A.C., *Methods of Analysis of the Association of Official Analytical Chemists*, Washington, D.C., Association of Official Analytical Chemists, 1965.
3. A.S.T.M., *Manual on Industrial Water*, Philadelphia, Pa., American Society for Testing and Materials, 1953.
4. COLLINS, A. G., PEARSON, C., ATTAWAY, D. H., and EBREY, T. G., *U.S. Bur. Mines Rept. Invest.* 6087 (1962).
5. DILLON, V. S., *Assay Practice on the Witwatersrand*, Johannesburg, Transvaal and Orange Free State Chamber of Mines, 1955.
6. FISHMAN, M. J., and SKOUGSTAD, M. W., *Anal. Chem.* **35**, 146–9 (1963).
7. GULLIKSON, D. M., CARAWAY, W. H., and GATES, G. L., *U.S. Bur. Mines Rept. Invest.* 5737 (1961).
8. HAMER, P., JACKSON, J., and THURSTON, E. F., *Industrial Water Treatment Practice*, London, Butterworth's Scientific Publications in association with I.C.I., 1961.

9. KLEIN, L., *River Pollution. 1. Chemical Analysis*, London, Butterworth, 1959.
10. LAKIN, H. W., ALMOND, H., and WARD, F. N., Washington, D.C., *Geol. Survey Circ.* 161 (1952).
11. MOSKOWITZ, A., DASHER, J., and JAMISON, H. W., *Anal. Chem.* **32**, 1362–4 (1960).
12. RAINWATER, F. H., and THATCHER, L. L., Washington, D.C., *U.S. Geol. Survey, Water Supply Paper* 1454 (1960).
13. RIEHL, M. L., *Hoover's Water Supply and Treatment*, Washington, D.C., National Lime Association, 1957.
14. SACHDEV, S. L., and WEST, P. W., *Anal. Chim. Acta* **44**, 301–7 (1969).
15. SILVEY, W. D., and BRENNAN, R., *Anal. Chem.* **34**, 784–6 (1962).
16. SOLVAY TECHNICAL AND ENGINEERING SERVICE, *Water Analysis*, Bulletin No. 11, 4th ed., 2nd printing, New York, Solvay Process Division, Allied Chemical and Dye Corporation, 1957.
17. STANTON, R. E., *Rapid Methods of Trace Analysis* (For Geochemical Application), London, Edward Arnold, 1966.
18. TAYLOR, E. W., *The Examination of Waters and Water Supplies*, (Thresh, Beale, and Suckling), 7th ed., London, J. and A. Churchill, 1958.

3. Xanthates

From time to time most laboratories serving the mining industries are required to determine xanthates. Several methods may be employed: oxidation of xanthate by standard iodine to di-xanthogen [3, 6]; decomposition of xanthate by an excess of standard acid and titration of this excess with standard alkali [1, 2]; precipitation of xanthate with a cuprous salt, filtration and determination of the excess copper; and a polarographic determination in a supporting electrolyte of NaOH–KCl [5].

Another procedure for xanthates, particularly for cyanide solutions from the gold mining industry, is based on the formation and extraction of cupric xanthate, and final measurement of its absorbance at 305 nm [4]. Normally, cupric xanthate decomposes into cuprous xanthate and di-xanthogen, but in this method the decomposition is delayed by the addition of sodium chloride in the presence of nickel and potassium cyanide.

A. ACID-BASE TITRATION

Transfer 5 g of the finely-ground xanthate to a 250 ml Erlenmeyer flask with about 100 ml of acetone. Stopper the flask, shake until the sample has dissolved, allow to stand at least 5 minutes, filter into a 250 ml volumetric flask, wash with acetone, and make up to the mark with the latter. Withdraw a 25-ml aliquot to a beaker containing 35 ml of 0·1 N hydrochloric acid, place the latter in a water bath at 45–55°C for 15 minutes, stir occasionally, and cool to room temperature. Add 2–3 drops of methyl red indicator and titrate with 0·1N sodium hydroxide to the orange-yellow endpoint.

$$\%\ \text{available xanthate} = \frac{\text{mol. wt. of xanthate (35–ml of 0·1 N NaOH)}}{50}$$

The molecular weights of xanthates in common use are:

	Mol. wt.
Sodium ethyl xanthate	144·2
Sodium isopropyl xanthate	158·2
Sodium secondary butyl xanthate	172·3
Sodium amyl xanthate	186·2
Potassium ethyl xanthate	160·3
Potassium isopropyl xanthate	174·3
Potassium normal butyl xanthate	188·3
Potassium secondary butyl xanthate	188·3
Potassium amyl xanthate	202·4
Potassium hexyl xanthate	216·3

B. Determination with Cuprous Salt

(1) *Available collecting power*

One of the chief decomposition products of xanthates is di-xanthogen, which is formed by:

$$2\left[\begin{array}{l} S{=}\underset{\displaystyle SK}{\underset{|}{C}}{-}OC_2H_5 \end{array}\right] + O + CO_2 \rightarrow \begin{array}{c} S{=}C{-}OC_2H_5 \\ | \\ S \\ | \\ S \\ | \\ S{=}C{-}OC_2H_5 \end{array} + K_2CO_3$$

potassium ethyl xanthate di-xanthogen

Because di-xanthogens are also collectors, some concentrator staffs prefer to include them in the estimation of xanthates. Di-xanthogens, unlike xanthates, are soluble in ether; the loss in weight from an ether extraction gives the di-xanthogen content.

Weigh 2 g of the sample into a small porcelain dish, add 15 ml ether, and grind thoroughly. Filter through a weighed Gooch crucible and wash thoroughly with a number of 5-ml portions of ether. Dry the residue in an oven at 70°C for 15 minutes, cool, and weigh; the loss in weight represents di-xanthogen.

Transfer the residue and asbestos pad to a 400-ml beaker, diluteto 200 ml with water, and when all the xanthate is in solution filter through Whatman No. 31 paper, washing thoroughly with cold water. Precipitate the xanthate with copper as outlined under (2) below. The xanthate found in the ether-insoluble precipitate, plus the di-xanthogen content, gives the total collecting power of the material.

(2) *Available xanthate*

Dissolve 2 g of the sample in cold water and filter through Whatman No. 31 paper, washing well with cold water. If di-xanthogen has been determined by an ether extraction, the ether-insoluble residue may be dissolved in water, filtered, and washed. Make up to a definite volume.

Pipette 10 ml of a solution containing 1 g cuprous chloride in 100 ml of 25% hydrochloric acid solution into a 600-ml beaker containing 200 ml water. Add from a burette or pipette 10 ml of the filtered xanthate solution to this and stir thoroughly. Allow to stand for 15 minutes with occasional stirring. One g of copper or 1·6 g cuprous chloride will be required to precipitate 2·5 g of potassium ethyl xanthate. Cupric salts cannot be used, for part of the xanthate is then converted to di-xanthogen which is not precipitated by copper.

Filter off the copper xanthate precipitate through a Whatman No. 40 paper into a 250-ml volumetric flask, wash with cold water, and make up to the mark. Pipette out 100 ml or other suitable aliquot and determine the copper in it by iodometric or electrolytic procedures. The difference in the copper content before and after the addition of the xanthate is a measure of the latter.

$$2KC_3H_5OS_2 + Cu_2Cl_2 \rightarrow Cu_2(C_3H_5OS_2)_2 + 2KCl$$

One g of copper = 2·5215 g potassium ethyl xanthate or 2·2681 g sodium ethyl xanthate.

REFERENCES

1. AMERICAN CYANAMID COMPANY, New York, personal communication, 1960.
2. DOW CHEMICAL COMPANY, Great Western Division, San Francisco, personal communication, 1946.
3. LINCH, A. L., *Anal. Chem.* **23**, 293–6 (1951).
4. POHLANDT, C., COOK, E. B. T., and STEELE, T. W., *Talanta* **16**, 1129–35 (1969).
5. SUN, S., and HOLZMANN, R. T., *Anal. Chem.* **29**, 1298–1300 (1957).
6. SUTHERLAND, K. L., and WARK, I. W., *Principles of Flotation*, Melbourne, Australasian Institute of Mining and Metallurgy, 1955.

4. Pine Oil

A colorimetric method for determining 1–30 ppm of pine oil in water has been published [1]. It has application in the mining industry for the re-use or discharge of concentrator waters. The method depends on the stable bluish-green colour produced when pin eoil is reacted with vanillin–hydrochloric acid reagent at 60°C for 25 minutes. Some waters may require a special preparation in which ion-exchange resins are used to remove interferences.

REFERENCE

1. SUTTON, J. A., *U.S. Bureau of Mines Rept. Invest.* 4990 (1953).

5. Cyanogen

Cyanides are widely used in the mining and metallurgical industries for gold extraction, ore flotation, electroplating, heat treatment of metals, and other purposes; the determination of cyanogen compounds is consequently required in many laboratories serving these industries.

When sulphides are present in the solution used for the analysis of cyanides, they must be removed by precipitation with lead carbonate or barium nitrate, and cyanides determined on the filtered solution.

A. Free Cyanide

The term "free cyanide" is used to indicate the equivalent, in terms of potassium cyanide or sodium cyanide, of all the cyanogen present as simple cyanides of the alkalies and the alkaline earth metals, such as K, Na, NH_4, Ca, and Ba cyanides. It does not include cyanogen present in the form of double cyanides or as hydrogen cyanide.

Weigh out a sample or measure sufficient filtered solution so that the content of sodium cyanide or its equivalent is not more than 0·2 g. Add a few drops of a 5% solution of potassium iodide. Titrate with standard silver nitrate solution over a black background with constant stirring to the first appearance of a permanent yellow turbidity or opalescence [3, 4, 5, 7].

$$AgNO_3 + 2KCN \rightarrow KAg(CN)_2 + KNO_3$$
$$KAg(CN)_2 + AgNO_3 \rightarrow 2\ AgCN + KNO_3$$
$$AgNO_3 + KI \rightarrow AgI + KNO_3$$

The endpoint of the reaction can be determined by the formation of insoluble silver cyanide when excess silver nitrate reacts with soluble $KAg(CN)_2$; the addition of potassium iodide, however, renders the endpoint much sharper by the formation of the insoluble compound silver iodide.

1 ml 0·1 N $AgNO_3$ = 0·01302 g KCN or 0·0098 g NaCN.

When many determinations are made, it is advisable to have a silver nitrate solution containing 6·524 g silver nitrate per litre, in which case 1 ml $AgNO_3$ = 0·005 g KCN. To convert to NaCN, multiply the KCN value by 0·753. For instance, a solution containing 8·664 g silver nitrate per litre will have a value of 1 ml = 0·005 g NaCN. Alternatively, a standard solution containing 9·825 g silver nitrate per litre can be used, and 100 ml of the cyanide solution taken for titration; 1 ml $AgNO_3$ = 0·01% NaCN.

Among the ion-selective electrodes now commercially available is one selective for the free cyanide ion.

B. Total Cyanide

The expression "total cyanide" in industrial work is used to indicate, in terms of potassium or sodium cyanide, all of the cyanogen existing in the form of simple cyanides, hydrogen cyanide, and the double cyanides of zinc. The determination depends on the fact that when sodium hydroxide is added to a cyanide solution the hydrocyanic acid and zincocyanide are converted to sodium cyanide, and the total sodium cyanide content of the solution is measured by titration with standard silver nitrate [3]. The addition of sodium hydroxide does not extract cyanide from ferrocyanide, thiocyanate, or cyanate. In practice, "total cyanide" can be done on the same sample as "free cyanide"; after the latter has been determined, sodium hydroxide is added so that about 0·5% is present, with a little potassium iodide, and the titration continued.

Measure out a suitable quantity of sample or cyanide solution, and add 10 ml of sodium hydroxide–potassium iodide solution. The latter consists of 4 g NaOH and 1 g KI per 100 ml of water. Titrate with standard silver nitrate solution as described above for free cyanide.

Most commercial cyanides and cyanogen compounds are water-soluble. If the sample contains an insoluble cyanide, place in a distilling flask, add 20 ml of 7% mercuric chloride solution, 10 ml of 50% $MgCl_2.6H_2O$ solution, 200 ml of 1 : 4 sulphuric acid, and distil about 150 ml into a 4% sodium hydroxide solution in a receiver. Sulphides, of course, must be removed previously. Distillation readily converts the simple cyanides into hydrogen cyanide; it also converts most cyanide complexes such as those of Fe, Cd, Cu, Ni, Ag, and Zn when mercuric and magnesium salts are present. Complete decomposition of the cobalticyanide complex does not occur even after distilling the sample all day.

C. Determination of Small Quantities of Cyanide

The cyanide ion, CN^-, is very toxic; the measurement of small quantities of cyanide is important in pollution studies. Because many complex cyanides may decompose under certain conditions to give CN^-, it is customary to distil the sample of water or industrial waste, as outlined above, to collect all cyanides except cobalticyanide. If the cyanide content is greater than about 1 mg per litre, a titration procedure with silver nitrate and the indicator *p*-dimethylaminobenzalrhodanine is employed. This indicator, turning from a yellow to pink when silver is in excess, is more sensitive for small quantities than potassium iodide [1].

If the cyanide content in the distillate is less than 1 mg per litre, a colorimetric procedure is indicated [1]. The cyanide is converted to cyanogen chloride by reaction with chloramine-T at a *p*H less than 8 without hydrolysing to the cyanate. Cyanogen chloride forms a blue dye with pyridine-pyrazolone reagent, which is measured at 620 nm. For low CN^- values, the colour may be extracted with butyl alcohol and the absorbance measured at 630 nm. Details of this specialized determination may be obtained from authoritative standard methods [1].

Determinations of cyanide in effluents [6], and a fluorometric method for cyanide [4] have been published.

Free hydrogen cyanide in river water has been determined by extracting with 1,1,1-trichloroethane, transferring to sodium pyrophosphate solution, and measuring photometrically [8].

REFERENCES

1. AMERICAN PUBLIC HEALTH ASSOCIATION, *Standard Methods for the Examination of Water and Wastewater*, New York, American Public Health Association, 1965.
2. BARK, L. S., and HIGSON, H. G., *Analyst* **88**, 751–60 (1963).
3. DILLON, V. S., *Assay Practice on the Witwatersrand*, Johannesburg, Transvaal and Orange Free State Chamber of Mines, 1955.
4. GUILBAULT, G. G., and KRAMER, D. N., *Anal. Chem.* **37**, 918 (1965).
5. HANSON–VAN WINKLE–MUNNING CO., *Simple Methods for Analyzing Plating Solutions*, Matawan, N.J., Hanson–Van Winkle–Munning, 1958.
6. HIGSON, H. G., and BARK, L. S., *Analyst* **89**, 338–45 (1964).
7. IMPERIAL CHEMICAL INDUSTRIES LTD., *Analysis of Cyanidation Solutions and Gold Precipitate*, London, I.C.I., 1963.
8. MONTGOMERY, H. A. C., GARDINER, D. K., and GREGORY, J. G. G., *Analyst* **94**, 284–91 (1969).

ANALYTICAL REFERENCE BOOKS FOR THE MINING AND METALLURGICAL INDUSTRIES

ADLER, I., *X-Ray Emission Spectrography in Geology*, Amsterdam and New York, Elsevier, 1966.

AHRENS, L. H., and TAYLOR, S. R., *Spectrochemical Analysis*, 2nd ed., Reading, Mass., Addison-Wesley, 1964.

ALLPORT, N. L., and BROCKSOPP, J. E., *Colorimetric Analysis*, Vol. 2, 2nd ed., London, Chapman and Hall, 1963.

ALUMINUM COMPANY OF AMERICA, *Chemical Analysis of Aluminum*, New Kensington, Pa., Aluminum Company of America, 1950.

AMBROSE, D., and AMBROSE, B. A., *Gas Chromatography*, London, Newnes, 1961.

AMERICAN PUBLIC HEALTH ASSOCIATION, *Standard Methods for the Examination of Water and Wastewater*, New York, American Public Health Association, 1965.

ANGINO, E. E., and BILLINGS, G. K., *Atomic Absorption Spectrophotometry in Geology*, New York, Elsevier, 1967.

ANONYMOUS, *Analytical Chemistry of Uranium*, London, Oldbourne Press, 1964.

A.O.A.C., *Methods of Analysis of the Association of Official Analytical Chemists*, Washington, D.C., Association of Official Analytical Chemists, 1965.

A.S.T.M., *Chemical Analysis of Metals. Sampling and Analysis of Metal Bearing Ores*, Philadelphia, Pa., American Society for Testing and Materials, 1969.

A.S.T.M., *Manual on Industrial Water*, Philadelphia, Pa., American Society for Testing and Materials, 1953.

A.S.T.M., *Methods for Emission Spectrochemical Analysis*, 5th ed., Philadelphia, Pa., American Society for Testing and Materials, 1968.

BALSTON, J. N., and TALBOT, B. E., *A Guide to Filter Paper and Cellulose Powder Chromatography*, London, H. Reeve Angel, 1952.

BARKSDALE, J., *Titanium; Its Occurrence, Chemistry and Technology*, 2nd ed., New York, Ronald Press, 1964.

BEAMISH, F. E., *The Analytical Chemistry of the Noble Metals*, Oxford, Pergamon, 1966.

BENNETT, H., and HAWLEY, W. G., *Methods of Silicate Analysis*, 2nd ed., New York, Academic Press, 1964.

BERTIN, E. P., *Principles and Practice of X-Ray Spectrometric Analysis*, New York, Plenum, 1970.

BIRKS, L. S., *X-Ray Spectrochemical Analysis*, 2nd ed., New York, Wiley-Interscience, 1969.

BLACK, C. A., ed., *Methods of Soil Analysis*, Part 2, Chemical and Microbiological Properties, Madison, Wisconsin, American Society of Agronomy, 1965.

BLOCK, R. J., LESTRANGE, R., and ZWEIG, G., *Paper Chromatography*, New York, Academic Press, 1952.

BLOKHIN, M. A., *Methods of X-Ray Spectroscopic Research*, Oxford, Pergamon, 1966.

BOLTZ, D. F., *Colorimetric Determination of Nonmetals*, New York, Interscience Publishers, 1958.

BRITISH ALUMINIUM COMPANY LTD., *Chemical Analysis of Aluminium and its Alloys*, London, British Aluminium Company, 1961.

BRITISH ALUMINIUM COMPANY LTD., *Spectrochemical Analysis of Aluminium and its Alloys*, London, British Aluminium Co, 1961.

BRITISH CAST IRON RESEARCH ASSOCIATION, *Chemical Analysis for Iron Foundries. Selected Methods*, London, Allen and Unwin, 1967.

BRITISH DRUG HOUSES LTD., *The B.D.H. Book of Organic Reagents for Analytical Use*, Poole, Dorset, British Drug Houses, 1958.

BRODE, W. R., *Chemical Spectroscopy*, New York, Wiley, 1943.

BURRELL CORPORATION, *Manual for Gas Analysts*, Pittsburgh, Pa., Burrell Corporation, 1951.

BURRIEL-MARTI, F., and RAMIREZ-MUNOZ, J., *Flame Photometry*, Amsterdam, Elsevier, 1957.

BUTTS, A., *Silver, Economics, Metallurgy and Use*, Princeton, N.J., Van Nostrand, 1967.

CALDER, A. B., *Photometric Methods of Analysis*, London, Adam Hilger, 1969.

CHAMOT, E. M., and MASON, C. W., *Handbook of Chemical Microscopy*, 3rd ed., New York, Wiley, 1958.

COATES, V. J., NOEBELS, H. J., and FAGERSON, I. S., *Gas Chromatography*, New York, Academic Press, 1958.

CODELL, M., *Analytical Chemistry of Titanium Metals and Compounds*, New York, Interscience Publishers, 1959.

CROW, D. R., and WESTWOOD, J. V., *Polarography*, New York, Barnes and Noble, 1968.

DEAN, J. A., *Flame Photometry*, New York, McGraw-Hill, 1960.

DEAN, J. A., and RAINS, T. C., *Flame Emission and Atomic Absorption Spectrometry*, Vol. 1, Theory, New York, Dekker, 1969.

DILLON, V. A., *Assay Practice on the Witwatersrand*, Johannesburg, Transvaal and Orange Free State Chamber of Mines, 1955.

DOLEZAL, J., POVONDRA, P., and SULCEK, Z., *Decomposition Techniques in Inorganic Analysis*, London, Iliffe, 1968.

DOZINEL, C. M., *Modern Methods of Analysis of Copper and its Alloys*, 2nd ed., London, Elsevier, 1963.

ELWELL, W. T., and GIDLEY, J. A. F., *Atomic-Absorption Spectrophotometry*, 2nd ed., Oxford, Pergamon, 1966.

ELWELL, W. T., and SCHOLES, I. R., *Analysis of Copper and its Alloys*, Oxford, Pergamon, 1967.

ELWELL, W. T., and WOOD, D. F., *The Analysis of Titanium, Zirconium and their Alloys*, London, Wiley, 1961.

ELWELL, W. T., and WOOD, D. F., *The Analysis of the New Metals Titanium, Zirconium, Hafnium, Niobium, Tantalum, Tungsten and their Alloys*, Oxford, Pergamon, 1966.

FEIGL, F., *Spot Tests in Inorganic Analysis*, 5th ed., Amsterdam, Elsevier, 1958.

FLASCHKA, H. A., *EDTA Titrations*, 2nd ed., Oxford, Pergamon, 1964.

FRANCIS, W. *Boiler House and Power Station Chemistry*, 3rd ed., London, Edward Arnold, 1955.

FURMAN, N. H., ed., *Scott's Standard Methods of Chemical Analysis*, 6th ed., Vol. 1, Princeton, N.J., Van Nostrand, 1962.

GORDON, A. H., and EASTOE, J. E., *Practical Chromatographic Techniques*, London, Newnes and Pearson, 1964.

HAIS, I. M., and MACEK, K., *Paper Chromatography*, New York, Academic Press, 1963.

HAMER, P., JACKSON, J., and THURSTON, E. F., *Industrial Water Treatment Practice*, London, Butterworth's Scientific Publications in Association with I.C.I., 1961.

HASSLER, J. W., *Activated Carbon*, New York, Chemical Publ. Co., 1963.

HAYWOOD, F. W., and WOOD, A. A. R., *Metallurgical Analysis by Means of the Spekker Photoelectric Absorptiometer*, 2nd ed., London, Hilger and Watts, 1956.

HEFTMANN, E., *Chromatography*, 2nd ed., New York, Reinhold, 1967.

HERRMANN, R., and ALKEMADE, C. T. J., *Chemical Analysis by Flame Photometry*, New York, Wiley, 1963.

HEYROVSKY, J., and KUTA, J., *Principles of Polarography*, New York, Academic Press, 1966.

HILLEBRAND, W. F., LUNDELL, G. E. F., BRIGHT, H. A., and HOFFMAN, J. I., *Applied Inorganic Analysis*, 2nd ed., New York, Wiley, 1953.

HOPKIN and WILLIAMS LTD., *Organic Reagents for Metals*, Chadwell Heath, England, Hopkin and Williams, Vol. 1, 1955, Vol. 2, 1964.

IMPERIAL CHEMICAL INDUSTRIES LTD., *Analysis of Cyanidation Solutions and Gold Precipitate*, London, I.C.I., 1963.

IMPERIAL CHEMICAL INDUSTRIES LTD., *The Analysis of Titanium and its Alloys*, 3rd ed., London, I.C.I., 1959.

ISRAEL PROGRAM FOR SCIENTIFIC TRANSLATIONS, Jerusalem. Translation of the volumes in the series Analytical Chemistry of the Elements, published by the Academy of Sciences of the U.S.S.R.— Moscow, as they appear. To 1970 the following volumes have been published: Uranium, Thallium, Thorium, Ruthenium, Molybdenum, Boron, Potassium, Zirconium and Hafnium, Cobalt, Plutonium, Beryllium, and Nickel.

JACOBS, M. B., *Analytical Chemistry of Industrial Poisons, Hazards, and Solvents*, 2nd ed., New York, Interscience Publishers, 1949.

JACOBS, M. B., *The Chemical Analysis of Air Pollutants*, New York, Wiley-Interscience, 1960.

JEFFERY, P. G., *Chemical Methods of Rock Analysis*, Oxford, Pergamon, 1970.

KEULEMANS, A. I. M., *Gas Chromatography*, 2nd ed., New York, Reinhold, 1959.

KLEIN, L., *River Pollution. 1. Chemical Analysis*, London, Butterworth, 1959.

KOLTHOFF, I. M., and BELCHER, R., *Volumetric Analysis*, Vol. III, New York, Interscience Publishers, 1957.

KOLTHOFF, I. M., and ELVING, P. J., *Treatise on Analytical Chemistry, Part II, Analytical Chemistry of the Elements*, Volumes 1–10, New York, Interscience Publishers, 1961–1969.

KOLTHOFF, I. M., and LINGANE, J. J., *Polarography*, New York, Interscience Publishers, 1952.

KOLTHOFF, I. M., and STENGER, V. A., *Volumetric Analysis*, New York, Interscience Publishers, Vol. I, 1942, Vol. II, 1947.

KORKISCH, J., *Modern Methods for the Separation of Rarer Metal Ions*, Oxford, Pergamon, 1969.

LAMBIE, D. A., *Techniques for the Use of Radioisotopes in Analysis. A Laboratory Manual*, London, Spon, 1964.

LANGFORD, K. E., *Analysis of Electroplating and Related Solutions*, 2nd ed., Teddington, England, Robert Draper, 1959.

LEWIS, C. L., and OTT, W. L., *Analytical Chemistry of Nickel*, Oxford, Pergamon, 1970.

LEWIS, C. L., OTT, W. L., and SINE, N. M., *The Analysis of Nickel*, Oxford, Pergamon, 1966.

LIEBHAFSKY, H. A., PFEIFFER, H. G., WINSLOW, E. H., and ZEMANY, P. D., *X-Ray Absorption and Emission in Analytical Chemistry*, New York, Wiley, 1960.

LINGANE, J. J., *Electroanalytical Chemistry*, 2nd ed., New York, Interscience Publishers, 1958.

LITTLEWOOD, A. B., *Gas Chromatography*, New York, Academic Press, 1962.

LOTHIAN, G. F., *Absorption Spectrophotometry*, 3rd ed., London, Adam Hilger, 1969.

LYON, W. S., *Guide to Activation Analysis*, London, Van Nostrand, 1964.

MAGILL, P. L., HOLDEN, F. R., and ACKLEY, C., *Air Pollution Handbook*, New York, McGraw-Hill, 1956.

MAVRODINEANU, R., and BOITEUX, H., *Flame Spectroscopy*, New York, Wiley, 1965.

MAXWELL, J. A., *Rock and Mineral Analysis*, New York, Wiley-Interscience, 1968.

MCKAY, H. A. C., *Solvent Extraction Chemistry of Metals*, London, Macmillan, 1965.

MEITES, L., *Polarographic Techniques*, New York, Interscience Publishers, 1955.

MEITES, L., *Handbook of Analytical Chemistry*, New York, McGraw-Hill, 1963.

MIKES, O., *Laboratory Handbook of Chromatographic Methods*, London, Van Nostrand, 1966.

MILNER, G. W. C., *The Principles and Applications of Polarography*, London, Longmans Green, 1957.

MITCHELL, R. L., *The Spectrochemical Analysis of Soils, Plants and Related Materials*, Farnham Royal, England, Commonwealth Agricultural Bureau, 1964.

MORRISON, G. H., and FREISER, H., *Solvent Extraction in Analytical Chemistry*, New York, Wiley, 1957.

MOSHIER, R. W., *Analytical Chemistry of Niobium and Tantalum*, London, Pergamon, 1964.

MULLEN, P. W., *Modern Gas Analysis*, New York, Interscience Publishers, 1955.

NAVOSELOVA, A. V., and BATSANOVA, L. R., *Analytical Chemistry of Beryllium*, Hartford, Conn., Daniel Davy, 1968.

NORWITZ, G., *Rapid Analysis of Nonferrous Metals and Alloys*, New York, Chemical Publ. Co., 1958.

PESHKOVA, V. M., and SAVOSTINA, N. M., *Analytical Chemistry of Nickel*, Hartford, Conn., Daniel Davy, 1967.

PUNGOR, E., *Flame Photometry Theory*, London, Van Nostrand, 1967.

PYATNITSKIĬ, I. V., *Analytical Chemistry of Cobalt*, Hartford, Conn., Daniel Davy, 1965.

RAMIREZ-MUNOZ, J., *Atomic-Absorption Spectroscopy and Analysis by Atomic-Absorption Flame Photometry*, New York, Elsevier, 1968.

REYNOLDS, R. J., and ALDOUS, K., *Atomic Absorption Spectroscopy*, London, Charles Griffin, 1970.

RIEHL, M. L., *Hoover's Water Supply and Treatment*, Washington, D.C., National Lime Association, 1957.

RIEMAN, W., and WALTON, H. F., *Ion Exchange in Analytical Chemistry*, Oxford, Pergamon, 1970.

ROBINSON, J. W., *Atomic Absorption Spectroscopy*, London, Edward Arnold, 1966.

RODDEN, C. J., *Analysis of Essential Nuclear Reactor Materials*, Washington, D.C., U.S. At. Energy Comm., 1964.

RYABCHIKOV, D. I., and GOL'BRAIKH, E. K., *The Analytical Chemistry of Thorium*, Oxford, Pergamon, 1963.

SAMUELSON, O., *Ion Exchange Separations in Analytical Chemistry*, New York, Wiley, 1963.

SANDELL, E. B., *Colorimetric Determination of Traces of Metals*, 3rd ed., New York, Interscience Publishers, 1959.

SCHOELLER, W. R., and POWELL, A. R., *The Analysis of Minerals and Ores of the Rarer Elements*, 3rd ed., London, Charles Griffin, 1955.

SHAPIRO, L., and BRANNOCK, W. W., *Rapid Analysis of Silicate, Carbonate, and Phosphate Rock*, Washington, D.C., *U.S. Geol. Survey Bull.* 1144–A, 1962.

SLAVIN, W., *Atomic Absorption Spectroscopy*, New York, Interscience Publishers, 1968.

SMITH, I., *Chromatographic and Electrophoretic Techniques*, Vol. 1 and 2, New York, Interscience Publishers, 1960.

SNELL, F. D., and HILTON, C. L., *Encyclopedia of Industrial Chemical Analysis*, New York, Interscience Publishers, Volumes 1–7 have appeared to 1968; Volumes 8– by F. D. Snell and L. Ettre, from 1969 onwards.

SNELL, F. D., and SNELL, C. T., *Colorimetric Methods of Analysis*, 3rd ed., New York, Van Nostrand, Vol. I, 1948; Vol. II, 1949; Vol. IIA 1959.

STANTON, R. E., *Rapid Methods of Trace Analysis (for Geochemical Application)*, London, Edward Arnold, 1966.

STERN, A. C., *Air Pollution*, Vol. 2, 2nd ed., New York, Academic Press, 1968.

STOCK, R., and RICE, C. B. F., *Chromatographic Methods*, 2nd ed., London, Chapman and Hall, 1967.

STROUTS, C. R. N., GILFILLAN, J. H., and WILSON, H. N., *Analytical Chemistry, The Working Tools*, 2 vols., London, Oxford University Press, 1955.

TAYLOR, D., *Neutron Irradiation and Activation Analysis*, London, Newnes, 1964.

TAYLOR, E. W., *The Examination of Water and Water Supplies*, (Thresh, Beale, and Suckling), 7th ed., London, J. and A. Churchill, 1958.

TEXAS GULF SULPHUR CO., *Sulphur Manual*, New York, Texas Gulf Sulphur Co., 1961.

TREADWELL, F. P., and HALL, W. T., *Analytical Chemistry*, 2 vols., New York, Wiley, 1942.

VERSAGI, F. J., *Routine Analysis of Copper-Base Alloys*, New York, Chemical Publ. Co., 1960.

VICKERY, R. C., *Analytical Chemistry of the Rare Earths*, Oxford, Pergamon, 1961.

VOGEL, A. I., *A Text-book of Quantitative Inorganic Analysis Including Elementary Instrumental Analysis*, 3rd ed., London, Longmans Green, 1961.

WEAST, R. C., *Handbook of Chemistry and Physics*, 50th ed., Cleveland, O., Chemical Rubber Co., 1969.

WELCHER, F. J., *The Analytical Uses of Ethylenediaminetetraacetic Acid*, New York, Van Nostrand, 1958.

WEST, T. S., *Complexometry with EDTA and Related Reagents*, 3rd ed., Poole, England, B.D.H. Chemicals, 1969.

WESTWOOD, W., and MAYER, A., *The Chemical Analysis of Cast Iron and Foundry Materials*, 2nd ed., London, Allen and Unwin, 1960.

WILSON, C. L., and WILSON, D. W., *Comprehensive Analytical Chemistry*, London, Elsevier, Vols. IA, B, C, IIA, B, 1959–68.

YOUNG, R. S., *The Analytical Chemistry of Cobalt*, Oxford, Pergamon, 1966.

CONVERSION TABLE

In the analysis of rocks, results for many elements are usually expressed in terms of an oxide. The following table gives conversions to the forms which are generally reported in rock analyses.

Element	*Conversion*	*Form reported*
Aluminium	Al × 1·890	Al_2O_3
Barium	Ba × 1·116	BaO
Beryllium	Be × 2·776	BeO
Cadmium	Cd × 1·142	CdO
Calcium	Ca × 1·399	CaO
	CaO × 1·785	$CaCO_3$
Carbon	C × 3·664	CO_2
Chromium	Cr × 1·462	Cr_2O_3
Cobalt	Co × 1·272	CoO
Copper	Cu × 1·252	CuO
Iron	Fe × 1·286	FeO
	× 1·430	Fe_2O_3
	× 1·382	Fe_3O_4
Lead	Pb × 1·077	PbO
Lithium	Li × 2·153	Li_2O
Magnesium	Mg × 1·658	MgO
	MgO × 2·092	$MgCO_3$
Manganese	Mn × 1·291	MnO
Molybdenum	Mo × 1·500	MoO_3
Nickel	Ni × 1·273	NiO
Niobium	Nb × 1·431	Nb_2O_5
Phosphorus	P × 2·292	P_2O_5
Potassium	K × 1·205	K_2O
Silicon	Si × 2·139	SiO_2
Sodium	Na × 1·348	Na_2O
Sulphur	S × 2·497	SO_3
Tantalum	Ta × 1·221	Ta_2O_5
Thorium	Th × 1·138	ThO_2
Tin	Sn × 1·270	SnO_2
Titanium	Ti × 1·668	TiO_2
Tungsten	W × 1·261	WO_3
Uranium	U × 1·134	UO_2
	× 1·202	UO_3
	× 1·179	U_3O_8
Vanadium	V × 1·471	V_2O_3
Zinc	Zn × 1·245	ZnO
Zirconium	Zr × 1·351	ZrO_2

ATOMIC WEIGHTS

Abridged from the 1961 Table of Atomic Weights, based on carbon-12, issued by the International Union of Pure and Applied Chemistry and Butterworth Scientific Publications.

	Symbol	*Atomic weight*		*Symbol*	*Atomic weight*
Aluminium	Al	26·98	Molybdenum	Mo	95·94
Antimony	Sb	121·75	Neodymium	Nd	144·24
Argon	Ar	39·95	Neon	Ne	20·18
Arsenic	As	74·92	Nickel	Ni	58·71
Barium	Ba	137·34	Niobium	Nb	92·91
Beryllium	Be	9·01	Nitrogen	N	14·01
Bismuth	Bi	208·98	Osmium	Os	190·2
Boron	B	10·81	Oxygen	O	16·0
Bromine	Br	79·91	Palladium	Pd	106·4
Cadmium	Cd	112·40	Phosphorus	P	30·97
Calcium	Ca	40·08	Platinum	Pt	195·09
Carbon	C	12·01	Potassium	K	39·10
Cerium	Ce	140·12	Praseodymium	Pr	140·91
Caesium	Cs	132·91	Rhenium	Re	186·2
Chlorine	Cl	35·45	Rhodium	Rh	102·91
Chromium	Cr	52·00	Rubidium	Rb	85·47
Cobalt	Co	58·93	Ruthenium	Ru	101·07
Copper	Cu	63·54	Samarium	Sm	150·35
Dysprosium	Dy	162·50	Scandium	Sc	44·96
Erbium	Er	167·26	Selenium	Se	78·96
Europium	Eu	151·96	Silicon	Si	28·09
Fluorine	F	19·00	Silver	Ag	107·87
Gadolinium	Gd	157·25	Sodium	Na	22·99
Gallium	Ga	69·72	Strontium	Sr	87·62
Germanium	Ge	72·59	Sulphur	S	32·06
Gold	Au	196·97	Tantalum	Ta	180·95
Hafnium	Hf	178·49	Tellurium	Te	127·60
Helium	He	4·00	Terbium	Tb	158·92
Holmium	Ho	164·93	Thallium	Tl	204·37
Hydrogen	H	1·008	Thorium	Th	232·04
Indium	In	114·82	Thulium	Tm	168·93
Iodine	I	126·90	Tin	Sn	118·69
Iridium	Ir	192·22	Titanium	Ti	47·90
Iron	Fe	55·85	Tungsten	W	183·85
Krypton	Kr	83·80	Uranium	U	238·03
Lanthanum	La	138·91	Vanadium	V	50·94
Lead	Pb	207·19	Xenon	Xe	131·30
Lithium	Li	6·94	Ytterbium	Yb	173·04
Lutetium	Lu	174·97	Yttrium	Y	88·91
Magnesium	Mg	24·31	Zinc	Zn	65·37
Manganese	Mn	54·94	Zirconium	Zr	91·22
Mercury	Hg	200·59			

AUTHOR INDEX

SUBJECT INDEX

Date Due